Integrability and Nonintegrability
in Geometry and Mechanics

Mathematics and Its Applications (*Soviet Series*)

Managing Editor:

M. HAZEWINKEL
Centre for Mathematics and Computer Science, Amsterdam, The Netherlands

Editorial Board:

A. A. KIRILLOV, *MGU, Moscow, U.S.S.R.*
Yu. I. MANIN, *Steklov Institute of Mathematics, Moscow, U.S.S.R.*
N. N. MOISEEV, *Computing Centre, Academy of Sciences, Moscow, U.S.S.R.*
S. P. NOVIKOV, *Landau Institute of Theoretical Physics, Moscow, U.S.S.R.*
M. C. POLYVANOV, *Steklov Institute of Mathematics, Moscow, U.S.S.R.*
Yu. A. ROZANOV, *Steklov Institute of Mathematics, Moscow, U.S.S.R.*

A. T. Fomenko
Department of Mathematics and Mechanics,
Moscow State University, Moscow, U.S.S.R.

Integrability and Nonintegrability in Geometry and Mechanics

KLUWER ACADEMIC PUBLISHERS
DORDRECHT / BOSTON / LONDON

Library of Congress Cataloging in Publication Data

Fomenko, A. T.
  Integrability and nonintegrability in geometry
and mechanics.

  (Mathematics and its applications (Soviet series))
  Includes index.
  1. Hamiltonian systems. 2. Differential equations.
3. Symplectic manifolds. I. Title. II. Series:
Mathematics and its applications (Kluwer Academic
Publishers). Soviet series.
QA614.83.I65 1988   514'.7   88-23724

ISBN 90-277-2818-6

Published by Kluwer Academic Publishers,
P.O. Box 17, 3300 AA Dordrecht, The Netherlands.

Kluwer Academic Publishers incorporates
the publishing programmes of
D. Reidel, Martinus Nijhoff, Dr W. Junk and MTP Press.

Sold and distributed in the U.S.A. and Canada
by Kluwer Academic Publishers,
101 Philip Drive, Norwell, MA 02061, U.S.A.

In all other countries, sold and distributed
by Kluwer Academic Publishers Group,
P.O. Box 322, 3300 AH Dordrecht, The Netherlands.

Translated by M. V. Tsaplina

All Rights Reserved
© 1988 by Kluwer Academic Publishers
No part of the material protected by this copyright notice may be reproduced or
utilized in any form or by any means, electronic or mechanical
including photocopying, recording or by any information storage and
retrieval system, without written permission from the copyright owner.

Printed in The Netherlands

SERIES EDITOR'S PREFACE

Approach your problems from the right end and begin with the answers. Then one day, perhaps you will find the final question.

'The Hermit Clad in Crane Feathers' in R. van Gulik's *The Chinese Maze Murders*.

It isn't that they can't see the solution. It is that they can't see the problem.

G.K. Chesterton. *The Scandal of Father Brown* 'The point of a Pin'.

Growing specialization and diversification have brought a host of monographs and textbooks on increasingly specialized topics. However, the "tree" of knowledge of mathematics and related fields does not grow only by putting forth new branches. It also happens, quite often in fact, that branches which were thought to be completely disparate are suddenly seen to be related.

Further, the kind and level of sophistication of mathematics applied in various sciences has changed drastically in recent years: measure theory is used (non-trivially) in regional and theoretical economics; algebraic geometry interacts with physics; the Minkowsky lemma, coding theory and the structure of water meet one another in packing and covering theory; quantum fields, crystal defects and mathematical programming profit from homotopy theory; Lie algebras are relevant to filtering; and prediction and electrical engineering can use Stein spaces. And in addition to this there are such new emerging subdisciplines as "experimental mathematics", "CFD", "completely integrable systems", "chaos, synergetics and large-scale order", which are almost impossible to fit into the existing classification schemes. They draw upon widely different sections of mathematics. This programme, Mathematics and Its Applications, is devoted to new emerging (sub)disciplines and to such (new) interrelations as exempla gratia:

- a central concept which plays an important role in several different mathematical and/or scientific specialized areas;
- new applications of the results and ideas from one area of scientific endeavour into another;
- influences which the results, problems and concepts of one field of enquiry have and have had on the development of another.

The Mathematics and Its Applications programme tries to make available a careful selection of books which fit the philosophy outlined above. With such books, which are stimulating rather than definitive, intriguing rather than encyclopaedic, we hope to contribute something towards better communication among the practitioners in diversified fields.

Until about 20 years ago, 1967 to be precise, the year of the discovery of the inverse spectral transform, the number of interesting (model) equations which could be solved completely was four. Now there are 40 or so. They include some of the more important equations of mathematical physics such as the Korteweg-de Vries equation, the Sine-Gordon equation and the nonlinear (or cubic) Schrödinger equation and most of them are the first members of infinite hierarchies of such socalled completely integrable equations. It is in fact not a total surprise that these completely integrable equations turn up often in applications: there is a good solid argument that says that in a wide variety of circumstances when dealing with wave phenomena the next step after the linear one is likely to involve the KdV or cubic Schrödinger equation.

Since its birth, the field of integrable systems has grown enormous; it now intertwines with most of the other fields in mathematics and accounts for many hundreds, quite possibly over a thousand, articles a year in mathematics, physics, chemistry, biology, geology and other journals. There is fair selection of introductory and basic books on the subject and a large collection of proceedings volumes. The field is now certainly far too large to be treated comprehensively in one volume; even 10 would probably not suffice to survey it in all its aspects.

Thus the time has come for a second generation of books on the topic, which concentrate in depth on certain main topics within the field. This is one of the first such books and by an author who has made fundamental contributions. Some of the unique features of the present volume are: a comprehensive treatment of surgery for completely integrable systems, a thorough treatment of how to recognize, analyse, and prove nonintegrability, and a discussion of 'noncommutative integrability'. The book contains a substantial number of results from the author's seminar at Moscow state university which are not, or not readily, available elsewhere. All in all it is a book that seems to fit the concept of this series very well and which should be of great interest to a large variety of mathematicians and physicists.

The unreasonable effectiveness of mathematics in science ...

    Eugene Wigner

Well, if you know of a better 'ole, go to it.

    Bruce Bairnsfather

What is now proved was once only imagined.

    William Blake

As long as algebra and geometry proceeded along separate paths, their advance was slow and their applications limited.

But when these sciences joined company they drew from each other fresh vitality and thenceforward marched on at a rapid pace towards perfection.

    Joseph Louis Lagrange.

Bussum, September 1988

Michiel Hazewinkel

TABLE OF CONTENTS

Preface — xiii

Chapter 1. Some Equations of Classical Mechanics and Their Hamiltonian Properties — 1

§1. Classical Equations of Motion of a Three-Dimensional Rigid Body — 1
   1.1. The Euler–Poisson Equations Describing the Motion of a Heavy Rigid Body around a Fixed Point — 1
   1.2. Integrable Euler, Lagrange, and Kovalevskaya Cases — 6
   1.3. General Equations of Motion of a Three-Dimensional Rigid Body — 10

§2. Symplectic Manifolds — 12
   2.1. Symplectic Structure in a Tangent Space to a Manifold — 12
   2.2. Symplectic Structure on a Manifold — 17
   2.3. Hamiltonian and Locally Hamiltonian Vector Fields and the Poisson Bracket — 20
   2.4. Integrals of Hamiltonian Fields — 30
   2.5. The Liouville Theorem — 32

§3. Hamiltonian Properties of the Equations of Motion of a Three-Dimensional Rigid Body — 34

§4. Some Information on Lie Groups and Lie Algebras Necessary for Hamiltonian Geometry — 39
   4.1. Adjoint and Coadjoint Representations, Semisimplicity, the System of Roots and Simple Roots, Orbits, and the Canonical Symplectic Structure — 39
   4.2. Model Example: $SL(n, \mathbb{C})$ and $sl(n, \mathbb{C})$ — 44
   4.3. Real, Compact, and Normal Subalgebras — 46

Chapter 2. The Theory of Surgery on Completely Integrable Hamiltonian Systems of Differential Equations — 55

§1. Classification of Constant-Energy Surfaces of Integrable Systems. Estimation of the Amount of Stable Periodic Solutions on a Constant-Energy Surface. Obstacles in the Way of Smooth Integrability of Hamiltonian Systems — 55
   1.1. Formulation of the Results in Four Dimensions — 55

- 1.2. A Short List of the Basic Data from the Classical Morse Theory — 68
- 1.3. Topological Surgery on Liouville Tori of an Integrable Hamiltonian System upon Varying Values of a Second Integral — 70
- 1.4. Separatrix Diagrams Cut out Nontrivial Cycles on Nonsingular Liouville Tori — 73
- 1.5. The Topology of Hamiltonian-Level Surfaces of an Integrable System and of the Corresponding One-Dimensional Graphs — 78
- 1.6. Proof of the Principal Classification Theorem 2.1.2 — 91
- 1.7. Proof of Claim 2.1.1 — 91
- 1.8. Proof of Theorem 2.1.1. Lower Estimates on the Number of Stable Periodic Solutions of a System — 92
- 1.9. Proof of Corollary 2.1.5 — 97
- 1.10 Topological Obstacles for Smooth Integrability and Graphlike Manifolds. Not each Three-Dimensional Manifold Can be Realized as a Constant-Energy Manifold of an Integrable System — 98
- 1.11. Proof of Claim 2.1.4 — 99

§2. Multidimensional Integrable Systems. Classification of the Surgery on Liouville Tori in the Neighbourhood of Bifurcation Diagrams — 103
- 2.1. Bifurcation Diagram of the Momentum Mapping for an Integrable System. The Surgery of General Position — 103
- 2.2. The Classification Theorem for Liouville Torus Surgery — 109
- 2.3. Toric Handles. A Separatrix Diagram is Always Glued to a Nonsingular Liouville Torus T^n Along a Nontrivial $(n-1)$-Dimensional Cycle T^{n-1} — 111
- 2.4. Any Composition of Elementary Bifurcations (of Three Types) of Liouville Tori Is Realized for a Certain Integrable System on an Appropriate Symplectic Manifold — 116
- 2.5. Classification of Nonintegrable Critical Submanifolds of Bott Integrals — 123

§3. The Properties of Decomposition of Constant-Energy Surfaces of Integrable Systems into the Sum of Simplest Manifolds — 126
- 3.1. A Fundamental Decomposition $Q = m\mathrm{I} + p\mathrm{II} + q\mathrm{III} + s\mathrm{IV} + r\mathrm{V}$ and the Structure of Singular Fibres — 126
- 3.2. Homological Properties of Constant-Energy Surfaces — 129

Chapter 3. Some General Principles of Integration of Hamiltonian Systems of Differential Equations — 143

§1. Noncommutative Integration Method — 143
- 1.1. Maximal Linear Commutative Subalgebras in the Algebra of Functions on Symplectic Manifolds — 143
- 1.2. A Hamiltonian System Is Integrable if Its Hamiltonian is Included in a Sufficiently Large Lie Algebra of Functions — 146

1.3. Proof of the Theorem	149
§2. The General Properties of Invariant Submanifolds of Hamiltonian Systems	157
2.1. Reduction of a System on One Isolated Level Surface	157
2.2. Further Generalizations of the Noncommutative Integration Method	160
§3. Systems Completely Integrable in the Noncommutative Sense Are Often Completely Liouville-Integrable in the Conventional Sense	165
3.1. The Formulation of the General Equivalence Hypothesis and its Validity for Compact Manifolds	165
3.2. The Properties of Momentum Mapping of a System Integrable in the Noncommutative Sense	167
3.3. Theorem on the Existence of Maximal Linear Commutative Algebras of Functions on Orbits in Semisimple and Reductive Lie Algebras	171
3.4. Proof of the Hypothesis for the Case of Compact Manifolds	173
3.5. Momentum Mapping of Systems Integrable in the Noncommutative Sense by Means of an Excessive Set of Integrals	173
3.6. Sufficient Conditions for Compactness of the Lie Algebra of Integrals of a Hamiltonian System	176
§4. Liouville Integrability on Complex Symplectic Manifolds	178
4.1. Different Notions of Complex Integrability and Their Interrelation	178
4.2. Integrability on Complex Tori	181
4.3. Integrability on $K3$-Type Surfaces	182
4.4. Integrability on Beauville Manifolds	184
4.5. Symplectic Structures Integrated without Degeneracies	186
Chapter 4. Integration of Concrete Hamiltonian Systems in Geometry and Mechanics. Methods and Applications	**187**
§1. Lie Algebras and Mechanics	187
1.1. Embeddings of Dynamic Systems into Lie Algebras	187
1.2. List of the Discovered Maximal Linear Commutative Algebras of Polynomials on the Orbits of Coadjoint Representations of Lie Groups	189
§2. Integrable Multidimensional Analogues of Mechanical Systems Whose Quadratic Hamiltonians are Contained in the Discovered Maximal Linear Commutative Algebras of Polynomials on Orbits of Lie Algebras	207
2.1. The Description of Integrable Quadratic Hamiltonians	207
2.2. Cases of Complete Integrability of Equations of Various Motions of a Rigid Body	210
2.3. Geometric Properties of Rigid-Body Invariant Metrics on Homogeneous Spaces	216

§3. Euler Equations on the Lie Algebra so(4)   220

§4. Duplication of Integrable Analogues of the Euler Equations by Means of Associative Algebra with Poincaré Duality   231

    4.1. Algorithm for Constructing Integrable Lie Algebras   231

    4.2. Frobenius Algebras and Extensions of Lie Algebras   236

    4.3. Maximal Linear Commutative Algebras of Functions on Contractions of Lie Algebras   243

§5. The Orbit Method in Hamiltonian Mechanics and Spin Dynamics of Superfluid Helium-3   250

Chapter 5. Nonintegrability of Certain Classical Hamiltonian Systems   256

§1. The Proof of Nonintegrability by the Poincaré Method   256

    1.1. Perturbation Theory and the Study of Systems Close to Integrable   256

    1.2. Nonintegrability of the Equations of Motion of a Dynamically Nonsymmetric Rigid Body with a Fixed Point   260

    1.3. Separatrix Splitting   261

    1.4. Nonintegrability in the General Case of the Kirchhoff Equations of Motion of a Rigid Body in an Ideal Liquid   266

§2. Topological Obstacles for Complete Integrability   267

    2.1. Nonintegrability of the Equations of Motion of Natural Mechanical Systems with Two Degrees of Freedom on High-Genus Surfaces   267

    2.2. Nonintegrability of Geodesic Flows on High-Genus Riemann Surfaces with Convex Boundary   272

    2.3. Nonintegrability of the Problem of n Gravitating Centres for $n > 2$ 275

    2.4. Nonintegrability of Several Gyroscopic Systems   277

§3. Topological Obstacles for Analytic Integrability of Geodesic Flows on Non-Simply-Connected Manifolds   281

§4. Integrability and Nonintegrability of Geodesic Flows on Two-Dimensional Surfaces, Spheres, and Tori   287

    4.1. The Holomorphic 1-Form of the Integral of a Geodesic Flow Polynomial in Momenta and the Theorem on Nonintegrability of Geodesic Flows on Compact Surfaces of Genus $g > 1$ in the Class of Functions Analytic in Momenta   287

    4.2. The Case of a Sphere and a Torus   291

    4.3. The Properties of Integrable Geodesic Flows on the Sphere   294

Chapter 6. A New Topological Invariant of Hamiltonian Systems of Liouville-Integrable Differential Equations. An Invariant Portrait of Integrable Equations and Hamiltonians   300

§1. Construction of the Topological Invariant   300

§2. Calculation of Topological Invariants of Certain Classical Mechanical Systems   311

§3. Morse-Type Theory for Hamiltonian Systems Integrated by Means of Non-Bott Integrals   324

References   326

Subject Index   341

PREFACE

In recent years, a new branch of science has appeared which originates in the classical theoretical mechanics, mathematical physics, the theory of Hamiltonian systems, and symplectic geometry. This branch may be conditionally outlined as follows: new methods of integration of Hamiltonian systems on symplectic manifolds. New profound relations between integrability of many systems and their implicit algebraic properties have been revealed. Among these properties, of primary importance are "system symmetries" which are not merely understood as groups of system invariance but in a more general sense as a set of algebraic properties of a system of differential equations which enable this system to be naturally "embedded," for instance, in the Lie algebra of a certain Lie group with preservation of its dynamical structure (for instance, with preservation of its Hamiltonian properties).

These mechanisms turned out to control integrals of many interesting Hamiltonian systems of equations which arise in geometry as well as in mechanics and physics. It is known that finding integrals for a concrete system is not a simple task. Furthermore, systems in "general position" do not generally have a sufficient number of integrals to integrate the equations. Therefore, sufficiently effective methods should be worked out to help search for rare integrable cases in the boundless ocean of all Hamiltonian systems (the "majority" of which are certainly nonintegrable). One of the aims of the present book is to acquaint the reader with several algorithms for finding integrals. Special attention is given to "formula-type" algorithms, that is, those algorithms which make it possible to efficiently represent integrals (whenever they are found) in an explicit form, for instance, in the form of polynomials or rational functions. This sometimes yields formulae explicitly for solutions (that is, integral trajectories) of a Hamiltonian system.

It is clear that we cannot omit the general mechanism responsible for nonintegrability of systems in general position. We believe it is instructive to present the methods of integration and the methods of proving nonintegrability of Hamiltonian systems in one book. Therefore, in the second part of the book, we give substantial attention to nonintegrability problems, especially to the qualitative aspect of the discussed effects, leaving out calculations and replacing them by references to the corresponding literature.

The book is aimed at acquainting a wide range of readers with geometric and algebraic mechanisms of integrability and nonintegrability. This aim determines the style of presentation, which is made as simple as possible. For the reader's convenience, we introduce the necessary information from related fields of science, such as the theory of Lie groups, symplectic geometry, topology, etc.

Thus, the main topics elucidated in the book are formulated as follows: 1) some mechanical systems and the corresponding Hamiltonian systems of differential equations; 2) fundamentals of symplectic geometry; 3) qualitative topological theory of integrable systems, classification of constant-energy surfaces of integrable systems, and Morse theory of integrable systems; 4) classification of Liouville torus surgery at the moment of intersection of critical energy levels; 5) general principles for integrating Hamiltonian systems: commutative and noncommutative integration, and applications; 6) integration of some concrete dynamical systems on symplectic manifolds. General methods and applications; 7) general mechanisms of nonintegrability of Hamiltonian systems.

This book is based on a two-year lecture course delivered by the author to students at the Department of Mechanics and Mathematics at Moscow State University in 1983–1985. In addition, the results of recent studies are included. Among the original material involved are the results obtained by the author and by participants of the seminar "Modern Geometric Methods," headed by the author at Moscow University in 1984–1987.

The author planned the program of the study of maximal involutive sets of functions on the orbits of Lie algebras. Investigations in this direction were started and first discussed at the above-mentioned seminar in 1980. The most interesting results obtained are also included in the book.

The book is intended for a wide range of readers, for those who are interested in applications of modern geometry to Hamiltonian mechanics and in the theory of integration of differential equations.

In the framework of a comparatively small book, it is practically impossible to give an exhaustive review of the modern state of integrability and nonintegrability of Hamiltonian systems of differential equations. The list of refereces includes some papers which will help the reader to study the problem. In particular, we recommend the papers by the following authors.

I. M. Gel'fand, S. P. Novikov, L. D. Faddeev, R. Bott, M. F. Atiyah, V. I. Arnold, J. Moser, V. P. Maslov, M. Adler, van Moerbeke, H. P. McKean, F. Calogero, O. I. Bogoyavlensky, A. M. Perelomov, A. V. Brailov, V. V. Trofimmov, A. M. Vinogradov, B. A. Kupershmidt, M. Vergne, C. S. Gardner, C. Godbillon, Yu. I. Manin, V. G. Drinfeld, V. E. Zakharov, B. Kostant, A. G.Reyman, D. Kazhdan, S. Sternberg, V. V. Koslov, P. D. Lax, J. Dixmier, D. Mumford, J. Marsden, A. Weinstein, M. D. Kruskall, F. Margi, M. A. Olshanetsky, M. A. Semenov-Tian-Shansky, T. Ratiu, A. V. Bolsinov, Ya. V. Tatarinov, A. Thimm, S. J. Takiff, S. L. Ziglin, V. Guillemin, M. Rais, M. Duflo, C. Conley, E. Zehnder, D. Ebin. G. Wilson, M. V. Karasev, B. A. Dubrovin, V. L. Golo, A. P. Veselov.

I express my gratitude to S. P. Novikov, V. I. Arnold, D. V. Anosov, Yu. I. Manin, H. Zieschang, V. V. Golo, O. I. Bogoyavlensky, Ya. V. Tatarinov, S. V. Matveev, V. V. Trofimmov, A. V. Brailov, and A. V. Bolsinov for stimulating scientific discussions which resulted in improvement and extension of some sections of the book.

I would like to thank the Reidel Publishing Company and Professor M. Hazewinkel and Dr. D. J. Larner personally for their support and interest in my work.

I am also grateful to M. V. Tsaplina for much effort put in the translation of the book.

CHAPTER 1

SOME EQUATIONS OF CLASSICAL MECHANICS AND THEIR HAMILTONIAN PROPERTIES

§1 Classical Equations of Motion of a Three-Dimensional Rigid Body

1.1 *The Euler–Poisson Equations Describing the Motion of a Heavy Rigid Body around a Fixed Point*

Let us consider a rigid body located in a gravity field and moving around a fixed point 0. To described this motion, it is convenient to use two Cartesian coordinate systems. The *first* one is *fixed* in an enveloping three-dimensional Euclidean space $\mathbb{R}^3$, and the coordinates of the point with respect to the system are denoted as x_1, y_1, z_1. The *second, moving* coordinate system is rigidly connected with the body ("frozen-in" in it), and the coordinates of the point with respect to this system are denoted by x, y, z. It is convenient to assume that the axes of the moving ("frozen-in" in the body) coordinate system are directed along the principal axes of inertia of the body for the point 0. Consider a unit vector e directed along the z_1-axis fixed in the space of the coordinate system. Let the coordinates of this vector be α, β, γ [with respect to the moving coordinate system (rotating with the body)]. It is clear that these coordinates are the functions of time t (Fig. 1).

Let us denote the centre of mass of the rigid body by P and let (x_0, y_0, z_0) be the coordinates of the body with respect to the moving coordinate system. Since this system is rigidly connected with the body, the numbers x_0, y_0, z_0 do not depend on time (the centre of mass does not change its position in the body with respect to the frozen-in coordinate axes). Let m be *the mass of the rigid body* and A, B, C the *principal moments of inertia of the body with respect to the point* 0. By virtue of the choice of the moving coordinate system, the inertia tensor of the rigid body may be written as the following quadratic diagonal matrix $\begin{pmatrix} A & 0 & 0 \\ 0 & B & 0 \\ 0 & 0 & C \end{pmatrix}$. The tensor is taken in the diagonal form because the axes of the frozen-in coordinate system are directed along the principal axes of inertia of the body. Let ω be the vector of the instantaneous angular velocity of the body, whose coordinates with respect to

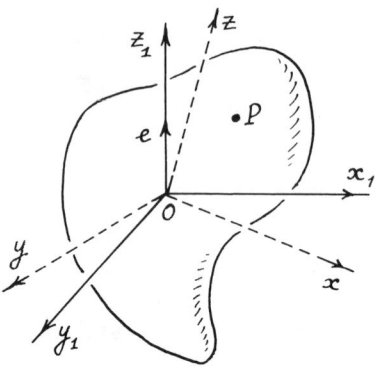

Figure 1

the moving coordinate system will be denoted by (p, q, r). It is clear that p, q, r are functions of time. Thus, the motion of a rigid body is described by six functions $p(t), q(t), r(t), \alpha(t), \beta(t), \gamma(t)$. The free fall acceleration will be denoted by g. It is obvious that if d is a gravitational vector applied in the centre of mass of the body, then d is directed vertically downward (along the z_1- axis) and has the form $d = -\text{mge}$.

From classical mechanics it is known that if K is the angular momentum of the body with respect to the point 0 and M is the moment of external forces (in our case, the moment of the gravitational force d) then, according to the theorem on the change of angular momentum, the following relation $\frac{dK}{dt} = M$, written with respect to the fixed coordinate system (x_1, y_1, z_1), holds. The quantities K and M can be explicitly calculated; however, the calculations are omitted. Projecting this vector equation onto the axes of the moving coordinate system (x, y, z), we can derive the following three scalar equations which are usually called *Euler dynamic equations*:

$$A\frac{dp}{dt} + (C - B)qr = mg(z_0\beta - y_0\gamma),$$
$$B\frac{dq}{dt} + (A - C)rp = mg(x_0\gamma - z_0\alpha),$$
$$C\frac{dr}{dt} + (B - A)pq = mg(y_0\alpha - x_0\beta).$$

These equations contain six unknown functions of time: $p, q, r, \alpha, \beta, \gamma$ and six constant quantities: A, B, C, x_0, y_0, z_0 characterizing body mass distribution with respect to the principal axes of inertia for the point 0. To make the system of equations closed, one should add three more equations which will imply that the velocity of the endpoint of the unit vector e in the fixed coordinate system (x_1, y_1, z_1) is equal to zero (the vector is fixed). Projecting the obtained vector equation onto the axes of the moving coordinate system, we can derive the three missing scalar

equations
$$\frac{d\gamma}{dt} = r\beta - q\gamma, \quad \frac{d\beta}{dt} = p\gamma - r\gamma, \quad \frac{d\gamma}{dt} - q\alpha - p\beta.$$
These equations are called the *Poisson equations*. The complete system of all six differential equations is called the *Euler–Poisson equations*. Their integration is a rather nontrivial problem.

The position of a rigid body in $\mathbf{R}^3$ at each time moment is uniquely determined by the position of the moving coordinate system (x, y, z) "frozen-in" in the body with respect to the fixed coordinate system (x_1, y_1, z_1). To set this position, it suffices to set the so-called *Euler angles* θ, φ, ψ (Fig. 2).

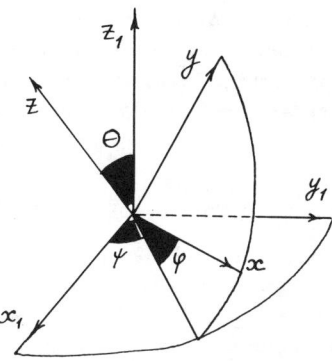

Figure 2

The angle θ is usually called the *nutation angle* and ψ the *precession angle*. The Euler angles can be expressed by the functions $p, q, r, \alpha, \beta, \gamma$ introduced above. By a direct calculation it can be found that

$$\theta = \arccos(\gamma), \quad \varphi = \arctan\frac{d}{\beta}, \quad \dot\psi = \frac{p\alpha + q\beta}{\alpha^2 + \beta^2},$$

It is seen that if the solutions of the Euler–Poisson equations

$$p(t), q(t), r(t), \alpha(t), \beta(t), \varphi(t)$$

are found in one or another way, then the above formulae give us the variation laws for the *nutation angles* $\theta(t)$ and for the *proper rotation* $\varphi(t)$ of the rigid body. The variation law for the *precession angle* $\psi(t)$ is determined by the quadrature from the last formula (see above). Thus, to describe the motion of a heavy rigid body, one has to solve the Euler–Poisson equations. From the theory of differential equations, it is known that finding a first integral of a system makes it possible to lower the order of the system by unity. If k functionally independent first integrals are found, the order of the system may be lowered by an integer k. One of the most important problems in the theory of motion is therefore the search for a sufficient number of independent integrals. For complete integration of Euler–Poisson equations, it suffices to find five independent first integrals which do not contain time and then to perform yet another simple quadrature. Since the latter operation does not encounter essential difficulties. the problem is reduced mainly to searching for five independent integrals. Three of them prove to be easily specified.

1) Since the functions $\alpha(t), \beta(t), \gamma(t)$ are coordinates of the unit vector e, it is obvious that $\alpha^2, \beta^2, \gamma^2 = 1$.

2) The *energy integral* $Ap^2 + Bq^2 + Cr^2 + 2mg \times (x_0\alpha + y_0\beta + z_0\gamma) = $ const.

3) The *area integral* $Ap\alpha + Bq\beta + Cr\gamma = $ const.

These integrals are written down in the coordinates of the moving system (x, y, z). Hence, for a complete integration of the Euler–Poisson equations it suffices to find *two more independent first integrals*.

Actually, it suffices to specify *only one more* independent (fourth) integral (which does not depend on time), after which the system can be completely integrated. Let us discuss this important circumstance in more detail.

Recall the known *Jacobi theorem* from the theory of differential equations. Let the differential equation

$$\frac{dx_1}{X_1(x_1, x_2)} = \frac{dx_2}{X_2(x_1, x_2)}$$

be given, which can be rewritten as $X_2 dx_1 - X_1 dx_2 = 0$. It is well known that to find the first integral of this equation, one should find such a function (the so-called *Jacobi multiplier*) $M(x_1, x_2)$, multiplication by which of our equation would transform its left-hand side into a complete differential of a certain function $f(x_1, x_2)$, i.e., such a function that the equality: $M(X_2 dx_1 - X_1 dx_2) = df$ holds. Indeed, if such a multiplier does exist, then $df = 0$ by virtue of our differential equation, i.e., $f(x_1, x_2) = $ const is just the required integral. The *integrating multiplier* (the Jacobi multiplier) can be found by way of solving a certain differential equation.

Indeed, since $df = \frac{\partial f}{\partial x_1} dx_1 + \frac{\partial f}{\partial x_2} dx_2$, then from the definition of M, it follows that $\frac{\partial f}{\partial x_1} = MX_2$ and $\frac{\partial f}{\partial x_2} = -MX_1$. From this and from the identity

$$\frac{\partial^2 f}{\partial x_1 \partial x_2} = \frac{\partial^2 f}{\partial x_2 \partial x_1},$$

it follows that $\frac{\partial}{\partial x_1}(MX_1) + \frac{\partial}{\partial x_2}(MX_2) = 0$. If the factor M is known in advance, then the first integral can be found by the formula

$$f(x_1, x_2) = \int_h^{x_1} M(x_1, x_2) X_2(x_1, x_2) dx_1 - \int_h^{x_2} M(h, x_2) X_1(h, x_2) dx_2 = \text{const}$$

This implies that to find the first integral $f(x_1, x_2) = $ const of the original equation, it suffices to solve the equation $\frac{\partial}{\partial x_1}(MX_1) + \frac{\partial}{\partial x_2}(MX_2) = 0$ (for the integrating factor), with any particular solution $M(x_1, x_2)$ of this equation being sufficient.

A similar construction is also fit for a system of n first-order equations

$$\frac{dx_i}{dt} = X_i(x_1, \ldots, x_n), 1 \leqslant i \leqslant n.$$

§1. Classical Equations of Motion

Let us represent this system in the form

$$\frac{dx_1}{X_1} = \frac{dx_2}{X_2} = \cdots = \frac{dx_n}{X_n}.$$

The function $M(x_1,\ldots,x_n)$, which is one of the solutions of the equation $\frac{\partial}{\partial x_1}(MX_1) + \cdots + \frac{\partial}{\partial x_n}(MX_n) = 0$, will be called the *last (integrating) Jacobi multiplier*.

JACOBI THEOREM. *If for a system of differential equations $\frac{dx_1}{X_1} = \cdots = \frac{dx_n}{X_n}$ the last multiplier $M(x_1,\ldots,x_n)$ and $n-2$ first integrals of the system are known, that is, $f_3(x_1,\ldots,x_n) = \text{const},\ldots,f_n(x_1,\ldots,x_n) = \text{const}$, then using the change of variables $y_1 = x_1, y_2 = x_2, y_3 = f_3,\ldots,y_n = f_n$ one reduces the original system of equations to the equation $\frac{dy_1}{Y_1} = \cdots = \frac{dy_n}{Y_n}$ by means of the known integrating factor $M(y_1,\ldots,y_n)$ which is algorithmically expressed by the known $n-2$ first integrals $f_3,\ldots,f_n$ and by the function M.*

We omit here the explicit expression because we will not need it hereafter. Thus, the general case is reduced to the case of an equation for two variables, which is integrated using the above technique. Now go back to the analysis of the Euler–Poisson equations. Integration of these equations is apparently equivalent to integration of the system

$$\frac{dp}{dt} = p', \quad \frac{dq}{dt} = q', \quad \frac{dr}{dt} = r', \quad \frac{d\alpha}{dt} = \alpha', \quad \frac{d\beta}{dt} = \beta', \quad \frac{d\gamma}{dt} = \gamma',$$

where

$$p' = \frac{1}{A}\left(mg(z_0\beta - y_0\gamma) - (C-B)qr\right),$$
$$q' = \frac{1}{B}\left(mg(x_0\gamma - z_0\alpha) - (A-C)rp\right),$$
$$r' = \frac{1}{C}\left(mg(y_0\alpha - x_0\beta) - (B-A)pq\right),$$
$$\alpha' = r\beta - q\gamma,$$
$$\beta' = p\gamma - r\alpha,$$
$$\gamma' = q\alpha - p\beta.$$

This system can be represented in the following canonical form

$$\frac{dp}{p'} = \frac{dq}{q'} = \frac{dr}{r'} = \frac{d\alpha}{\alpha'} = \frac{d\beta}{\beta'} = \frac{d\gamma}{\gamma'} = dt.$$

As mentioned above, this system always possesses three independent first integrals

$$\begin{cases} Ap^2 + Bq^2 + Cr^2 + 2mg(x_0\alpha + y_0\beta + z_0\gamma) + C_1 & (=\text{const}) \\ Ap\alpha + B_q\beta + Cr\gamma = C_2 & (=\text{const}) \\ \alpha^2 + \beta^2 + \gamma^2 = 1 & (=C_3 = \text{const}) \end{cases}$$

Now set an equation for the last integrating factor M:

$$\frac{\partial}{\partial p}(Mp') + \frac{\partial}{\partial q}(Mq') + \frac{\partial}{\partial r}(Mr') + \frac{\partial}{\partial \alpha}(M\alpha') + \frac{\partial}{\partial \beta}(M\beta') + \frac{\partial}{\partial \gamma}(M\gamma') = 0.$$

From the explicit formulae which set the functions $p', q', r', \alpha', \beta', \gamma'$ (see above) as functions of the variables $p, q, r, \alpha, \beta, \gamma$, we easily obtain

$$\frac{\partial p'}{\partial p} = \frac{\partial q'}{\partial q} = \frac{\partial r'}{\partial r} = \frac{\partial \alpha'}{\partial \alpha} = \frac{\partial \beta'}{\partial \beta} = \frac{\partial \gamma'}{\partial \gamma} = 0.$$

Therefore, the equation for M is certainly satisfied if we set $M \equiv 1$. Thus, we have found one particular solution of this equation, and $M \equiv 1$ may be considered to be precisely the last multiplier of the Euler–Poisson equations.

Hence, according to the Jacobi theorem, to reduce these equations to quadratures, it suffices to know only four (out of the five) first integrals, that is, one should add to the three already known quadratic integrals an additional fourth integral.

Many papers of outstanding mathematicians are devoted to the search for this *fourth integral*. Although *it has not been found in the general case*, nonetheless many of its properties can be revealed, and in certain particular cases the fourth integral can be manifestly specified. The fourth integral can be sought for among different classes of functions, such as analytic ones, smooth ones, and others.

Those particular cases where *the fourth integral can be indicated* are characterized by limitations on the shape of a rigid body, that is, on constants A, B, C (moments of inertia) and on x_0, y_0, z_0 (coordinates of the centre of mass of a body). We list these particular cases.

1.2 Integrable Euler, Lagrange, and Kovalevskaya Cases

1) The *Euler case*. The case is characterized by the fact that a body of an arbitrary shape is supposed to be fixed *in its centre of mass*, that is, $x_0 = y_0 = z_0 = 0$. In this case, the Euler equations take the form

$$A\frac{dp}{dt} + (C-B)qr = 0, \quad B\frac{dq}{dt} + (A-C)rp = 0, \quad C\frac{dr}{dt} + (B-A)pq = 0.$$

Multiplying them respectively by Ap, Bq, Cr and summing up, we obviously obtain the required fourth integral $A^2p^2 + B^2q^2 + C^2r^2 = C_4 (= \text{const})$. The Euler equations are interpreted as a smooth vector field in the Euclidean space $\mathbb{R}^3$ of the variables p, q, r. Since p, q, r are coordinates of the vector ω of instantaneous angular velocity in the body, then the integral trajectories of the Euler equations describe the time evolution of the angular velocity vector (with respect to the coordinate system "frozen-in" in the body).

The Euler equations have two first integrals: $Ap^2 + Bq^2 + Cr^2 = \text{const}$ and $A^2p^2 + B^2q^2 + C^2r^2 = \text{const}$. To investigate the trajectories of the system, it is convenient to consider the *vector of the angular momentum* K which is connected with the vector ω of the *angular velocity* by a simple relation $K = h(\omega)$, where $h = \begin{pmatrix} A & 0 & 0 \\ 0 & B & 0 \\ 0 & 0 & C \end{pmatrix}$ is the *inertia operator*, that is, a diagonal matrix whose eigenvalues coincide with the *moments of inertia*. Then the Euler equations can be written in terms of the angular momentum as $\frac{dK}{dt} = [K, \omega]$, where $K = h(\omega)$ and $[\ ,\ ]$ stands for the vector production of vectors in $\mathbb{R}^3$. This differential equation may, of course, be interpreted as an equation for angular velocity vector ω, because the operator h is invertible (in the case of general position when all the three principal moments of inertia are nonzero). Since $\omega = h^{-1}(K)$, then $h(\dot{\omega}) = [h(\omega), \omega]$, or $\dot{\omega} = h^{-1}[h(\omega), \omega]$.

If the Cartesian coordinates of the vector of the angular momentum K are denoted by (k_1, k_2, k_3), then the Euler equations take the form

$$\frac{dk_1}{dt} = a_1 k_2 k_3, \qquad \frac{dk_2}{dt} = a_2 k_3 k_1, \qquad \frac{dk_3}{dt} = a_3 k_1 k_2,$$

where

$$a_1 = \frac{B-C}{BC}, \qquad a_2 = \frac{C-A}{CA}, \qquad a_3 = \frac{A-B}{AB}.$$

Then the indicated two first quadratic integrals of the Euler equations are written as

$$\frac{k_1^2}{A} + \frac{k_2^2}{B} + \frac{k_3^2}{C} = \text{const}$$

and

$$k_1^2 + k_2^2 + k_3^2 = \text{const}.$$

The first of them implies the *energy conservation law*, and the second implies the angular momentum conservation law. Thus, moving along integral trajectories of the vector field described by the Euler equations, the angular momentum vector $K(t)$ moves actually along the trajectories obtained as intersections of a sphere and an ellipsoid (Fig. 3).

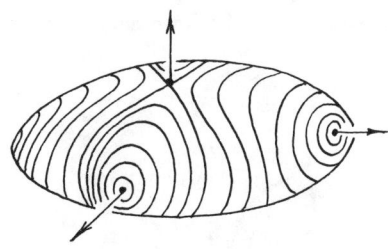

Figure 3

For definiteness, one may fix the ellipsoid and vary the sphere radius. The qualitative picture of the trajectories is demonstrated in Fig. 3. Among them one should distinguish six points which are endpoints of six semiaxes of the ellipsoid. These are separate trajectories of the Euler equations. To each of these six positions of the angular momentum vector, there corresponds a constant value of the angular velocity vector directed along one of the axes of inertia of the body, the angular velocity ω remaining all the time collinear to the angular momentum vector. Thus, we deal with *a body rotating at a constant angular velocity around a fixed axis of inertia of the body*. The motion of a body where the angular velocity vector is constant is called the *stationary rotation* in classical mechanics. Thus the existence of the indicated two first integrals results in the following: *a rigid body fixed in its centre of mass always admits stationary rotation around any of its three axes of inertia*.

These rotations differ from one another in their stability. One can verify that the stationary solutions $K(t)$ corresponding to the major and the minor axes of inertia (in the case where all the moments of inertia are pairwise different) are *Lyapunov-stable*, and the solution $K(t)$, which corresponds to the mean axis, is *unstable*.

Indeed, as is seen in Fig. 3, under small pertubation of initial data near the *major and minor axes*, the integral trajectory of the Euler equations remains a small closed curve (circle) and, therefore, the stationary solution $K(t)$ is stable. In the case of the *mean axis*, the small perturbation of the stationary solution makes the endpoint of the angular momentum vector $K(t)$ move along a closed trajectory of "large diameter," as a result of which the vector $K(t)$ quickly moves away from the mean ellipsoid axis (Fig. 3).

Thus, in the Euler case we have a fairly complete description of the body motion around a fixed centre of mass. We shall make a few remarks that will be instructive for the following section. In the Euler case, a rigid body is a mechanical system whose configuration space is a group of orthogonal proper rotations of the space $\mathbb{R}^3$, that is, a group $SO(3)$. As is seen hereafter, this situation is a particular case of more general "group systems" which are multidimensional analogues of the Euler equations and also admit complete integration.

2) The *Lagrange case*. The case is characterized by the fact that $A = B$ and $x_0 = y_0 = 0$. This means that we deal with an *axisymmetric rigid body*, and on account of this, the Lagrange case is usually called the case of *symmetric top*. In other words, the *symmetric top* (or the *Lagrange top*) is a rigid body whose inertia ellipsoid is an ellipsoid of rotation and whose centre of gravity lies on the rotation axis (Fig. 4).

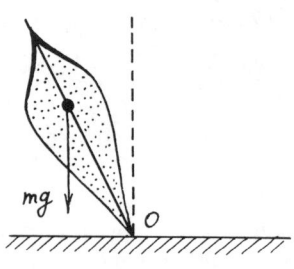

Figure 4

In this case, the last of the Euler equations takes the form $\frac{dr}{dt} = 0$. The fourth required integral is therefore of particularly simple form, $r = $ const. The motion of the Lagrange top can be described in detail, for instance, in terms of the Euler angles θ, φ, ψ. In particular, the slope θ of the Lagrange top axis periodically changes with time between two limit values $\theta_{\min}$ and $\theta_{\max}$ (Fig. 5).

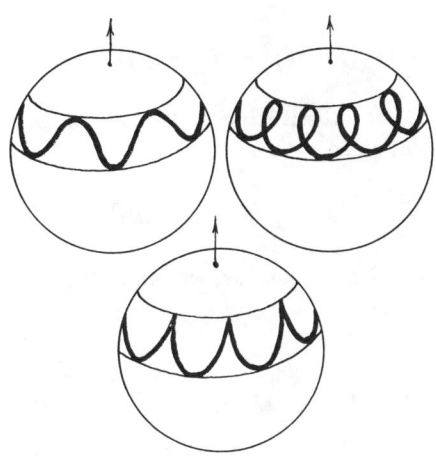

Figure 5

If one considers two parallels corresponding to $\theta_{\min}$ and $\theta_{\max}$, then between them there appears a strip in which the top axis moves. A periodic change of the angle θ is called *nutation*. A change of the angle φ is called *precession*. Consequently, the resultant motion of the Lagrange top consists of rotation, nutation, and precession (for the geometric properties of these motions, see, for instance, Ref. 3).

3) The *Kovalevskaya case*. This important case of integrability has lately become (like the Euler case) of increasing interest because its multidimensional analogues have been discovered. It is relevant to note here an exceeding attention given in recent years to the works by Kovalevskaya and to her biography (for instance, the fundamental monograph by Koblitz [154]).

The Kovalevskaya case is characterized by the fact that $A = B = 2C$ and $y_0 = z_0 = 0$. Here the Euler equations take the form

$$2\frac{dp}{dt} = qr, \qquad 2\frac{dq}{dt} + rp = c\gamma, \qquad \frac{dr}{dt} = -c\beta,$$

where the constant c is given by $c = \frac{1}{C}mgx_0$ (C is the third moment of inertia). Multiplying the second equation by i and adding it to the first one, we get $2\frac{d}{dt}(p + iq) = -ir(p - iq) + ic\gamma$. Now let us consider the first two Poisson equations: $\frac{d\alpha}{dt} = r\beta - q\gamma$, $\frac{d\beta}{dt} = p\gamma - rd$. Multiplying the second equation by i and adding it to the first one, we get $\frac{d}{dt}(\alpha + i\beta) = -ir(\alpha + i\beta) + i\gamma(p + iq)$. Now we eliminate the quantity γ from the two derived complex equations. To this end, we multiply the first equation by $p+iq$ and the second equation by $-c$ and add the results together. Thus, we have

$$\frac{d}{dt}\left[(p+iq)^2 - c(\alpha+i\beta)\right] = ir\left[(p+iq)^2 - c(\alpha+i\beta)\right].$$

This equation is obviously rewritten to become

$$\frac{d}{dt}\ln\left[(p+iq)^2 - c(\alpha+i\beta)\right] = -ir.$$

In a similar manner, but substituting multiplication by $-i$ for the multiplication by i, we derive the equation

$$\frac{d}{dt}\ln\left[(p-iq)^2 - c(\alpha-i\beta)\right] = ir.$$

By adding together the two last equations, we obtain the identity

$$\frac{d}{dt}\ln\left[(p+iq)^2 - c(\alpha+i\beta)\right] \cdot \left[(p-iq)^2 - c(\alpha-i\beta)\right] = 0$$

Thus, *we have found the fourth integral*, which can be written as

$$\left[(p+iq)^2 - c(\alpha+i\beta)\right] \cdot \left[(p-iq)^2 - c(\alpha-i\beta)\right] = \text{const}$$

or

$$(p^2 - q^2 - c\alpha)^2 + (2pq - c\beta)^2 = \text{const}.$$

In all three cases (Euler, Lagrange, Kovalevskaya) the problem reduces to quadratures. The general solution of the Euler–Poisson equations is expressed in the Euler and Lagrange cases in terms of *elliptic functions*, and in the Kovalevskaya case in terms of *hyperelliptic functions* (see, for instance, [156]).

The integrable cases also include the so-called Goryachev–Chaplygin case, where integrability takes place *on one level surface*.

1.3 *General Equations of Motion of a Three-Dimensional Rigid Body*

The motion of a rigid body in $\mathbb{R}^3$ under the action of various forces is described in many problems of classical mechanics by the following general equations $\dot{K} = [K,\omega] - [e,u], \dot{e} = [e,\omega]$, where K is the *angular momentum of the body*, ω is the *angular velocity of the body*, and the vectors e and u determine the physical content of the problem and depend on the conditions of motion of the rigid body. It is convenient to assume that $\omega = \frac{\partial H}{\partial K}$ and $u = \frac{\partial H}{\partial e}$, where $H(K,e)$ is a certain known function on the Euclidean space $\mathbb{R}^6 = \mathbb{R}^3(K) \times \mathbb{R}^3(e)$. Here each of the vectors K and e changes in the three-dimensional space $\mathbb{R}^3(K)$ and $\mathbb{R}^3(e)$, respectively. As a particular case, we obtain the above *equations of motion of a rigid body with a fixed point in the gravity field*. Indeed, let e be a unit vector directed along the vertical z-axis and let $u = mr$ be the product of the body mass m by the radius-vector of the centre of mass. The *total energy*, that is, the function H then has the following form:

$$H = \frac{1}{2}\langle K, h^{-1}(k)\rangle + m\langle r, e\rangle.$$

Here $\langle \, , \, \rangle$ is an ordinary *Euclidean scalar product* in $\mathbb{R}^3$, and h^{-1} is *a positive definite self-conjugate operator*. The equations of motion of the body will now be written as

$$h(\dot\omega) = [h(\omega), \omega] + m[e, r], \dot e = [e, \omega].$$

Comparing these equations with those of Subsection 1.1, we see that they are exactly the *Euler–Poisson equations* of motion of a heavy rigid body with a fixed point. Since h is a self-conjugate operator, in a certain orthogonal reper connected with the rotating body, it is then reduced to a diagonal form and then its matrix becomes

$$\begin{pmatrix} h_1 & 0 & 0 \\ 0 & h_2 & 0 \\ 0 & 0 & h_3 \end{pmatrix}.$$

The proper directions of the matrix h are called the *axes of inertia of the body*, and the eigenvalues h_1, h_2, h_3 are called *principal moments of inertia of the body*. In this case, the equations of motion contain six parameters: h_1, h_2, h_3 and mx_0, my_0, mz_0, where $r = (x_0, y_0, z_0)$ is the *radius-vector of the centre of mass of the body*.

We shall give another important example. Take as the function H the following positive definite quadratic form on a six-dimensional space $\mathbb{R}^3 = \mathbb{R}^3(K) \times \mathbb{R}^3(e)$:

$$H(K, e) = \frac{1}{2}\langle AK, K\rangle + \langle BK, e\rangle + \frac{1}{2}\langle Ce, e\rangle.$$

Then the general equations written above are transformed into the so-called *Kirchhoff equations* which describe the *motion of a rigid body in an ideal boundless liquid*. In this case, the vectors e and $u = \frac{\partial H}{\partial e}$ are usually called an *impulsive force* and an *impulsive momentum*, respectively.

Since the matrices A, B, C are symmetric, it follows that without loss of generality, the matrix A may be assumed to have been already reduced to a diagonal form, that is

$$A = \begin{pmatrix} a_1 & 0 & 0 \\ 0 & a_2 & o \\ 0 & 0 & a_3 \end{pmatrix}.$$

Of course, the remaining matrices B and C need not be diagonalized. Hence, in the general case, the quadratic form H (the total energy of the body) depends on fifteen real parameters, that is, the coefficients of the matrices A, B, C. If we impose various limitations of the type of symmetry on the body, then the number of parameters will, of course, decrease. For instance, if a body has three mutually orthogonal symmetry planes (this property is inherent, for instance, in a triaxial ellipsoid), then

$$B = 0, \quad C = \begin{pmatrix} c_1 & 0 & 0 \\ 0 & c_2 & 0 \\ 0 & 0 & c_2 \end{pmatrix}.$$

§2 Symplectic Manifolds

2.1 Symplectic Structure in a Tangent Space to a Manifold

A highly important role in geometry in played by *Riemannian manifolds*, that is, smooth n-dimensional manifolds M^n supplied with a *Riemannian metric* g_{ij}. It is known that setting this metric is equivalent to setting in each tangent space $T_x M^n$ at a point $x \in M$ a *bilinear symmetric nondegenerate positive definite scalar product* which smoothly depends on the point x. If $a, b \in T_x M$ is an arbitrary pair of tangent vectors, this scalar product can be written in the form

$$\langle a, b \rangle = \sum_{i,j=1}^{n} g_{ij} a_i b_j,$$

where

$$g_{ij}(x) = g_{ji}(x); \qquad a = (a_1, \ldots, a_n), \qquad b = (b_1, \ldots, b_n).$$

It is well known that a Riemannian metric can always be set on any finite-dimensional smooth manifold. If a point x is fixed on M, then in the tangent space $T_x M^n$, the form $\langle a, b \rangle$ can always be reduced through a nondegenerate linear transformation to the form

$$\langle a, b \rangle = \sum_{i=1}^{n} a_i b_i,$$

that is, it can transform into an ordinary Euclidean scalar product. Thus, a Riemannian metric determines in each tangent space a Euclidean geometry (after an appropriate choice of coordinates in the tangent plane).

In line with this theory, a *symplectic geometry* in a linear space $T_x M$ is constructed whose properties are determined by setting a *skew-symmetric (!) nondegenerate scalar product* (,). This cannot be done in all cases but only when the dimension of the manifold M is even. Then an even-dimensional space $T_x M^{2n}$ transforms into a *symplectic space* L^{2n}. In what follows, we shall assume for convenience that a symplectic space L^{2n} is *modelled* on a Euclidean space $\mathbb{R}^{2n}$ (on which, therefore, two forms are simultaneously given, namely, symmetric and skew-symmetric).

DEFINITION 1.2.1: We say that in $\mathbb{R}^{2n}$ a *linear symplectic structure* is given if a bilinear skew-symmetric nondegenerte scalar product (,) is given, that is, $(a, b) = -(b, a)$ for any vectors $a, b \in \mathbb{R}^{2n}$ and $(a, b) = 0$ for any $a \in \mathbb{R}^{2n}$ if and only if $b = 0$. The space $\mathbb{R}^{2n}$ with such a structure will be called a *symplectic space*.

If $e_1, \ldots, e_{2n}$ is the basis in $\mathbb{R}^{2n}$, then the scalar product (,) can be written as follows:

$$(a, b) = \sum_{i,j=1}^{2n} \omega_{ij} a_i b_i,$$

where $\Omega = (\omega_{ij})$ is a nondegenerate skew-symmetric matrix, $\omega_{ij} = -\omega_{ji}$. Any nondegenerate linear transformation of a basis leaves the matrix Ω skew-symmetric. The *simplest symplectic structure* in $\mathbb{R}^{2n}$ is constructed in the following way. Divide Euclidean coordinates into two groups: $(p_1, \ldots, p_n, q_1, \ldots, q_n)$ and set the symplectic structure by the formulae

$$(a, b) = \sum_{i=1}^{n} p_i q'_i - q_i p'_i,$$

where

$$a = (p_1, \ldots, p_n, q_1, \ldots, q_n), \qquad b = (p'_1, \ldots, p'_n, q'_1, \ldots, q'_n).$$

In this basis, the matrix Ω will be written as $\begin{pmatrix} 0 & -E \\ E & 0 \end{pmatrix} = I$, where E is a unit matrix of order n. From linear algebra, it is known that if Ω is an arbitrary nondegenerate skew-symmetric matrix, it can be reduced to the above-mentioned canonical form through a linear nondegenerate transformation of the basis.

A basis of $2n$ vectors $\alpha_1, \ldots, \alpha_n, \beta_1, \ldots, \beta_n$, in which the structure Ω takes the canonical form I, will be called *symplectic*. It is clear that the matrix of pairwise scalar products of basis vectors will have the form

$$\begin{pmatrix} 0 & E \\ -E & 0 \end{pmatrix} = \begin{pmatrix} & & & 1 & & \\ & 0 & & & \ddots & \\ & & & & & 1 \\ -1 & & & & & \\ & \ddots & & & 0 & \\ & & -1 & & & \end{pmatrix}$$

that is,

$$(\alpha_i, \alpha_j) = (\beta_i, \beta_j) = 0, \qquad (\alpha_i, \beta_i) = 1, \qquad (\alpha_i, \beta_j) = 0$$

when $i \neq j$. Note that the *scalar square of each vector is equal to zero*, that is, all vectors in a symplectic space are *isotropic*. The vectors α_i and β_i will be called *dual* because $(\alpha_i, \beta_i) = 1$. By analogy with a symmetric scalar product, we say that two vectors a and b are *skew-orthogonal* if their scalar product is equal to zero, that is, $(a, b) = 0$. In particular, *in a symplectic space each vector is skew-orthogonal to itself*. In spite of these distinctions between Euclidean and symplectic scalar products, they have many common features.

A set of all vectors, skew-orthogonal to all vectors from a certain plane Π^k in $\mathbb{R}^{2n}$, is called a *skew-orthogonal complement* of the plane Π in $\mathbb{R}^{2n}$. In particular, a skew-orthogonal complement of one vector a is a $(2n-1)$-dimensional hyperplane in $\mathbb{R}^{2n}$, which contains the *vector a*. This follows from the nondegeneracy of symplectic structure, because a nonzero vector a cannot be skew-orthogonal to the entire space $\mathbb{R}^{2n}$. It is also clear that if T is a skew-orthogonal complement of the plane Π^k in $\mathbb{R}^{2n}$, then the dimension of T is equal to $2n - k$.

A subspace K in $\mathbb{R}^{2n}$ will be called *symplectic* if the restriction imposed on it by the symplectic structure $(\ ,\ )$ is nondegenerate.

It follows from the linear algebra (see above) that in each *symplectic space* $\mathbb{R}^{2n}$, *there always exists a symplectic basis*. There are many such bases and, besides, as the first vector of a symplectic basis one may take any nonzero vector $a \in \mathbb{R}^{2n}$.

We will prove this instructive assertion using induction by n. When $n = 1$, for a vector e_2 which complements a vector $e_1 = a$ to a symplectic basis in $\mathbb{R}^2$, it suffices to take a vector which is not skew-orthogonal to a in $\mathbb{R}^2$ and to choose its length so that $(b, a) = 1$. Such a vector b exists by virtue of nondegeneracy of the form $(\ ,\ )$.

Now let $n > 1$ and $a = e_1 \neq 0$. As for the first stage, we shall choose the vector $e_2 \in \mathbb{R}^{2n}$, such that $(e_2, e_1) = 1$. Let us consider a two-dimensional plane $\sqcap^2$ spanned by these vectors and let K be a skew-orthogonal complement of $\sqcap^2$ in $\mathbb{R}^{2n}$. It is clear that the dimension of K is equal to $2n - 2$ and, besides, neither of the vectors e_1 and e_2 lies in K. Indeed, if, for instance, e_1 lay in K, then e_1 would be skew-orthogonal to e_2, which is impossible by virtue of the choice of e_2. Now let us show that a $(2n - 2)$-dimensional plane K is a symplectic subspace in $\mathbb{R}^{2n}$, i.e., that the restriction imposed by the symplectic structure on K is nondegenerate. Indeed, suppose the converse and let a certain vector $h \in K$ be skew-orthogonal to the entire plane K. But in this case, it is skew-orthogonal to the entire space $\mathbb{R}^{2n}$, because the plane K is skew-orthogonal to the plane $\sqcap^2$, and the sum of K and $\sqcap^2$ gives $\mathbb{R}^{2n}$. Since by the induction hypothesis one can choose a symplectic basis in a symplectic space K^{2n-2}, then complementing it with two vectors e_2 and e_1, we just obtain a symplectic basis in the entire space $\mathbb{R}^{2n}$, and the assertion follows.

DEFINITION 1.2.2: We say that a plane $\sqcap^k$ in a symplectic space $\mathbb{R}^{2n}$ is *isotropic* if it is skew-orthogonal to itself, that is to say, a skew-scalar product of any two vectors of the plane is equal to zero. If k is equal to n (that is, to half the dimension of $\mathbb{R}^{2n}$ then an isotropic plane will be called a *Lagrangian plane*.

Lagrangian planes are isotropic planes of maximal possible dimension. Since, for the sake of convenience, we model a symplectic structure on $\mathbb{R}^{2n}$, this permits setting a skew-symmetric product in a distinct form. If

$$\langle a, b \rangle = \sum_i p_i p'_i + q_i q'_i,$$

where

$$a = (p_1, \ldots, q_n), \qquad b = (p'_1, \ldots, q'_n)$$

in $\mathbb{R}^{2n}$, then the product (a, b) can be rewritten as $(a, b) = \langle \Omega a, b \rangle$, where the operator $\Omega : \mathbb{R}^{2n} \to \mathbb{R}^{2n}$ sets a certain linear nondegenerate transformation. Clearly, the operator Ω is skew-symmetric.

We shall assume the symplectic and Euclidean structures to be *in correspondence* in $\mathbb{R}^{2n}$ in the sense that the vectors of the symplectic basis $\alpha_1, \ldots, \alpha_n, \beta_1, \ldots, \beta_n$ are orthonormalized, that is, they are mutually orthogonal, and their Euclidean length is equal to unity. In a symplectic orthonormalized basis, the operator Ω is given by the matrix

$$I = \begin{pmatrix} 0 & -E \\ E & 0 \end{pmatrix},$$

in particular, $\Omega^2 = -E$. The action of the operator Ω is as follows: in each of n two-dimensional planes spanned by the vectors α_1, β_1, there occurs rotation by an angle $\pi/2$.

LEMMA 1.2.1. *A plane Π in a symplectic space $\mathbb{R}^{2n}$ is isotropic if and only if the planes Π and $\Omega\Pi$ are mutually orthogonal (in the Euclidean sense).*

PROOF: By virtue of the consistence between the symplectic and Euclidean structures, we have $(a, b) = \langle \Omega a, b \rangle$. Therefore, $(a, b) = 0$ if and only if $\langle \Omega a, b \rangle = 0$, as required.

LEMMA 1.2.2. *The dimension of an isotropic plane Π^k in a symplectic space $\mathbb{R}^{2n}$ is never higher than n.*

PROOF: By virtue of Lemma 1.2.1, the planes Π and $\Omega\Pi$ are mutually orthogonal, which is equivalent to isotropy of Π. Since the operator Ω is nondegenerate, the dimensions of the planes Π and $\Omega\Pi$ are equal. Consequently, $k+k \leqslant 2n$, as required.

DEFINITION 1.2.3: A linear transformation $g : \mathbb{R}^{2n} \to \mathbb{R}^{2n}$ of a symplectic space $\mathbb{R}^{2n}$ is called *symplectic* if and only if it preserves its symplectic structure, that is, $(ga, gb) = (a, b)$ for any $a, b \in \mathbb{R}^{2n}$. The sum total of all symplectic transformations of $\mathbb{R}^{2n}$, which leave the origin in its place, forms a group called a *real symplectic group* and denoted by $\mathrm{Sp}(n, \mathbb{R})$.

We should verify the correctness of Definition 1.2.3, that is, prove that the symplectic transformations form a group. It is obvious that the composition of symplectic transformations is itself a symplectic transformation. We shall prove that symplectic transformations are nondegenerate.

Recall the relation between skew-symmetric scalar products and *exterior differential 2-forms*. Consider the form

$$\omega = \sum_{i=1}^{n} dp_i \wedge dq_i.$$

It determines the skew-symmetric scalar product

$$(a, b) = \sum_{i=1}^{n} p_i q'_i - p'_i q_i$$

under a natural identification of $\mathbb{R}^{2n}$ with the tangent space $T\mathbb{R}^{2n}$. In other words, the product of the vectors $a = (p_1, \ldots, q_n)$ and $b = (p'_1, \ldots, q'_n)$ can be written in the form $(a, b) = \omega(a, b)$, since

$$(dp_i \wedge dq_i)(a, b) = p_i q'_i - p'_i q_i.$$

LEMMA 1.2.3. *The determinant of any linear symplectic transformation $g : \mathbb{R}^{2n} \to \mathbb{R}^{2n}$ is equal to unity.*

PROOF: A mapping g is symplectic if and only if it preserves the exterior 2-form ω. This characteristic of symplectic transformations will be often used hereafter. Let us consider the nth exterior degree of the form ω and put $\tau = \omega \wedge \cdots \wedge \omega$ (n times). The general properties of exterior forms imply that $\tau = \lambda(x) dx_1 \wedge \cdots \wedge dx_{2n}$, where $\lambda(x)$ is some smooth function. We assert that

$$\lambda = \sqrt{\det(\omega_{ij})} = \sqrt{\det \Omega},$$

where $\Omega = (\omega_{ij})$ is the matrix of coefficients of the form ω. If A is a nondegenerate linear change of coordinates, then $A\Omega' A^T = \Omega$, where Ω' is the matrix of the form ω in the new coordinate system. From this, we obtain $\det \Omega = \det \Omega' \cdot (\det A)^2$. If the chosen coordinates are canonical, i.e., Ω' is canonical, then $\det \Omega' = 1$ and $\det \Omega = (\det A)^2$. With the change of coordinates $(x) \to (p, q)$, the form τ transforms by the following law:

$$\tau = (\det A)\tau' = (\det A)dx_1 \wedge \cdots \wedge dx_{2n}.$$

Since $\det A = \sqrt{\det \Omega}$, then $\sqrt{\det \Omega} = \lambda$, and the result follows.

Thus, the form $\tau = \omega \wedge \cdots \wedge \omega$ (n times) is interpreted as the form of $2n$-dimensional volume in $\mathbb{R}^{2n}$. Therefore, preserving the form ω, the transformation g automatically preserves its nth exterior degree, that is, preserves the form τ and, consequently, preserves the element of a $(2n)$-dimensional volume in $\mathbb{R}^{2n}$. This implies that $\det g = 1$, as required.

COROLLARY 1.2.1. *A skew-scalar product $(a, b) = \omega(a, b)$ is nondegenerate at a point $x \in \mathbb{R}^{2n}$ if and only if the form $\tau = \omega \wedge \cdots \wedge \omega$ (n times) is nonzero at this point.*

From Lemma 1.2.3 it immediately follows that a set of symplectic transformations does actually *form a group*. What has been proved also implies that a symplectic transformation maps any symplectic basis again into a symplectic basis. The inverse is also true: if some linear transformation maps a symplectic basis into a symplectic basis, the transformation is symplectic. Therefore, *any two symplectic bases can be made coincident by means of a symplectic transformation*.

In a Euclidean space, all planes of equal dimensions are equivalent in the sense that any of them can be mapped into another one by means of isometry of $\mathbb{R}^{2n}$, i.e., by means of parallel mapping and orthogonal transformation. In a symplectic space, the situation is different.

We have already proved above that any nonzero vector of symplectic space may be taken as a first vector of the symplectic basis. Therefore, any nonzero vector of a symplectic space can be carried into any other (nonzero) vector by a symplectic transformation. In other words, a *symplectic group* $\mathrm{Sp}(n, \mathbb{R})$ *is transitive on the set of all nonzero vectors of a symplectic space*, which come from the origin. But already for two-dimensional planes this property does not hold. There exist pairs of two-dimensional planes which cannot be made coincident through a symplectic transformation.

LEMMA 1.2.4. *A symplectic transformation maps any isotropic plane again into an isotropic plane. In particular, an image of a Lagrangian plane is a Lagrangian plane.*

The proof is obvious.

Let us consider in more detail the properties of the group $\mathrm{Sp}(1, \mathbb{R})$. It consists of linear transformations which are homogeneous and preserve in a two-dimensional plane a skew-symmetric from $(a, b) = a_1 b_2 - a_2 b_1$, where $a = (a_1 a_2)$, $b = (b_1, b_2)$.

LEMMA 1.2.5. *The group $\mathrm{Sp}(1, \mathbb{R})$ is isomorphic to a group of real matrices of order 2 with a determinant equal to unity, that is, to a group $\mathrm{SL}(2, \mathbb{R})$.*

PROOF: A skew-scalar product of a pair of vectors a, b is equal to the area of the parallelogram spanned by the vectors a and b. It is known that if $\gamma = \begin{pmatrix} p & q \\ r & s \end{pmatrix}$ is an arbitrary linear homogeneous transformation, then the area of the parallelogram spanned by the vectors $g(a)$ and $g(b)$ is obtained from the initial area through multiplication by the transformation determinant g. The assertion of the Lemma follows.

Along with the group $\mathrm{Sp}(n, \mathbb{R})$, we will also need a group $\mathrm{Sp}(n, \mathbb{C})$ of symplectic transformations of a complex linear space. The *Hermitian form* in $\mathbb{C}^{2n}$ is given as $\sum_{i=1}^{2n} a_i \bar{b}_i$, where the "bar" sign stands for complex conjugation. The symplectic structure in $\mathbb{C}^{2n}$ is given by

$$(a, b) = \sum_{i=1}^{n} a_i b_{n+i} - a_{n+i} b_i.$$

DEFINITION 1.2.4: A nondegenerate complex linear transformation $g : \mathbb{C}^{2n} \to \mathbb{C}^{2n}$ is called *symplectic* if it preserves its symplectic structure in $\mathbb{C}^{2n}$, that is, preserves the exterior 2-form

$$dz_1 \wedge dz_{n+1} + \cdots + dz_n \wedge dz_{2n}.$$

The set of all such homogeneous transformations forms a complex symplectic group $\mathrm{Sp}(n, \mathbb{C})$.

2.2 Symplectic Structure on a Manifold

We have introduced a symplectic linear structure in the tangent space $T_x M^{2n}$ to an even-dimensional smooth manifold. In some cases, one can set a symplectic *smoothly point-dependent structure* on the entire manifold M^{2n}. In other words, it may happen that on M^{2n} a smooth skew-symmetric tensor field ω_{ij} of rank 2 is defined, which is nondegenerate at each point of the manifold M^{2n} and satisfying a certain "closedness" condition. In this case, M^{2n} is called a *symplectic manifold*. It can be shown that such a structure cannot be given at all on each even-dimensional smooth manifold.

DEFINITION 1.2.5: A smooth even-dimensional manifold M^{2n} is called *symplectic* if on this manifold is given an exterior differential 2-form (that is, form of degree 2)

$$\omega = \sum_{i<j} \omega_{ij} dx_i \wedge dx_j$$

which possesses the following properties:

1) This *form is nondegenerate*, that is, the matrix of its coefficients $\Omega(x) = (\omega_{ij}(x))$ is nondegenerate at each point of the manifold.

2) This *form is closed*, that is, its exterior differential is equal to zero: $d\omega = 0$. Such a form is called a *symplectic structure* on M^{2n}.

Here $x_1, \ldots, x_{2n}$ are local regular coordinates on M^{2n}. At each point $x \in M^{2n}$, such a 2-form apparently determines the linear symplectic structure

$$\omega(a, b) = (a, b) = \sum_{i<j} \omega_{ij} a_i b_j,$$

where $a, b \in T_x M$. For any fixed point x, the form ω determines the linear symplectic structure in $T_x M$.

We shall recall that each point $x \in M$ is naturally connected not only with the tangent space $T_x M$, but also with the cotangent space $T_x^* M$ conjugate to the space $T_x M$. The space $T_x^* M$ consists of all the covectors, that is, linear forms (linear functionals) on the tangent space. Generally speaking, in the case of an arbitrary smooth manifold M, the linear spaces $T_x M$ and $T_x^* M$ cannot be identified in such a way that this identification by invariant under the change of coordinates, i.e., be of a tensor character. However, if on a manifold, for instance, a nondegenerate tensor field q_{ij} of rank 2 is given, then these linear spaces can be identified. To this end, we set a mapping $q : T_x M \to T_x^* M$ and let

$$(q(a))_i = \sum_{j=1}^{2n} q_{ij} a_j,$$

where $a = (a_1, \ldots, a_{2n}) \in T_x M$. Since the matrix (q_{ij}) is invertible, one can obviously determine the inverse linear mapping. Therefore, q sets an isomorphism between T_x and T_x^* invariant under a change of coordinates.

The *procedure of canonical identification of tangent and cotangent spaces* is especially important when $q_{ij} = g_{ij}$ sets a Riemannian metric on M or when $q_{ij} = \omega_{ij}$ sets a symplectic structure on M. Let us consider a smooth function $f(x)$ on M. The gradient $\operatorname{grad} f(x)$ of this function is a covector, that is, an element of the space $T_x^* M$ at each point x. If the manifold M is Reimannian then, using the identification of the tangent and cotangent spaces by means of the metric g_{ij}, one may view the gradient of the function as a vector field on M. For simplicity, this field will be denoted by the same symbol $\operatorname{grad} f(x)$. If M is a symplectic manifold, then the identification of the tangent and cotangent spaces by means of the symplectic structure ω transforms the covector field $\operatorname{grad} f$ into a vector field which we denote as $\operatorname{sgrad} f$ and which is, of course, distinct from the vector field $\operatorname{grad} f$.

It is clear that the vector field $\operatorname{grad} f$ is uniquely defined by the equality $\langle v, \operatorname{grad} f \rangle = v(f)$, where v runs through all the vector fields on M, $v(f)$ is a derivative of the function f along the field v, and $\langle \, , \, \rangle$ is a scalar product corresponding to the Riemannian metric g_{ij}.

DEFINITION 1.2.6: Let f be a smooth function on M^{2n} and ω a symplectic structure on M. A smooth vector field on M which is uniquely defined by the relation $\omega(v, \operatorname{sgrad} f) = v(f)$, where v runs through a set of all smooth vector fields on M and $v(f)$ is the value of the field v on the function f (that is, a derivative of the function f in the direction of the field v), is called a *skew-symmetric gradient* $\operatorname{sgrad} f$ of the function f (a *skew gradient*).

We see that the definition of $\operatorname{sgrad} f$ copies the definition of $\operatorname{grad} f$. The only difference is that, instead of the *symmetric* tensor g_{ij}, one considers the *skew-symmetric* tensor ω_{ij}.

As has already been mentioned, the Riemannian metric g_{ij} can always be reduced at one point to a canonical diagonal form δ_{ij} by a choice of appropriate local coordinates on the manifold. Precisely in the same manner, the symplectic structure ω_{ij} can always be reduced at one point to the canonical form

$$\Omega(x) = \begin{pmatrix} 0 & E \\ -E & 0 \end{pmatrix}.$$

Here we rely on the information from the algebra presented in §2.1. Thus, in this respect, the Riemannian and symplectic structures are similar.

However, when one passes from one isolated point (and a separate tangent space) to a whole open neighbourhood of this point, one immediately finds a drastic distinction between the Riemannian metric and the symplectic structure.

It is well known that, *in the general case, the Riemannian metric is not reduced by a local change of coordinates to a diagonal form simultaneously in the whole neighbourhood of a point.* This may be hampered by the fact that the Riemann curvature tensor is nonzero. Namely, if at a given point at least one of the tensor components is nonzero, then no local changes of coordinates can reduce the metric to a diagonal form simultaneously at all points of any arbitrarily small open neighbourhood of this point.

The symplectic structure behaves quite differently. It turns out that by an appropriate change of coordinates, it can always be reduced to the canonical form $\begin{pmatrix} 0 & E \\ -E & 0 \end{pmatrix}$ simultaneously at all the points of a certain (maybe small) open neighbourhood of any point on a symplectic manifold. This fact constitutes the content of the known *Darboux theorem* which has no analogue (in the indicated sense) in Riemannian geometry.

THEOREM 1.2.1 (DARBOUX). *Let ω_{ij} be a symplectic structure on M^{2n}. Then, for any point $x \in M$, there always exists an open neighbourhood with local regular coordinates $p_1, \ldots, p_n, q_1, \ldots, q_n$, such that the form ω in them is written in the simplest canonical form $\sum dp_i \wedge dq_i$, that is, at each point of this neighbourhood the matrix (ω_{ij}) is of the simplest form $\begin{pmatrix} 0 & E \\ -E & 0 \end{pmatrix}$.*

DEFINITION 1.2.7: *Local coordinates $p_1, \ldots, p_n, q_1, \ldots, q_n$ on a symplectic manifold are called symplectic if in these coordinates the form ω is written in the canonical form.*

So, for each point x and M one can point out an open neighbourhood with symplectic coordinates. Covering M with such neighbourhoods, we obtain a symplectic atlas. We omit the proof of the Darboux theorem.

Let a symplectic system of coordinates $p_1, \ldots, p_n, q_1, \ldots, q_n$ in the neighbourhood of a certain point of a symplectic manifold be given. How will a vector field sgrad f be written in these coordinates? Since the matrix of the form looks like $\begin{pmatrix} 0 & E \\ -E & 0 \end{pmatrix}$, and since the gradient is given

$$\left(\frac{\partial f}{\partial p_1}, \ldots, \frac{\partial f}{\partial q_n} \right) = \operatorname{grad} f,$$

it follows that the skew gradient is written as

$$\operatorname{sgrad} f = \left(\frac{\partial f}{\partial q_1}, \ldots, \frac{\partial f}{\partial q_n}, -\frac{\partial f}{\partial p_1}, \ldots, -\frac{\partial f}{\partial p_n} \right).$$

An example of a symplectic manifold is a Euclidean space $\mathbb{R}^{2n}(p_1,\ldots,q_n)$ on which a nondegenerate closed 2-form $\omega = \sum dp_i \wedge dq_i$ is given. Another example may be smooth two-dimensional orientable Riemannian manifolds. A symplectic structure is given here by a 2-form of Riemannian volume (area). A more general example is a cotangent bundle T^*M^n of an arbitrary smooth manifold M^n. This is a $2n$- dimensional smooth manifold on which there exists a naturally defined closed and nondegenerate 2-form [3].

2.3 Hamiltonian and Locally Hamiltonian Vector Fields and the Poisson Bracket

We will denote by $V(M)$ an infinite-dimensional linear space *of all smooth vector fields* on a manifold M. If M^{2n} is a symplectic manifold, then in $V(M)$ there exists a linear subspace $H_{\text{loc}}(M)$ of vector fields of a special form, called *locally Hamiltonian fields*, and the subspace $H_{\text{loc}}(M)$ contains, in turn, another linear subspace $H(M)$ consisting of the so-called *Hamiltonian fields*. Thus, $V(M) \supset H_{\text{loc}}(M) \supseteq H(M)$.

DEFINITION 1.2.8: A smooth vector field v on a symplectic manifold M, which has the form $v = \text{sgrad}\, F$, is called a *Hamiltonian field* on M if the smooth function F is defined on the entire manifold M. The function F is then called a *Hamiltonian*.

Thus, Hamiltonian fields are in some sense analogues of the gradient (or potential) fields $v = \text{grad}\, F$, which play an important role in Riemannian geometry. The role of these fields in symplectic geometry is not smaller. Note that in modern geometry and mechanics there exist systems in which a *Hamiltonian is not a single-valued function* on a manifold (it is multivalued). Therefore the following definition is meaningful.

DEFINITION 1.2.9: A smooth vector field v on a symplectic manifold M is called *locally Hamiltonian* if for any point $x \in M$, there exists such an open neighbourhood $U(x)$ of the point x and such a smooth function F_U defined on this neighbourhood that $v = \text{sgrad}\, F_U$, that is, the field v is Hamiltonian in the neighbourhood U with the local Hamiltonian F_U.

It is clear that locally Hamiltonian fields from a linear subspace $H_{\text{loc}}(M)$ in $V(M)$ and that $H_{\text{loc}}(M) \supseteq H(M)$ inasmuch an any *Hamiltonian field is obviously locally Hamiltonian*. The converse is false. This means that there exist fields that admit in the neighbourhood of each point x on M a representation in the form $v|_u = \text{sgrad}\, F_U$ but do not admit a unique representation in the form $v = \text{sgrad}\, F$, where F is a single-valued function defined at all points of M. In other words, in such cases, the local Hamiltonians defined on separate neighbourhoods U cannot be "sewed" into one smooth single-valued function defined on the entire M.

The simplest example: in a Euclidean plane without a point (we discard the origin 0), we consider a vector field

$$\left(\frac{x}{x^2+y^2}, \frac{y}{x^2+y^2}\right) = v,$$

where x, y are Cartesian coordinates. This field is a velocity field on a liquid flow from the origin along radian rays (Fig. 6).
It is readily seen that this field is locally Hamiltonian on a symplectic manifold $\mathbb{R}^2 \setminus 0$ supplied with a 2-form $dx \wedge dy$. Indeed, as the local Hamiltonian F_U, it suffices to take the function of the polar angle $\varphi = \arctan\left(\frac{y}{x}\right)$, which is a smooth function in

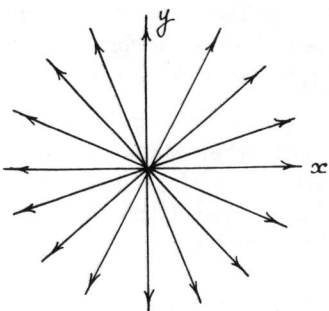

Figure 6

a sufficiently small neighbourhood of any point $(x, y) \neq (0,0)$ and defined up to a constant. Its usual gradient is of the form

$$\left(\frac{-y}{x^2 + y^2}, \frac{x}{x^2 + y^2}\right).$$

Since we have chosen the symplectic structure $dx \wedge dy$ on $\mathbb{R}^2 \setminus 0$ to be canonical, the vector sgrad f is obtained from the vector grad $f = (f_x, f_y)$ as follows: sgrad $f = (-f_y, f_x)$. From this, we have

$$\text{sgrad } F_U = \left(\frac{x}{x^2 + y^2}, \frac{y}{x^2 + y^2}\right) = v(x, y).$$

However, this field is *not a globally Hamiltonian* one because the polar angle $\varphi(x, y)$ is a multivalued function on a plane without a point, and the above-mentioned local Hamiltonians cannot be "sewed" into one smooth single-valued function. This results from the fact that the symplectic manifold $\mathbb{R}^2 \setminus 0$ is *not simply-connected*.

Let us go back to Hamiltonian fields. They can also be described in the language of such one-parameter groups of diffeomorphisms of the manifold M which are produced by these fields. Let v be a Hamiltonian field and $\mathfrak{G}^v$ a *one-dimensional group of diffeomorphisms* of the manifold M which is represented on M shifts along integral trajectories of the field v. This means that the group $\mathfrak{G}^v$ consists of smooth transformations g_t acting on M as follows: $g_t(x) = y$, where $x = \gamma(0), y = \gamma(t)$, and γ is an integral curve of the field v which passes through the point x at $t = 0$. In other words, the diffeomorphism g_t shifts each point x by a time t along a trajectory $\gamma(t)$. Since the 2-form ω which sets the symplectic structure on M is carried by the diffeomorphism g_t into some new form $g_t^* \omega$ (defined as $(g_t^* \omega)(x) = \omega(g_t(x))$), the derivative $\frac{d\omega}{dt}$ of the form ω along the vector field v is well defined, i.e.,

$$\frac{d\omega(x)}{dt} = \frac{d}{dt}(g_t^* \omega)|_{t=0}.$$

It can be shown that the symplectic structure ω is invariant under the group of diffeomorphisms $\mathfrak{G}^v$ generated by the Hamiltonian field v. We shall prove the general assertion.

THEOREM 1.2.2. *A smooth vector field v on a symplectic manifold M with a symplectic structure ω is locally Hamiltonian if and only if it preserves its symplectic structure, that is, if the derivative of the form ω in the direction of the field v is zero or, in other words, if $g_t^*\omega \equiv \omega$ at all t.*

PROOF: By virtue of the Darboux theorem (see Theorem 1.2.1) it suffices to prove this assertion only for the case where M is the simplest symplectic manifold $\mathbb{R}^{2n}$ with the canonical structure $\omega = \sum dp_i \wedge dq_i$. Let the form ω be preserved by a one-parameter group g_t, that is, $\frac{d}{dt}\omega(\gamma(t)) \equiv 0$, where $\gamma(t)$ are integral trajectories of the field $v = (X_i, Y_i)$. Thus, we have

$$\frac{d}{dt}\omega(\gamma(t)) = \frac{d}{dt}\sum dp_i(t) \wedge dq_i(t)$$

$$= \sum d\left(\frac{dp_i}{dt}\right) \wedge dq_i + dp_i \wedge d\left(\frac{dq_i}{dt}\right)$$

$$= \sum dX_i \wedge dq_i + dp_i \wedge dY_i$$

$$= \sum_{i,k}\left(\frac{\partial X_i}{\partial p_k} + \frac{\partial Y_k}{\partial q_i}\right) dp_k \wedge dq_i$$

$$+ \sum_{k<i}\left(\frac{\partial X_i}{\partial q_k} - \frac{\partial X_k}{\partial q_i}\right) dq_k \wedge dq_i$$

$$+ \sum_{i<k}\left(\frac{\partial Y_i}{\partial p_k} - \frac{\partial Y_k}{\partial p_i}\right) dp_i \wedge dp_k \equiv 0.$$

This implies zero coefficients before the independent monomials, that is,

$$-\frac{\partial X_i}{\partial p_k} = \frac{\partial Y_k}{\partial q_i}, \quad \frac{\partial X_i}{\partial q_k} = \frac{\partial X_k}{\partial q_i}, \quad \frac{\partial Y_i}{\partial p_k} = \frac{\partial Y_k}{\partial p_i}.$$

Let us consider a differential form

$$\alpha = \sum -Y_i dp_i + X_i dq_i.$$

It is closed.

Indeed,

$$d\alpha = \sum -\frac{\partial Y_i}{\partial p_k} dp_k \wedge dp_i - \frac{\partial Y_i}{\partial q_K} dq_k \wedge dp_i + \frac{\partial X_i}{\partial p_k} dp_k \wedge dq_i + \frac{\partial X_i}{\partial q_k} dq_k \wedge dq_i \equiv 0,$$

which becomes clear after reduction of like terms and the use of the above conditions on partial derivatives of the components of the vector field v. Closedness of the form α on $\mathbb{R}^{2n}$ means its precision (by virtue of the Poincaré lemma), that is, there exists a smooth function H such that $\alpha = dH$. From this we have

$$\alpha = \sum \frac{dH}{\partial p_i} dp_i + \frac{\partial H}{\partial q_i} dq_i = \sum -Y_i dp_i + X_i dq_i,$$

that is,
$$Y_i = -\frac{\partial H}{\partial p_i}, \qquad X_i = \frac{\partial H}{\partial q_i},$$

that is, $v = \operatorname{sgrad} H$, as required. The inverse assertion is obtained by repeating the arguments in the inverse order. This proves the theorem.

In the case of a two-dimensional Riemannian manifold M^2 with a Riemannian metric g_{ij} and with the form of the Riemannian area $\omega = \sqrt{\det(g_{ij})}\,dx \wedge dy$ as a symplectic structure (see above), the condition that the group $\mathfrak{G}$ of diffeomorphisms g_t preserve the form ω is equivalent to the condition that the domain areas be preserved on the surface M^2 when these domains are shifted by the diffeomorphisms g_t. Thus, the shifts along integral trajectories of a Hamiltonian field on a two-dimensional symplectic manifold preserve the domain areas.

A similar assertion is valid in a multidimensional case as well. Let ω be a symplectic structure on a manifold M^{2n}. Then the nth exterior degree of the 2-form ω may be regarded as the form of the Riemannian volume on M^{2n}, since the form $\tau = \omega \wedge \cdots \wedge \omega$ (n times) transforms exactly in the same way as the form of the Riemannian volume (it is additionally multiplied by the determinant of the Jacobi matrix of change of coordinates). Since the locally Hamiltonian field preserves the 2-form ω, the same field also preserves the exterior $(2n)$-form τ, that is, it preserves domain volumes on the manifold M^{2n}.

In the case of a two-dimensional symplectic manifold, the condition of the locally Hamiltonian character of the field admits another vivid geometrical interpretation. Let g_{ij} be a Riemannian metric on M^2 and let $\omega = \sqrt{\det(g_{ij})}\,dx \wedge dy$ be the form of the Riemannian area. By virtue of the Darboux theorem, one can always choose local coordinates p and q such that the form ω be written in the canonical form $dp \wedge dq$. Here p and q are certain functions of x and y (and vice versa). Let v be a locally Hamiltonian field $v = (P(x,y), Q(x,y))$, where P and Q are coordinates of the field in the local system of coordinates p and q. Let us interpret the field v as a velocity field of the flow of liquid of constant density (equal to unity) on the surface M^2. Let us investigate the variation of the mass of the liquid bounded by an infinitesimal rectangle on the surface when it is shifted along integral trajectories of the field v. It is clear that the mass of this liquid is equal to the area of the rectangle. Therefore, the mass of the liquid contained in a bounded (sufficiently small) domain on M^2 is equal to the area of the domain.

Let D be a bounded domain with a piecewise smooth boundary in M^2, and let $D(t)$ be a domain obtained from the domain D through its shift by a time t along integral trajectories of the field v. In other words, $D(t) = g_t(D)$, i.e., $D(t)$ is the image of the domain D under the diffeomorphism $g_t : M \to M$. Let us denote by $\operatorname{div}(v) = \frac{\partial P}{\partial p} + \frac{\partial Q}{\partial q}$ the divergence of the field v and let $S(t)$ be the area of the domain $D(t)$, i.e., the mass of the liquid contained in this domain.

PROPOSITION 1.2.1. *For any bounded domain $D(t)$ on the surface M^2, there exists the relation*
$$\frac{d}{dt}S(t) = \iint\limits_{D(t)} \operatorname{div}(v)\,dp \wedge dq.$$

A flow of liquid is called *incompressible* if the change in its mass in any domain is zero. From Proposition 1.2.1, an instructive assertion follows.

COROLLARY 1.2.1. *A flow v of liquid of constant density on the surface M^2 is incompressible if and only if* $\mathrm{div}(v) = 0$.

PROOF OF PROPOSITION 1.2.1: It suffices to prove the assertion for a domain D which has the form of an infinitesimal rectangle. It is well known that the change of the rectangle area under a shift is measured by the determinant of the Jacobi transformation matrix. Consequently, it suffices to calculate this determinant for an infinitesimal transformation

$$(p,q) \to (p + \varepsilon P(p,q), q + \varepsilon Q(p,q)),$$

where ε is an infinitesimal parameter. It is clear that the Jacobi matrix has the form

$$\begin{pmatrix} 1 + \varepsilon \frac{\partial P}{\partial p} & \varepsilon \frac{\partial P}{\partial q} \\ \varepsilon \frac{\partial Q}{\partial p} & 1 + \varepsilon \frac{\partial Q}{\partial q} \end{pmatrix}.$$

The determinant of this matrix (up to ε^2) is as follows:

$$1 + \varepsilon \,\mathrm{div}(v) = 1 + \varepsilon \left(\frac{\partial P}{\partial p} + \frac{\partial Q}{\partial q} \right).$$

Thus, the distortion of the area of an infinitesimal domain is measured exactly by the divergence of the field v, which proves Proposition 1.2.1.

Thus, in the case of a two-dimensional symplectic manifold, the locally Hamiltonian vector fields are exactly the flows of incompressible liquid, that is, the vector fields with zero divergence. In other words, the condition for the local Hamiltonian properties of the field v in the two-dimensional case is equivalent to the condition $\mathrm{div}(v) = 0$.

We have given an example of a locally Hamiltonian vector field on $\mathbb{R}^2 \setminus 0$ which is not a Hamiltonian field. The symplectic manifold was not compact here. But such kind of examples may be constructed on compact closed (i.e., without a boundary) manifolds as well. Take, for instance, an ordinary flat torus T^2 supplied with a Euclidean metric $g_{ij} = \delta_{ij}$. Then the 2-form of the area on this torus will be written in Cartesian coordinates x, y as follows: $dx \wedge dy$, i.e., this 2-form is canonical. On the torus, consider a vector field $v = (1, 0)$ given by a uniform liquid motion along its parallels (Fig. 7).

Torus parallels are integral trajectories of this field. We will prove that this field is not Hamiltonian on the entire torus (although it is locally Hamiltonian, which is obvious). Indeed, let us assume that there exists a smooth single-valued function $f(x,y)$ defined on a torus, that is, doubly periodic on $\mathbb{R}^2(x,y)$, namely $f(x + m, y + n) = f(x,y)$ for any integer m, n, and such that $v = \mathrm{sgrad}\, f$, that is, $(1,0) = (-f_y, f_x)$. Then $f_x = 0, f_y = -1$. From this it follows that a vector field $w = (0, -1)$ on the torus is potential, that is, $w = \mathrm{grad}\, f$. This field is *conjugate* to the field v, that is, their integral trajectories are *mutually orthogonal*. Therefore, integral trajectories of the field w are meridians of the flat torus (Fig. 7). In

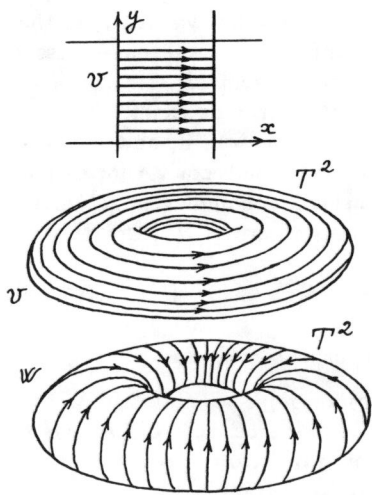

Figure 7

particular, all the integral trajectories of the field w are closed. It remains only to recall the following general statement. If a smooth vector field w on a manifold M has at least one closed trajectory γ without singular points of the field w, this field cannot be potential. Indeed, assume the existence of a potential f, that is, such a function that there exists the equality $w = \mathrm{grad}\, f$. Parametrize the trajectory γ by a parameter t, so that $\gamma(0) = \gamma(1)$. Integrate the 1-form df along the closed trajectory γ. And we obtain

$$\int_0^1 df = \int_0^1 \sum_i \frac{\partial f}{\partial x_i} dx_i(t) = f(1) - f(0) \neq 0$$

inasmuch as the function monotonously increases along the trajectory γ. Thus, the potential f cannot be a single-valued function on M.

This contradiction proves the absence on nonsingular closed trajectories in potential fields. Thus, the vector field w on a torus is not potential and, therefore, the initial field v is not Hamiltonian on the entire torus, and the result follows.

We remind the reader that a manifold is called *simply-corrected* if each closed path (loop) is contracted along it to a point. The differential k-form τ is called *closed* on M if $d\tau = 0$, and it is called *exact* if $\tau = d\rho$ for a certain $(k-1)$-form ρ. Any exact form is closed. The converse is, generally, false.

The obstacle which prevents an arbitrary local Hamiltonian field from being a Hamiltonian one is that the *manifold is not simply-connected*. In our first example,t he manifold $\mathbb{R}^n \backslash 0$ is not simply-connected because it punctures the point out of the plane (a closed detour of this point is not contracted in the point on $\mathbb{R}^n \backslash 0$). In our second example, the torus T^2 is non-simply-connected. Its parallel and meridian set generators in the nonzero fundamental group $\pi_1(T^2) = \mathbf{Z} \oplus \mathbf{Z}$.

From the proof of Theorem 1.2.2, we see that the local Hamiltonian properties of the field v are equivalent to the closedness of the differential 1-form $\alpha = \sum -Y_i dp_i + X_i dq_i$ on the manifold M. For a field to be globally Hamiltonian, it is sufficient that this form be exact. For instance, this will always be the case on $\mathbf{R}^{2n}$ (the *Poincaré lemma*). If, however, a symplectic manifold is not simply-connected, then closed but not exact 1-forms may exist on it. This will be so if a group of one-dimensional cohomologies $H^1(M, \mathbf{R})$ is nonzero. In both our examples, we deal with a nonzero group $H^1(M^2, \mathbf{R})$, namely

$$H^1(\mathbf{R}^2 \backslash 0, \mathbf{R}) = \mathbf{R} \quad \text{and} \quad H^1(T^2, \mathbf{R}) = \mathbf{R} \oplus \mathbf{R} = \mathbf{R}^2.$$

In particular, the closed 1-forms $\alpha = dp$ and $\alpha' = dq$ on a torus are not exact.

PROPOSITION 1.2.2. *Let a symplectic manifold M have a zero first group of real cohomologies $H^1(M, \mathbf{R})$ (for instance, this will always be the case with a simply-connected manifold). Then any locally Hamiltonian vector field on the manifold will be at the same time globally Hamiltonian.*

If a manifold is simply-connected, then we always have $H^1(M, \mathbf{R}) = 0$. The converse is, generally, not true. There exist non-simply-connected manifolds for which $H^1(M, \mathbf{R}) = 0$. In fact, in these cases, the fundamental group $\pi_1(M)$ of the manifold M "is not very large" in the sense that is factor *by the commutant*, i.e., the group $\pi_1/[\pi_1, \pi_1]$ (which coincides with the one-dimensional group of integer-valued homologies $H_1(M, \mathbf{Z})$) has no infinite-order elements.

Now let us proceed to describe an important operation of calculating a *Poisson bracket* of two functions on a symplectic manifold.

DEFINITION 1.2.10.: A function $\{f, g\}$ given by the formula

$$\{f, g\} = \omega(\text{sgrad } f, \text{sgrad } g) = \sum \omega_{ij} (\text{sgrad } f)_i (\text{sgrad } g)_j$$

is called the *Poisson bracket* of smooth function f and g on a symplectic manifold M with a form ω.

Here $\omega = \sum \omega_{ij} dx_i \wedge dx_j$ and sgrad $f = ((\text{sgrad } f)_1, \ldots, (\text{sgrad } f)_{2n})$. Thus, the *Poisson bracket of the functions f and g is a skew-symmetric scalar product (with respect to the form ω) of their skew gradients*. If we denote by ω^{ij} the coefficients of the matrix reciprocal to the matrix (ω_{ij}), then the Poisson bracket can be written in the local coordinates $x_1, \ldots, x_{2n}$ on M as

$$\{f, g\} = \sum \omega^{ij} \frac{\partial f}{\partial x_i} \frac{\partial g}{\partial x_j}.$$

Our argument is based on the fact that the gradient of the function and its skew gradient are related as

$$(\text{sgrad } f)_i = \sum_j \omega^{ij} \frac{\partial f}{\partial x_j}.$$

Thus, the Poisson bracket can be calculated if the gradients of the functions f and g are known.

This formula may be taken as the *definition of the Poisson bracket*. The formula may be also extended to the case where the matrix (ω^{ij}) is *degenerate*. In a certain exact sense, the *Poisson bracket is a more general and primary notion that the symplectic structure*. Of course, setting a symplectic structure (in its classical interpretation) makes it possible to define well (and unambiguously) the operation of calculating the Poisson bracket. The Poisson bracket may, however, be defined for a much broader range of cases in which the corresponding structure ω is degenerate. But, for the present, we will keep to the definition of the Poisson bracket in terms of the symplectic structure ω.

The Poisson bracket has a simple interpretation. There exists an important relation: $\{f,g\} = (\operatorname{sgrad} f)g$, that is, the *Poisson bracket coincides with the derivative of the function g in the direction of the vector field* $(\operatorname{sgrad} f$. By virtue of skew symmetry of the form ω, the equality $\{f,g\} = -(\operatorname{sgrad} g)f$ is valid, too. This follows immediately from the definition of the operation $\operatorname{sgrad} g$, namely,

$$\omega(v, \operatorname{sgrad} g) = v(g) = \sum v_i \frac{\partial g}{\partial x_i},$$

where v is an arbitrary smooth vector field on M.

PROPOSITION 1.2.3. *The operation of calculating the Poisson bracket $f, g \to \{f,g\}$ has the following properties:*
1) *bilinearity, that is, $\{\alpha f + \beta g, h\} = \alpha\{f,h\} + \beta\{g,h\}$, where α and β are constants;*
2) *skew symmetry, that is, $\{f,g\} = -\{g,f\}$;*
3) *there exists the Jacobi identity, that is, $\{h,\{f,g\}\} + \{g,\{h,f\}\} + \{f,\{g,h\}\}$ $= 0$ for any smooth functions f, g, h on M;*
4) *the Poisson bracket considered as a differential operation satisfies the Leibniz identity, that is, $\{h, fg\} = \{h,f\}g + f\{h,g\}$, where fg is the product of the functions f and g.*

PROOF: Properties 1 and 2 obviously follow from the definition of the Poisson bracket. Verification of property 3 requires some calculations. It is clear that

$$\{h,\{f,g\}\} = \sum \omega^{ij} \frac{\partial \omega^{pq}}{\partial x_j} \frac{\partial h}{\partial x_i} \frac{\partial f}{\partial x_p} \frac{\partial g}{\partial x_q}$$
$$+ \sum \omega^{ij} \omega^{pq} \frac{\partial h}{\partial x_i} \frac{\partial^2 f}{\partial x_j \partial x_p} \frac{\partial g}{\partial x_q}$$
$$+ \sum \omega^{ij} \omega^{pq} \frac{\partial h}{\partial x_i} \frac{\partial f}{\partial x_p} \frac{\partial^2 g}{\partial x_j \partial x_q}.$$

By permuting cyclically h, f, g and summing up the results obtained, we discover that the first sum vanishes because the form ω is closed, that is,

$$d\omega = \sum \frac{\partial \omega_{ij}}{\partial x_\alpha} dx_\alpha \wedge dx_i \wedge dx_j = 0.$$

The second and third sums will vanish due to the symmetry of second partial derivatives and because the commutator of the two vector fields is again a vector field, that is, a first-(but not second-) order differential operator. Eventually, property 4 follows from the fact that

$$\{h, fg\} = (\operatorname{sgrad} h)(fg) = ((\operatorname{sgrad} h)f)\, g + f(\operatorname{sgrad} h)g = \{h, f\}g + f\{h, g\},$$

and the result follows.

Proposition 1.2.3 has an important corollary which has no analogue in Riemannian geometry based on the concept of the *symmetric* scalar product.

PROPOSITION 1.2.4. *Such a linear infinite-dimensional space $C^\infty(M)$ of smooth functions f on a symplectic manifold M, which is endowed with the operation of calculating the Poisson bracket $f, g \to \{f, g\}$, naturally transforms into some infinite-dimensional Lie algebra (over the field $\mathbb{R}$).*

Recall that the *Lie algebra is a linear space endowed with a bilinear skew-symmetric operation satisfying the Jacobi identity*. We should emphasize that the appearance of the structure of Lie algebra in the function space on a symplectic manifold is entirely due to the presence of the symplectic structure on M. It can be readily seen that the operation $f \to \operatorname{grad} f$ given by a symmetric (i.e., Riemannian) scalar product does not give rise to the natural structure of Lie algebra on the function space. Here lies the essential difference between the Riemannian (symmetric) and symplectic (skew-symmetric) geometries on a manifold.

There exists a close connection between the Lie algebra $C^\infty(M)$ (with a Poisson bracket as the operation) and the Lie algebra $V(M)$ of all smooth vector fields on the manifold M (with an ordinary vector-field commutator as the operation). Let us construct a natural mapping α of the function space $C^\infty(M)$ into the vector-field space $V(M)$, setting it by the formula $\alpha(f) = \operatorname{sgrad} f$.

PROPOSITION 1.2.5.

1) *the mapping $\alpha : f \to \operatorname{sgrad} f$ set a homomorphism of the Lie algebra $C^\infty(M)$ into the Lie algebra $V(M)$, that is, $\operatorname{sgrad}\{f, g\} = [\operatorname{sgrad} f, \operatorname{sgrad} g]$. In other words, under the mapping α, the Poisson bracket $\{f, g\}$ is transformed into an ordinary vector-field commutator $\operatorname{sgrad} f, \operatorname{sgrad} g$.*

2) *the image of the Lie algebra $C^\infty(M)$ under the homomorphism α is a subalgebra $H(M)$ consisting of all Hamiltonian vector fields.*

3) *if the manifold is connected, then there holds an isomorphism: $H(M) \approx C^\infty(M)/\mathbb{R}^1$, where $\mathbb{R}^1$ stands for the ideal in $C^\infty(M)$ which consists of constant functions on M (that is, of constants).*

4) *locally Hamiltonian fields form a Lie subalgebra $H_{\operatorname{loc}}(M)$ in the Lie algebra $H(M)$ of all vector fields.*

5) *Hamiltonian fields form the ideal in the Lie algebra of locally Hamiltonian vector fields.*

PROOF: From the definition of the operation sgrad we have:

$$(\operatorname{sgrad}\{f,g\})h = \{\{f,g\},h\} = \quad \text{(by virtue of the Jacobi identity)}$$
$$= -\{\{h,f\},g\} - \{\{g,h\},f\}$$
$$= \{f,\{g,h\}\} - \{g,\{f,h\}\}$$
$$= (\operatorname{sgrad} f)(\operatorname{sgrad} g)h - (\operatorname{sgrad} g)(\operatorname{sgrad} f)h$$
$$= [\operatorname{sgrad} f, \operatorname{sgrad} g]h,$$

which proves item 1. To prove item 2, it suffices to check that the ordinary commutator of two Hamiltonian fields sgrad f and sgrad g is again a Hamiltonian field. This follows immediately from item 1 because $[\operatorname{sgrad} f, \operatorname{sgrad} g] = \operatorname{sgrad}\{f,g\}$. In particular, the Hamiltonian of the commutator of Hamiltonian fields is the Poisson bracket of Hamiltonians of these fields. Thus, the space of Hamiltonian fields is a subalgebra in the space of all vector fields on a manifold. Let us prove item 3. Let f be such a smooth function on a connected manifold that sgrad $f = 0$. Since the form ω is nondegenerate, sgrad $f = 0$. Since the manifold is connected, f is constant on the entire M, as required. Let us prove item 4. Let v_1 and v_2 be locally Hamiltonian vector fields. This means that for any point on the manifold, there exists such a sufficiently small neighbourhood U in which $v_1 = \operatorname{sgrad} f_1^U$, $v_2 = \operatorname{sgrad} f_2^U$. Then their commutator $[v_1, v_2]$ in this neighbourhood admits the representation

$$[v_1, v_2] = [\operatorname{sgrad} f_1^U, \operatorname{sgrad} f_2^U] = \operatorname{sgrad}\{f_1^U, f_2^U\};$$
$$\{f_1^U, f_2^U\} = \omega(\operatorname{sgrad} f_1^U, \operatorname{sgrad} f_2^U) = \omega(v_1, v_2).$$

Thus, the commutator $[v_1, v_2]$ is a globally defined Hamiltonian field, because the smooth function $\omega(v_1, v_2)$ is well defined *on the entire manifold* (and does not depend on the choice of the neighbourhood U).

Thus, we have proved not only item 4, but item 5 as well. Indeed, as has become clear, the commutator of two locally Hamiltonian vector fields v_1 and v_2 is a globally Hamiltonian field. Therefore, *Hamiltonian fields form the ideal in the space of locally Hamiltonian fields*. This completes the proof of the theorem.

Note that the Lie subalgebras $H(M)$ and $H_{\mathrm{loc}}(M)$ depend on the choice of the form ω on M. With a change of their symplectic structure, these subalgebras, in general, will *change* in the Lie algebra of all functions.

For applications, it is instructive to know the explicit expression for the Poisson bracket of two functions in canonical symplectic coordinates. Let $p_1, \ldots, p_n, q_1, \ldots, q_n$ be canonical coordinates in a symplectic space $\mathbf{R}^{2n}$. Then

$$\{f, g\} = \omega(\operatorname{sgrad} f, \operatorname{sgrad} g)$$
$$= \sum_{i=1}^{n} \frac{\partial f}{\partial p_i}\frac{\partial g}{\partial q_i} - \frac{\partial g}{\partial p_i}\frac{\partial f}{\partial q_i}.$$

Here we have used the fact that the form $\omega = \sum dp_i \wedge dq_i$ has the matrix $\begin{pmatrix} 0 & E \\ -E & 0 \end{pmatrix}$.

2.4 Integrals of Hamiltonian Fields

It should be recalled that a smooth function f on a manifold M is called an *integral of a vector field* v if it is constant along the field v, that is, $v(f) = 0$. This is equivalent to the fact that the function f is constant on all integral trajectories of the field v. It is well known that the presence of an integral in a vector field $v(x) = (v_1(x), \ldots, v_N(x))$, which corresponds to a system of ordinary differential equations

$$\dot{x}_i = v_i(x_1, \ldots, x_N), 1 \leqslant i \leqslant N \quad \text{on} \quad M^N,$$

enables the *order of the initial system to be lowered by unity*. The point is that integral trajectories of the system lie on level surfaces of the integral f. These surfaces are (at the points of general position) smooth hypersurfaces in M^N. If several integrals $f_1, \ldots, f_r$ of the field (equation) v are known and if they are functionally independent on some open domain in the manifold, the *order of the system (on this domain) can be lowered by an integer* r.

Thus, the search for a maximal possible number of independent integrals of a vector field is an important task. For Hamiltonian systems, the search for integrals is sometimes facilitated by specific properties of these systems.

THEOREM 1.2.3.
1) *let* $v = \text{sgrad } f$ *be a Hamiltonian field with the Hamiltonian* f *on a symplectic manifold* M. *Then a function* g *on* M *is the integral for the field* v *if and only if its Poisson bracket with the Hamiltonian* f *is zero, that is,* $\{f, g\} \equiv 0$.
2) *the Hamiltonian* f *is always an integral of the Hamiltonian field* $v = \text{sgrad } f$.
3) *if* h *and* g *are two integrals of the Hamiltonian field* $v = \text{sgrad } f$, *then their Poisson bracket* $\{h, g\}$ *is also an integral of this field. This integral may generally turn out to depend ont he integrals* h *and* g.

PROOF: A function g is an integral of the field $v = \text{sgrad } f$ if and only if $v(g) \equiv 0$, that is, $(\text{sgrad } f)g = \{f, g\} = 0$, which precisely proves item 1. If $g = f$ then by virtue of skew symmetry of the Poisson bracket we obtain $\{f, f\} = 0$, which proves item 2. If h and g are two integrals, then from the Jacobi identity we have

$$\{\{h, g\}, f\} + \{\{f, h\}, g\} + \{\{g, f\}, h\} = 0$$

and since $\{f, h\} = \{g, f\} = 0$ (the functions h and g are integrals), it follows that $\{\{h, g\}, f\} = 0$, that is, $\{h, g\}$ is also an integral, and the theorem follows.

Thus, knowing some set of integrals of a Hamiltonian field, one may construct a series of other integrals by calculating Poisson brackets of the initial integrals. It is clear that not all of the integrals obtained in this way will be essentially new, i.e., independent of the preceding ones. The point is that on a $2n$- dimensional symplectic manifold, there cannot exist more than n functionally independent (on open domains) functions. But, in some cases, this procedure may actually give a new integral independent of the initial ones.

Let us consider a Hamiltonian field $v = \text{sgrad } f$ on M. The Hamiltonian f may be interpreted as an element of the Lie algebra $C^\infty(M)$ of smooth functions on M. Then we may analyze a linear subspace $G(f)$ of all functions $g \in C^\infty(M)$ which commute with the element f with respect to the Poisson bracket, i.e., functions

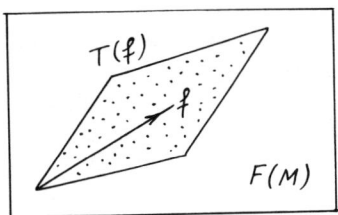

Figure 8

which satisfy the equation $\{f, g\} = 0$. Clearly, the subspace $G(f)$ is a set of all integrals of the Hamiltonian field $v = \operatorname{sgrad} f$. Obviously, $f \in G(f)$ (Fig. 8). Thus, each Hamiltonian f uniquely defines a certain subspace $G(f)$ of functions commuting with it, i.e., a space of all integrals of the field $v = \operatorname{sgrad} f$. From Theorem 1.2.3, it immediately follows that the space $G(f)$ of all integrals is a subalgebra in the Lie algebra $C^\infty(M)$ of functions on M. The subspace $G(f)$ may have an *infinite dimension*, but one can choose in it *not more than n* distinct functionally independent integrals of the field $v = \operatorname{sgrad} f$ on a $2n$-dimensional manifold M. The rest of the elements of the space $G(f)$ will be functionally expressed by these independent integrals.

We will present a useful theorem which generalizes Noether's result.

THEOREM 1.2.4. *Let a Hamiltonian field $v = \operatorname{sgrad} f$ be given on a symplectic manifold M. Let the Hamiltonian f be invariant under a one-parameter group $\mathfrak{G}^\tau$ of transformations $g_t : M \to M$ due to some other Hamiltonian field $\tau = \operatorname{sgrad} g$. Then the Hamiltonian g is an integral of the field $v = \operatorname{sgrad} f$.*

PROOF: Since the Hamiltonian f is invariant under the transformation g_t, $\tau(f) \equiv 0$, i.e., $(\operatorname{sgrad} g)f = 0$, i.e. $\{g, f\} = 0$, as required.

There exist other generalizations of *Noether's theorem*. Let a Hamiltonian field $v = \operatorname{sgrad} f$ and another Hamilton field $w = \operatorname{sgrad} g$ commuting with v be given. This means that $[v, w] = 0$. Recall that this is equivalent to the fact that two corresponding one-parameter groups of transformations $\mathfrak{G}^v$ and $\mathfrak{G}^w$ commute as diffeomorphisms of the manifold m. Commutation of the fields v and w implies that

$$0 = [v, w] = [\operatorname{sgrad} f, \operatorname{sgrad} g] = \operatorname{sgrad}\{f, g\}.$$

From proposition 1.2.5, it follows that the Poisson bracket $\{f, g\}$ is a locally constant function, and if the manifold M is connected, then the function is constant on the entire manifold M. Thus, we have proved that two Hamiltonian field v and w commute (like vector fields) if and only if the Poisson bracket of their Hamiltonians, $\{f, g\}$, is locally constant (or constant, in case the manifold is connected).

2.5 The Liouville Theorem

If for a vector field v on a manifold M^N one can find r independent (almost everywhere) integrals $f_1, \ldots, f_r$, then the order of the corresponding system of differential equations decreases by an integer r. Geometrically this corresponds to the fact that integral trajectories of the system v move along common level surfaces of the integrals $f_1, \ldots, f_r$ which (in the case of general position) are $(N - r)$-dimensional smooth submanifolds in M^N. Therefore, for a complete integration of a system, one should have $N - 1$ independent integrals. Then their common level surfaces will be one-dimensional and will coincide with the integral trajectories of this system. But such an ideal case is very rate. Most often one cannot find a complete set of integrals (or such a set does not exist at all). One has therefore to be satisfied with various versions of "partial integrability." And it is desirable that partial integrability look approximately as follows.

It is required to find a set of independent (almost everywhere) integrals $f_1, \ldots, f_r$, such that *their common level surfaces be sufficiently simple*, for instance, that all of them (in the case of general position) be diffeomorphic to a same *simple manifold*. Besides, it is also desirable that, being restricted to this common level surface, the initial system be transformed on it into a "simply organized" system, i.e., that the *integral trajectories admit a simple description*.

It is remarkable that in many cases this program can be realized.

For some Hamiltonian systems, there exists an extremely advisable partial integrability, for which the common level surface of integrals turns out to be a *torus*, and the restriction to it of the initial system sets a *conditionally-periodic motion along the torus*.

We will conditionally refer to this situation as the case of *commutative integrability of a Hamiltonian system* and will examine it in more detail.

We will say that *two functions f and g are in involution* on a symplectic manifold M^{2n} if their Poisson bracket is exactly zero. It turns out that for Hamiltonian systems it suffices to find only n independent integrals in involution, in order that the description of the motion of integral trajectories of a system could be complete enough. In this case, each integral "is taken for two integrals." More precisely, each such integral makes it possible to lower the order of the system not by unity, as in the general case, but by two unities. We will present the known Liouville theorem which solves this integration problem.

THEOREM 1.2.5. *Let a set of n smooth functions $f_1, \ldots, f_n$ in involution be given on a symplectic manifold M^{2n}. We will denote by M_ξ the common level surface given by the system of equations $f_1(x) = \xi_1, \ldots, f_n(x) = \xi_n$. Suppose that on this surface all the functions f_i are functionally independent (this is, the gradients of the functions or their skew gradients are linearly independent at each point on the level surface). Then there exist the following assertions.*

1) *The level surface M_ξ is a smooth n-dimensional submanifold invariant with respect to each vector field sgrad $f_i = v_i$, that is, all these fields are tangent to the level surface M_ξ.*

2) *If the level surface M_ξ is connected and compact, it is diffeomorphic to an n-dimensional torus T^n. In the general case, if M_ξ is connected (but not necessarily compact) and if all vector fields v_i are complete on the level sur-*

face, then M_ξ is diffeomorphic to a certain "cylinder," that is, a factor of the Euclidean space $\mathbb{R}^n$ by a certain lattice (whose rank does not exceed n).

3) If the level surface M_ξ is connected and compact (in this case, it is a torus T^n), then in a certain open neighbourhood of this surface, one may introduce regular curvilinear coordinates $s_1, \ldots, s_n; \varphi_1, \ldots, \varphi_n$ called "action-angle," which have the following properties:

a) The functions $s_1(x), \ldots, s_n(x)$ set coordinates in the directions transversal to the torus T^n and are functionally expressed through the integrals $f_1, \ldots, f_n$. In these coordinates, the equation of the torus is given as $s_1 = \cdots = s_n = 0$.

b) The functions $\varphi_1(x), \ldots, \varphi_n(x)$ set angular coordinates on the torus T^n (and therefore, on the tori close to this one, which are level surfaces of the integrals $f_1, \ldots, f_n$). In other words, φ_i change from 0 to 2π and set the representation of the torus T^n in the form of a standard direct product of n circles, where φ_i is the angular coordinate on the ith circle.

c) Written in the neighbourhood of the torus T^n in "action-angle" coordinates, the symplectic structure ω acquires the canonical form $\omega = \sum ds_i \wedge d\varphi_i$. This is equivalent to the fact that the pairwise Poisson brackets of "action-angle" coordinates have the simplest form:

$$\{s_i, s_j\} = \{\varphi_i, \varphi_j\} = 0, \qquad \{s_i, \varphi_j\} = \delta_{ij}.$$

d) Written in the "action-angle" coordinates, each vector field $v = \text{sgrad } f$, where f is any function of the set $f_1, \ldots, f_n$ or any function which can be functionally expressed through the functions $f_1, \ldots, f_n$ (the field is tangent to the torus), acquires a simple form

$$\dot{\varphi}_i = q_i(s_1 = 0, \ldots, s_n = 0) \doteq \text{const}.$$

Here the equations $s_i = 0$ are equivalent to the equations $f_i = \xi_i$, where $\xi_1, \ldots, \xi_n$ are constants defining the torus T^n as the level surface $f_1 = \xi_1, \ldots, f_n = \xi_n$. Thus, the vector field v has the simplest form on the torus T^n (in coordinates $\varphi_1, \ldots, \varphi_n$): its components are constant, and its integral trajectories set a rectilinear winding of the torus, that is to say, a conditionally periodic motion along the torus. At each point of the torus, the vectors $v_i = \text{sgrad } f_i$ form a basis in a tangent plane to the torus.

e) The functions $q_1, \ldots, q_n$ from the previous item d) are also defined in a certain neighbourhood of the torus T^n. Therefore, on the level surfaces which are close to the torus (and are also tori), we also have $\dot{\varphi}_i = q_i(s_1^0, \ldots, s_n^0)$, where the equations $s_1 = s_1^0, \ldots, s_n = s_n^0$ set a close level surface. Thus, the initial system $v = \text{sgrad } f$ is written in the neighbourhood of the torus in the coordinates

$$s_1, \ldots, s_n, \varphi_1, \ldots, \varphi_n$$

in the form

$$\dot{s}_1 = 0, \ldots, \dot{s}_n = 0, \qquad \dot{\varphi}_i = q_i(s_1, \ldots, s_n), 1 \leqslant i \leqslant n.$$

Thus, if the integrals $f_1, \ldots, f_n$ are independent, then *nonsingular compact common level surfaces are unions of tori.* Embedding each of these tori into an enveloping symplectic manifold may be fairly complicated. It is given by the functions $f_1, \ldots, f_n$, and the more complicated the functions, the more complicated the manifold. Nonetheless, knowing the functions $f_1, \ldots, f_n$, one can say much about the character of the torus location in M^{2n}. On the torus itself, the vector field $v = \operatorname{sgrad} f$ which we now investigate is constructed extremely simply. In the coordinates $\varphi_1, \ldots, \varphi_n$, it becomes a field with *constant* components, i.e., *it is completely defined by setting the velocity vector at a certain point of the torus*. All the rest of the velocity vectors are obtained from it by parallel transport in the coordinates $\varphi_1, \ldots, \varphi_n$.

The Liouville theorem has become highly important in modern geometry, mechanics, and physics because many mechanical systems (and their multidimensional analogues) appear to posses a set of integrals in involution which enable system to be integrated in the indicated sense. We will say that a Hamiltonian system $v = \operatorname{sgrad} f$ on a symplectic manifold M^{2n} is *completely Liouville-integrable* (or *admits a complete commutative integration*) if for this system there exists a set of n functions $f_1, \ldots, f_n$ in involution, the functions $f_1 = f$ and $f_1, \ldots, f_n$ satisfying the conditions of the Liouville theorem. The set of functions $f_1, \ldots, f_n$ will be called a *complete commutative (involutive) set of functions*.

Integrating a given system in the sense of Liouville means involving the system's Hamiltonian f in the family of functions which are in involution and are such that among them one can choose n independent functions, where n is *half the dimension of the enveloping manifold*. If such a set of functions is found, then (under the assumptions of Theorem 1.2.5) the trajectories of the system move along tori of half the dimension and set on these tori a conditionally periodic motion in appropriate coordinates.

From the point of view of the Lie algebra $C^\infty(M)$ of functions on M^{2n}, an integration of a Hamiltonian system is equivalent to a search for a commutative subalgebra $G(f)$ which contains the Hamiltonian f of a given system and possesses the property that one can choose an additive basis of n independent functions in $G(f)$.

§3 Hamiltonian Properties of the Equations of Motion of a Three-Dimensional Rigid Body

It should be recalled that the motion of a rigid body in $\mathbf{R}^3$ is described by the equations (see §1.3)
$$\dot K = [K, \omega] + [e, u], \qquad \dot e = [e, \omega],$$
where K is the *angular momentum of the body*, ω the *angular velocity of the body*, the vectors e and u are determined by the *physical contents of the problem* and describe the action of *external forces*. These general equations have the following three integrals: $f_1 = H$, where the total energy H has the form
$$H = \frac{1}{2}\langle K, h^{-1}(K)\rangle + m\langle r, e\rangle$$
(see §1.3), $f_2 = \langle K, e\rangle$, and $f_3 = \langle e, e\rangle$. Examine, for instance, the *motion of a heavy rigid body around a fixed point*. In this problem, e is a unit vector directed along the vertical z-axis, and therefore, the third integral f_3 is of the form $f_3 = \langle e, e\rangle = 1$.

§3. Hamiltonian Properties of the Equations

The general equations of motion of a rigid body set a vector field on the *six-dimensional* space $\mathbf{R}^6 = \mathbf{R}^3(K) \times \mathbf{R}^3(e)$.

This field turns out to be Hamiltonian on the common level surface of two integrals $f_2 = c_2 = $ const. *and* $f_3 = c_3 = $ const. with respect to a certain natural Poisson bracket. This makes it possible to apply to this system the rich techniques used for the study of the general Hamiltonian systems.

Let us consider the common level surface M_{23} of two integrals f_2 and f_3, that is, $M_{23} = \{f_2 = c_2, f_3 = c_3 > 0\}$. This surface is a four-dimensional submanifold in $\mathbf{R}^6$. Furthermore, the topological structure of this surface can be easily described.

LEMMA 1.3.1. *The level surface M_{23} is diffeomorphic to the (co)tangent bundle of a two-dimensional sphere.*

PROOF: Let us write two quadratic equations in $\mathbf{R}^6(K, e)$:

$$\langle e, e \rangle = c_3 = \text{const.} \rangle 0 \quad \text{and} \quad \langle K, e \rangle = c_2 = \text{const.}$$

Since the vector e varies in $\mathbf{R}^3(e)$, the first equation implies that the end-point of this vector runs through all the points of a standard two-dimensional sphere of radius $\sqrt{c_3}$. The second equation can be interpreted as follows.

Let us identify $\mathbf{R}^3(K)$ with $\mathbf{R}^3(e)$, then the equation $\langle K, e \rangle = c_2$ transforms for each fixed e into an equation of two-dimensional Euclidean plane orthogonal to the vector e and moved away from the origin by one and the same quantity (independent of e). See Fig. 9.

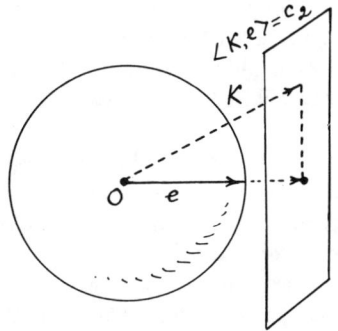

Figure 9

By varying e, we obtain a set of planes, each tangent to a two-dimensional sphere of one and the same radius. Therefore, the two vectors K, e, which satisfy the above equations, define uniquely a certain vector tangent to a two- dimensional sphere. This precisely proves the lemma since the bundle tangent to a sphere S^2 is defined as a four-dimensional manifold whose points are pairs of the form (e, ξ), where e is a point on the sphere, and ξ is an arbitrary vector tangent to the sphere at the point e.

Now let us define ont he space $\mathbf{R}^6(K,e)$ a certain operation $\{\,,\,\}$ of which we then prepare a Poisson bracket on the level surface M_{23}. We assume, by definition, that if $K=(K_1,K_2,K_3)$ and $e=(e_1,e_2,e_3)$, then

$$\{K_1,K_2\}=-K_3, \quad \{K_3,K_1\}=-K_2, \quad \{K_2,K_3\}=-K_1,$$
$$\{K_1,e_1\}=\{K_2,e_2\}=\{K_3,e_3\}=0, \quad \{K_1,e_2\}=-e_3, \quad \{K_1,e_3\}=e_2,$$
$$\{K_3,e_1\}=-e_2, \quad \{K_3,e_2\}=e_1, \quad \{K_2,e_1\}=e_3,$$
$$\{K_2,e_3\}=-e_1;$$
$$\{e_i,e_j\}=0.$$

The corresponding square table which sets the operation $\{\,,\,\}$ on the generators K_1 and e_i is of the form

	K_1	K_2	K_3	e_1	e_2	e_3
K_1	0	$-K_3$	K_2	0	$-e_3$	e_2
K_2	K_3	0	$-K_1$	e_3	0	$-e_1$
K_3	$-K_2$	K_1	0	$-e_2$	e_1	0
e_1	0	$-e_3$	e_2	0	0	0
e_2	e_3	0	$-e_1$	0	0	0
e_3	$-e_2$	e_1	0	0	0	0

It is readily seen that this operation sets on the space $\mathbf{R}^6(K,e)$ the *structure of the Lie algebra*, which is isomorphic to the Lie algebra of the group of motions of the three-dimensional space $\mathbf{R}^3$. In other words, it is isomorphic to a semidirect sum of two Lie subalgebras: the three-dimensional Lie subalgebra $\mathbf{R}^3(K)$ isomorphic to the *Lie algebra of the group* SO(3) *of rotations* in $\mathbf{R}^3$ and the three-dimensional Abelian Lie subalgebra $\mathbf{R}^3(e)$ isomorphic to the *Lie algebra of the group of shifts (translations) in* $\mathbf{R}^3$. The Lie subalgebra $\mathbf{R}^3$ is represented on the subalgebra $\mathbf{R}^3(e)$ in a standard manner.

Assuming now the operation $\{\,,\,\}$ to be bilinear, skew-symmetric, and satisfying the *Leibniz rule* of function product differentiation, we may extend it to the space of all smooth functions given on $\mathbf{R}^6(K,e)$. One can check that the operation defined in this way satisfies the *Jacobi identity*.

Thus, we can now compute the "Poisson bracket" of any pair of functions on $\mathbf{R}^6(K,e)$. It can be checked that the equations of motion of a rigid body are now represented int he following form:

$$\dot{K}_i=\{K_i,H\}, \quad \dot{e}_i=\{e_i,H\}, \quad i=1,2,3.$$

However, the operation constructed above has one defect, namely, *it is degenerate*. Indeed, the functions (integrals) f_2 and f_3 commute (in the sense of our bracket) with any smooth functions $g(K,e)$. To make sure of this, it suffices to check that the functions

$$f_2=\langle K,e\rangle=\sum K_i e_i \quad \text{and} \quad f_3=\langle e,e\rangle=\sum e_i e_i$$

commute with all the generators K_i, e_i. Let us compute, for instance, the bracket

$$\{f_2, K_1\} = \sum_{i=1}^{3}\{K_i e_i, K_1\}$$
$$= \sum_{i=1}^{3}\{K_i, K_1\}e_i + K_i\{e_i, K_1\}$$
$$= K_3 e_2 - K_2 e_3 + K_2 e_3 - K_3 e_2 \equiv 0$$

One may similarly check that $\{f_2, K_2\} = \{f_2, K_3\} = 0$. Next we compute

$$\{f_2, e_1\} = \sum_{i}\{K_i e_i, e_1\}$$
$$= \sum_{i}\{K_i, e_1\}e_i + K_i\{e_i, e_1\}$$
$$= \sum_{i}\{K_i, e_1\}e_i = e_3 e_2 - e_2 e_3 = 0$$

Similarly,
$$\{f_2, e_2\} = \{f_2, e_3\} = 0.$$

Next,

$$\{f_3, e_j\} = \sum_{i}\{e_i^2, e_j\} = 0;$$
$$\{f_3, K_1\} = \sum_{i}\{e_i^2, K_1\}$$
$$= \sum_{i} 2e_i\{e_i, K_1\}$$
$$= 2e_2 e_3 - 2e_3 e_2 = 0;$$

and similarly
$$\{f_3, K_2\} = \{f_3, K_3\} = 0.$$

Hence, the degenerate Poisson bracket constructed above can be restricted (limited) to common level surfaces M_{23} of the integrals f_2 and f_3. One can check that as a result, a *nondegenerate Poisson bracket* $\{\ ,\ \}$ arises already on the space of the functions defined on the manifold M_{23}.

The computation of this bracket may be described in another way. Let f and g be smooth functions on the level surface M_{23}. Let them continue smoothly up to the functions f' and g' given on the entire space $\mathbb{R}^6(K, e)$. Suppose, by definition, that $\{f, g\}'(x) = \{f', g'\}(x)$, where $x \in M_{23}$. The bracket $\{\ ,\ \}'$ is thus well defined. that is to say, such a definition does not depend on how the functions f and g continue up to the functions f' and g'. Hence, the Poisson bracket $\{\ ,\ \}'$ sets the structure of a *symplectic manifold* on the surface M_{23}.

THEOREM 1.3.1 (NOVIKOV AND SHMELTSER [97]). *The equations of motion of a rigid body $\dot{K} = [K, w] + [e, u], \dot{e} = [e, \omega]$ can be represented on the common level surface M_{23} of two integrals f_2 and f_3 in the Hamiltonian form $\dot{f} = \{f, h\}'$, where h is the restriction of the function H to the surface M_{23}.*

In the case the second integral f_2 is zero, this construction acquires a particularly simple form on the surface $M_{23}^0 = (f_2 = 0, f_3 = \text{const.})$. We will describe it following Kozlov [160].

Let $f_2 = 0$ and put $K = [p, e]$. If $f_3 > 0$ and $f_2 = \langle K, e \rangle = 0$, then the vector p exists and is uniquely defined up to the shifts along the vector e. Next, we set $M(p, e) = H([p, e], e)$.

THEOREM 1.3.2 (SEE [160]). *The functions $p(t)$ and $e(t)$ satisfy the canonical Hamiltonian equations*

$$\dot{p} = -\frac{\partial M}{\partial e}, \qquad \dot{e} = \frac{\partial M}{\partial p}.$$

The vectors e, p, K admit a simple interpretation, namely, *e is the radius vector of a point in $\mathbb{R}^3$, p the momentum of this point,* and K the *angular momentum* (taken with an opposite sign). In case $f_2 \neq 0$, the change of the variables $(K, e) \to (p, e)$ should be made in a more delicate way.

PROOF OF THEOREM 1.3.2: Let us find

$$\dot{e} = \frac{\partial M}{\partial p} = \frac{\partial H}{\partial K}\frac{\partial K}{\partial p} = [e, \omega].$$

Since $K = [p, e]$, it follows that

$$\dot{K} = [\dot{p}, e] + [p, \dot{e}] = -\left[\frac{\partial M}{\partial e}, e\right] + [p, [e, \omega]],$$

$$\frac{\partial M}{\partial e} = \frac{\partial H}{\partial e} + \frac{\partial H}{\partial K}\frac{\partial K}{\partial e} = u + [\omega, p].$$

From this we have

$$\dot{K} = -[u, e] + [e, [\omega, p]] + [p, [e, \omega]] = [K, \omega] + [e, u],$$

which complete the proof.

The *Hamiltonian properties of the equations of motion of a rigid body on a level surface M_{23} of integrals f_2 and f_3* has a deep algebraic meaning. In the chapters which follow, we will see that this is also true for various *multidimensional analogues of the equations of motion of a rigid body* and that it is this fact that helps integrate, in the sense of Liouville, the corresponding multidimensional equations of motion.

§4 Some Information on Lie Groups and Lie Algebras Necessary for Hamiltonian Geometry.

4.1 *Adjoint and Coadjoint Representations, Semisimplicity, the System of Roots and Simple Roots, Orbits, and the Canonical Symplectic Structure*

As we will see below, in modern problems of Hamiltonian mechanics and geometry there often appear systems which in general possess certain *noncommutative symmetry groups*. In this connection, we will need some known facts from the theory of Lie groups and Lie algebras. For the reader's convenience, we have collected all such facts in this section. We omit the proofs, of course, and refer the reader to the manuals (for instance, [28]).

DEFINITION 1.4.1: Let $\mathfrak{G}$ be a smooth manifold on which the group structure is given. Then $\mathfrak{G}$ is called a *Lie group* if the mapping $\mathfrak{G} \times \mathfrak{G} \to \mathfrak{G}$, given by the formula $(a,b) \to ab^{-1}$, is smooth.

Let $v \in T_e\mathfrak{G}$, and let us transfer the vector v, using *left shifts*, about the entire group $\mathfrak{G}$. Thus we obtain the vector field on $\mathfrak{G}$. More precisely, for $a \in \mathfrak{G}$, we put $L_a(g) = ag$. Since $L_a^{-1} = L_{a^{-1}}$, this is a diffeomorphism of the group $\mathfrak{G}$. Assume that $\xi_a = (dL_a)_e(v); L_a : \mathfrak{G} \to \mathfrak{G}$. The constructed vector field is *left-invariant*:

$$\xi_{ab} = (dL_a)_e(\xi_b) \qquad \text{for all} \quad a, b \in \mathfrak{G}.$$

A *linear space of all left-invariant vector fields is a Lie algebra G with respect to the vector-field commutator*. This Lie algebra is finite-dimensional, and its dimension is equal to the dimension of the group. It is called the Lie algebra of the Lie group $\mathfrak{G}$.

If ξ is a left-invariant vector field on $\mathfrak{G}$, then ξ generates a certain globally defined *group of diffeomorphisms*. A smooth homomorphism $\xi : \mathbb{R}^1 \to \mathfrak{G}$ is called a *one-parameter subgroup* in the Lie group $\mathfrak{G}$. It can be easily verified that the left-invariant field ξ generates a one-parameter group in $\mathfrak{G}$.

Thus, the Lie algebra of a Lie group may be defined in one of the four equivalent ways:

1) one-parameter subgroups
2) tangent vectors at a unit element
3) left-invariant vector fields
4) left-invariant $\mathbb{R}^1$-actions

One can define a universal mapping $\exp : G \to \mathfrak{G}$, under which $\exp(tX) : \mathbb{R}^1 \to \mathfrak{G}$ set a one-parameter subgroup with a velocity vector X at a unit element.

DEFINITION 1.4.2: The group $\mathfrak{G}$ acts on itself by means of *conjugation* $(a,b) \to aba^{-1}$. This action induces already a linear action of the group on its Lie algebra $\mathrm{Ad} : \mathfrak{G} \to GL(G)$, the so-called *adjoint representation of the group* $\mathfrak{G}$, that is, $\mathrm{Ad}_g \xi = d(\mathrm{Ad}_g)_e \xi$.

The *adjoint representation of a Lie group induces the adjoint representation of its Lie algebra*, namely, $ad = d(\mathrm{Ad})_e : G \to \mathrm{Hom}(G)$. The following equality holds: $ad_X Y = [X, Y]$, where $[X, Y]$ is the commutator in a Lie algebra G.

For *Hamiltonian mechanics* another representation is needed. Let G^* be a space *dual* to G, that is, a space of linear mappings $f : G \to \mathbb{R}$. Let us define the *coadjoint action* $\mathrm{Ad}^* : \mathfrak{G} \to GL(G^*); (\mathrm{Ad}_g^* f)X = f(\mathrm{Ad}_{g^{-1}} X)$. The differential of this representation is called the *coadjoint representation of a Lie algebra*: $\mathrm{ad}^* : G \to \mathrm{Hom}(G^*)$. The following equality holds: $(\mathrm{ad}_X^* f)Y = f([X,Y]); X, Y \in G, f \in G^*$. Sometimes we will denote $\mathrm{ad}_X^* f$ by $\{X, f\}$. Here $X \in G, f \in G^*$. Not to be confused with the Poisson bracket!

We shall now deal with the *orbits of coadjoint representation*. On these orbits, there is a *natural symplectic structure*, and any *homogeneous symplectic manifold* may be realized in the from of the orbit of the representation Ad^* in a certain Lie algebra.

Note that in the general case the orbits of the representations Ad^* and Ad are *distinct*. We shall give the simplest example illustrating this phenomenon. Let $\mathfrak{G}$ be the group of affine transformations of a straight line $x \to ax + b, a \neq 0; a, b \in \mathbf{R}$. This group admits matrix realization

$$\mathfrak{G} = \left\{ \begin{pmatrix} a & b \\ 0 & 1 \end{pmatrix}, a \neq 0 \right\}.$$

It can be easily verified that

$$G = T_e \mathfrak{G} = \begin{pmatrix} \xi_1 & \xi_2 \\ 0 & 0 \end{pmatrix}$$

and

$$\mathrm{Ad}_a \xi = a\xi a^{-1}, \; a \in \mathfrak{G}, \xi \in G.$$

If

$$a = \begin{pmatrix} a_1 & a_2 \\ 0 & 1 \end{pmatrix}, \quad \xi = \begin{pmatrix} \xi_1 & \xi_2 \\ 0 & 0 \end{pmatrix}$$

then

$$\mathrm{Ad}_a \xi = \begin{pmatrix} \xi_1 & a_1\xi_2 - \xi_1 a_2 \\ 0 & 0 \end{pmatrix} = \begin{pmatrix} \eta_1 & \eta_2 \\ 0 & 0 \end{pmatrix}.$$

From the explicit form of the adjoint representation Ad, we have

$$\eta_1 = \xi_1, \quad \eta_2 = \xi_2 a_1 - \xi_1 a_2.$$

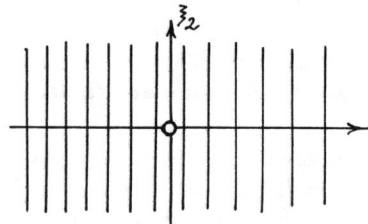

Figure 10

Hence, the orbits are organized as shown in Fig. 10. Let us now investigate the orbits of coadjoint representation. In the algebra G, we choose the basis

$$e_1 = \begin{pmatrix} 1 & 0 \\ 0 & 0 \end{pmatrix}, \quad e_2 = \begin{pmatrix} 0 & 1 \\ 0 & 0 \end{pmatrix},$$

and in the algebra G^*, the dual basis $f_1, f_2 : f_i(e_j) = \delta_{ij}$. In this basis, we have coordinates x_1, x_2. Simple calculations show that

$$\mathrm{Ad}^*_{\begin{pmatrix} a_1 & a_2 \\ 0 & 1 \end{pmatrix}}(x_1, x_2) = (x_1 - x_2 a_2, x_2 a_1),$$

and therefore the orbits are organized as shown in Fig. 11. In particular, the representations Ad and Ad* of this group are *not equivalent*.

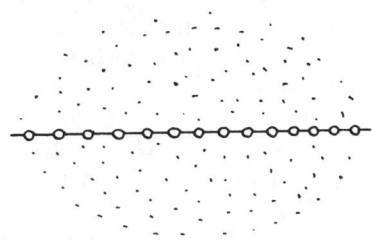

Figure 11

If on the algebra G there exists a symmetric nondegenerate scalar product $\langle X, Y \rangle$ such that $\langle \mathrm{Ad}_g X, \mathrm{Ad}_g Y \rangle = \langle X, Y \rangle$, then the adjoint and coadjoint representations are equivalent, that is, they have, in particular, identical orbits.

Hereafter, we denote a scalar product by $\langle \, , \, \rangle$, which is not to be confused with the notation of the previous section!

In particular, all so-called *semisimple Lie algebras* satisfy this condition. The bilinear form $\mathrm{tr}(ad_X ad_Y)$ (or $\mathrm{Re\,tr}\, ad_X ad_Y$) is called the *Killing form*. Let us denote it simply by $\langle X, Y \rangle$. Semisimple Lie algebras are characterized by the fact that the form $\langle X, Y \rangle$ is nondegenerate.

Recall the structural theory of *complex semisimple Lie algebras*. The *maximal Abelian subalgebra* $T \subset G$, such that for all $h \in T$, ad_h is a semisimple linear transformation of G, is called the *Cartan subalgebra*. If $X \in G$ is an arbitrary element, then by $G(X, 0)$ we denote the subspace of elements commuting with X in G. An element $X \in G$ is called *regular* provided that $\dim G(X, 0)$ is *minimal*. If X is a regular element, then $G(X, 0)$ is a Cartan subalgebra in G. This subalgebra will be denoted by $T(X)$. *Regular elements from an open and everywhere dense subset in G. If G is a semisimple Lie algebra, then any Cartan subalgebra is commutative.*

Let G be a semisimple Lie algebra over a field of complex numbers $\mathbb{C}$. Let us fix a certain Cartan subalgebra T. The linear form $\alpha(h)$ on T is called a *root* if there exists an element $E_\alpha \in G, E_\alpha \neq 0$, such that $[h, E_\alpha] = \alpha(h) E_\alpha$ for any $h \in T$. Let G_α be a proper subspace corresponding to α. Then $G = T \oplus \sum_{\alpha \neq 0} G_\alpha$, where the sign $\oplus$ stands for a direct sum of linear subspaces. *In the semisimple Lie algebra G, all subspaces G_α are one-dimensional when $\alpha \neq 0$* (over the field $\mathbb{C}$).

It is known that $[G_\alpha, G_\beta] \subset G_{\alpha+\beta}$, that is, $[E_\alpha, E_\beta] = N_{\alpha\beta} \cdot E_{\alpha+\beta}$. If $\alpha+\beta \neq 0$, then E_α and E_β are *orthogonal* with respect to the form $\langle X, Y \rangle$. The vectors E_α and $E_{-\alpha}$ are, on the contrary, *not orthogonal*. The restriction of the form $\langle X, Y \rangle$ to T is nondegenerate. If $r = \dim_\mathbb{C} T$ (the number r is called the *rank of the algebra G*), then there exist r linearly independent roots of the algebra G with respect to T. The total number of roots is, generally, greater than r. Therefore, the *set of all roots is not linearly independent*. If $\alpha, \beta, \alpha+\beta$ are nonzero roots, then $[G_\alpha, G_\beta] = G_{\alpha+\beta}$. The only roots proportional to the root $\alpha \neq 0$ are the roots $0, \pm\alpha$. The roots α can be represented as the vectors $H'_\alpha \in T$. Since the form $\langle X, T \rangle$ is nondegenerate on T, it follows that for each $\alpha \in T^*$ (where T^* is a space dual to T), there exists a single element $H'_\alpha \in T$, such that $\alpha(h) = \langle h, H'_\alpha \rangle$ for all $h \in T$. Then, if $\alpha \neq 0$, it follows that $[E_\alpha, X] = \langle E_\alpha, X \rangle H'_\alpha$ for $X \in G_{-\alpha}$ and $\langle \alpha, \alpha \rangle \neq 0$. We will denote by $T_0 \subset T$ the subspace in T generated by all vectors H'_α with rational coefficients. Clearly, T_0 is the "real part" of T. It turns out that $\dim_\mathbb{Q} T_0 = \dim_\mathbb{C} T = \frac{1}{2} \dim_\mathbb{R} T$. Here $\mathbb{Q}$ is the field of rational numbers. Next, the restriction on T_0 of the form $\langle h, h' \rangle$ is positive definite and takes on rational values $(h, h' \in T_0)$; $\alpha(h') \in \mathbb{Q}$ for any $h' \in T_0$. In particular, $\alpha(h')$ is a real number if $h' \in T_0$. In what follows, we will denote by Δ a *set of nonzero roots of the algebra G*. Let $H_1, \ldots, H_r$ be a certain fixed basis in T_0. If λ, μ are two linear forms on T_0, then λ is said to be greater than μ, provided that $\lambda(H_i) = \mu(H_i)$ when $i = 1, 2, \ldots, n$ and $\lambda(H_{n+1}) > \mu(H_{n+1})$. Recall that if λ, μ are roots, then $\lambda(h'), \mu(h')$ are real numbers for any $h' \in T_0$. Thus, in the set Δ there arises *linear ordering*. The root $\alpha \in \Delta$ is called *positive* if $\alpha > 0$, that is, $\alpha(H_i) = 0$ when $i = 1, 2, \ldots, n$ and $\alpha(H_{n+1}) > 0$. Positiveness of the root α thus implies that its first nonzero coordinate is positive. The linear ordering is introduced not uniquely. We assume hereafter that the basis $H_1, \ldots, H_r$ (where $r = \operatorname{rank} G$) *is fixed*. We will denote the *set of positive roots* by Δ^+. Then $\Delta = \Delta^+ \cup \Delta^-$, where $\Delta^+ \cap \Delta^- = \emptyset$, and there exists a one-to-one correspondence between Δ^+ and Δ^- established by the involution $\alpha \to -\alpha$. Clearly, if $\alpha \in \Delta^+$, then $(-\alpha) \in \Delta^-$. A positive root α is called *simple* if it cannot be represented as a sum of two positive roots. If $r = \operatorname{rank} G = \dim_\mathbb{C} T$, then there exist exactly r simple roots $\alpha_1, \ldots, \alpha_r$ which form the basis in T over $\mathbb{C}$ and the basis in T_0 over $\mathbb{Q}$. Besides, each root $\beta \in \Delta$ can be represented in the form $\beta = \sum m_i \alpha_i$, where $m_i \in \mathbb{Z}$ are integers of the same sign. If $m_i \geqslant 0$, then $\beta \in \Delta^+$, and if $m_i \leqslant 0$, then $\beta \in \Delta^-$. The *system of simple roots* $\alpha_1, \ldots, \alpha_r$ is usually denoted by Π. The system Δ can be uniquely established from the system Π.

Set

$$V^+ = \sum_{\alpha > 0} G_\alpha, \quad V^- = \sum_{\alpha < 0} G_\alpha,$$

then $G = T \oplus V^+ \oplus V^-$. For future purposes, let us fix the basis of a special form in G. An arbitrary basis $H_1, \ldots, H_r$ in T (over $\mathbb{C}$) generates to the basis H' in T_0 (over $\mathbb{Q}$) and in T (over $\mathbb{Q}$). We will complement this basis with the vectors $E_\alpha \in G_\alpha, \alpha \neq 0, \alpha \in \Delta$. The vectors E_α may be chosen so that $\langle E_\alpha, E_{-\alpha} \rangle = 1$. Then the commutation operation in the algebra G is given by $[h, E_\alpha] = \alpha(h) E_\alpha, h \in T$

(where $\alpha(h) \in \mathbb{Q}$ if $h \in T_0$); $[E_\alpha, E_{-\alpha}] = -H'_\alpha$;

$$[E_\alpha, E_\beta] = \begin{cases} N_{\alpha\beta} \cdot E_{\alpha+\beta}, & \alpha+\beta \neq 0 \text{ is a root} \\ 0, & \alpha+\beta \neq 0 \text{ is not a root} \end{cases} \quad \langle h, H'_\alpha \rangle = \alpha(h), h \in T.$$

The vectors $E_\alpha \in G_\alpha$ may be so chosen that $N_{\alpha\beta} = N_{-\alpha,-\beta}$. The constants $N_{\alpha\beta}$ may be regarded as *real* (after the corresponding normalization of the vectors E_α).

Let us consider the coadjoint representation Ad^* of the group $\mathfrak{G}$, i.e., $\text{Ad}^* : \mathfrak{G} \to GL(G^*)$, where G^* is a space dual to the Lie algebra G. If $f \in G^*$, then the set $0(f) = \{\text{Ad}_g^* f, g \in \mathfrak{G}\}$ is called an *orbit* passing through the point $f \in G^*$.

It is known that *on the orbits of coadjoint representation of any Lie group, there exists a natural symplectic structure (the structure of Kirillov) invariant with respect to the coadjoint representation.*

Let $0(f)$ be the orbit of the representation Ad^* of the Lie group $\mathfrak{G}$ which passes through the point $f \in G^*$. Then

$$T_f 0(f) = \{\text{ad}_\xi^* f, \xi \in G\} \subset G^*.$$

Indeed, any tangent vector to the orbit $0(f)$ at the point $f \in G^*$ has the form

$$v = \frac{d}{dt}\Big|_{t=0} \text{Ad}^*_{\exp(t\xi)} f \in T_f 0(f)$$

for a certain $\xi \in G$. Let $e_1, \ldots, e_n$ be a basis in G and $e^1, \ldots, e^n$ a dual basis to e_i in G^*, that is, $e^i(e_j) = \delta^i_j$. If $f \in G^*$, then $f = \sum f_i e^i$, f_i is a linear function on G^*. Thus

$$v_i = \frac{d}{dt}\Big|_{t=0} f_i(\text{Ad}^*_{\exp(t\xi)} f) = \frac{d}{dt}\Big|_{t=0}(\text{Ad}^*_{\exp(t\xi)} f)e_i$$

$$= f(\frac{d}{dt}\Big|_{t=0} \text{Ad}_{\exp(-t\xi)} e_i) = f(-[\xi, e_i]) = -(\text{ad}_\xi^* f)e_i.$$

Hence, $v = -\text{ad}_\xi^* f$, as required.

DEFINITION 1.4.3: Let $\xi, \eta \in T_f 0(f)$. The, as proved above, we may assume that

$$\xi = \text{ad}^*_{\xi_1} f, \qquad \eta = \text{ad}^*_{\eta_1} f.$$

We assume by definition, that $\omega(\xi, \eta) = f([\xi_1, \eta_1])$.

Let us check the correctness of the definition. Generally speaking, ξ_1, η_1 are *not uniquely* defined. Let $f \in G^*$. Then the subspace $\text{Ann } f = \{X \in G : \text{ad}_X^* f = 0\}$ is called the *annihilator of the covector f*. As is readily seen, the subspace $\text{Ann } f$ is a *subalgebra*. Obviously,

$$\xi = \text{ad}^*_{\xi_1} f = \text{ad}^*_{\xi_2} f$$

if and only if $\xi_1 - \xi_2 \in \text{Ann } f$. To check the correctness, one should therefore show that

$$f([\xi_1 + \xi_0, \eta_1 + \eta_0]) = f([\xi_1, \eta_1])$$

where $\xi_0, \eta_0 \in \mathrm{Ann}\, f$, which follows from the linearity of f and from the definition of the annihilator.

It is asserted that ω is a *nondegenerate exterior 2-form on* $T_f 0(f)$. This means that either a) $\det(\omega_{ij}) \neq 0$ or b) if $\omega(\xi, \eta) = 0$ for any $\xi \in T_f 0(f)$, then $\eta = 0$. Let us check the second statement. Let $\omega(\xi, \eta) = f([\xi_1, \eta_1)$ for any ξ_1, where

$$\xi = \mathrm{ad}^*_{\xi_1} f, \qquad \eta = \mathrm{ad}^*_{\eta_1} f.$$

Then $(\mathrm{ad}^*_{\eta_1} f)\xi_1 = 0$ for any $\xi_1 \in G$, and therefore $\mathrm{ad}^*_{\eta_1} f = 0$, and hence $\eta = \mathrm{ad}^*_{\eta_1} f = 0$.

In particular, the *dimension of each orbit of coadjoint representation is even*. It is easily seen that the *symplectic form* ω *is invariant under the coadjoint action*, that is, $\mathrm{Ad}^*_h \omega = \omega, h \in \mathfrak{G}$. It also turns out that $d\omega = 0$.

Thus, *on each orbit* $0(f)$, *we have constructed a nondegenerate exterior 2-form* ω (*the form of Kirillov*). In order that ω can define a *symplectic structure* it suffices that $d\omega = 0$ (see above).

Let a nondegenerate differential 2-form ω be given on a manifold M. Then, with the help of this form one can define the Poisson bracket setting

$$\{f, g\} = \sum \omega^{ij} \frac{\partial f}{\partial x_i} \frac{\partial g}{\partial x_j}$$

for $f, g \in C^\infty(M)$. Here $\omega = (\omega_{ij})$ and $(\omega^{ij}) = (\omega_{ij})^{-1}$. It is asserted that $d\omega = 0$ if and only if the bracket $\{f, g\}$ satisfies the Jacobi identity

$$\{f, \{g, h\}\} + \{g, \{h, f\}\} + \{h, \{f, g\}\} = 0.$$

4.2 Model Example: $\mathrm{SL}(n, \mathbb{C})$ *and* $\mathrm{sl}(n, \mathbb{C})$

All basic effects connected with the root system of semisimple Lie algebras manifest themselves fully on the example of a semisimple Lie algebra $\mathrm{sl}(n, \mathbb{C})$. For this reason and for the reader's convenience, we will illustrate the basic properties of the root system on the example of this algebra. After the acquaintance with this example, the interested reader will be able to confidently and without difficulty on root systems in any other semisimple Lie algebra.

If a Lie algebra G is a *matrix one* (like $\mathrm{sl}(n, \mathbb{C})$), then the commutation operation in it is given by $[X, Y] = XY - YX$, where XY and YX are ordinary products of matrices X and Y from the Lie algebra G. For the algebra $\mathrm{sl}(n, \mathbb{C})$, the *Killing form* is as follows $\langle X, Y \rangle = \mathrm{tr}\, X \cdot Y$. We recall that the algebra $\mathrm{sl}(n, \mathbb{C})$ consists of complex square $n \times n$ matrices X, such that $\mathrm{tr}\, X = 0$. As the Cartan subalgebra T, one may take the family of diagonal matrices

$$\begin{pmatrix} a_1 & & 0 \\ & \ddots & \\ 0 & & a_n \end{pmatrix}, \quad a_1 + \cdots + a_n = 0.$$

The element $h \in T$ is regular if and only if all eigenvalues of the operator h are distinct, that is, $a_i \neq a_j$ when $i \neq j$.

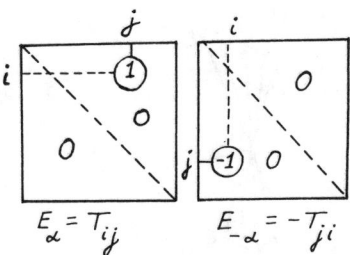

Figure 12

We will denote by T_{ij} an elementary $n \times n$ matrix, in which only one element in the place (i,j) is nonzero; here i is the row number and j is the column number. The Cartan subalgebra has been described above. The vectors E_α, which are eigenvectors for all transformations $\mathrm{ad}_h : G \to G$, have the form T_{ij} if $i < j$ and $-T_{ij}$ if $i > j$ (Fig. 12).

Indeed, by calculating $\mathrm{ad}_h T_{ij}$, we obtain $[h, T_{ij}] = a_i - a_j$, that is, $\alpha(h) = a_i - a_j$. Thus, the roots α of the algebra $\mathrm{sl}(n, \mathbb{C})$ are indexed by a pair of subscripts i, j. We shall now write $\alpha = \alpha_{ij}; \alpha_{ij}(h) = a_i - a_j$ (Fig. 13). Since $(-\alpha_{ij})(h) = a_j - a_i = \alpha_{ji}(h)$, it follows that $-\alpha_{ij} = \alpha_{ji}$ (Fig. 13). The algebra $\mathrm{sl}(n, \mathbb{C})$ is therefore represented in the form $T \oplus \sum \mathbb{C} T_{ij}$. Thus, as the root decomposition of the algebra, one may take its standard decomposition into a direct sum of one-dimensional subspaces $\mathbb{C} T_{ij}$ and the plane T.

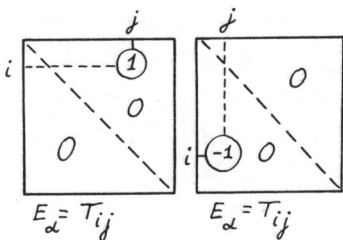

Figure 13

Note that if desired, the Cartan subalgebra T may be interpreted as a proper subspace of transformations ad_T which corresponds to zero eigenvalue. The multiplicity of zero eigenvalue is equal to r, that is, to the rank of the algebra G of the dimension of the Cartan subalgebra (Fig. 14).

Let $G = \mathrm{sl}(n, \mathbb{C})$. Since $E_\alpha = T_{ij}$, it follows that by commuting the matrices $E_\alpha = T_{ij}$ and $E_\beta = T_{pq}$, we obtain the relation $[T_{ij}, T_{pq}] = N_{\alpha\beta} E_{\alpha+\beta}$, because $N_{\alpha\beta} = 0$ if all the subscripts i, j, p, q are distinct and $[T_{ij}, T_{jq}] = T_{iq}$. The explicit expression of the Killing form on the algebra $\mathrm{sl}(n, \mathbb{C})$ is as follows: $\langle X, Y \rangle = \mathrm{tr}\, X \cdot Y$. Therefore, if $\alpha + \beta \neq 0$, then $\langle E_\alpha, E_\beta \rangle = \mathrm{tr}\, T_{ij} T_{pq} = 0$. Thus, E_α and E_β are orthogonal when $\alpha + \beta \neq 0$. If $\alpha + \beta = 0$, then the vectors $E_\alpha = T_{ij}$ and $E_{-\alpha} = -T_{ji}$

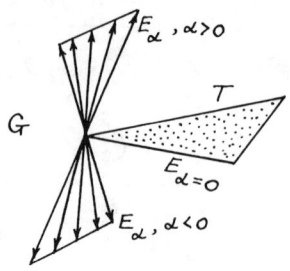

Figure 14

are not orthogonal, because $\langle E_\alpha, E_{-\alpha}\rangle = \operatorname{tr} T_{ij}(-T_{ji}) = -1$. Finally, the equality $[G_\alpha, G_\beta] = G_{\alpha+\beta}$, where, for instance, $\alpha = \alpha_{ij}, i < j, \beta = \alpha_{pq}, p < q$, may take place if and only if two among the four indices i, j, p, and q coincide: either $j = p$ or $i = q$. This is equivalent to the condition that the sum of the roots $\alpha + \beta$ is also a root. Indeed,

$$\alpha_{ij}(h) + \alpha_{pq}(h) = (\alpha + \beta)h = a_i - a_j + a_p - a_q$$

is equal to $a_i - a_q$ if $j = p$ and is equal to $a_p - a_j$ if $i = q$. This mechanism is exhibited in Fig. 15, where the root $\alpha_{ij}(h) = a_i - a_j$ for $i < j$ is shown by an arrow indicating that a_j is subtracted from a_i.

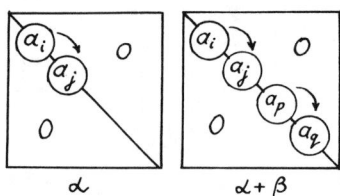

Figure 15

In this case $\mathrm{sl}(n, \mathbb{C})$, the root H'_α, where $\alpha = \alpha_{ij}, i < j$, is expressed by the matrix pictured in Fig. 16.

If $\alpha = \alpha_{ij}, i > j$, then for the root H'_α, see Fig. 17.

Thus, $H'_{\alpha=\alpha_{ij}} = T_{ii} - T_{jj}$. From this it follows that the linear subspace generated by the matrices $T_{ii} - T_{jj}$ coincides with the subalgebra $T = \sum \mathbb{C} T_{ii}$.

4.3 Real, Compact, and Normal Subalgebras

Up to this moment we have dealt with complex semisimple Lie algebras. But an important role is also played by various real subalgebras contained in complex algebras. One of them is especially remarkable, since the corresponding Lie group is compact.

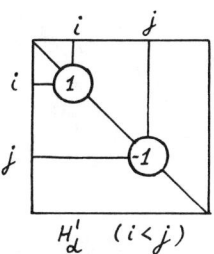

Figure 16

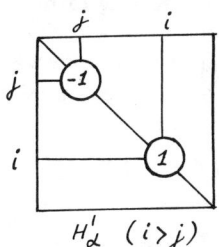

Figure 17

DEFINITION 1.4.4: Let G be a complex semisimple Lie algebra. A real subalgebra G_0 of the Lie algebra G (interpreted as a Lie algebra over a field $\mathbb{R}$) is called the *real form of the complex Lie algebra* G if the canonical mapping of the complex extension $G_0^{\mathbb{C}} = G_0 \otimes_{\mathbb{R}} \mathbb{C}$ of the algebra G_0 into the algebra G is an isomorphism. In this case, we have the equality $\dim_{\mathbb{R}} G_0 = \dim_{\mathbb{C}} G$.

This means that complexifying the real algebra G_0, that is to say, examining linear combinations of its elements with complex coefficients, we obtain the entire enveloping algebra G.

Let G_0 be a certain real form of a complex semisimple Lie algebra G. Then any element of the algebra G can be uniquely represented in the form $X + iY$, where $X, Y \in G_0$. This decomposition of the algebra G gives rise to a natural involution σ which maps the algebra G into itself. Namely $\sigma(X + iY) = X - iY$. This involution depends on the subalgebra G_0 and possesses the following obvious properties:

$$\sigma^2 = 1 = \mathrm{id}, \sigma X = X \quad \text{if} \quad X \in G_0;$$
$$\sigma(A + B) = \sigma A + \sigma B, \sigma(\lambda A) = \bar{\lambda}\sigma A, \sigma[A, B] = [\sigma A, \sigma B].$$

LEMMA 1.4.1. *On a Lie algebra G let an involution σ be given which possesses the properties specified above. Then this involution defines a certain subalgebra G_0 in the algebra G, which is the real form.*

PROOF: Denote by G_0 a set of fixed points of the involution σ on G. From the properties of σ it follows that G_0 is the real subalgebra in G. On the other hand, any element A from the algebra G can be represented in the from $X + iY$, where $X, Y \in G_0$. Indeed,

$$A = \frac{1}{2}(A + \sigma A) + i\left(\frac{A - \sigma A}{2i}\right),$$

where

$$X = \frac{1}{2}(A + \sigma A) \in G_0, \qquad Y = \frac{1}{2i}(A - \sigma A) \in G_0$$

because

$$\sigma X = \sigma(A + \sigma A) = X, \qquad \sigma Y = Y.$$

This proves the lemma.

In the algebra G, consider the Killing form $\langle\ ,\ \rangle_G$. It can be restricted to the real subalgebra G_0. Denote this restriction of the form as $\langle\ ,\ \rangle'_{G_0}$. On the other hand, on the subalgebra G_0, its own Killing form $\langle\ ,\ \rangle_{G_0}$ is defined. The question is whether these two forms coincide (up to a scalar nonzero factor). Generally, the restriction of the Killing form of an enveloping algebra to an arbitrary subalgebra does not coincide with the Killing form of this subalgebra. In the case of real forms, the situation is, however, more favourable.

LEMMA 1.4.2. *If G_0 is the real form of the semisimple Lie algebra G, then two above-mentioned forms coincide (up to the nonzero factor).*

DEFINITION 1.4.5: A real Lie algebra is called *compact* if its Killing from is *negative definite*, that is, if the corresponding quadratic form satisfies the inequality $\langle X, X \rangle < 0$ for $X \neq 0$.

DEFINITION 1.4.6: The real form G_0 of a complex Lie algebra G is called a *compact real form of the algebra G* if G_0 is a compact real algebra.

The term "compact algebra" is reasoned by the fact that a Lie group with a compact Lie algebra turns out itself to be a *compact topological space*.

LEMMA 1.4.3. *In order that the real form G_0 of a Lie algebra G be compact, it is necessary and sufficient that the Hermitian form $\langle A, \sigma A \rangle$ on the algebra G be negative definite.*

PROOF: Let G_0 be a σ compact Lie algebra and $A = X + iY \in G$, then

$$\langle A, \sigma A \rangle = \langle X + iY, X - iY \rangle = \langle X, X \rangle + \langle YY \rangle < 0.$$

Inversely, if the form $\langle A, \sigma B \rangle$ is negative definite, then for $A \in G_0$, $A \neq 0$, we have $\sigma A = A$ and $\langle A, A \rangle < 0$, and the lemma follows.

In the sequel, the compact form will be denoted by G_c.

The *classification of all real forms of the Lie algebra G reduces to the classification of all non-equivalent involutions on semisimple algebra*. The compact form is defined by a special involution, which we shall now describe.

First we consider our model example $\mathrm{sl}(n, \mathbb{C})$. Consider an involution $\sigma A = \bar{A}$ on G, that is, an operation of complex conjugation. The set of fixed points of this involution coincides, obviously, with the subalgebra of real matrices, which is the Lie algebra $\mathrm{sl}(n, \mathbb{R})$. It is clear that the Killing form on the algebra $\mathrm{sl}(n, \mathbb{R})$ is a real form tr $X^2 = \sum x_{ij}x_{ij}$. Obviously, it is not negative definite (it is indefinite). Therefore, sl$(n, \mathbb{R})$ is a *real, but not a compact form of the complex Lie algebra* sl$(n, \mathbb{C})$.

The compact form G_c is constructed here as follows. Consider an involution $\tau : G \to G, \tau A = -\bar{A}^T$, where T implies transposition. The fixed points of this involution are skew-Hermitian matrices with a zero trace. It is readily seen that this space is the Lie algebra of the compact group $su(n)$. Indeed, calculating the Killing form on this real form G_0, we immediately obtain that it is negative definite.

Examining the case of the algebra sl$(n, \mathbb{C})$, it turns out that we have, in fact, modelled the general situation for all complex simple Lie algebras.

THEOREM 1.4.1. *Each semisimple complex Lie algebra G possesses a compact real form G_c.*

PROOF: We merely present in an explicit form the embedding of a compact subalgebra G_c into an algebra G. Consider the Weyl basis in the algebra G (for the definition of this bases, wee above). Consider on G an involution σ given on the Weyl basis as follows: $\sigma E_\alpha = E_{-\alpha}$ if $\alpha \neq 0, \sigma h = -h$ for any vector $h \in T_0$, where $T_0 \subset T$ is the "real part" of the Cartan subalgebra T. We will assume that $\sigma(\lambda X) = \bar{\lambda}\sigma X$. Thus, the mapping σ acts as shown in Fig. 18, that is,

$$\sigma : V^+ \to V^-, \sigma : V^- \to V^+, \sigma : T_0 \to -T_0, \sigma : iT_0 \to iT_0$$

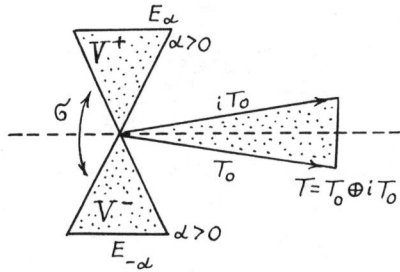

Figure 18

From the properties of the Weyl basis, it immediately follows that σ is an automorphism of the algebra G. Now let us find the real form corresponding to this involution.

From the explicit form of σ, it follows that as a basis (over $\mathbb{R}$) in the subalgebra of fixed points of the involution σ, one may take the vectors

$$E_\alpha + E_{-\alpha}; i(E_\alpha - E_{-\alpha}); iH'_\alpha.$$

It is asserted that this is precisely the compact subalgebra in the algebra G. Indeed, since $\langle E_\alpha, E_\alpha \rangle = 0$ and $\langle E_\alpha, E_{-\alpha} \rangle = -1$, it suffices to calculate the following scalar products:

$$\langle E_\alpha + E_{-\alpha}, E_\alpha + E_{-\alpha}\rangle = -2,$$
$$\langle i(E_\alpha - E_{-\alpha}), i(E_\alpha - E_{-\alpha})\rangle = 2\langle E_\alpha, E_{-\alpha}\rangle = -2,$$
$$\langle E_\alpha + E_{-\alpha}, i(E_\alpha - E_{-\alpha})\rangle = 0,$$
$$\langle iH'_\alpha, iH'_\alpha\rangle = -\alpha(H'_\alpha) < 0,$$

because $\alpha(H'_\alpha) > 0$, and the vector H'_α is dual to the linear form α. Hence, the Killing form is actually negative definite throughout the subalgebra of fixed points. This completes the proof of the theorem.

Now let us analyze in more detail the embedding of the compact form G_c into the algebra G. In G one may obviously choose the following basis (over $\mathbb{R}$):

$$E_\alpha + E_{-\alpha}, \ i(E_\alpha - E_{-\alpha}), \ E_\alpha - E_{-\alpha}, \ i(E_\alpha + E_{-\alpha}), \ H'_\alpha, \ iH'_\alpha.$$

This means that, along with the root decomposition of the algebra $G = T \oplus V^+ \oplus V^-$ (over $\mathbb{C}$), there exists another natural decomposition (over $\mathbb{R}$):

$$G = T_0 \oplus iT_0 \oplus W^+ \oplus W^-, \quad \text{where } W^+ = \{E_\alpha + E_{-\alpha}, i(E_\alpha - E_{-\alpha})\},$$
$$W^- = \{E_\alpha - E_{-\alpha}, i(E_\alpha + E_{-\alpha})\},$$
$$T_0 = \{H'_\alpha\}, \qquad iT_0 = \{iH'_\alpha\}$$

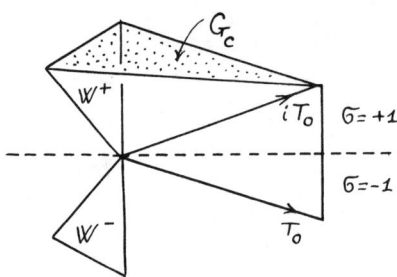

Figure 19

In braces (not to be confused with the Poisson bracket!) we have shown the vectors which form a basis in a corresponding plane (Fig. 19). It is clear that $\sigma = +1$ on the plane $W^+ \oplus iT_0$ and $\sigma = -1$ on the plane $W^- \oplus T_0$. The plane $W^+ \oplus iT_0$ is therefore a subalgebra (as distinct from the plane $W^- \oplus T_0$). This subalgebra of fixed points coincides with the subalgebra G_c in the algebra G, that is, $W^+ \oplus iT_0 = G_c$. Thus, the compact Lie algebra G_c in the complex algebra G is spanned by the following vectors:

$$G_c = \{E_\alpha + E_{-\alpha}, i(E_\alpha - E_{-\alpha}), iH'_\alpha\} = W^+ \oplus iT_0.$$

§4 Information on Lie Groups and Lie Algebras

Now consider an adjoint action of the Lie algebra G_c on itself, that is, examine the action of transformation of the form $\mathrm{ad}_h : G_c \to G_c$, where $h \in iT_0$. Since the element h lies in the Cartan subalgebra, it follows that the transformation ad_h carries into itself the plane W^+ orthogonal to the plane iT_0. We make use of the fact that the operators ad_h are skew-symmetric with respect to the Killing form and therefore preserve the orthogonal complement by carrying it into itself.

Let $h = iq$, where $q \in T_0$. Clearly,

$$\mathrm{ad}_h(E_\alpha + E_{-\alpha}) = i\,\mathrm{ad}_q(E_\alpha + E_{-\alpha}) = i\alpha(q)(E_\alpha - E_{-\alpha}),$$

where $\alpha(q)$ is a real number. Hence,

$$\mathrm{ad}_h(E_\alpha + E_{-\alpha}) = \alpha(q)(i(E_\alpha - E_{-\alpha})).$$

Similarly,

$$\mathrm{ad}_h\, i(E_\alpha - E_{-\alpha}) = -\alpha(q)(E_\alpha + E_{-\alpha}).$$

Thus, the operator ad_h carries the two-dimensional real plane spanned by the vectors $E_\alpha + E_{-\alpha}, i(E_\alpha - E_{-\alpha})$ into itself and is given in this plane by the following skew-symmetric second-order matrix

$$\mathrm{ad}_h = \begin{pmatrix} 0 & \alpha(q) \\ -\alpha(q) & 0 \end{pmatrix},$$

where $h = iq$. Thus, we see a distinction of the action of the operator ad_h on a compact algebra from a similar action on a complex algebra.

In the complex case, this operator is reduced to a diagonal form in a basis composed of root vectors, whereas in the real compact case, this operator does not have real eigenvectors in the orthogonal complement of the Cartan subalgebra iT_0. It is reduced to a box-diagonal form, that is, written as a matrix with (2×2) boxes on the diagonal. Each of these boxes corresponds to one root and is written by means of the above-mentioned matrix.

In our model example, the compact algebra $G_c = su(n)$, where $G_c \subset G = sl(n, \mathbb{C})$, is decomposed into the direct sum of the following subspaces:

$$su(n) = W^+ \oplus iT_0,$$

where

$$iT_0 = \begin{pmatrix} i\varphi_1 & & 0 \\ & \ddots & \\ 0 & & i\varphi_n \end{pmatrix}, \sum \varphi_i = 0, \varphi_i$$

are real numbers, $W^+ = \mathrm{Re}\, W^+ \oplus \mathrm{Im}\, W^+$, where

$$\mathrm{Re}\, W^+ = \{E_\alpha + E_{-\alpha}\} = \left\{ \begin{pmatrix} 0 & & 1 \\ & \ddots & \\ -1 & & 0 \end{pmatrix} \right\},$$

$$\mathrm{Im}\, W^+ = \{i(E_\alpha - E_{-\alpha})\} = \left\{ \begin{pmatrix} 0 & & i \\ & \ddots & \\ i & & 0 \end{pmatrix} \right\}$$

Here, the two-dimensional invariant subspace spanned by the pair of vectors $E_\alpha + E_{-\alpha}, i(E_\alpha - E_{-\alpha})$ has the form

$$\begin{pmatrix} 0 & & a+ib \\ & \ddots & \\ -a+ib & & 0 \end{pmatrix},$$

where a, b are real.

It turns out that the *compact form is unique in G up to the automorphism of the algebra G*.

We have investigated the root decomposition of the algebra $\mathrm{sl}(n, \mathbb{C})$. Therefore, we can now easily write in an explicit form the canonical embedding of the compact form G_c into the algebra $\mathrm{sl}(n, \mathbb{C})$.

The root vectors E_α coincide with the elementary matrices $T_{pq}, p < q$, if $\alpha > 0$, and with the matrices $-T_{pq}, p > q$, if $\alpha < 0$. Hence,

$$E_\alpha + E_{-\alpha} = T_{pq} - T_{qp}; i(E_\alpha - E_{-\alpha}) = iT_{pq} + iT_{qp},$$

and finally, $iH'_\alpha = i(T_{pp} - T_{qq})$ if $\alpha(h) = a_p - a_q, p < q$ (Fig. 20). Thus, the subspace iT_0 coincides with the subspace of all diagonal purely imaginary zero-trace matrices. The subspace $\{E_\alpha + E_{-\alpha}\}$ coincides with the subspace of all real skew-symmetric matrices. The subspace $\{i(E_\alpha - E_{-\alpha})\}$ coincides with the subpsace of all symmetric purely imaginary matrices with zeros in the diagonal.

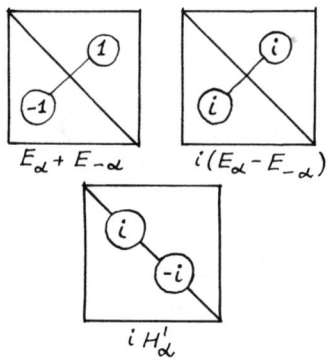

Figure 20

The subalgebra $G_c = W^+ \oplus iT_0$ coincides with the subspace of all skew-Hermitian zero-trace matrices, that is, with the Lie algebra of the group $\mathrm{SU}(n)$. Thus, we have, in fact, proved the following assertion.

LEMMA 1.4.4. *The standard embedding of the subalgebra* $\mathrm{su}(n)$ *into the Lie algebra* $\mathrm{sl}(n, \mathbb{C})$ *coincides with the canonical embedding of the compact form* G_c *into* G.

Along with the canonical compact form, *each semisimple complex Lie algebra possesses a canonical noncompact form sometimes* called a *normal noncompact form*.

§4   *Information on Lie Groups and Lie Algebras*   53

Let us again consider the Weyl basis in the algebra G and construct the following involution $\tau : G \to G$, where $\tau E_\alpha = E_\alpha, \tau H'_\alpha = H'_\alpha$, that is, the mapping τ is identical on the real parts of the planes V^+ and V^-, as well as on the real part T_0 of the Cartan subalgebra T. But this involution is not at all identical on the entire algebra because $\tau(\lambda X) = \bar{\lambda}\tau X$. Therefore, $\tau(iT_0) = -iT_0, \tau(iE_\alpha) = -iE_\alpha$.

The set of fixed points of the involution τ coincides with the plane $T_0 \oplus \operatorname{Re} V^+ \oplus \operatorname{Re} V^-$, that is, with the linear span (over **R**) of the vectors $E_\alpha, E_{-\alpha}, H'_\alpha$. Clearly, this is a subalgebra.

DEFINITION 1.4.7: The real subalgebra $\{E_\alpha, E_{-\alpha}, H'_\alpha\}$ is called a *normal noncompact form* of the Lie algebra G.

In our model example, this subalgebra coincides with the subalgebra of real zero-trace matrices, that is, with $\operatorname{sl}(n, \mathbf{R})$. The involution τ coincides here with the involution $\tau A = \bar{A}$, where the "bar" symbol implies complex conjugation.

The same as the compact form, the *noncompact normal form is uniquely defined up to the automorphism of an enveloping complex algebra*. For the interposition of the compact and noncompact forms in the algebra G, see Fig. 21.

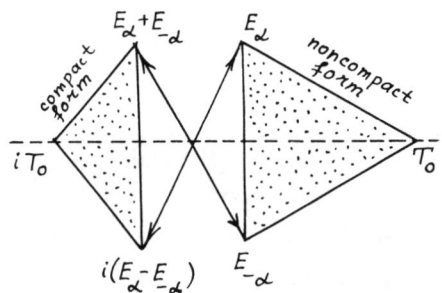

Figure 21

A more detailed scheme is given in Fig. 22. This decomposition is shown over the field **R**, and therefore real and imaginary planes are presented separately.

The compact real form G_c admits another invariant characteristic. This subalgebra is a *maximal compact subalgebra* in the complex Lie algebra G.

We will point out another compact subalgebra in the complex semisimple algebra G, which frequently appears in concrete problems of Hamiltonian geometry and mechanics. Let us examine the two involutions described above, namely σ, which determines the compact form, and τ, which determines the noncompact normal form. Let us consider a set G_n of points fixed under both involutions. Since $\sigma = 1$ on $W^+ \oplus iT_0$ and $\tau = 1$ on $T_0 \oplus \{E_\alpha\} \oplus \{E_{-\alpha}\}$, it follows that G_n is spanned by the vectors $E_\alpha + E_{-\alpha}$, because it is only these vectors that remain in their places under a simultaneous action of both involutions. Hence, G_n is a *compact subalgebra*. It coincides with the intersection in the algebra G of its two real forms: the *compact form* G_c and the *noncompact normal form*.

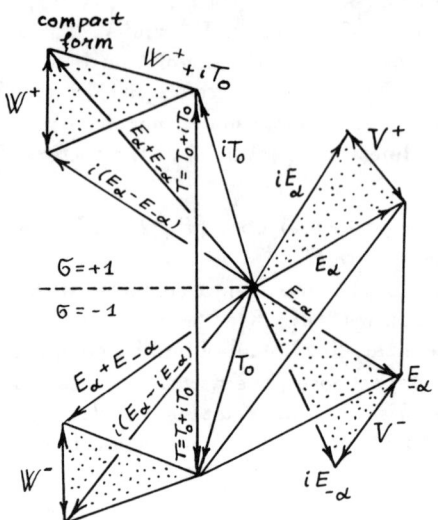

Figure 22

Note that the subalgebra G_n is not at all a real form of the algebra G, because its complexification does not coincide with G. The subalgebra G_n is sometimes called the *normal compact subalgebra* (but not to be confused with the form!).

In our model example of the algebra $\mathrm{sl}(n, \mathbb{C})$, we obtain:

$$G_n = G_c \cap \text{ (the normal noncompact form)} = \mathrm{su}(n) \cap \mathrm{sl}(n, \mathbb{R}) = \mathrm{so}(n),$$

because skew-Hermitian real matrices are skew-symmetric.

CHAPTER 2

THE THEORY OF SURGERY ON COMPLETELY INTEGRABLE HAMILTONIAN SYSTEMS OF DIFFERENTIAL EQUATIONS

§1 Classification of Constant-Energy Surfaces of Integrable Systems. Estimation of the Amount of Stable Periodic Solutions on a Constant-Energy Surface. Obstacles in the Way of Smooth Integrability of Hamiltonian Systems

1.1 *Formulation of the Results in Four Dimensions*

We shall present mainly the new results obtained by the author recently. We begin with remarks of qualitative character.

As will be shown in Ch. 5, systems of "general position" are, as a rule, nonintegrable. Since of particular importance for applications are cases where systems are integrable (for instance, in the sense of Liouville), it becomes clear how difficult it is to find such rare integrable systems in the vast ocean of systems, the "majority" of which are nonintegrable. Beginning with this chapter, we investigate the integrability phenomenon itself and seek efficient integrability conditions.

It is natural to start with the following question: *How is the fact of integrability of a Hamiltonian system associated with the topology of the phase (or configuration) manifold?*

Complete Liouville integrability of a Hamiltonian system is traditionally assumed to provide a more or less detailed qualitative description of the behaviour of integral trajectories of the system. In principle, this is certainly the case. But one often ignores the fact that for such a description one should effectively find the action-angle variables (with respect to which the trajectories of the system become rectilinear torus windings) in the neighbourhood of Liouville tori. Embeddings of Liouville tori into an enveloping phase manifold may be rather complicated (soon we will see examples). It is clear that the complicacy of torus embedding increases as the system integrals become more complicated. Since integrals are most frequently polynomials (with increasing degrees), a concrete analysis of tori and action-angle variables is often hampered by the necessity to solve nontrivial algebraic and analytic problems.

It is therefore relevant to ask the following questions. How are Liouville tori arranged in the phase space? How do they adjoin one another, fill open domains, how are they transformed in the neighbourhood of critical surfaces, etc? In other words, what is the way of constructing a qualitative theory of topological arrangements and interaction of Liouville tori (and thereby of the arrangement of integral trajectories of a system), for instance, on a constant-energy surface of a system?

In the present chapter, we answer some of these questions.

For the sake of simplicity, we begin with the four-dimensional case, although we have proved almost all the results formulated here in the multidimensional case as well. This will be discussed in §2.

In recent years many new results have been obtained concerning integration of Hamiltonian systems on symplectic manifolds. See, for instance, the survey in [143], [188]. Of particular interest in this connection is the discovery of stable closed integral trajectories of integrable systems. Such trajectories correspond to *periodic motions of the system*.

We prove that in some cases one may guarantee the existence of *at least two stable periodic solutions* on three-dimensional constant-energy surfaces (isoenergy surfaces) of an integrable system by proceeding merely from the data on one-dimensional homologies of these surfaces (Theorem 2.1.1) or from the data on the fundamental group. Such stable solutions may be effectively found.

This result follows from the general Theorem 2.1.3 *on complete topological classification and on canonical representation of constant-energy surfaces of integrable systems* (on four-dimensional manifolds of the three simplest types). In particular, this yields a simple classification of all nonsingular constant-energy surfaces for integrable systems.

We assume a Hamiltonian system to posses a second smooth integral which will be called a *"Bott"* integral. This is an integral whose critical points may be degenerate but are of necessity organized as nondegenerate smooth (critical) submanifolds.

With this purpose in mind, we have developed a new specific *"Morse-type theory" of integrable Hamiltonian systems* which is distinct from the ordinary Morse theory and from the Bott theory of functions with degenerate singularities [172].

In particular, we also develop some ideas suggested by Anosov in [171], Novikov in [95], [97], Smale in [127], and Kozlov in [61].

Next we ask another question. *Do nonsingular constant-energy surfaces of integrable systems posses specific properties which distinguish them among all smooth three-dimensional manifolds?* It is natural to expect that in Bott's (Morse's) as well as analytic and algebraic cases not all manifolds may play such a role. In other words, it is obvious that *not all manifolds are realized in the form of a constant-energy surface of an integrable system*.

In Corollary 2.1.4 we prove this hypothesis for Bott's case. From this we obtain, in particular, some *new topological obstacles for integrability of Hamiltonian systems* in the class of smooth Bott functions (Corollary 2.1.2). Examining geodesic flows on a two-dimensional sphere (and proving nonintegrability of flows of general position in the above-mentioned class of integrals), we follow the paper by Anosov [171] and also by Klingenberg and Takens [170].

As mentioned above, many ways of proving analytic nonintegrability of systems of general position are now available (see, for instance, the papers [160], [161], [164]–[167], and [169]). Our results contribute to these studies, for they make it possible to efficiently answer the question of the existence or nonexistence (in many cases) of smooth Bott integrals. This class of integrals is distinct from the class of analytic integrals.

We should emphasize that, in our opinion, special attention should be given to *integrability on a separate fixed constant-energy surface*. The reason is that in mechanics and physics one often has to deal with a Hamiltonian system integrable only on *one constant-energy surface* and nonintegrable on the others. Analytic systems are usually integrable either simultaneously on all Hamiltonian level surfaces or on none of them. For this reason, we think it to be instructive to consider smooth systems and smooth integrals which admit a simultaneous presence of both "integrable" and "nonintegrable" constant-energy surfaces.

In this connection, the following general problem arises. Let a Hamiltonian system with a Hamiltonian H be given. One should establish whether among the constant-energy surfaces of this system there exists at least one on which the system is integrable.

It is natural to hypothesize that many Hamiltonian systems (nonintegrable globally) do have such a surface. Clearly, a smooth (or analytic) integral f, which integrates a system $v = \text{sgrad}\, H$ only on one Hamiltonian level surface, satisfies an equation weaker than the ordinary equation $\{H, f\} = 0$. Namely, this integral surface itself, whereas outside this surface, it does not yet have to commute with H. In the simplest form, this condition can be written as follows: $\{H, f\} = \lambda(H)$, where the function $\lambda(H)$ is such that $\lambda(0) = 0$ and $\lambda'(0) = 0$. We assume that the isolated Hamiltonian level surface of interest is given by the equation $H = 0$. Thus, the general equation $\{H, f\} = \lambda(H)$ deserves most thorough examination. On a Hamiltonian level surface, consider two vector fields $\text{sgrad}\, H$ and $\text{sgrad}\, f$. For these vector fields to commute on a level surface, it suffices that the gradient of the function $\{H, f\}$ vanish on the surface $H = 0$. Indeed, we know that $[\text{sgrad}\, H, \text{sgrad}\, f] = \text{sgrad}\{H, f\}$. Thus, the function $\lambda(H)$ must be quadratic in H in the neighbourhood of the value $H = 0$. In particular, it would be possible first to investigate the general properties of the equation $\{H, f\} = \varepsilon H^2$, where ε is a nonzero constant.

The problem listed above are particular cases of the general noncommutative integration problem which is, in fact, traced back to Cartan and has been analyzed from different points of view by Fomenko and Trofimov in the survey [188], and Marsden and Weinstein in [78].

From the Liouville theorem it follows that all nonsingular two-dimensional compact level surfaces of a second integral f (on a constant-energy manifold), that is, common level surfaces of both integrals H and f are unions of tori. It turns out that the *structure of singular constant-energy surfaces can also be completely described*. They prove to be homeomorphic to two-dimensional complexes obtained by a *special (easily described) gluing of two-dimensional tori*. This description, which we have obtained, provides in particular, a *complete description of the behaviour of integral trajectories of a system* on singular integral levels.

In §2 of the present chapter, we construct a new multidimensional topological theory of integrable systems. As one of its applications, we suggest a *complete classification of surgery on Liouville tori of general position in the neighbourhood of bifurcation diagrams of momentum mapping of integrable Hamiltonian systems.* It turns out that the canonical surgery on tori can be explicitly and efficiently described. This permits the description of events happening to Liouville tori and to integral trajectories of a system on critical energy levels (that is, at the moment when a Liouville torus "runs into" the critical value of the energy integral).

Let M^4 be a four-dimensional smooth symplectic manifold on which a Hamiltonian system $v = \text{sgrad } H$ is given, with H being a smooth Hamiltonian. *Equilibrium positions* x_0 of the system v are *critical points of the function* H. Since H is an integral of the system v, it follows that the field v may be restricted to an *invariant three-dimensional constant-energy surface* Q, that is, $Q = \{x \in M : H(x) = \text{const}\}$. Being a symplectic manifold, M^4 is orientable, and therefore the manifold Q is also orientable.

Consider *noncritical level surfaces* Q, that is, such surfaces on which grad $H \neq 0$. For a complete Liouville-integrability of the system v on the manifold M, it suffices to find *one more* additional (second) integral f which is independent of the integral H (almost everywhere) and is in involution with it. Let such an integral exist. We restrict it to the surface Q and obtain a smooth function. As has already been mentioned, we will consider integrability of the system only *on one separate constant-energy surface*.

DEFINITION 2.1.1: We will call a smooth integral f a *Bott integral* on a level surface Q if the critical points of the function f and Q form nondegenerate critical smooth submanifolds.

The general properties of such functions were studied in the well-known papers by Bott (see, for example, [172]). Based on this, it makes sense to call such functions *Bott functions*. We do not confuse them with the ordinary Morse functions, because we do not frequently employ the latter in the course of our analysis (such cases will be stipulated). Recall that the critical submanifold for a function f is called *nondegenerate* if the Hessian d^2f of the function f is nondegenerate on normal planes to the submanifold. In our case, nondegenerate critical submanifolds for the function f on Q may be *zero-dimensional, one-dimensional, and two-dimensional.* We shall be interested in the existence of the Bott integrals in a Hamiltonian system. The point is that the experience of investigating concrete systems (see, for instance, [177]–[179]) has shown that the overwhelming majority of the discovered integrals are Bott integrals in the sense just mentioned. The class of Bott integrals which we have introduced therefore seems natural.

DEFINITION 2.1.2: Let γ be a closed integral trajectory of a system v on a surface Q (that is, a periodic solution). We say that the trajectory γ is *stable* if its certain tubular neighbourhood is fully fibred into two-dimensional tori which are invariant with respect to the system v, that is, all integral trajectories close to γ "are located" on invariant two-dimensional tori whose common axis is the circle γ (Fig. 23).

Stability of a trajectory implies that a normal two-dimensional disk (of small radius) is fibred fully, without gaps, into *concentric* circles. Since we are primarily concerned with integrable systems, the above definition of stability coincides with the traditional notion of strong stability. The fact is that nonsingular level surfaces of a second integral of such a system are two-dimensional Liouville tori, and there-

§1   *Classification of Constant-Energy Surfaces*   59

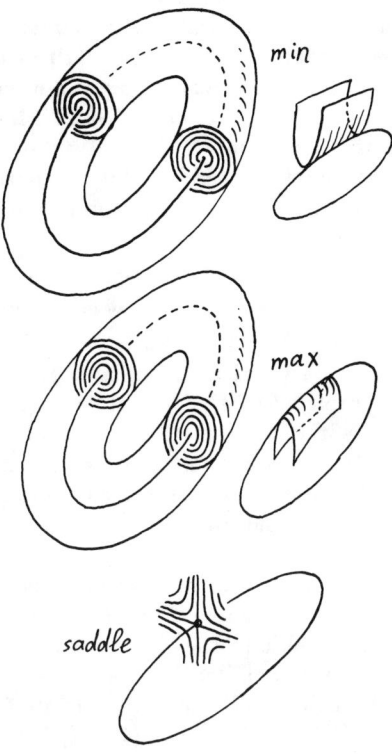

Figure 23

fore stable trajectories are those on which a second (additional) integral attains the local minimum or maximum.

A system may be integrable but have not a single closed stable trajectory (although it may have many closed trajectories). A simple example is a geodesic flow of a two-dimensional flat torus, that is, a torus T^2 with a metric $g_{ij} = \delta_{ij}$. It is easy to see that this geodesic flow has an additional linear integral *but that all closed trajectories of the system are unstable*(!).

As proved below, there exists a simple qualitative connection among the following three objects:

a) an additional Bott integral f on a constant-energy surface Q (that is, the system is completely integrable).

b) stable periodic solutions of the system on this surface,

c) a group of one-dimensional integer-valued homologies $H_1(Q, \mathbb{Z})$ (or a fundamental group $\pi_1(Q)$) of this surface.

So, we are considering one constant-energy surface Q, on which a second additional integral f is given. If it is defined not only on the surface $Q = \{H = 0\}$

but also in a certain neighbourhood of this surface, then f satisfies the equation $\{H, f\} = \lambda(H)$ (weaker than the ordinary involutivity condition $\{H, f\} = 0$).

Consider the critical nondegenerate submanifolds T of the integral f on Q. Each of them possesses a separatrix diagram $P(T)$ (see, for instance, [172],[143]). The reader will recall that a *separatrix diagram* is a union of integral trajectories of the field grad f incoming in T and outcoming from T. In view of this, we shall speak of an *incoming separatrix diagram* $P_-(T)$ and an *outcoming separatrix diagram* $P_+(T)$. In the neighbourhood of a submanifold T, both in- and out-diagrams are two-dimensional smooth manifolds with a common boundary T. They may be either *orientable* or *nonorientable*.

DEFINITION 2.1.3: We say that a Bott integral f on a surface Q is *orientable* if all of its critical submanifolds are *orientable*. If at least one of its critical submanifolds is nonorientable, we say that the integral f is *nonorientable*.

An isoenergy surface on which there exists a Bott integral will be sometimes referred to as an "integrable surface", for short.

It turns out that, without a substantial loss of generality, it suffices to investigate *only orientable integrals* f on Q. The reason is that one is considering the isoenergy surfaces Q up to a two-sheeted covering (over them one may always assume the integral f to be orientable).

Claim 2.1.1. Let Q^3 be a nonsingular compact constant-energy surface in M^4 and f a Bott nonorientable integral on Q. Then all its nonorientable critical submanifolds are homeomorphic to the Klein bottle, with the integral f attaining either its minimum or maximum (local) on these submanifolds. Let $U(Q)$ be a sufficiently small tubular neighbourhood of the surface Q in M. Then there exists a two-sheeted covering $\pi : (\tilde{U}(\tilde{Q}), \tilde{H}, \tilde{f}) \to (U(Q), H, f)$ (with a fibre $\mathbb{Z}_2$), where $\tilde{U}(\tilde{Q})$ is a symplectic manifold with a Hamiltonian system $\tilde{v} = \text{sgrad } \tilde{H}$ (the Hamiltonian $\tilde{H}$ is of the form $\tilde{H} = \pi^*(H)$) which is integrable on $\tilde{Q} = \pi^{-1}(Q)$ by means of a Bott orientable (!) integral $\tilde{f} = \pi^*(f)$. In this case, all the critical Klein bottles in Q "unfold" into critical tori T^2 in $\tilde{Q}$ (maxima or minima of the integral $\tilde{f}$). The manifold $\tilde{U}(\tilde{Q})$ is a tubular neighbourhood of $U(Q)$.

From this it follows that if f is a nonorientable integral on Q, then $\pi_1(Q) \neq 0$, and the group $\pi_1(Q)$ contains a subgroup of index two. If, for instance, Q is homeomorphic to the sphere S^3 (a particular case in mechanics), then any Bott integral f on S^3 is *always orientable*.

Hereafter $m = m(Q)$ denotes the number of stable periodic solutions of the system v on Q. Let, next, $r = r(Q)$ be the number of such critical submanifolds of the integral f on Q which are homeomorphic to the Klein bottle. If the integral is orientable, then $r = 0$.

THEOREM 2.1.1 (FOMENKO). *Let M^4 be a smooth symplectic four-dimensional manifold (compact or noncompact) and let $v = \text{sgrad } H$ be a Hamiltonian vector field on M^4, where H is a smooth Hamiltonian. Suppose the system is Liouville-integrable on a certain nonsingular compact three-dimensional level surface Q of the Hamiltonian H, with an additional smooth integral f commuting with H on Q being a Bott integral on Q. Then the number $m = m(Q)$ of stable periodic*

solutions of the system v on the surface Q is estimated from below through the topological invariants of the surface Q as follows:
1) In case the integral f is orientable on Q, we have:
   a) $m \geqslant 2$ if the homology group $H_1(Q, \mathbb{Z})$ is finite;
   b) $m \geqslant 2$ if the fundamental group $\pi_1(Q) = \mathbb{Z}$.
2) In case the integral f is nonorientable on Q, we have:
   a) $m + r \geqslant 2$ if the homology group $H_1(Q, \mathbb{Z})$ is finite;
   b) $m \geqslant 2$ if $H_1(Q, \mathbb{Z}) = 0$ (the group $\pi_1(Q)$ may be infinite);
   c) $m \geqslant 1$ if $H_1(Q, \mathbb{Z})$ is a finite cyclic group;
   d) $m \geqslant 1$ if $\pi_1(Q) = \mathbb{Z}$ or if $\pi_1(Q)$ is a finite group;
   e) $m \geqslant 2$ if $H_1(Q, \mathbb{Z})$ is a finite cyclic group and if the surface Q does not belong to a small series of manifolds $Q_0 = (S^1 \times D^2) + sA^3 + rK^3$ which are easily described in an explicit form (see the description below).

In cases 1 and 2, the integral f reaches a local minimum or maximum on each of these stable periodic solutions of the system v (or on Klein bottles). If the homology group $H_1(Q, \mathbb{Z})$ is infinite, that is, rank $H_1 \geqslant 1$, then the system v may have no stable periodic solution on the surface Q.

The criterion obtained is rather efficient because it is usually not difficult to check whether an integral is a Bott one and to calculate the rank of a one-dimensional homology group. In many mechanical integrable systems, constant-energy surfaces are often diffeomorphic either to a sphere S^3 or to a projective space $\mathbb{R}P^3$ or $S^1 \times S^2$. For instance, for the equations of motion of a heavy rigid body at high velocities, one may assume after some factorization (see Ch. 1), that the surfaces Q are homeomorphic to $\mathbb{R}P^3$. Besides, if a Hamiltonian H has an isolated minimum or maximum (an isolated equilibrium position of the system) on M^4, then all sufficiently close level surfaces $Q = \{H = \text{const}\}$ are three-dimensional spheres S^3. We denote by $L_{p,q}$ the so-called lens spaces (factors of the sphere S^3 under the action of the cyclic group). The cases of interest for Hamiltonian mechanics are singled out as a special assertion.

PROPOSITION 2.1.1. *Let a Hamiltonian system $v = \text{sgrad}\, H$ be integrable by means of a Bott integral f on some separate constant-energy surface Q homeomorphic to one of the following three-dimensional manifolds: $S^3, \mathbb{R}P^3, S^1 \times S^2, L_{p,q}$.*
1) *If the integral f is orientable, we always have $m \geqslant 2$, that is, the system v necessarily has at least two stable periodic solutions on each of these surfaces.*
2) *If the integral f is nonorientable, we have $m \geqslant 2$ for S^3, and for the manifolds $\mathbb{R}P^3, S^1 \times S^2, L_{p,q}$ we have $m \geqslant 1$. In particular, on the sphere S^3 an integrable system always has at least two stable periodic solutions for any Bott integral.*

Thus, an integrable system has two stable periodic solutions not only on three-dimensional small spheres close to an isolated equilibrium position (minimum or maximum of the energy H), but also on all extending level surfaces of the function H as long as they are homeomorphic to the sphere S^3. To the best of our knowledge, this qualitative result is also new.

The criterion of Theorem 2.1.1 is exact in the following sense. Some cases are known where an integrable system v on a surface $Q \approx \mathbb{R}P^3$ or $Q \approx S^3$ has *exactly* two (and not more!) stable periodic solutions [61].

From the results of Anosov, Klingenberg, and Takens, it follows that in the set of all geodesic flows on smooth Riemannian manifolds there exists an open everywhere dense subset of flows without closed stable integral trajectories [170], 171]. This means that the *property of a geodesic flow to have no stable trajectories is the property of general position*. Recall once again that we mean "strong stability" (see Definition 2.1.2).

COROLLARY 2.1.1. *Let a two-dimensional smooth surface be homeomorphic to a sphere and have a Riemannian metric of general position which means that on the surface there is not a single stable closed geodesic line. Then the geodesic flow corresponding to the metric is nonintegrable (on each nonsingular constant-energy surface) in the class of smooth Bott integrals.*

By the *rank of a fundamental group* $\pi_1(Q)$ is meant the least possible number of generators (in the corepresentation of this group).

So, if rank $\pi_1(Q) = 1$, then a system v integrated on Q by means of a Bott integral f will necessarily have *at least one stable periodic solution* on Q. For instance, in several integrable cases of inertial motion of a four-dimensional rigid body with a fixed point (see [143]), some three-dimensional nonsingular constant-energy surfaces are diffeomorphic to $S^1 \times S^2$ (the remarks of Brailov), that is, $\pi_1(S^1 \times S^2) = \mathbb{Z}$ and $R = 1$. Similarly, as is well known, in the integrable case of Kovalevskaya (for a three-dimensional heavy body), certain constant-energy surfaces are also homeomorphic to $S^1 \times S^2$ after an appropriate factorization.

PROPOSITION 2.1.2. *Let a Hamiltonian system v on a nonsingular compact three-dimensional constant-energy surface Q be integrated by means of a Bott integral f. Now, if the system v on Q has no stable periodic solutions then:*
1) *the group $H_1(Q, \mathbb{Z})$ is not a finite cyclic group,*
2) *rank $\pi_1(Q) \geq 2$, with at least one of the generators in the group $\pi_1(Q)$ having an infinite order.*

Consider a geodesic flow of a flat two-dimensional torus, that is, a torus with a locally Euclidean metric. This flow is integrable in the class of Bott integrals and obviously has no closed stable trajectories. By virtue of Proposition 2.1.2, we must have: rank $\pi_1(Q) \geq 2$. Indeed, the nonsingular surfaces Q are diffeomorphic here to a three-dimensional torus T^3, for which $H_1(T^3, \mathbb{Z}) = \mathbb{Z} \oplus \mathbb{Z} \oplus \mathbb{Z}$.

COROLLARY 2.1.2. *Let $v = \operatorname{sgrad} H$ be a smooth Hamiltonian system on M^4 and let Q be a certain nonsingular compact three-dimensional constant-energy surface. Let the following two conditions be fulfilled:*
1) *the system v on Q has no stable periodic solutions;*
2) *the group $H_1(Q, \mathbb{Z})$ is a finite cyclic group or rank $\pi_1(Q) \geq 1$. Then the system v is nonintegrable in the class of smooth Bott integrals on a given surface Q.*

An application of this assertion is seen, for instance, from Corollary 2.1.1 on nonintegrability of geodesic flows of general position on a two-dimensional sphere. In the case of a geodesic flow of a flat torus T^2, we have $Q = T^3, H_1(T^3, \mathbb{Z}) = \mathbb{Z}^3, R = 3$ (that is, the conditions of Corollary 2.1.2 are not fulfilled), and although the flow has no closed stable trajectories on Q, it is nonetheless integrable in the class of Bott integrals.

In [90] were proved the *theorems on noncommutative integrability of Hamiltonian systems with symmetries*. The theorem extends the classical Liouville theorem to the case where the Hamiltonian H of a system is included in a certain finite-dimensional Lie algebra of functions, G which satisfies the condition $\dim G + \operatorname{rank} G = \dim M$. Although in this case the functions from the Lie algebra G need not yet be integrals of the field sgrad H, such systems nevertheless turn out also to possess invariant tori (of the dimension equal to the rank of G) on which the trajectories of the system set a conditionally periodic motion [90], [143].

COROLLARY 2.1.3. *Let $v = \operatorname{sgrad} H$ be a Hamiltonian system integrable on M^{2n} in the noncommutative sense, and let G be a Lie algebra of functions on M^{2n} (with respect to the Poisson bracket) with functionally independent (generally, noncommuting) generators $f_1, \ldots, f_k$, where $f_1 = H, k = \dim G$, and $\dim G + \operatorname{rank} G = \dim M$. Let $\operatorname{rank} G = 2$. Suppose that among the functions f_i there exists at least one function f_α such that its restriction to a certain common three-dimensional compact level surface Q of the remaining functions f_j, where $j \neq \alpha$, is a Bott function. Then for the system v on Q, we have all the assertions of Theorem 2.1.1.*

Recall that the rank of a homology group (as distinct from the rank of a fundamental group) is the number of independent infinite-order generators. Above, we have formulated the question: Can any three-dimensional closed manifold be a Hamiltonian level surface of an integrable system? The answer is given by the following theorem.

COROLLARY 2.1.4. *Not nearly each three-dimensional smooth compact closed orientable manifold may play the role of a constant-energy surface of a Hamiltonian system integrated by means of a smooth Bott integral.*

Thus, the *obstacle for integrability of a Hamiltonian system may be the topology of the constant-energy surface.* The words "not nearly each" are explained below. The topological obstacle preventing the "overwhelming majority" of three-dimensional manifolds from being realized in the form of constant-energy surfaces of integrable systems will be presented in an explicit form. One may construct concrete examples of sufficiently simple three-dimensional manifolds which cannot be such surfaces.

The results listed above are actually consequences of the general theorem of the author on the topological classification of constant-energy surfaces of integrable systems. Before formulating the theorem, we shall describe five types of simplest three-dimensional manifolds that appear as those "elementary bricks," of which an arbitrary constant-energy surface of an integrable system is glued together.

Type 1. The direct product $S^1 \times D^2$ will be called a *full torus*. Its boundary is one torus T^2.

Type 2. The direct product $T^2 \times D^1$ will be called a *cylinder*. Its boundary consists of two tori T^2.

Type 3. The direct product $N^2 \times S^1$ will be called an *oriented saddle* or, more descriptively, "trousers," where N^2 is a two-dimensional sphere with three removed disks. This manifold is homotopy equivalent to a figure eight, that is, to a bouquet (union) of two circles. Its boundary consists of three tori T^2.

Type 4. We shall realize the manifold N^2 in the form of a disk D^2 with two holes which we fix for further purposes and number by figures 1 and 2. Consider a nontrivial fibre bundle $A^3 \xrightarrow{N^2} S^1$ with a circle as the base S^1 and with a fibre N^2 (a disk with two holes). It is clear that over the circle there exist only two nonequivalent fibre bundles with a fibre N^2. These are the direct product $N^2 \times S^1$ (see Type 3 above) and the fibre bundle A^3. The fibre bundle is characterized by the fact that have been transferred along the base S^1, the fibre N^2 returns to the initial place only with exchanged places of holes 1 and 2. Because N^2 is homotopy equivalent to the figure eight composed of the two circles 1 and 2, types 3 and 4 may also be represented as follows. In type 3, we have a direct product of the figure eight by a circle, while in type 4, the figure eight moves along a circle in such a way that after a revolution the two circles 1 and 2 exchange places (the figure eight reverses). A small neighbourhood of the circle (base) S^1 is homeomorphic in this case to two Möbius strips which intersect transversally along their common axis. The boundary of the manifold A^3 consists of two tori T^2. It is clear that A^3 may be realized in $\mathbb{R}^3$ and then has the following form. Consider a full torus restricted to a standardly embedded torus, inside which we drill a thin full torus, which winds twice around the generator of the large full torus (Fig. 24). The manifold A^3 will be called a nonorientable saddle. It is clear that A^3 is the space of the oriented skew product $N^2 \tilde{\times} S^1$. From the topological point of view the manifold A^3 in not new, however. It is obtained by gluing a full torus and trousers through a torus diffeomorphism. This may be conditionally written as follows:

$$A^3 = \text{I} + \text{III} = (S^1 \times D^2) + (N^2 \times S^1).$$

The proof will be given later on.

Type 5. We denote by K^2 a Klein bottle and by K^3 the space of an oriented skew product of K^2 by a segment, that is, $K^3 = K^2 \tilde{\times} D^1$. The boundary of K^3 is a torus T^2. From the topological point of view, the manifold K^3 is not essentially new either, because (see the proof below) it is represented as the following gluing:

$$K^3 = \text{I} + \text{IV} = (S^1 \times D^2) + A^3 = 2\,\text{I} + \text{III} = 2(S^1 \times D^2) + (N^2 \times S^1).$$

Thus, out of the five types of manifolds listed above, only the first three are topologically independent. The last two are decomposed into combinations of manifolds of types 1, 2, 3. But the manifolds A^2 and K^3 are of great interest for the analysis of trajectories of the system v, because they correspond to peculiar and interesting motions of a mechanical system.

THEOREM 2.1.2. (FOMENKO. THE TOPOLOGICAL CLASSIFICATION THEOREM FOR THREE-DIMENSIONAL CONSTANT-ENERGY SURFACES OF INTEGRABLE SYSTEMS).

Let M^4 be a smooth symplectic manifold (compact or noncompact) and let $v = \operatorname{sgrad} H$ be a Hamiltonian system that is Liouville-integrable on a certain nonsingular compact three-dimensional constant-energy surface Q by means of a Bott integral f. Let m by the number of such periodic solutions of the system v on the surface Q on which the integral f attains a strictly local minimum or maximum (then the solutions are stable). Next, let p be the number of two-dimensional

§1 *Classification of Constant-Energy Surfaces* 65

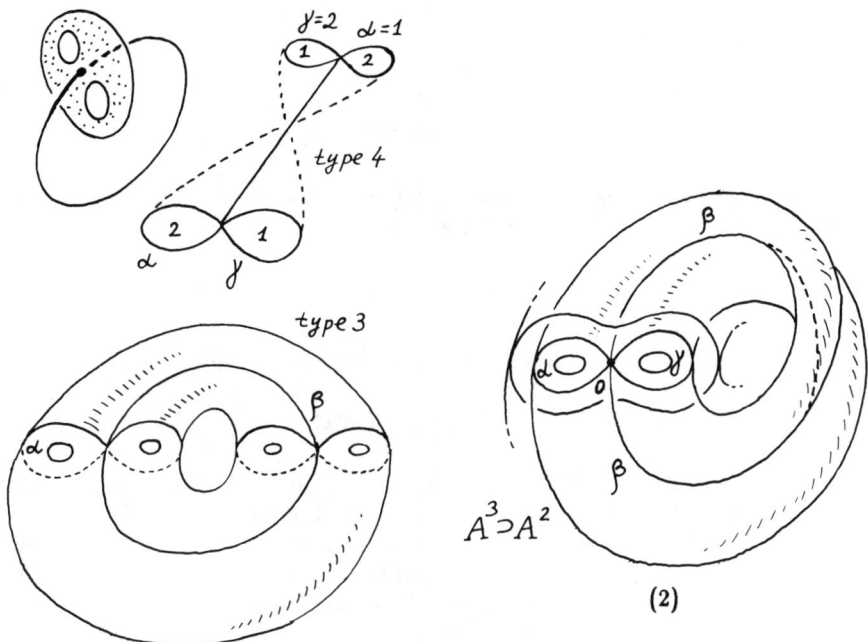

Figure 24

critical tori of the integral f (the minima or maxima of the integral); q the number of critical circles of the integral f (unstable trajectories of the system) with an orientable separatrix diagram; s the number of critical circles of the integral f (unstable trajectories of the system) with a nonorientable separatrix diagram; r the number of critical Klein bottles. This is the exhaustive list of all possible critical submanifolds of the integral f on Q. Then the manifold Q is represented in the form of gluing (by some diffeomorphisms of boundary tori) of the following "elementary bricks":

$$Q = m\ \mathrm{I} + p\ \mathrm{II} + q = \mathrm{III} + s\ \mathrm{IV} + r\ \mathrm{V} = m(S^1 \times D^2) \\ + p(T^2 \times D^1) + q(N^2 \times S^1) + sA^3 + rK^3.$$

If the integral f is orientable, then the last summand is missing, that is, $r = 0$.

Thus, in the above canonical representation of the manifold Q, all nonnegative integers m, p, q, s, r admit a clear interpretation, namely, they let us know how many critical submanifolds of each type a given integral f has on a given manifold Q. If we ignore this interpretation of the numbers m, p, q, s, r and require a simpler topological representation of the isoenergy surface Q, our requirement is met by the following theorem.

THEOREM 2.1.3. *Let Q be a compact nonsingular constant-energy surface of a Hamiltonian system $v = $ sgrad on Q integrated by a Bott integral f. Then Q admits the following representation:*

$$Q = m' \text{ I} + p' \text{ II} + q' \text{ III} = m'(S^1 \times D^2)$$
$$+ p'(T^2 \times D^1) + q'(N^2 \times S^1) = m'' \text{ I} + q'' \text{ III},$$

where m', p', q' are certain nonnegative integers. These numbers are related to the numbers m, p, q, s, r from Theorem 2.1.2 as follows:

$$m' = m + s + 2r, \quad p' = p, \quad q' = q + s + r, \quad m'' = m' + p', \quad q'' = q' + p'.$$

Thus, for an isoenergy surface Q^3, there occur two decompositions:

$$Q = m \text{ I} + p \text{ II} + q \text{ III} + s \text{ IV} + r \text{ V} \text{ and } Q = m' \text{ I} + p' \text{ II} + q' \text{ III}.$$

The former will be called a Hamiltonian decomposition of isoenergy surface, and the latter a topological decomposition. It is clear that the Hamiltonian decomposition is more "detailed." It "remembers" the structure of the critical submanifolds of the Bott integral. The topological decomposition is rougher. Its elementary blocks have already partially "forgotten" the original Hamiltonian picture. In the sequel, we will use one or the other decompositions subject to the problem of interest.

Now we may suggest a complete classification of all types of surgery on Liouville tori for varying values of the integral f. (When H and f exchange places, one can speak of bifurcation of Liouville tori as soon as they pass through the critical energy level when the second integral f is fixed).

We shall examine the following five types of surgery on a torus T^2 which correspond to the manifolds I, II, III, IV, V specified above. We realize the torus T^2 as one of the components of the boundary of a corresponding manifold. Then, carried away by the change of the integral f, the torus T^2 transforms into a union of Liouville tori which are the remaining components of the boundary. This surgery acquires the following form.

1) A torus T^2 is contracted to the axial circle of a full torus and then "vanishes" from the level surface of the integral f. Denote this surgery as $T^2 \to S^1 \to 0$.
2) Two tori T^2 move towards each other along a cylinder, flow together into one torus, and "vanish." The notation is: $2T^2 \to T^2 \to 0$.
3) A torus T^2 splits into two tori as it passes through the centre of the trousers (oriented saddle) when they "stay" on the level surface of the integral f. The notation is: $T^2 \to 2T^2$.
4) A torus T^2 spirals twice round a torus T^2 (following the topology of the nonoriented saddle A^3) and then stays on the level surface of the integral f. The notation is: $T^2 \xrightarrow{2} T^2$.
5) A torus T^2 transforms into a Klein bottle (covering it twice) and "vanishes" from the level surface of the integral f. The notation is: $T^2 \to K^2 \to 0$.

The five types of surgery obtained by reversing the arrows will not be treated as new.

THEOREM 2.1.4. (FOMENKO. CLASSIFICATION OF BIFURCATIONS OF TWO-DIMENSIONAL LIOUVILLE TORI).
Let f be a Bott integral on a nonsingular constant-energy surface Q^3. Then any surgery on general position of a Liouville torus, which occurs when the torus passes through the critical level surface of the integral f, is a composition of the elementary types 1-5 of the surgery (modifications) specified above. Furthermore, among these five types, only the first three are independent (from the topological point of view). The modifications 4 and 5 split into compositions of the modifications 1 and 3.

The general classification theorem for bifurcations of multidimensional Liouville tori is formulated below. For more detailed information, see Ch. 6.

Now let us consider four classes of three-dimensional compact orientable closed manifolds.

1) Class (H) consists of constant-energy (isoenergy) surfaces Q^3 of integrable Hamiltonian systems (integrated by a Bott integral f).
2) Class (Q) consists of all manifolds decomposed into the sum of "elementary bricks" of types 1, 2, 3, that is,

$$m'(S^1 \times D^2) + p'(T^2 \times D^1) + q'(N^2 \times S^1).$$

As has already been proved by the author (Theorem 2.1.2), class (H) is contained in class (Q).

3) Proceeding from the intrinsic problems of three-dimensional topology, Waldhausen [184] introduced the class (Γ) of three-dimensional manifolds W, which he called Graphenmannigfaltigkeiten. They are defined as follows. It is required that in W there exists a set of nonintersecting two-dimensional tori whose removal gives us a manifold each of whose connectedness components is fibred with a circle as typical fibre over the two-dimensional manifold (maybe with a boundary).
4) Developing the ideas of Fomenko described in the present chapter, Matveev and Burmistrova [310] have recently analyzed class (S) of three-dimensional manifolds, on which there exists a smooth function g with all critical points organized into nondegenerate circles, and on which all nonsingular level surfaces of the function g are union of two-dimensional tori. Brailov and Fomenko have proved that for classes (H) and (Q) we have an inverse inclusion: $(Q) \subset (H)$, that is, in the end, classes (H) and (Q) coincide.

CLAIM 2.1.2 (BRAILOV, FOMENKO [285]). . We have an equality: $(H) = (Q)$, that is, any manifold obtained by gluing full tori, cylinders, and trousers may be realized in the form of a constant-energy surface Q^3 of a certain Hamiltonian system integrated (by a Bott integral) on an appropriate symplectic manifold M^4.

Next, Fomenko and Zieschang [286], [289], [297], [298] have proved the following equality: $(\Gamma) = (Q)$. Finally, Matveev and Burmistrova have proved that $(S) = (Q)$. Gathering all these results, we formulate the theorem.

CLAIM 2.1.3. . All the four classes of three-dimensional manifolds described above coincide: $(H) = (Q) = (\Gamma) = (S)$.

Next, it turns out that the number of critical submanifolds of the integral f on Q can be estimated from below by a universal constant which depends only on the first homology group $H_1(Q, \mathbb{Z})$. Let $\beta = \operatorname{rank} H_1(Q, \mathbb{Z})$ (i.e., a one-dimensional Betti number), and let ε be the number of elementary multipliers in the finite part $\operatorname{Tor}(H_1)$ of the group $H_1(Q, \mathbb{Z})$. If $\operatorname{Tor}(H_1)$ is decomposed into an ordered sum of subgroups, where the order of any subgroup divides the order of the preceding one, then ε is the number of such summands.

CLAIM 2.1.4 (FOMENKO, ZIESCHANG [289], [298]). . Let $Q \in (H)$, that is, Q^3 is a constant-energy surface of an integrable system (by a Bott integral f). Let m be the number of stable periodic solutions of the system, s the number of unstable periodic solutions with nonorientable separatrix diagram, and r the number of critical Klein bottles. Then we always have the following inequalities:

$$m' = m + s + 2r \geqslant \varepsilon - 2\beta + 1, q' \geqslant m' - 2 \quad for \quad m + r + s + q > 0.$$

If $m = r = s = q = 0$, then $\varepsilon - 2\beta \leqslant 0$. Next, $q \geqslant m + r - 2$ (where q', q are defined above). If the integral f is orientable and so are all the separatrix diagrams of its critical submanifolds, then we have $s = r = 0$, that is, in this case we estimate from below the number m of stable periodic solutions of the system: $m \geqslant \varepsilon - 2\beta + 1$, and also $q \geqslant m - 2$.

COROLLARY 2.1.5. Let an integral f be orientable, and let all separatrix diagrams of its critical submanifolds on Q be also orientable. Then the manifold Q admits the following representation. Let m be the number of such stable periodic solutions of the system on Q on which f attains its local minimum or maximum. Consider a two-dimensional closed connected compact orientable manifold M_g^2 of genus g, where $g \geqslant 1$ (that is, a sphere with g handles), and a direct product $M_g^2 \times S^1$. Let us choose in M_g^2 an arbitrary finite set of nonintersecting and nonself-intersecting smooth circles α_i, among which there are exactly m contracted circles (the rest of the circles are nonhomotopic to zero in M_g^2). In the direct product $M_g^2 \times S^1$, the circles α_i define two-dimensional tori $T_i^2 = \alpha_i \times S^1$. We cut $M_g^2 \times S^1$ in all the tori T_i^2, after which we again identify the same tori by means of generally nonidentical diffeomorphisms. As a result we arrive at a new three-dimensional manifold. The connected constant-energy surface Q turns out to be a manifold of precisely this type.

Above, we have considered the case where the common level surface of the integrals H and f is compact. But it is not difficult to formulate and prove similar assertions for the noncompact case as well. We leave the details to the reader, since they can be easily worked out.

1.2 A Short List of the Basic Data From the Classical Morse Theory

It is well known that if a smooth function f with *nondegenerate critical points*, i.e., a *Morse function*, is given on a smooth manifold Q, then knowing these points and *their indices* allows us to say much about the topology of the manifold Q. It will be shown in the present chapter that an analogue of this theory exists also in the case where on a symplectic manifold a set of independent functions in involution is given, the number of which is equal to half the dimension of the manifold.

Recall that a point $x_0 \in Q$ is termed *critical* for a function f if $\operatorname{grad} f(x_0) = 0$. A critical point is termed *nondegenerate* if the matrix of the second differential

$d^2 f$, that is, the matrix $\left(\frac{\partial^2 f}{\partial x_i \partial x_j}\right)$, is nondegenerate at the point x_0. The maximal dimension of the linear subspace in a tangent plane $T_{x_0}Q$, on which the bilinear symmetric form $d^2 f$ is negative definite, is called the *index of the critical point* x_0. In other words, the form $d^2 f$ may always be reduced to the diagonal form at the critical point x_0, and then the index of the critical point is the number of negative squares in the diagonal notation of this form. The function f is called the *Morse function* if all of its critical points are nondegenerate. It is fairly well known (see, for instance, [28], [198]) that *on any smooth compact manifold there always exist Morse functions. These functions are dense everywhere in the space of all smooth functions on the manifold*. Each Morse function on a compact manifold Q has only a *finite* number of critical points, in particular, all these points are isolated. In the set of all Morse functions, there exists an everywhere dense subset consisting of functions, such that to each critical value of the function, there corresponds one and only one critical point on the manifold. To say it differently, *on each critical level of such a function there exists exactly one critical point*.

An important stage in the construction of the ordinary Morse theory is the well-known *Morse lemma*. It asserts that in some open neighbourhood of a nondegenerate critical point x_0 of a Morse function f, there always exist local regular coordinates $y_1, \ldots, y_n$, such that the function f is written in the form

$$f(y) = -y_1^2 - \cdots - y_\lambda^2 + y_{\lambda+1}^2 + \cdots + y_n^2,$$

where λ is the index of the critical point.

Let f be a Morse function on Q. Introduce the notation: $f_a = f^{-1}(a)$ is the level surface of the function f corresponding to the value a. Let $Q_a = \{x \in Q : f(x) \leqslant a\}$, that is, Q_a consists of all points x at which the values of f do not exceed a. Clearly, $\partial Q_a = f_a$.

Let a segment $[a, b]$ (where $a < b$) contain no critical values of the function f, that is, the set $f^{-1}[a, b]$ lying in the manifold Q contains no critical points of the function f. Then the manifolds f_a and f_b are diffeomorphic and, besides, the manifolds Q_a and Q_b are diffeomorphic, too. Here Q_a is a deformation retract of Q_b.

Let there exist exactly one critical point of index λ in a fibre $f^{-1}[a, b] = Q_b \setminus Q_a$. Then the manifold Q_b is homotopy equivalent to the finite cell complex obtained from the manifold Q_a by gluing one cell σ^λ of dimension to the boundary $f_a = \partial Q_a$. The direct product of two indices $D^\lambda \times D^{n-\lambda}$ is called a handle H_λ^n of dimension n and index λ. The boundary of the handle has the form:

$$\partial H_\lambda^n = (S^{\lambda-1} \times D^{n-\lambda}) \cup (D^\lambda \times S^{n-\lambda-1}).$$

Let $S^{\lambda-1} \subset f_a$ be a smoothly embedded sphere, such that its sufficiently small tubular neighbourhood $N_\varepsilon S^{\lambda-1}$ of radius $\varepsilon > 0$ be represented as the direct product $S^{\lambda-1} \times D^{n-\lambda}$, where $D^{n-\lambda}$ is a normal disk of dimension $n - \lambda$ and radius ε. Then one may construct a new smooth manifold Q_a with a boundary by gluing the manifold Q_a with the handle H_λ^n via the diffeomorphism

$$g : S^{\lambda-1} \times D^{n-\lambda} \to N_\varepsilon S^{\lambda-1} \approx S^{\lambda-1} \times D^{n-\lambda}.$$

We have an important assertion. The manifold Q_b turns out to be obtained from the manifold Q_a using precisely this operation, that is, Q_b results from gluing the handle H_λ^n of index λ to the boundary of the manifold Q_a. In other words, $\tilde{Q}_a \approx Q_b$.

As shown below, we have an important analogue of these assertions if the function f is replaced by the symplectic manifold *momentum mapping induced by a complete set of commuting integrals*. In particular, one may examine one integral on a three-dimensional constant-energy surface of an integrable system.

Now we are in a position to prove the theorems of §1.1.

1.3 Topological Surgery on Liouville Tori of an Integrable Hamiltonian System upon Varying Values of a Second Integral

In what follows, we think of all the assumptions of Theorem 2.1.2 as fulfilled.

LEMMA 2.1.1. *A smooth Bott integral f cannot have isolated critical points on a nonsingular compact constant-energy surface Q.*

The proof follows from the fact that Q^3 contains no critical points of the function H. Therefore, from each critical point x_0 of the function f on Q, there goes a nondegenerate integral trajectory of the field v which consists entirely of the critical points of the function f.

LEMMA 2.1.2. *The critical points of a smooth Bott integral f on a compact nonsingular surface Q fill either isolated smooth critical circles or smooth two-dimensional tori, or Klein bottles.*

PROOF: If a nondegenerate integral trajectory of the field v, which goes from a critical point x_0 of the integral f, is closed, then this trajectory is a circle. If a trajectory is open, then its closure P is a two-dimensional connected subset consisting of critical points of the integral. Therefore, P lies on a certain two-dimensional critical common level surface L of the integrals H and f. Generally speaking, a *singular level surface of an integral is not necessarily a manifold at all*. Since f is a Bott integral, its critical points are organized into nondegenerate submanifolds. Hence, P lies in a two-dimensional critical submanifold P' for the function f. It is asserted that the intersection of L with a sufficiently small open neighbourhood of P' in Q coincides with P'. Since along the normal to P' the function f is nondegenerate, it either strictly increases or strictly decreases on both sides of the normal to P'. A close nonsingular level surface P is, therefore, a two-sheeted covering of P'. By virtue of the integrability of the system v, the nonsingular compact level surface $\tilde{P}$ (on Q) must be Liouville tori and, therefore, P' are homeomorphic to a torus or to a Klein bottle. The point is that on the manifold P', there is a nonzero vector field sgrad H. Such a field may be only on the torus T^2 and on the Klein bottle K^2. On the projective plane $\mathbb{R}P^2$, there is no such field. This proves the lemma.

In Lemma 2.1.2, we have dealt with a critical level surface L which is, generally, not homeomorphic to a union of nonsingular tori for it may have singularities (although far from P'). The nonsingular level surfaces of the integral f and Q are compact and by the Liouville theorem are unions of tori.

§1 Classification of Constant-Energy Surfaces

Let S^1 be a critical circle of the integral f on Q. We ascribe to it *index* $0, 1$, or 2 depending on the index of the function f restricted to a two-dimensional disk normal to the circle. It is clear that the index of S^1 does not depend on the choice of the point on S^1. The circles of index 0 are local minima of the integral f, those of index 2 are maxima, and of index 1 are saddle circles.

LEMMA 2.1.3. *Critical circles of the integral f on Q may have indices $0, 1, 2$, whereas critical tori and Klein bottles may have indices 0 and 1 only.*

The proof is immediate from Lemma 2.1.2.

Let us investigate the surgery on the level surfaces $B_a = \{x \in Q : f(x) = a\} = f^{-1}(a)$ of the function f as a increases. We put $C_a = \{x \in Q : f(x) \leqslant a\}$. It is clear that $B_a = \partial C_a$. If a is a noncritical value, then B_a is a union of Liouville tori.

DEFINITION 2.1.4: The direct product of a circle with a two-dimensional disk on whose boundary two connected nonintersecting arcs l_1 and l_2 are distinguished will be called a *round handle* (Fig. 25). A round handle (full torus) is a *thickened cylinder with feet (bases)* $l_1 \times S^1$ and $l_2 \times S^1$.

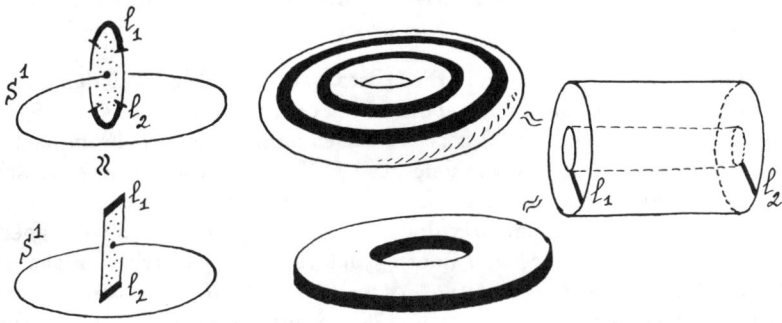

Figure 25

Define the *operation of gluing a round handle* to a three-dimensional manifold C_a with a boundary B_a. Let there lie two nonintersecting and non-self-intersecting circles γ_1 and γ_2 on B_a. Consider their small tubular neighbourhoods N_1 and N_2. By virtue of the orientability of B_a (union of tori), these neighbourhoods are homeomorphic to $S^1 \times D^1$, where D^1 is a segment. We will glue a round handle to B_a, identifying by means of homeomorphism the ring $l_1 \times S^1$ with a ring N_1, and the ring $l_2 \times S^1$ with N_2. Thus we are led to a new three-dimensional manifold. The circles γ_1 and γ_2 will be called the *axes of the feet of the round handle* (Fig. 26). The separatrix diagram of the critical circle S^1 will be denoted by $sd(S^1)$.

LEMMA 2.1.4. *Let a be a critical value of the integral f on Q. Suppose that on a critical level surface B_a there lies exactly one critical saddle circle S^1. Let $\varepsilon > 0$ be so small that on a segment $[a - \varepsilon, a + \varepsilon]$ there are no critical values of the function f other than a. 1) Let $sd(S^1)$ be orientable. Then $C_{a+\varepsilon}$ is obtained from*

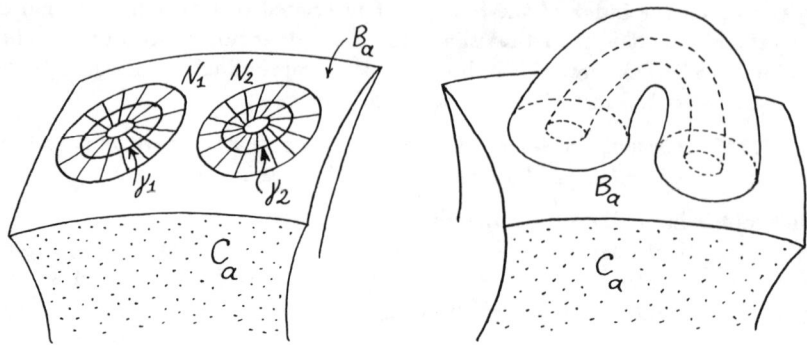

Figure 26

$C_{a-\varepsilon}$ by gluing a round handle to the boundary $B_{a-\varepsilon}$ of the manifold $C_{a+\varepsilon}$. The manifold $C_{a+\varepsilon}$ is homotopy equivalent to the manifold $C_{a-\varepsilon}$, to which the cylinder $S^1 \times D^1$ is glued with both boundaries. 2) Let $sd(S^1)$ be nonorientable. Then $C_{a+\varepsilon}$ is homotopy equivalent to $C_{a-\varepsilon}$ to which a Möbius strip is glued.

PROOF: Begin with case 1. Analyze the critical saddle circle S^1 of index 1. At each point x of this circle examine a normal disk $D^2(x)$ of small radius ε. Consider a field grad f on Q by introducing on Q a certain Riemannian metric. From each point $x \in S^1$, we send separatrices of the field grad f. The union of these separatrices is a *separatrix diagram of the saddle circle S^1*. By virtue of the nondegeneracy of the function f on $D^2(x)$, the separatrix diagram of each point $x \in S^1$ is hyperbolic. Varying the point x along S^1, we smoothly deform the separatrices in the normal disk $D^2(x)$. Consider the *"incoming" part $P_- = P_-^2$ of the separatrix diagram* contained in the fibre $a - \varepsilon \leqslant f \leqslant a$ (Fig 27). Since ε is small, it follows that P_- is a smooth two-dimensional submanifold containing S^1 and having the boundary homeomorphic either to a circle or to a disconnected union of two circles. One of them is the critical circle S^1. In the first case, the manifold P_- is homeomorphic to the *Möbius strip*, in the second case to the *cylinder $S^1 \times D^1$*.

Since, for the moment, the separatrix diagram of the critical circle S^1 is supposed to be orientable, the first case (the Möbius strip) is impossible here. The tubular neighbourhood of the surface P_- is therefore homeomorphic to a round handle. It is glued to $C_{a-\varepsilon}$ precisely in the fashion suggested by the definition of the round-handle -gluing operation (see above). The axes of both feet of the round handle are glued to two smooth circles γ_1 and γ_2 drawn on $B_{a-\varepsilon}$ by points A and B (Fig. 27) when the point x slides upon S^1.

If $sd(S^1)$ is nonorientable, then it is clear from the preceding consideration that instead of a thick cylinder (a full torus) a "thick Möbius strip" is glued. This completes the proof of the lemma.

Note that a round handle may be glued to $B_{a-\varepsilon}$ in only two ways: to one or two distinct tori. A "thick Möbius strip" may be glued only to one torus because

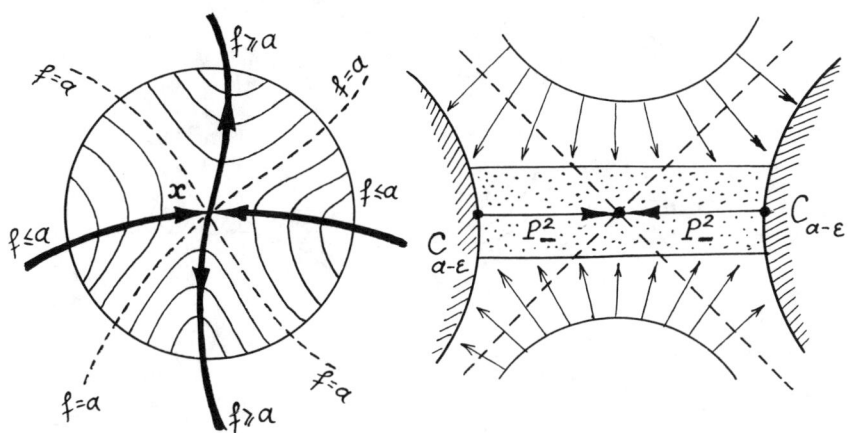

Figure 27

its boundary is connected (as distinguished from the boundary of a cylinder).

1.4 Separatrix Diagrams Cut Out Nontrivial Cycles on Nonsingular Liouville Tori

LEMMA 2.1.5. *Suppose on a critical level B_a there exists exactly one critical saddle circle S^1. 1) Let $sd(S^1)$ be orientable. Consider a round handle corresponding to the critical circle S^1 and glued with both feet to $B_{a-\varepsilon}$. Then each of the feet lies on a certain Liouville torus. It is asserted that the axis of each of the handle feet is a smooth non-self-intersecting circle realizing a nonzero element of the fundamental group of the torus. If both feet are glued to one and the same torus, then the axes γ_1 and γ_2 of both feet of the round handle do not intersect on the torus, they realize one and the same generator of the fundamental group of the torus, and are isotopic to each other on this torus. 2) Let now $sd(S^1)$ be nonorientable. Then the incoming separatrix diagram P_- homeomorphic to a Möbius strip is glued with its boundary circle to one Liouville torus on which this smooth non-self-intersecting circle realizes one of the generators of the fundamental group of the torus.*

The tubular neighbourhood of the separatrix diagram P_- in the nonorientable case will be called a *thickened (or thick) Möbius strip*.

LEMMA 2.1.6. *1) Let $sd(S^1)$ be orientable. If a round handle is glued to two distinct tori, then after having passed through the critical saddle circle S^1, which corresponds to the handle, these tori are deformed into one torus. If a handle is glued to one torus, then after having passed through the critical circle S^1, the torus splits into a union of two tori. 2) Let now $sd(S^1)$ be nonorientable. Then, after having passed through the critical saddle circle S^1, the Liouville torus to which a Möbius strip (in the form of the diagram P) is glued is again deformed into one Liouville torus.*

PROOFS OF LEMMAS 2.1.5 AND 2.1.6: Let $sd(S^1)$ be orientable. Consider the first case where a handle is glued to distinct tori T_1 and T_2. Let γ_1 and γ_2 be single

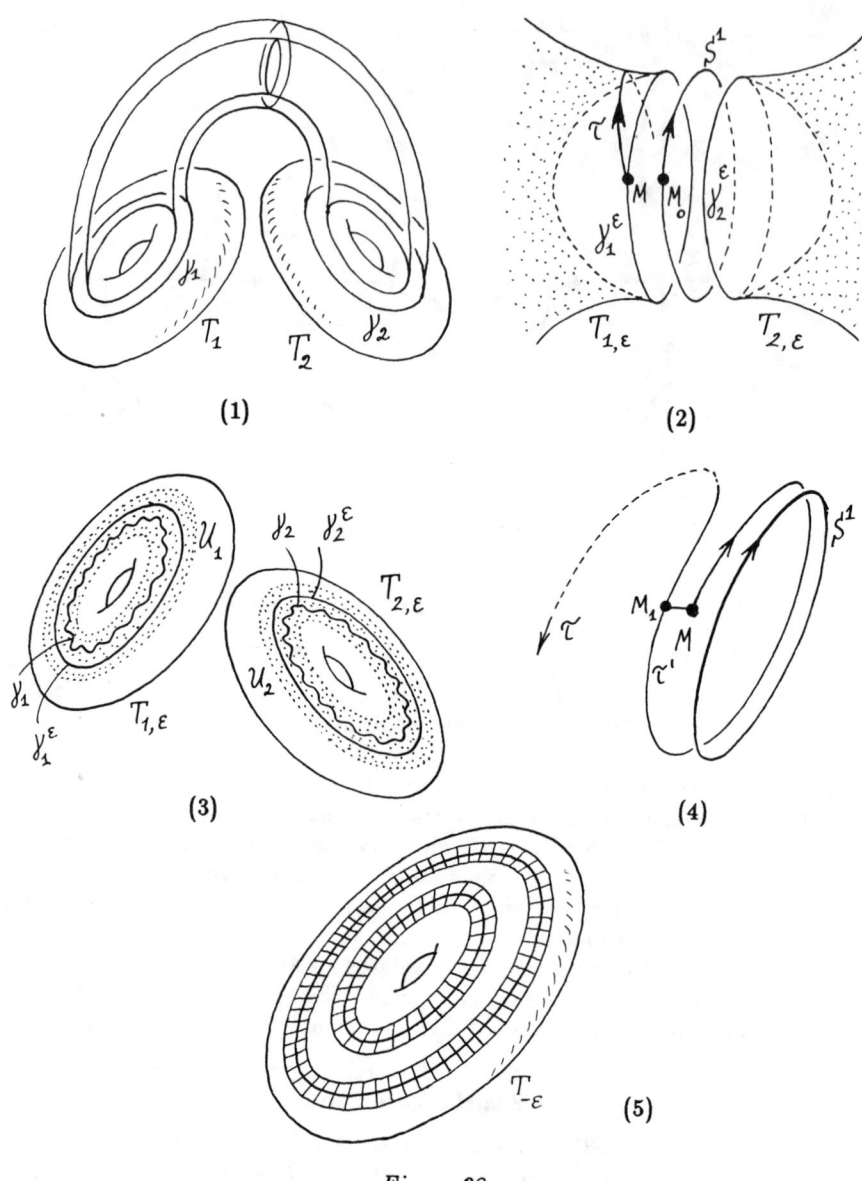

Figure 28

axes of the handle feet and let S^1 be a critical saddle circle on a level B_a. Consider two tori, $T_{1,\varepsilon}$ and $T_{2,\varepsilon}$, which lie on $B_{a-\varepsilon}$ and are two conectedness components of $B_{a-\varepsilon}$ (disregarding the other components of the level which are of no interest for the present). Let γ_1 and γ_2 be two circles cut out on $T_{1,\varepsilon}$ and $T_{2,\varepsilon}$ by the incoming separatrix diagram P_- (Figure 28). We may assume that $T_i = T_{i,\varepsilon_0}$ and $\gamma_i = \gamma_i^{\varepsilon_0}$, where $i = 1, 2$ and ε_0 is a sufficiently small fixed number. As ε decreases, the circles

γ_1 and γ_2 come closer together, of course, and for $\varepsilon = 0$, they flow into one circle S^1, which is a critical saddle circle. Since f is a Bott integral, the single circles γ_1^ε and γ_2^ε come close together in a hyperbolic way. Under the variation of ε, the tori $T_{1,\varepsilon}$ and $T_{2,\varepsilon}$ do not intersect and come closer together only in the neighbourhood of their circles γ_1^ε and γ_2^ε. One may assume, therefore, that on each of the fixed tori T_1 and T_2 two sufficiently small tubular neighbourhoods u_1 and u_2, respectively, are singled out of the circles γ_1 and γ_2, within which (neighbourhoods) there stir circles γ_1^ε and γ_2^ε that do not overstep the limits of U_1 and U_2, the circles γ_i^ε being isotopic to γ_i, $i = 1, 2$. It is clear that all tori $T_{i,\varepsilon}$ (with a variable ε) are canonically identified with the fixed torus $T_i = T_{i,\varepsilon_0}$, $i = 1, 2$, through a diffeomorphism along the integral trajectories of the field grad f. Examine a point M on the circle γ_1^ε, letting an integral trajectory τ of the field v go from this point. Two cases are possible: a) the trajectory is *closed*, b) the trajectory is *nonclosed*. In case a) the circle τ, the whole of which lies on the torus $T_{1,\varepsilon}$ is close to the closed trajectory S^1 if ε is sufficiently small. In this case, the circle τ is closed on a nonsingular torus $T_{1,\varepsilon}$. Since the system v is integrable, the torus $T_{1,\varepsilon}$ admits the assertions of the Liouville theorem. Therefore, on each of the tori $T_{1,\varepsilon}$ and $T_{2,\varepsilon}$, there exist regular curvilinear coordinates with respect to which the restriction of the field v on the torus determines a conditionally periodic motion. Since the field v on $T_{1,\varepsilon}$ appears to have a closed integral trajectory τ, all the other trajectories of this field $T_{1,\varepsilon}$ are closed as well. In this case, on making a revolution along the torus $T_{1,\varepsilon}$, the single trajectory τ realizes a nontrivial element of the fundamental group $\pi_1(T_{1,\varepsilon})$, since it is generated by the rectilinear winding of the torus. But the trajectory τ is arbitrarily close to γ_1^ε because both of them are close to the saddle circle S^1. Consequently, γ_1^ε realizes the nonzero element of the group $\gamma_1(T_{1,\varepsilon})$, and accordingly γ_1 realizes the nonzero element of the group $\pi_1(T_1)$. One can prove in a similar manner than in case a), the curve γ_2 realizes the nonzero element of the group $\pi_1(T_2)$. Consider case b), where the trajectory τ is open on $T_{1,\varepsilon}$. Since the decreasing ε the trajectory τ may be regarded as passing arbitrarily close to the integral trajectory S^1 (the saddle circle) on each a priori given (but fixed) time interval, it follows that if ε continues to decrease, then, within some time, the trajectory τ will return (for the first time!) to the small neighbourhood of the point M. Let $M_1 \in \tau$ be the return point located near the point M in Q. The trajectory τ has not left the neighbourhood U_1 of the circle γ_1 on the torus $T_{1,\varepsilon}$. Therefore, the trajectory τ made a revolution along the torus $T_{1,\varepsilon}$ and returned to the small neighbourhood of the initial point M on the torus. It is obvious that here we have used orientability of the separatrix diagram. Joining the points M and M_1 by a small geodesic line segment on the torus $T_{1,\varepsilon}$, we obtain from the trajectory τ a new closed trajectory τ' which wholly lies within the neighbourhood U_1 (Fig. 28).

The Liouville theorem implies that τ' realizes the nonzero element of the group $\pi_1(T_{1,\varepsilon})$, since it is obtained through the small closure of an almost-periodic trajectory which has a made a revolution along the torus and returned to the point close to the initial one (the first return).

Since the single trajectory τ' is arbitrarily close to the single trajectory γ_1^ε, it follows that γ_1^ε and, therefore, γ_1 realize nonzero elements in the groups $\pi_1(T_{1,\varepsilon})$ and $\pi_1(T_1)$, respectively. We have essentially used the fact that when the trajectory τ returns to the point M_1, it has a made a revolution along the torus $T_{1,\varepsilon}$ (Fig.

28).

Thus, if a handle is glued to two distinct tori, Lemma 2.1.5 follows for the orientable case. Let now a handle be glued to one torus (case 2). Now the tori $T_{1,\varepsilon}$ and $T_{2,\varepsilon}$ may be assumed (from the preceding reasoning) to coincide. Denote this torus by $T_{-\varepsilon}$. On the torus T_ε, we obtain two circles $\bar{\gamma}_1^\varepsilon$ and $\bar{\gamma}_2^\varepsilon$ which become slightly deformed within their primarily chosen (finite) neighbourhoods U_1 and U_2 and do not leave them as ε decreases. The neighbourhood U_1 and U_2 (rings) may apparently be chosen as nonintersecting on a torus T_ε, since as ε decreases, the circles $\gamma_1^{-\varepsilon}$ and $\bar{\gamma}_2^\varepsilon$ flow together in Morse's manner (Fig. 27). Let us consider cases a) and b) once again. In case a) (a closed integral trajectory τ), the proof is exactly the same. Case b) requires additional reasoning. Let the trajectory τ be non-closed. By decreasing ε, one can compel this trajectory to return (for the first time) within some time to a certain point M_1 lying in the small neighbourhood of the point M. One should prove the fact that the point M_1 is located near the point M from the viewpoint (in topology) of the torus $T_{-\varepsilon}$. Nearness of the points M and M_1 in the manifold Q does not yet guarantee that they are near in the topology of the torus $T_{-\varepsilon}$. From Lemma 2.1.4 it follows that the torus $T_{-\varepsilon}$ intersects the tubular neighbourhood of the circle S^1 in two neighbourhoods U_1 and U_2. One may assume (for a small ε) that $U_1 \cap U_2 \neq 0$ (because f is a Bott integral). But in this case, one revolution is not enough for the trajectory τ to leave U_1 and get into U_2, because during the first revolution the trajectory τ goes near the circle $\gamma_1^{-\varepsilon}$, remaining in its tubular neighbourhood U_1. In fact, we again rest upon orientability of the separatrix diagram, which has the consequence that after one revolution point B on γ_1 (Fig. 27) again takes its initial position but does not come to the point A. In other words, we have made use of the fact that the intersection of P_- and $B_{a-\varepsilon}$ consists exactly of two nonintersecting circles γ_1 and γ_2. Thus, in case 2, both circles $\gamma_1^{-\varepsilon}$ and $\gamma_2^{-\varepsilon}$ realize nontrivial cycles on one and the same torus $T_{-\varepsilon}$. These circles do not intersect, because the separatrices of the critical points do not intersect outside the critical points. The elementary properties of a two-dimensional torus imply that these circles are isotopic and realize one and the same cycle on the torus $T_{-\varepsilon}$.

Let now $sd(S^1)$ be nonorientable, that is, the separatrix diagram P_- is homeomorphic to a Möbius strip. After the first revolution along the circle S^1, the point A will come to the point B (Fig. 27), that is, the point M_1 will not be close to the point M in the topology of the torus $T_{-\varepsilon}$. In other words, the trajectory τ will make only half a revolution along the torus $\tau_{-\varepsilon}$. On making another revolution along S^1 and returning to a certain point M_2 close to the point M in the topology of the torus $T_{-\varepsilon}$, the trajectory τ concludes its motion on the torus $T_{-\varepsilon}$. Further consideration obviously repeats the argument of the orientable case. This completes the proof of Lemma 2.1.5.

Now we proceed to the proof of Lemma 2.1.6. We begin with the orientable case. Let a round handle corresponding to a saddle circle be glued to distinct tori T_1 and T_2 along rings whose axes are noncontractible (by virtue of Lemma 2.1.5) circles γ_1 and γ_2, respectively.

Cutting the tori along these circles, we obtain two rings out of each torus. Gluing the round handle and considering the boundary of the obtained three-dimensional manifold, we get one torus as the upper component of the boundary.

§1 Classification of Constant-Energy Surfaces

Reversing the consideration, we arrive at a modification of two Liouville tori into one. In the nonorientable case, both components P_+ and P_- of the separatrix diagram P of the critical circle S^1 are homeomorphic to a Möbius strip (separately). Their boundaries are therefore connected and homeomorphic to a circle. This implies that one torus $T_{-\epsilon}$ is modified to exactly one torus $T_{+\epsilon}$, and Lemma 2.1.6 follows.

In the proof of Lemmas 2.1.5 and 2.1.6, we have essentially used the fact that a saddle circle is an integral trajectory of the field v. If we restrict our examination only to a smooth Bott function f whose nonsingular level surfaces are two-dimensional tori (that is, if we disregard the condition that f is an integral of a Hamiltonian field), then the number of modifications of a set of Liouville tori into a set of tori immediately increases. We present a corresponding table, for example, for the orientable case. The following *four cases* of arrangement of two smooth non-self-intersecting and nonintersecting circles γ_1 and γ_2 on a torus (Fig 29) are possible.

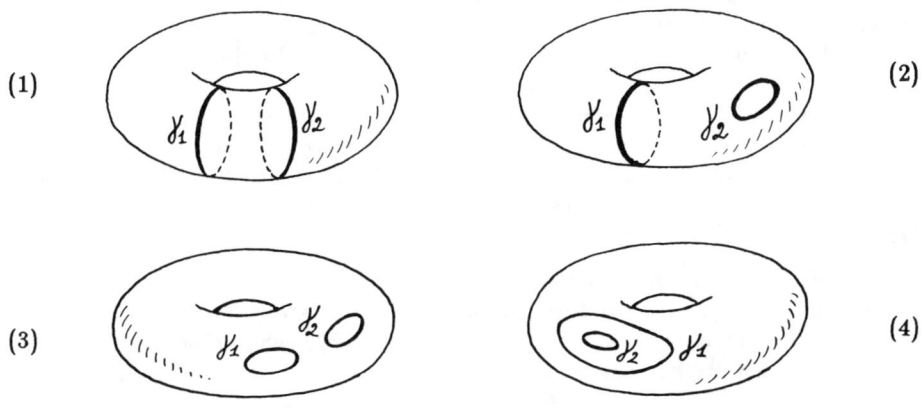

Figure 29

1) The circles γ_1 and γ_2 are *noncontractible*. Then they realize one and the same nontrivial cycle on the torus and are isotopic to each other.
2) The circle γ_1 is *noncontractible*, and the circle γ_2 is *contractible*.
3) The circles γ_1 and γ_2 are *contractible* and are located outside each other.
4) The circles γ_1 and γ_2 are *contractible*, and γ_2 is located within γ_1.

Gluing a round handle modifies one torus correspondingly into the following surfaces (in cases 1–4):

1) two tori, 2) one torus, 3) a cracknel (that is, a sphere with two handles) and a sphere, 4) two tori.

Thus, as compared with the assertion of Lemma 2.1.6, one could add two new torus modifications: a torus is transformed into a cracknel and a sphere. Cases 2, 3, 4 are forbidden by Lemmas 2.1.5 and 2.1.6, provided that $sd(S^1)$ is orientable.

Resting on the conditions of Theorem 2.1.3, we discard "75 percent" of all possible types of turn surgery and retain only case 1, that is, the "25 percent" allowed by the Liouville theorem.

1.5 The Topology of Hamiltonian-Level Surfaces of an Integrable System and of the Corresponding One-Dimensional Graphs

Consider an integral f on Q. Two cases are possible:
A) The function f has at least one critical saddle circle on the surface Q.
B) The function f does not have critical saddle circles.

We begin with case A) and assume $sd(S^1)$ to be orientable. Let S^1 be a saddle circle on Q. Consider two close noncritical level surfaces $B_{a+\varepsilon}$ and $B_{a-\varepsilon}$. According to Lemma 2.1.6, the circle S_1 gives rise either to splitting of a Liouville torus into two tori or confluence of two tori into one. Taking a function $-f$ instead of f, we may examine the splitting of one torus into two.

In a fibre $(a - \varepsilon \leqslant f \leqslant a + \varepsilon)$, single out a connected component $U(S^1)$ (a three-dimensional manifold with boundary) which contains S^1 and, as the boundary, has one torus $T_{-\varepsilon}$ in the composition of the "lower" surface $B_{a-\varepsilon}$ and two tori $T_{1,\varepsilon}$ and $T_{2,\varepsilon}$ in the composition of the "upper" surface $B_{a+\varepsilon}$ (Fig. 30). When passing through the critical value a, the torus $T_{-\varepsilon}$ splits into two tori, $T_{1,\varepsilon}$ and $T_{2,\varepsilon}$. The separatrix diagram P_- goes from the critical surface S^1 and, when descending, meets the torus $T_{-\varepsilon}$ along two circles γ_1 and γ_2 (see above). Lemmas 2.1.5 and 2.1.6 imply that γ_1 and γ_2 are boundaries of the ring which lies within the torus $T_{-\varepsilon}$.

Denote by K_1 and K_2 both rings, into the union of which the circles γ_1 and γ_2 split the torus $T_{-\varepsilon}$. Construct a new surface P_1 by adding to the separatrix diagram P_- the ring K_1 and a second surface P_2 by adding to P_- the ring K_2 (Fig. 30). It is clear that both surfaces P_1 and P_2 are homeomorphic to the torus (Fig. 31).

LEMMA 2.1.7. *Let $sd(S^1)$ be orientable. Then the tori P_1 and P_2 are isotopic in the manifold $U(S^1)$ to the tori $T_{1,\varepsilon}$ and $T_{2,\varepsilon}$, respectively. Circles γ_1, γ_2, S^1 lie on each of the tori P_1, P_2 and realize on them nontrivial generating cycles (they do not intersect and are not homotopic to zero).*

PROOF: Isotopy of the torus $T_{i,\varepsilon}$ onto the torus P_i may be constructed exploiting the argument of the ordinary Morse theory where the level surface is deformed along integral trajectories of a vector field until the surface gets into the small neighbourhood of a separatrix diagram, after which contraction proceeds along normals to the separatrix diagram. The other assertions follow from Lemmas 2.1.5 and 2.1.6 and complete the proof.

As we have assumed everywhere that on the critical level B_a there exists exactly one critical saddle circle. Now consider the general case, where on B_a there exist, in general, several such circles (always a finite number).

LEMMA 2.1.8. *It may always be assumed (in the study of surgery on Liouville tori) that on each critical level B_a there exists exactly one critical saddle circle. In other words, it may always be assumed that round handles or thick Möbius strips are glued successively and not simultaneously.*

PROOF: The analogue to this Lemma in the ordinary Morse theory is well known, but in our case the proof is more delicate, since here we deal with the *integral* f (and not merely with a smooth function), and therefore we should essentially use the conditons of Theorem 2.1.3. An arbitrary smooth perturbation of an integral

§1   *Classification of Constant-Energy Surfaces*   79

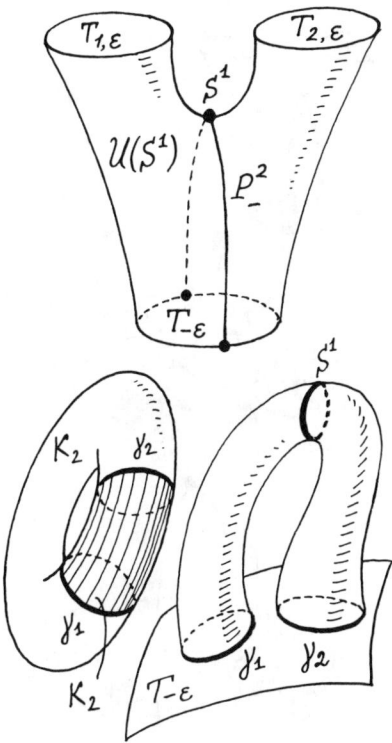

Figure 30

does not generally yield an integral. Consider a nonsingular (close to B_a) surface $B_{a-\varepsilon}$ which is a union of several tori. Suppose that several separatrix diagrams of saddle circles meet one of these tori $T_{-\varepsilon}$. Because f is a Bott integral, each separatrix diagram intersects the torus $T_{-\varepsilon}$ along a smooth non-self-intersecting circle, and the circles corresponding to diferent diagrams do not intersect. All the previous arguments can be applied separately to each of the circles. Consequently, each of the circles is the axis of a narrow ring along which the foot of one round handle or the foot of a thick Möbius strip is glued. Figure 32(1), illustrating for simplicity only the orientable case, shows that all round handles and thick Möbius strips may be thought of as glued to the torus $T_{-\varepsilon}$ irrespective of one another. Let the separatrix diagrams of several saddle circles 1–4 (shown by points in Fig. 32(1)) "come out," for instance, onto a torus $T_{-\varepsilon}$. First, one may perform surgery along circle 1. As a result, the torus $T_{-\varepsilon}$ splits into two tori represented in Fig 32(1) by dashed lines. Obviously, the intersection of these two new tori with the separatrix diagrams of circles 2, 3, and 4 is of the same topological type as before (noncontractible generators on a torus). The tori shown by dashed lines may be

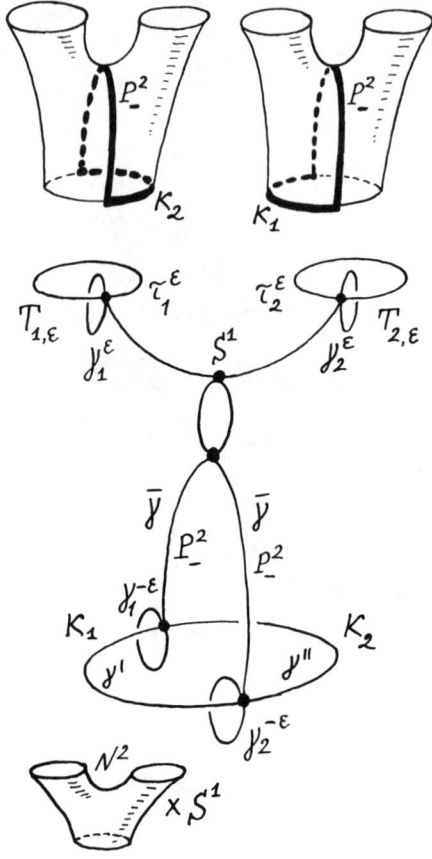

Figure 31

thought of as close to the torus $T_{-\varepsilon}$ which is united with the separatrix diagram P_- of the circle 1. Therefore, some isotopic generators (shown by open circles in Fig. 32) are cut out on these tori by the previous diagrams. Now one can carry out the next surgery along circle 2, etc. Thus, this may be interpreted as gluing only one round handle or one thick Möbius strip each time, that is, as passing through only one critical saddle circle. It is of importance that in our case round handles and thick Möbius strips are glued not to one another (which would confuse the picture), but to one and the same nonsingular torus (in different places of the torus), or to its small upward shift along integral trajectories of the field grad f. Due to this fact, we may commute the gluing of round handles and Möbius strips induced by critical saddle circles which correspond to one and the same critical value of the integral. Besides, nonintersecting and non-self-intersecting circles on a torus (in any number), which realize nonzero elements of the fundamental group, realize one and the same element, and therefore are pairwise isotopic.

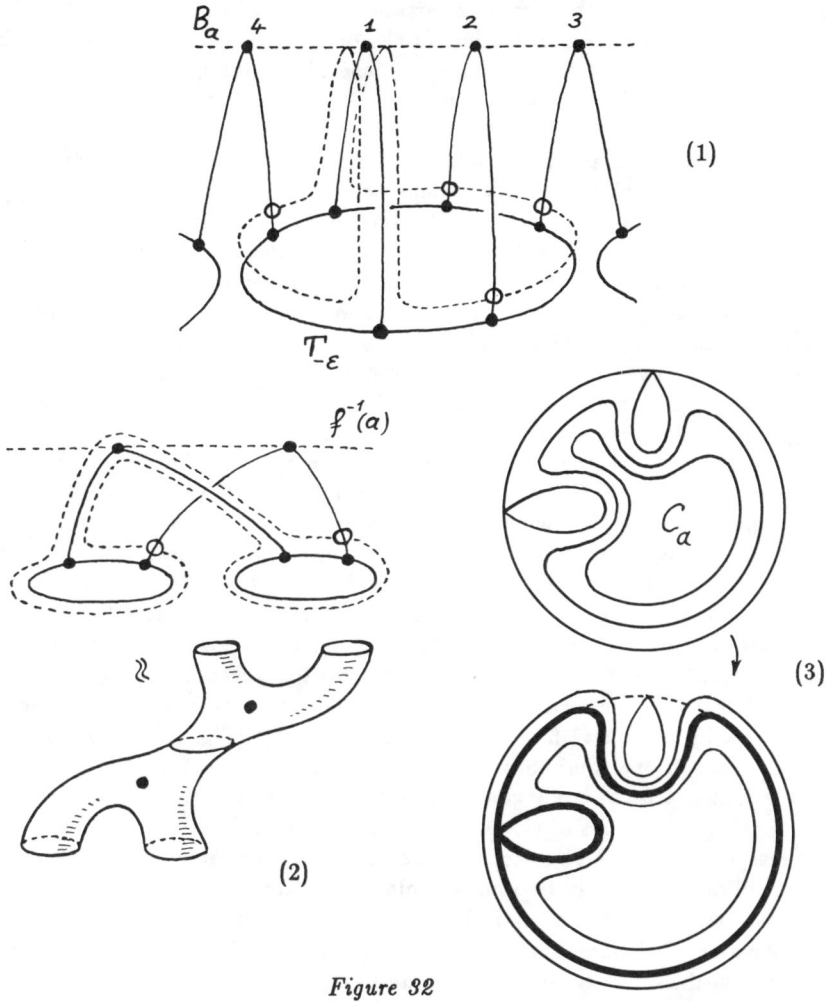

Figure 32

Now we may deform the function f so that on each of its critical levels there remains exactly one saddle circle. The smooth deformation of level surfaces of the function f, which allows us to set also a smooth deformation of the function itself, is shown in Fig 32(3). The figure exhibits the cross-section of a three-dimensional manifold C_a in the neighbourhood of its singular level surface B_a. As the function f deforms, it of course stops being an integral of the field v. But all the necessary properties of the integral have already been used in the proof of the important Lemmas 2.1.5 and 2.1.6. Therefore, we now may treat f as a usual smooth function. The lemma follows.

All local minima of the function f may be assumed to be located on one level $f^{-1}(0)$, and all local maxima of the integral on one level $f^{-1}(1)$. This can be achieved by varying the function f only in the neighbourhood of minimal and maximal submanifolds. New critical points do not occur in this case. A manifold Q is convenient to set in the form of a certain graph $\Gamma = \Gamma(Q, f)$. Let $0 \leqslant f \leqslant 1$ on the surface Q. All minima and maxima of the integral may be treated as absolute (see above). Since f is a Bott function on Q, the picture of splitting or annihilation of nonsingular Liouville tori near critical saddle circles is strictly definite (see Lemmas 2.1.5 and 2.1.6).

In Fig. 33 a black circle denotes the connected component of the singular level of the integral f, on which there lies exactly one critical saddle circle with an orientable separatrix diagram P_-. By virtue of Lemmas 2.1.5 and 2.1.6 this *singular level is homeomorphic to two tori glued along a nontrivial circle*. In other words, two tori have "touched" each other in a nonzero cycle. A light circle denotes the critical torus, that is, the minimum or maximum of the integral different from a circle. An ordinary black point stands for a nonsingular Liouville torus. An asterisk indicates such a connected component of a singular level of the integral which contains exactly one critical saddle circle with a nonorientable separatrix diagram P. A circle centered at a point is the critical Klein bottle. Thus, each noncritical level surface of the function is shown by a set of ordinary (black) points. Varying the values of f, we let these points shift vertically. As a result, we obtain a one-dimensional graph Γ which starts on a plane ($f = 0$) and terminates on a plane ($f = 1$) (see Fig. 33). It is clear that this graph may be realized in $\mathbf{R}^3$. Obviously there exists a continuous mapping h, carrying Q onto the graph $\Gamma(Q, f)$.

Each saddle circle (a black vertex in the graph) generates a "trefoil," where the edge of the graph, coming into the vertex from below, splits into two outgoing edges, provided that one torus splits into two tori (creation) [Fig. 34(1)]. In case two tori flow into one torus (annihilation), the two edges merging into one vertex of the graph (from below) flow into one outgoing edge [Fig. 34(1)]. Next, from each black vertex of the minimum, exactly one edge goes up from the graph, because each minimal critical circle generates exactly one nonsingular torus.

Exactly two edges of the graph go up from each light vertex (the minimum of the integral), since by virtue of Bott's character of the integral, the critical minimal torus splits into exactly two nonsingular tori. Replacing f by $-f$, we obtain the same assertions (and geometrical pictures) for the maxima of the integral.

An asterisk denotes the connected component of the singular level of an integral which contains exactly one critical saddle circle with a nonorientable separatrix diagram. Exactly one edge of the graph goes up and one down from each asterisk on the graph Γ(Fig. 34(1)); and exactly one edge of the graph exits from each circle with a dot. It is not accidental that we have labeled these different types of vertices of the graph Γ by the numbers 1, 2, 3, 4, 5 in Fig. 34(1). The point is that these five types of vertices of the graph Γ and the five types of the simplest three-dimensional manifolds enumerated in Theorem 2.1.2 are in one-to-one correspondence. It turns out that a sufficiently small "neighbourhood" of a vertex of the graph of type i (where $i = 1, 2, 3, 4, 5$) is homeomorphic (from the point of view of Q) to an elementary manifold of type i from Theorem 2.1.2.

Let us consider a critical submanifold L and nonsingular level surfaces $B_{a+\epsilon}$ and $B_{a-\epsilon}$ close to this submanifold. Let $U(L)$ be a connected component of the fibre between $B_{a+\epsilon}$ and $B_{a-\epsilon}$ which contains L.

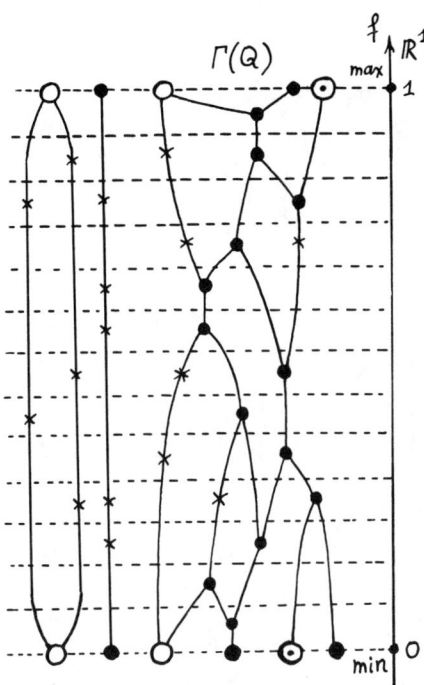

Figure 33

LEMMA 2.1.9. *1)* Let S^1 be a critical saddle circle and let its separatrix diagram P_- be orientable. Then a three-dimensional manifold $U(S^1)$ with the boundary $T_{1,\epsilon} \cup T_{2,\epsilon} \cup T_{-\epsilon}$ is homeomorphic to a direct product $N^2 \times S^1$, where N^2 is a two-dimensional sphere with three holes. In other words, the manifold N^2 is homeomorphic to a disk D^2 with two holes.

2) Let now the separatrix diagram P_- be nonorientable. Then $U(S^1)$ is homeomorphic to a manifold A^3, that is, to a nontrivial fibre bundle $A^3 \to S^1$ with a fibre N^2; $U(S^1) = N^2 \tilde{\times} S^1$ (a skew product).

3) Let $L = S^1$ be a maximal (or minimal) circle. Then $U(S^1) \approx S^1 \times D^2$ (a full torus).

4) Let $L = T^2$ be a maximal or minimal torus. Then $U(T^2) \approx T^2 \times D^1$ (a cylinder).

5) Let $L = K^2$ be a maximal (or minimal) Klein bottle. Then $U(K^2) \approx K^3 \approx K^2 \tilde{\times} D^1$ (a skew product).

A two-dimensional manifold N^2 may be treated as an *element cobordism* ("*trousers*") *containing exactly one critical point of index 1* (Fig. 35).

PROOF: We begin with the orientable case. Lemmas 2.1.5, 2.1.6, and 2.1.7 imply that the manifold $U(S^1)$ contains two tori, $P_1 = P_- \cup K_1$ and $P_2 = P_- \cup K_2$, which

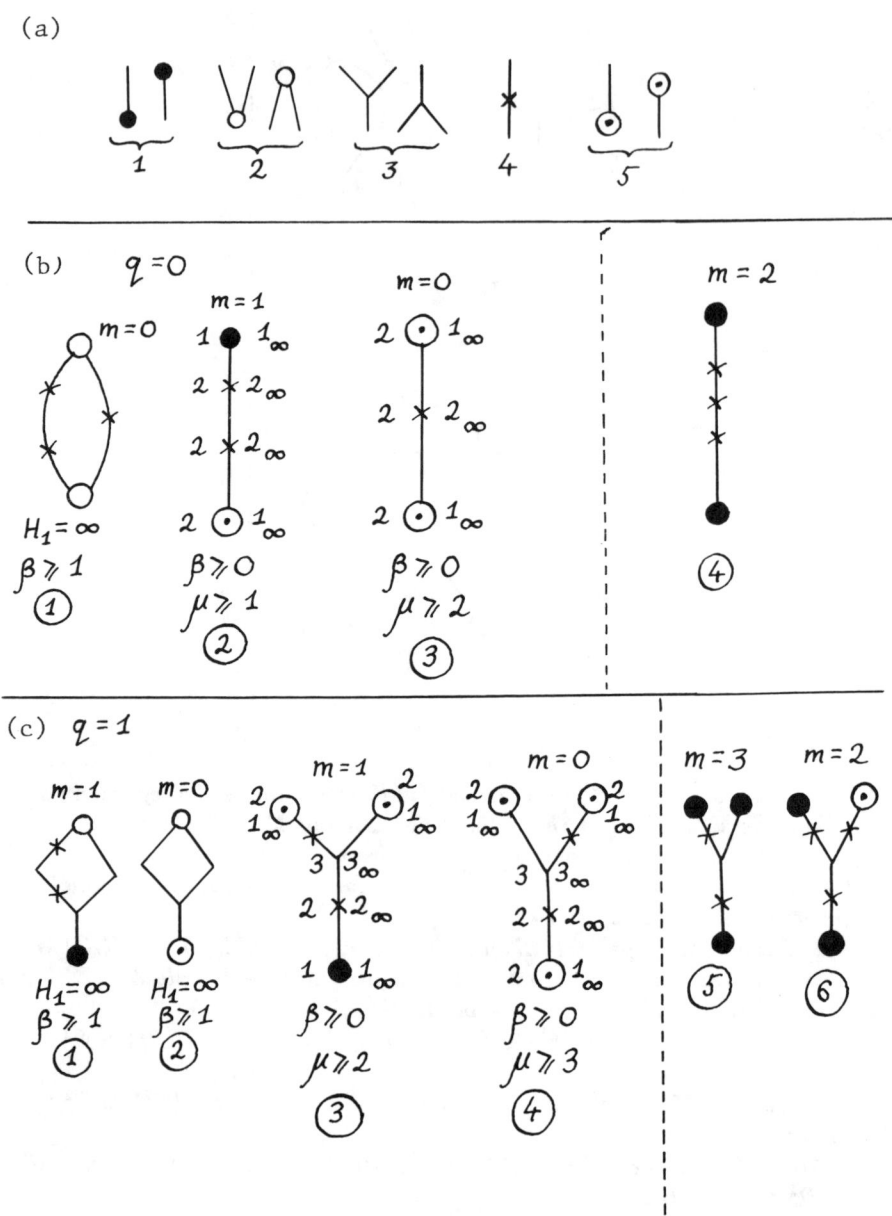

Figure 34

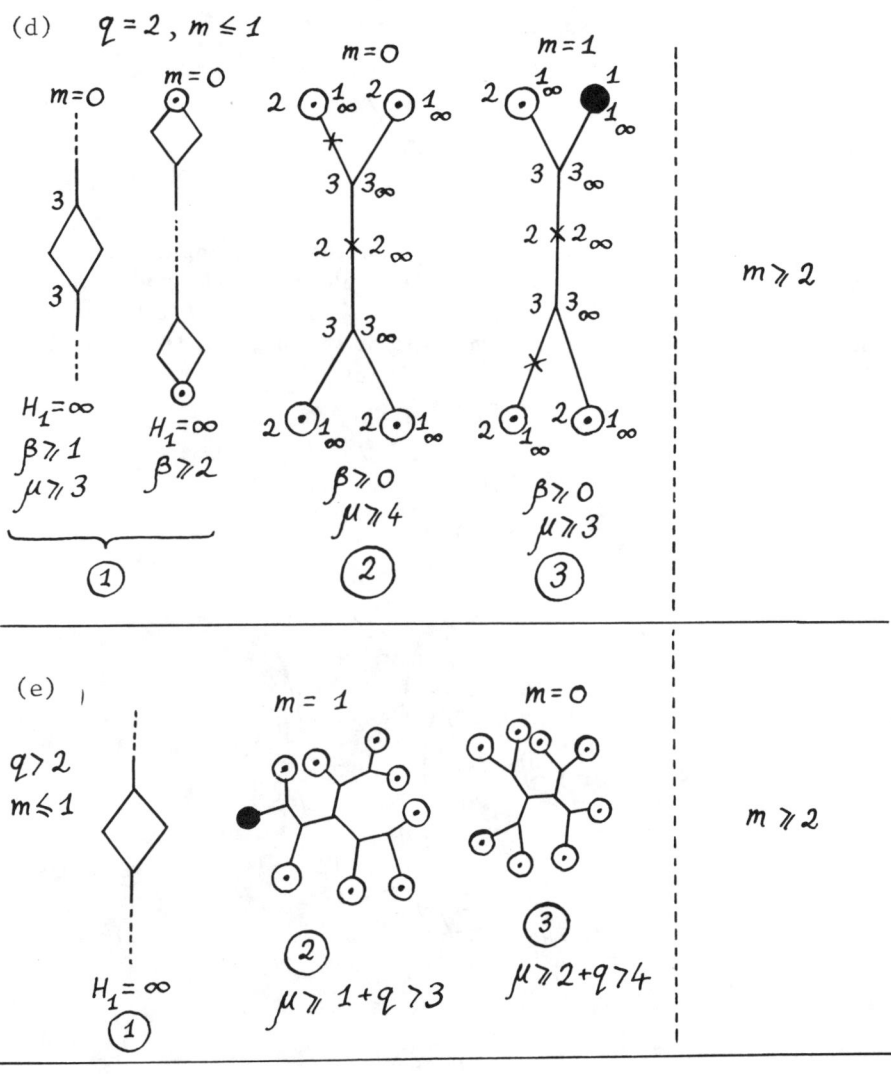

Figure 34 (cont.)

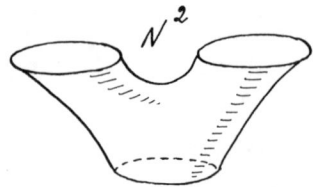

Figure 35

are isotopic respectively to tori $T_{1,\varepsilon}$ and $T_{2,\varepsilon}$. Let these be isotopies. The ascending part of the separatrix diagram of the circle S^1 along the field grad f cuts out a generator γ_1^ε on a torus $T_{1,\varepsilon}$ and a generator γ_2^ε on a torus $T_{2,\varepsilon}$ (Figs. 31, 36). Under isotopy of the torus $T_{1,\varepsilon}$ downward onto the torus P_1, the circle γ_1^ε turns into S^1.

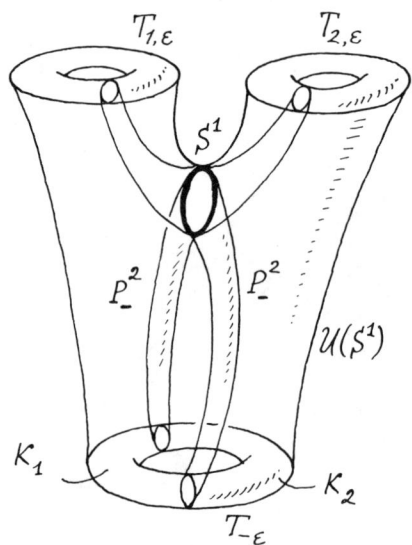

Figure 36

Similarly, under isotopy of the torus $T_{2,\varepsilon}$ onto the torus P_2, the circle γ_2^ε transforms into S^1. Any generator on a two-dimensional torus can always be complemented (not uniquely) with another circle, which is the second generator on the torus and intersects the first cycle exactly at one point. Let us choose a second generator on the torus $T_{1,\varepsilon}$ (Figs. 31, 36). Pulling it down under isotopy of the torus $T_{1,\varepsilon}$ onto P_1, we obtain on the torus P_1 its generator γ which is shown with a dashed line in Figs. 31 and 36 and is a complement of the circle (generator) S^1. The circle γ splits into two arcs: $\gamma = \bar{\gamma} \cup \gamma'$ in the ring K_1. It is clear that the cycles γ_1^ε and $\gamma_2^{-\varepsilon}$ are isotopic on the torus $T_{-\varepsilon}$ and split it into a union of the rings K_1 and K_2. Let us complement the arc $\bar\gamma$ with an arc γ'' in K_2 so that the circle $\bar\gamma \cup \gamma''$ on the torus P_2 will be a generator complementary of S_1, and that the circle $\tau^{-\varepsilon} = \gamma' \cup \gamma''$ on the torus $T_{-\varepsilon}$ will be a generator complementary of $\gamma_1^{-\varepsilon}$ (or of $\gamma_2^{-\varepsilon}$, which is equivalent). (See Figs. 31, 36 and 37.) This can be done because ε is sufficiently small and a part of the separatrix diagram P_- is diffeomorphic to a cylinder. Obviously, on the torus P_2 the cycle $\bar\gamma \cup \gamma''$ is a generator complementary of S^1. Under isotopy of the torus P_2 onto $T_{2,\varepsilon}$, its generator S^1 turns into the generator γ_2^ε on the torus $T_{2,\varepsilon}$, and the cycle $\bar\gamma \cup \gamma''$ turns into a curve τ_2^ε which is a generator of the torus $T_{2,\varepsilon}$, complementary of the generator γ_2^ε.

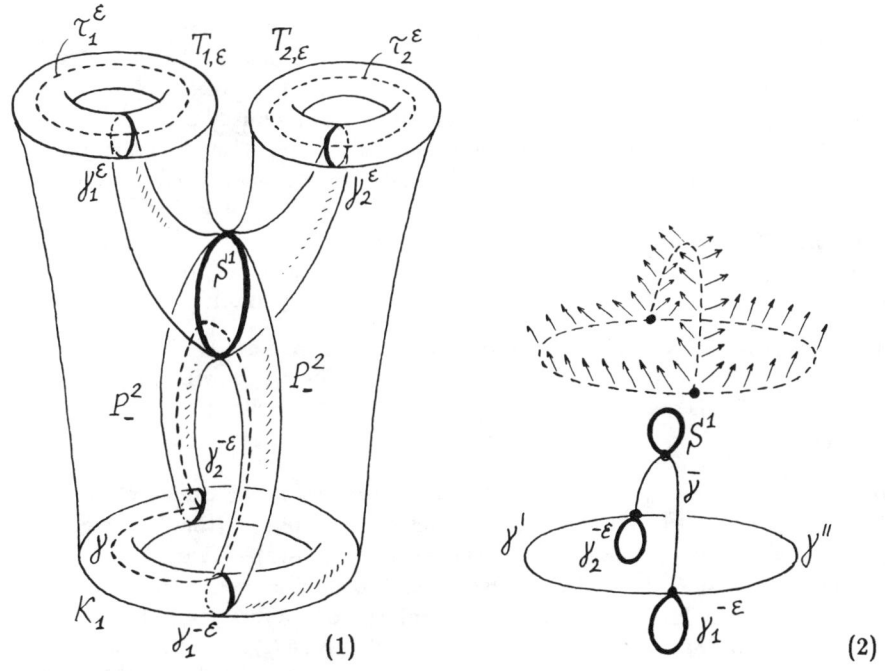

Figure 37

Thus, we have constructed the coordinate system in $U(S^1)$. As N^2 we take a surface formed by two circles $\bar\gamma \cup \gamma'$ and $\bar\gamma \cup \gamma''$ under isotopy of tori P_i onto tori $T_{i,\varepsilon}$, where $i = 1, 2$. It is clear that $\tau_1^\varepsilon = N^2 \cap T_{1,\varepsilon}$, and $N^2 \cap T_{2,\varepsilon} = \tau_2^\varepsilon$ with τ_2^ε being a generator on the torus $T_{i,\varepsilon}$, $i = 1, 2$. Next, $N^2 \cap T_{-\varepsilon} = \tau^{-\varepsilon} = \gamma' \cup \gamma''$, where $T_{-\varepsilon} = K_1 \cup K_2$. It is obvious that circles isotopic to S^1 under the above-mentioned isotopies of P_i onto $T_{i,\varepsilon}$, where $i = 1, 2$, set fibres of the direct product $N^2 \times S^1 = U(S^1)$. This proves the lemma *for the orientable case*.

Now consider the *nonorientable case*. The boundary circle of a separatrix diagram P_- (Möbius strip) is glued to a torus $T_{-\varepsilon}$ along some of its generators

$\gamma^{-\epsilon}$. We add to this generator a second generator on a torus $T_{-\epsilon}$, which we denote by $\tau^{-\epsilon}$. The second generator $\tau^{-\epsilon}$ is not uniquely defined, but this does not matter here.

Since the circle $\tau^{-\epsilon}$ lies on a nonsingular Liouville torus, a normal vector field grad f is defined at each point of this circle. This field carries away the circle $\tau^{-\epsilon}$. Obviously, a continuous deformation of the circle $\tau^{-\epsilon}$ induced by the deformation of the level surface may be well defined. Two versions are possible.

1) The circle $\tau^{-\epsilon}$ first turns into a figure eight (two points of the circle are glued on a singular level surface B_a), after which this figure eight breaks down and turns into a pair of nonintersecting circles.

2) The inverse process. The circle $\tau^{-\epsilon}$ and its duplicate, obtained after a single transport along the critical circle S^1, are first glued at one point and form a figure eight.

Further on, this figure eight is modified to become one circle.

In both versions of the surgery, it is obvious that a deformed circle $\tau^{-\epsilon}$ (or a doubled circle $\tau^{-\epsilon}$) describe a two-dimensional disk with two holes. Moreover, after one revolution along the circles S^1, this disk returns to its initial place, but the two holes of this disk exchange places. We have obtained none other than the manifold A^3 fibred over the base S^1 with fibre N^2. Let S^1 be a critical circle of index 2 or 0. Consider its tubular neighbourhood $U(S^1)$. Since f is a Bott function, it follows that $U(S^1)$ fibres into tori which contract onto the circle S^1 and are level surfaces of the function f. Since these are Liouville tori, the integral trajectories of the system v lie on them. Then the circle S^1 is obviously stable.

LEMMA 2.1.10. *We have the following homeomorphisms:* 1) $A^3 = N^2 \tilde{\times} S^1 = (S^1 \times D^2) + (N^2 \times S^1)$, *that is, A^3 results from gluing a full torus to trousers;*

2) $K^3 = K^2 \tilde{\times} D^1 = (S^1 \times D^2) + A^3 = 2(S^1 \times D^2) + (N^2 \times S^1)$, *that is, K^3 is obtained by gluing two full tori to trousers.*

PROOF: 1) Examine in Fig. 24(2) the point 0 which is the centre of the figure eight. Removing the tubular neighbourhood of the circle β (that is, removing a full torus), we obtain the manifold R shown in Fig. 38(1). Figure 38(2) exhibits fibration of the manifold R with a circle as the typical fibre over a two-dimensional manifold with boundary. This base is homeomorphic to a disk with two holes, that is, N^2. Figure 38(2) illustrates the corresponding procedure of obtaining N^2. By cutting the original ring in half and gluing the two parts according to the arrows, we obtain N^2. Gluings are determined by the character of fibring R into circles.

2) Examine the meridian h on the Klein bottle K^2 shown in Fig. 38(3). This meridian may be regarded as the base of the nontrivial product $K^2 \xrightarrow{S^1} h$ with the circle as the fibre. From this it follows that the manifold K^3 fibres over the circle h with a ring as fibre, that is, $S^1 \times D^1$. Thus, we have $p : K^3 \to h$. The boundary torus T^2 (where $T^2 = \partial K^3$) fibres under the projection P with a pair of circles (the ring boundary) as fibre. This fibre bundle is schematically illustrated in Fig. 38(3). It is clear that the Klein bottle K^2 may be embedded in the fibration space of K^3 as a "zero cross-section" (Fig. 38(3)). Therefore, in K^3 one can single out a circle $\tilde{h}$, which lies on the "zero cross-section" and is projected (homeomorphically) onto the base h under the projection p. Let us remove the

§1 *Classification of Constant-Energy Surfaces* 89

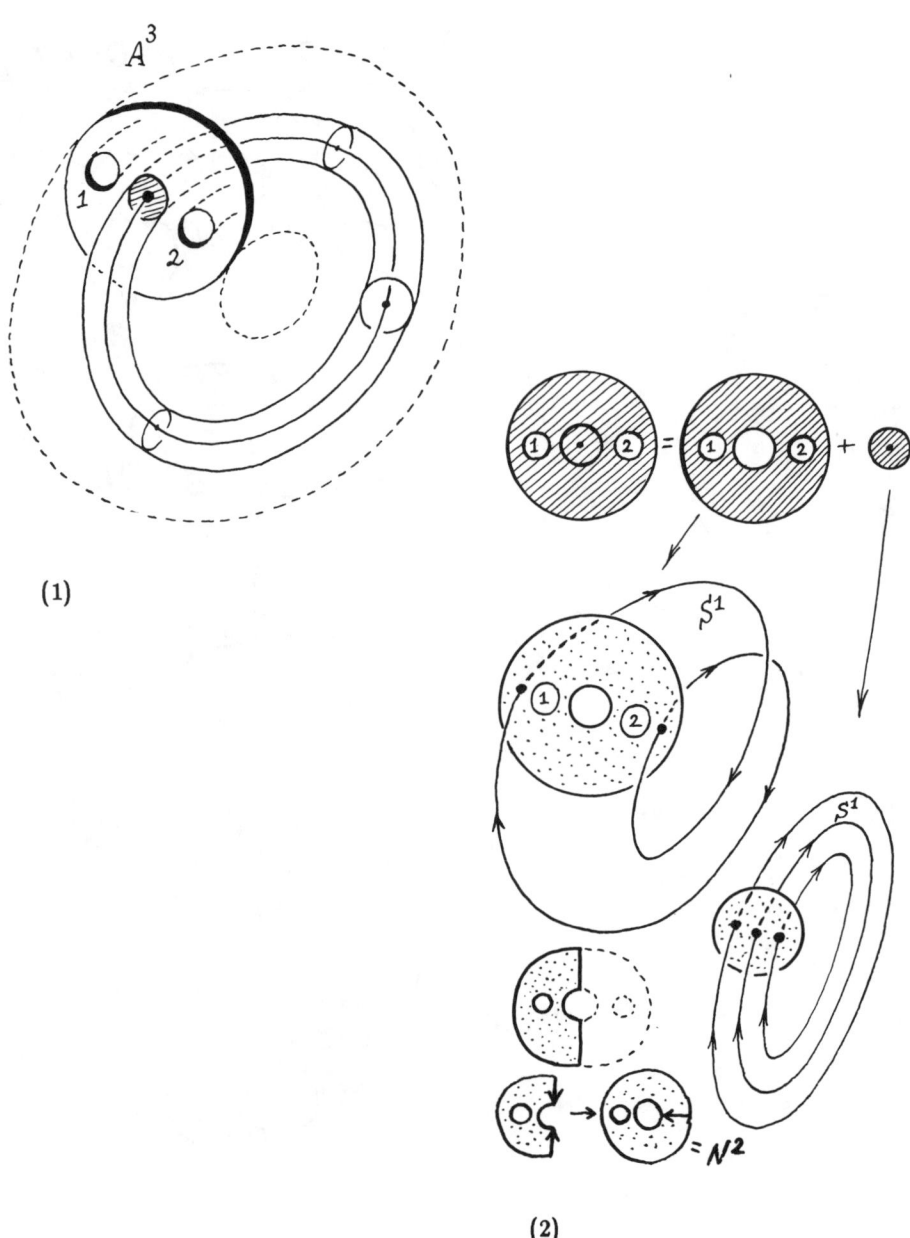

Figure 38

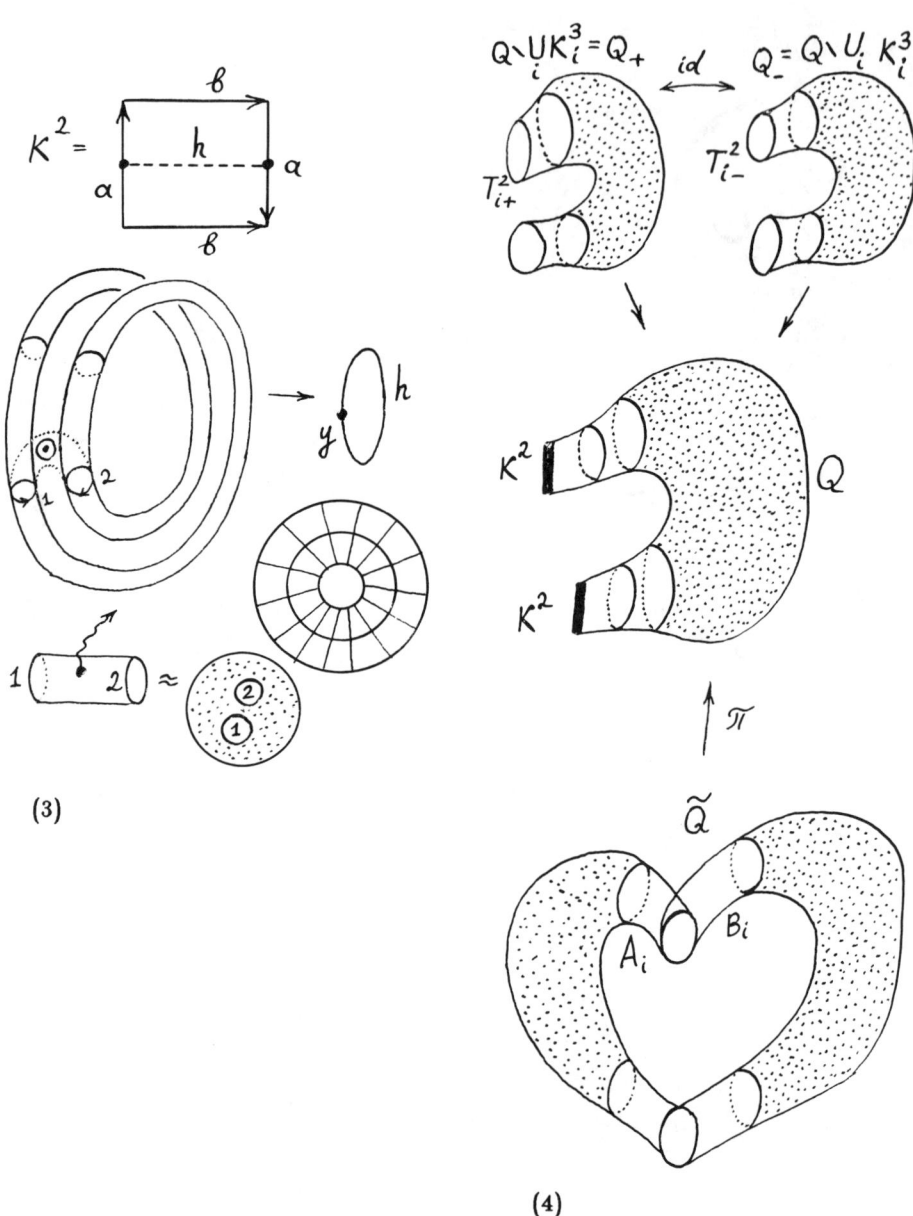

Figure 38 (cont.)

circle $\tilde{h}$ from K^3. As a result, we come to a new fibre bundle $\tilde{p}: K^3\setminus \tilde{h} \to h$ with a fibre N^2 (because by removing a point from the ring, we obtain N^2). It is also clear that the fibre bundles thus obtained is nontrivial, that is, $K^3\setminus \tilde{h} = A^3$. Thus, $K^3 = A^3 + (S^1 \times D^2) = 2(S^1 \times D^2) + (N^2 \times S^1)$, which is the desired conclusion.

1.6 Proof of the Principal Classification Theorem 2.1.2

Any constant-energy surface of an integrable system can be represented as gluing the simple three-dimensional manifolds of three types.

Let Q^3 be an "integrable" constant-energy surface. Single out in Q^3 all critical submanifolds of the integral f. Then have:

1) $m \geqslant 0$ maximal and minimal circles S^1;
2) $p \geqslant 0$ maximal and minimal tori T^2;
3) $q \geqslant 0$ saddle circles S_+^1 for which the separatrix diagram is orientable;
4) $s \geqslant 0$ circles S_-^1 for which the separatrix diagram is nonorientable;
5) $r \geqslant 0$ maximal and minimal Klein bottles. Then Q admits the representation $q = mU(S^1) + pU(T^2) + rU(K^2) + qU(S_+^1) + sU(S_-^1)$, where $U(L)$ denotes a connected regular tubular neighbourhood of the critical level surface containing the submanifold L (see above). In Lemma 2.1.9 all these manifolds are completely described:

$$U(S^1) = S^1 \times D^2, \quad U(T^2) = T^2 \times D^1, \quad U(S_+^1) = N^2 \times S^1,$$
$$U(S_-^1) = A^3, \quad U(K^2) = K^3.$$

This obviously implies the assertion of Theorem 2.1.2.

The proof of Theorem 2.1.3 is obtained from Theorem 2.1.2 and from Lemma 2.1.10.

1.7 Proof of Claim 2.1.1

Let $K_1^2, \ldots, K_r^2$ be a finite number of Klein bottles which are minima or maxima of an integral f. Level surfaces of the integral, which are close to K_i^2, are homeomorphic to a torus being a two-sheeted covering of the Klein bottle K_i^2. Cut out of Q^3 all the manifolds K_i^3 surrounding K_i^2. Recall that K_i^3 is a tubular neighbourhood of the Klein bottle K_i^2. Take two copies Q_+ and Q_- homeomorphic to $Q\setminus \cup_{i=1}^r K_i^3$. It is clear that the boundary $\partial(Q_+ \cup Q_-)$ consists of $2r$ tori T_{i+}^2, T_{i-}^2 obtained by doubling the tori ∂K_i^3, $1 \leqslant i \leqslant r$. Consider $2r$ copies A_i, B_i of cylinders $T^2 \times D^1$, assigning to each pair of tori T_{i+}^2, T_{i-}^2 two $2r$ cylinders: $A_i = T_{i+}^2 \times D^1$ and $B_i = T_{i-}^2 \times D^1$. In each of the cylinders, identify its base T_{i+}^2 (and T_{i-}^2, respectively), with the torus T_{i+}^2 (respectively, T_{i-}^2) in Q_+ (respectively, Q_-). Identify the two remaining bases of the cylinders in such a manner that the two-sheeted covering $A_i \cup B_i \to K_i^3$ will be well defined. See Fig. 38(4). As a result, we obtain a new manifold $\tilde{Q} = Q_+ \cup Q_- \cup (\cup_{i=1}^r A_i \cup B_i)$. Obviously, we have constructed the two-sheeted covering $\pi: \tilde{Q} \to Q$, where Q_+ and Q_- are identically projected onto $Q\setminus \cup_{i=1}^r K^3$. Since the tubular neighbourhood $U(Q)$ of the surface Q^3 in M^4 is homeomorphic to the direct product $Q \times D^1$, one can construct the two-sheeted covering $\pi: \tilde{U}(\tilde{Q}) \to U(Q)$.

Since the projection $\pi: \tilde{Q} \to Q$ may be thought of as smooth, it follows that the symplectic structure ω, the Hamiltonian field sgrad H, and the integral f, which are all given on Q, may be carried back onto $\tilde{Q}$. All these objects retain their characteristic properties. This implies our assertion.

1.8 Proof of Theorem 2.1.1. Lower Estimates on Number of Stable Periodic Solutions of a System

By virtue of Theorem 2.1.2, we have for Q^3 the following decomposition: $Q = m\,\mathrm{I} + p\,\mathrm{II} + q\,\mathrm{III} + sA^3 + rK^3$. The corepresentation of a finitely generated group G with generators $a_1, \ldots, a_n$ and relations $W_1, \ldots, W_m$ will be written as follows:

$$(a_1, \ldots, a_n | W_1 = 1, \ldots, W_m = 1).$$

It is easy to calculate the fundamental groups of these five elementary bricks. We have:

1) $\pi_1(S^1 \times D^2) = \mathbb{Z}$;
2) $\pi_1(T^2 \times D^1) = \mathbb{Z} \oplus \mathbb{Z}$;
3) $\pi_1(N^2 \times S^1) = F_2(a,b) \oplus \mathbb{Z}(c)$,

where F_2 is a free group; $H_1(N^2 \times S^1) = \mathbb{Z} \oplus \mathbb{Z} \oplus \mathbb{Z}$;

4) $\pi_1(A^3) = \pi_1(A^2) = F_2(a,b)/(ab^2 = b^2 a)$, $\qquad H_1(A^3) = \mathbb{Z} \oplus \mathbb{Z}$;
5) $\pi_1(K^3) = \pi_1(K^2) = (a, b | aba^{-1}b = 1)$, $\qquad H_1(K^3) = \mathbb{Z} \oplus \mathbb{Z}_2$.

Examine Q, an integral f, the corresponding graph $\Gamma(Q, f)$. Consider the following 4 cases: 1) $q = 0$ (no trousers), 2) $q = 1$, 3) $q = 2$, 4) $q > 2$. Describe all possible connected graphs $\Gamma(Q, f)$ in cases 1-3.

LEMMA 2.1.11. *Let $q = 0$. Then all possible connected graphs $\Gamma(Q, f)$ are depicted in Fig. 34b. Under the condition $m \leqslant 1$, the graphs $\Gamma(Q, f)$ are enumerated in Fig. 34b (1-3). The number of the copies A^3, that is, the number s of the asterisk on the edges of the graph may be arbitrary.*

The proof is reduced to simply going through all possible cases. In Fig. 34 at any of the nodes (vertices) on the graph Γ, we have given two numbers: the number of the free generators (with a ∞ sign) and the number of all generators in the group H_1 of the corresponding elementary manifold. For instance, near the full torus (the black vertex in the graph), there are numbers 1_∞ and 1. Let β be a one-dimensional Betti number, and μ the number of all generators in the homology group H_1.

LEMMA 2.1.12. *Let $q = 1$. All possible connected graphs are shown in Fig. 34c. The graphs Γ for which $m \leqslant 1$ are presented in Fig. 34c (1-4). The number s (that is, the number of asterisks) may be arbitrary.*

LEMMA 2.1.13. *Let $q = 2$. All possible connected graphs $\Gamma(Q, f)$ for which $m \leqslant 1$ are listed in Fig. 34d. The number of asterisks may be arbitrary.*

Now we proceed to the proof of those assertions of Theorem 2.1.1 which contain the homology group $H_1(Q, \mathbb{Z})$.

LEMMA 2.1.14. *1) Let the graph $\Gamma(Q,f)$ be connected and $q = 0$. For the graph Γ in Fig. 34b (1) the group H_1 is always infinite, that is, $\beta \geqslant 1$. For the graph Γ in Fig. 34b (2) the group H_1 has at least one (perhaps finite) generator, that is, $\mu \geqslant 1$ and $\beta \geqslant 0$. In particular, H_{-1} may be a cyclic finite group if $Q_0 = (S^1 \times D^2) + sA^3 + K^3$. For the graph Γ in Fig. 34b (3) the group H_1 has at least two independent (perhaps finite) generators, that is, $\mu \geqslant 2$, $\beta \geqslant 0$.*

2) Let the graph Γ be connected and $q = 1$. For the graphs Γ in Fig. 34c (1,2), the group H_1 is always infinite, that is, $\beta \geqslant 1$. For the graphs Γ in Fig. 34c (3,4), we have: $\beta \geqslant 0$, $\mu \geqslant 2$ [for Fig. 34c (3)] and $\beta \geqslant 0, \mu \geqslant 3$ [for Fig. 34c (4)]. In particular, in Fig. 34c (3,4), the group H_1 is always nonzero and cannot be finite cyclic.

3) Let the graph Γ be connected and $q = 2$, $m \leqslant 1$. If the graph Γ contains at least one cycle (see Fig. 34d (1)), then the group H_1 is always infinite, that is, $\beta \geqslant 1$. For the graph in Fig. 34d (2), we have $\beta \geqslant 0$, $\mu \geqslant 4$. For the graph in Fig. 34d (3), we have: $\beta \geqslant 0$, $\mu \geqslant 3$. In the latter cases, the group H_1 is always nonzero and cannot be finite cyclic.

4) Let the graph Γ be connected and $q > 2$. If the graph Γ contains at least one cycle, then the group H_1 is infinite. If it does not contain a single cycle, the graph Γ is a tree. In this case, we always have the inequality $\mu \geqslant 1 + q > 3$, and therefore the group H_1 is always nonzero and cannot be a finite cyclic group.

PROOF: Let a complex X be obtained by gluing two of its subcomplexes Y and T intersecting along a connected subcomplex R. Let β_Y (respectively, μ_Y) and β_T (respectively, μ_T) be Betti numbers (respectively, the number of all generators) for homology groups $H_1(Y,\mathbb{Z})$ and $H_1(T,\mathbb{Z})$. Let μ_R be the number of generators in the group $H_1(R,\mathbb{Z})$. Then it follows from Van Kampen's theorem that $\beta_X \geqslant \beta_Y + \beta_T - \beta_R$ and $\mu_X \geqslant \mu_Y + \mu_T - \mu_R$. Next, if the graph $\Gamma(Q,f)$ contains at least one closed cycle, then the same Van Kampen theorem (applied for the case of disconnected intersection R) implies that the group $H_1(X,\mathbb{Z})$ contains at least one infinite generator. Combining these two simple observations, one should go through all the cases listed in Fig. 34. Thus we have:

1) In Fig. 34b (1) the graph Γ contains a cycle, that is, $H_1 = \infty$. In Fig. 34b (2) we have:

$$\beta(Q) \geqslant \beta(S^1 \times D^2) + \beta(K^3) + s \cdot \beta(A^3) - (s+1)\beta(T^2) = 1 + 1 + 2s - 2(s+1) = 0.$$

that is, $\beta \geqslant 0$. Similarly,

$$\mu(Q) \geqslant \mu(S^1 \times D^2) + \mu(K^3) + s\mu(A^3) - (s+1)\mu(T^2) = 1 + 2 + 2s - 2(s+1) = 1.$$

It is easy to construct an example (choosing appropriate diffeomorphisms of two-dimensional boundary tori) when $H_1 = \mathbb{Z}_\alpha$. Hence, here the group H_1 may be a finite cyclic group, but in the case under consideration we have $m = 1$. This event happens if

$$Q_0 = (S^1 \times D^2) + sA^3 + K^3 = (s+4)(S^1 \times D^2) + (s+2)(N^2 \times S^1).$$

If we know beforehand that the manifold Q is not diffeomorphic to Q_0, then this special case (where H_1 is a finite cyclic group) is eliminated. All the other cases are considered in a similar manner. We present only the final results. For Fig. 34b (3) we have:

$$\beta(Q) \geqslant 2\beta(K^3) + s\beta(A^3) - (s+1)\beta(T^2) = 2 + 2s - 2(s+1) = 0,$$

that is,

$$\beta(Q) \geqslant 0; \quad \mu(Q) \geqslant 2\mu(K^3) + s\mu(A^3) - (s+1)\mu(T^2) = 4 + 2s - 2(s+1) = 2,$$

that is, $\mu(Q) \geqslant 2$.

2) For $q = 1$ the graphs in Fig. 34c (1,2) have cycles, and therefore $H_1 = \infty$. For Fig. 34c (3) we have:

$$\beta(Q) \geqslant 2\beta(K^3) + \beta(S^1 \times D^2) + \beta(N^2 \times S^1) + s\beta(A^3) - (s+3)\beta(T^2)$$
$$= 2 + 1 + 3 + 2s - 2(s+3) = 0; \quad \beta(Q) \geqslant 0;$$
$$\mu(Q) \geqslant 2\mu(K^3) + \mu(S^1 \times D^2) + \mu(N^2 \times S^1) + s\mu(A^3) - (s+3)\mu(T^2)$$
$$= 4 + 1 + 3 + 2s - 2(s+3) = 2; \quad \mu(Q) \geqslant 2.$$

For Figure 34c (4) we have:

$$\beta(Q) \geqslant 3\beta(K^3) + \beta(N^2 \times S^1) + s\beta(A^3) - (s+3)\beta(T^2) \geqslant 0;$$
$$\mu(Q) \geqslant 3\mu(K^3) + \mu(N^2 \times S^1) + s\mu(A^3) - (s+3)\mu(T^2) \geqslant 3.$$

3) For $q = 2$ and $m \leqslant 1$, the graphs in Fig. 34d (1) have a cycle, and therefore $H_1 = \infty$. Next, for Fig. 34d (2) we have:

$$\beta(Q) \geqslant 4\beta(K^3) + 2\beta(N^2 \times S^1) + s\beta(A^3) - (s+5)\beta(T^2) \geqslant 0;$$
$$\mu(Q) \geqslant 4\mu(K^3) + 2(\mu(N^2 \times S^1) + s\mu(A^3) - (s+5)\mu(T^2) \geqslant 4.$$

For Fig. 34d (3) we have:

$$\mu(Q) \geqslant 3\mu(K^3) + \mu(S^1 \times D^2) + 2\mu(N^2 \times S^1) + s\mu(A^3) - (s+5)\mu(T^2) \geqslant 3; \beta(Q) \geqslant 0.$$

4) Let $q > 2, m \leqslant 1$. If the graph Γ contains a cycle, then $H_1 = \infty$. Let Γ be a tree. Then only two cases are possible:

$$Q = (S^1 \times D^2) + q(N^2 \times S^1) + sA^3 + rK^3;$$
$$Q = q(N^2 \times S^1) + sA^3 + rK^3.$$

In the first case we have:

$$\mu(Q) \geqslant \mu(S^1 \times D^2) + q\mu(N^2 \times S^1) + s\mu(A^3) + r\mu(K^3) - (r+1+q-1+s)\mu(T^2)$$
$$= 1 + 3q + 2s + 2r - 2(r+q+s) = 1 + q \geqslant 3.$$

In the second case, we have:

$$\mu(Q) \geqslant q + 2 \geqslant 4.$$

This proves Lemma 2.1.14.

The proof of the homological part of Theorem 2.1.1 is concluded as follows. If the integral is orientable, then $r = 0$, and therefore, under the condition that the group H_1 is finite (in Fig. 34b,c,d), there remain only graphs located on the right of the vertical dashed line, that is, $m \geq 2$. If the integral is nonorientable, then the homological assertions of the theorem follow from Lemma 2.1.14. The inequality $m + r \geq 2$ (under the condition that H_1 is finite) is obviously equivalent to the inequality $m \geq 2$ (in the orientable case) if each K^3 is considered to be a full torus; see Lemma 2.1.14.

It remains to prove the assertions of Theorem 2.1.1 related to $\pi_1(Q)$. Let $\pi_1(Q) = \mathbb{Z}$ in the case of the orientable integral. From Lemma 2.1.14 (Fig. 34), it follows that for graphs of the form 34b (2,3), 34c (3,4), 34d (2,3), and 34e (2,3), the group $H_1(Q, \mathbb{Z})$ has at least two independent generators, and therefore $\pi_1(Q) \neq \mathbb{Z}$. Thus, we are not interested in these two graphs. We still have to check that if the graph $\Gamma(Q, f)$ contains at least one cycle, then the group $\pi_1(Q)$ contains at least two independent generators.

Consider the case of Fig. 34b (1), that is, the gluing of two cylinders. Examine the cases of Figs. 34b (1), 34c (1,2), 34d (1), 34e (1). In Fig. 34d (1), two saddles may be glued along two (out of three) of their boundary tori (see the left-hand side of Fig. 34d (1)). A calculation yields that here $\mu \geq 3$, and therefore $\pi_1(Q) \neq \mathbb{Z}$. If two saddles are glued in the form of a tree (on the right in Fig. 34d (1)), one should glue together the four boundary tori. Two versions are possible: gluing two cylinders (then $\beta \geq 2$ and $\pi_1 \neq \mathbb{Z}$) or gluing one cylinder and two full tori (but in this case $m = 2$).

Similar arguments show that in the case of Fig 34e (1), that is, $q \geq 2$, it is even more probable that $\pi_1 \neq \mathbb{Z}$. We have to analyze the cases shown on the left in Fig. 34c (1) and Fig. 34b (1). It is clear that it suffices to investigate the case of Fig. 34b (1), since in the left-hand side of Fig. 34c (1), gluing a full torus into trousers at the worst may give a commutative group with two generators, and this is precisely the case of the lower cylinder in Fig. 34b (1).

Now consider the case $Q = 2(T^2 \times D^1) + sA^3$ (see Fig. 34b (1)). The graph Γ contains one nontrivial cycle. The corresponding generator in the group $\pi_1(Q)$ will be denoted by t. Represent Q in the following form. Cut one edge in the graph Γ and obtain a cylinder $T^2 \times D^1$, in which several nonoriented saddles are glued. At first, assume that there are no such saddles, i.e., that $s = 0$. Let $G = \pi_1(T^2) = \mathbb{Z} \oplus \mathbb{Z}$ and let i_1, i_2 be two embeddings of the torus T^2 into the upper and lower boundaries of the cylinder $T^2 \times D^1$, respectively. We obtain homomorphisms $i_{1*} : G \to \pi_1(T^2 \times D^1) = \mathbb{Z} \oplus \mathbb{Z}$ and $i_{2*} : G \to \pi_1(T^2 \times D^1)$. It is clear that i_{1*}, and i_{2*} are isomorphisms of the groups. Denote by G_1 and G_2 the images of the group G under these homomorphisms. Then the group $\pi_1(Q)$ is obtained from the free product $\mathbb{Z}(t) * G$ by introducing the relations $ti_{1*}(h)t^{-1} = i_{2*}(h)$, where $h \in G$. The brief notion for the set of these relations will be $tG_1t^{-1} = G_2$. In our case, $G_1 \approx G_2 \approx G \approx \mathbb{Z} \oplus \mathbb{Z}$ because i_1 and i_2 are homeomorphisms of the tori. We could assume from the very start that $G_1 = G$, i.e., that the mapping i_1 is identical and i_2 is a certain homeomorphism. Then i_{2*} may be given by the unimodular matrix

$\begin{pmatrix} a & b \\ c & d \end{pmatrix}$ and the relations will take the form

$$t\alpha t^{-1} = \alpha^a \beta^c; \qquad t\beta t^{-1} \alpha^b \beta^d,$$

where α and β are generators of the group $G_1 = G = \mathbb{Z} \oplus \mathbb{Z}$. This implies that rank $\pi_1(Q) \geq 2$, and that at least one generator t has an infinite order. Let now exactly one nonoriented saddle be glued into the cylinder $T^2 \times D^1$. In line with the preceding scheme, we estimate from below the rank of $\pi_1(Q)$. Then we find $\pi_1(\widetilde{Q})$, where $\widetilde{Q}$ is obtained from Q by a cut along one Liouville torus, that is, one edge is cut in the graph Γ. It is clear that $\widetilde{Q}$ is homeomorphic to A^3, and A^3 is homotopy equivalent to the two-dimensional complex A^2 which is obtained as follows. We take a circle and a sufficiently small figure eight orthogonal to this circle. Then we move this figure eight along the circle in an orthogonal way and simultaneously rotate the figure eight, so that after one revolution it returns to the initial position and that two of its circles α and γ exchange places. From Fig. 24, it follows that

$$\pi_1(\widetilde{Q}) = F_2(\alpha, \gamma) * \mathbb{Z}(\beta)/(\beta\alpha\beta^{-1} = \gamma; \ \beta\gamma\beta^{-1} = \alpha)$$
$$\approx F_1(\alpha) * \mathbb{Z}(\beta)/(\beta^2\alpha\beta^{-2} = \alpha) \approx F_2(\alpha, \beta)/(\beta^2\alpha = \alpha\beta^2).$$

Here F_2 is a free group with two generators. Let us calculate the group $\pi_1(\widetilde{Q})$. Let i_1 and i_2 be again two embeddings of the torus T^2 onto the upper and lower boundaries of the manifold $\widetilde{Q} \approx A^3$. Assume i_1 to be an identical embedding. Here i_{1*} and i_{2*}, which map the group $\mathbb{Z} \oplus \mathbb{Z}$ into $\pi_1(A^3)$, are organized as follows. One may assume that under the homomorphism i_{1*}, the generators of the group $\pi_1(T^2) = \mathbb{Z} \oplus \mathbb{Z}$ are mapped into the elements α and β^2 of the group $\pi_1(A^3)$. It remains to understand what will become of the elements α and β after their detour along the generator t. From Fig. 24, we see that one should compare the generators on two Liouville tori located on different sides of a singular fibre homeomorphic to the two-dimensional complex A^2 examined above. We have $t\alpha \tau^{-1} = \alpha\beta\alpha\beta^{-1}, t\beta^2 t^{-1} = \beta$.

Indeed, the generators α and β may be interpreted (Fig. 24) as generators on the interior thin torus in A^3 which goes twice round the axis β of the large full torus. The generator α is here the meridian of the thin interior torus. Using the structure of Q, we draw the interior torus out of A^3 into Q and over to the exterior torus A^3. The element $t\alpha t^{-1}$ (that is, the image of α with this operation) will embrace the figure eight from the outside. It is clear that the figure eight has the form $\alpha(\beta\alpha\beta^{-1})$. We have checked the first assertion.

Similarly, the long parallel of the interior thin torus leaves A^3 and goes back on A^3 "from the outside" (Fig. 24). This parallel transforms into the element $t\beta^2 t^{-1}$ and unfolds, covering β once, as required. Thus,

$$\pi_1(Q) = F_3(\alpha, \beta, t)/(t\alpha t^{-1} = \alpha\beta\alpha\beta^{-1}, t\beta^2 t^{-1} = \beta).$$

This implies that rank $\pi_1(Q) \geq 2$, at least one of the generators, namely t, having an infinite order in the group $\pi_1(Q)$.

It is clear that an increase of the number of nonoriented saddles in the composition of Q cannot decrease the rank to less than two. Thus, Theorem 2.1.1 is proved in the orientable case. Let now the integral be nonorientable. We shall prove that $m \geq 1$ if $\pi_1(Q) = \mathbb{Z}$. Consider a two-sheeted covering $\pi : \tilde{Q} \to Q$ constructed in Claim 2.1.1. Since $\pi_1(\tilde{Q})$ is a subgroup of index two in the group $\pi_1(Q)$, it follows that $\pi_1(\tilde{Q}) = \mathbb{Z}$. But now we have an oriented integral on $\tilde{Q}$, and by the preceding reasoning, we have $\tilde{m} \geq 2$ for $\tilde{f} = \pi * f$. With a downward projection onto Q, two stable periodic trajectories may merge into one, and therefore $m \geq 1$. This completes the proof of Theorem 2.1.1.

It is readily seen that in the case of the graph in Fig. 34b (1), choosing an appropriate cylinder gluing, one can arrive at $H_1(Q, \mathbb{Z}) = \mathbb{Z}$ although the group $\pi_1(Q)$ remains at least with two generators. Indeed, we have written above the relations for the fundamental group $\pi_1(Q)$. Proceeding to homologies, we obtain the following relations: $\alpha = a\alpha + c\beta, \beta = b\alpha + d\beta$, that is, $\alpha(a-1) + \beta c = 0, \alpha b + \beta(d-1) = 0$. Consider the matrix of this system $I = \begin{pmatrix} a-1 & c \\ b & d-1 \end{pmatrix}$. If the determinant of this matrix is nonzero, then $H_1(Q, \mathbb{Z}) = \mathbb{Z}$ because in this case the system has only a zero solution, that is $\alpha = \beta = 0$. As an example we take the following unimodular matrix $\begin{pmatrix} a & b \\ c & d \end{pmatrix} = \begin{pmatrix} 0 & 1 \\ -1 & 0 \end{pmatrix}$. Then $I = \begin{pmatrix} -1 & -1 \\ 1 & 0 \end{pmatrix}$ and $\det I = 1 \neq 0$. Therefore, the manifold Q obtained from the cylinder $T^2 \times D^1$ by gluing two its boundary tori through diffeomorphism with the matrix $\begin{pmatrix} 0 & 1 \\ -1 & 1 \end{pmatrix}$ has the group $H_1(Q, \mathbb{Z}) = \mathbb{Z}$, that is, rank $H_1(Q, \mathbb{Z}) = 1$. This example rests only on topological considerations, but such a manifold Q may always be realized as a constant-energy surface of an integrable system.

It is of importance that manifolds A^3 and K^3 actually arise in concrete integrable systems. For instance, topological surgery, equivalent to A^3, in the dynamics of a three-dimensional heavy rigid body is presented in the work by Kharlamov [177].

Next, a geodesic flow of flat metric on a Klein bottle K^2 possesses an isoenergy (integrable) surface $Q^3 = 2K^3$ on which a Bott integral is defined. Under a two-sheeted covering both Klein bottles K^2 (minimum and maximum of the integral) unfold to become two-dimensional tori, which gives a manifold $\tilde{Q}^3 = 2(T^2 \times D^1)$ with an orientable integral.

1.9. Proof of Corollary 2.1.5

Let all the conditions of Corollary 2.1.5 be fulfilled. Let a system v have on Q exactly m periodic trajectories which are local minima or maxima of the integral f. By virtue of Theorem 2.1.3, the surface Q is glued of cylinders, orientable saddles, nonorientable saddles, and exactly m full tori. Since f is now assumed to be an oriented integral, then in the composition of Q there are actually no nonoriented saddles. The number of cylinders is equal to the number of maximal and minimal critical tori of the integral. In each of these "elementary bricks" we have introduced a coordinate system of the direct product, which we will fix. We will try to introduce a unique coordinate system of the direct product on the entire manifold Q. It is clear that, generally, this cannot be done. The point is that if two "bricks" are glued along a torus, then each of them induces on the torus its own pair of generators setting a coordinate system on the torus. Any pair of generators on the torus can be made coincident through a certain diffeomorphism of the torus.

Cutting Q along this torus and applying this diffeomorphism, we obviously make coincident the two above-mentioned coordinate systems. Continuing this process successively, we extend this unique coordinate system onto all other "bricks".

The process stops at the moment when the surface Q under surgery becomes a direct product $M_g^2 \times S^1$. It is clear that under this procedure the integral f changes and becomes a certain smooth function on the direct product, which is constant on the multiplier S^1, and Corollary 2.1.5 follows.

1.10 *Topological Obstacles for Smooth Integrability and Graphlike Manifolds. Not Each Three-Dimensional Manifold Can Be Realized as a Constant-Energy Manifold of an Integrable System*

Now we will prove Corollary 2.1.4. We will use the results of Waldhausen presented in [184]. This work was first kindly mentioned to me by Zieschang. Waldhausen examined a special class of three-dimensional manifolds W which he called "graphlike manifolds" (Graphenmannigfaltigkeiten). They are defined in the following way. In a manifold W there must exist a family T of non-intersecting tori T^2, discarding which we obtain a manifold each of whose connected components is fibred with circle S^1 as typical fibre over a two-dimensional manifold (maybe, with a boundary).

It is readily seen that *the manifolds W contain all constant-energy surfaces Q of Hamiltonian systems integrated by means of a smooth Bott integral.*

This follows from Theorem 2.1.3. Indeed, manifolds of first three types, namely (1) full tori $S^1 \times D^2$, (2) cylinders $T^2 \times D^1$, (3) oriented saddles $N^2 \times S^1$ are obviously fibred with a fibre S^1 over a two-dimensional manifold with a boundary. Examine a manifold of the fourth type, that is, A^3. It cannot, of course, be represented in the form of a fibre bundle $A^3 \xrightarrow{S^1} M^2$, where M^2 is a two-dimensional surface with a boundary. But A^3 may be cut into two pieces, A_1^3 and A_2^3, each of which admits the required representation.

Figure 38 illustrates such a partition. We cut out a small disk centred at a point O on N^2. Then this disk defines in A^3 a thin full torus which goes once round the axis of A^3. Removing this full torus from A^3, we obtain a manifold P which is fibred with a circle-fibre S^1 over a disk with three holes. Fig. 38 demonstrates such a fibre which goes twice round the axis of A^3 and punctures the disk with three holes at two points.

Thus, $A^3 = P + (S^1 \times D^2)$, where each of these manifolds is a "graph-like manifold" (see above).

Hence, the surfaces Q of integrable systems are of the form of gluing "elementary bricks" of three types, each of which splits into a union of skew products of two-dimensional manifolds (with a boundary) and the circle S^1, i.e. is a manifold of type W.

One can show, in fact, that *the class of manifolds of the form W exactly coincides with the class of constant-energy surfaces of integrable systems.*

Next, it is noted in [184] that *the class of manifolds of the form W does not coincide with the class of all three-dimensional manifolds*. This proves Corollary 2.1.4

The paper [184] gives examples of three-dimensional manifolds (with a boundary) which do not belong to the class W (and, therefore, to our class Q). Take a sphere S^3 and drill in it a full torus $U(K)$ whose axis is given by some knot K. If the obtained manifold $S^3 \setminus U(K)$ is of the type W, then the knot K must, as it turns out, be in a certain way connected with the toric knots. For the exact definition, which we will not present here, see [184]. Since such knots give only a "small percent" of all the knots, we see that *manifolds of the type W (and, therefore, the constant-energy surfaces of integrable systems) are "very rare," they are relatively few among all three-dimensional manifolds.*

Zieschang has noticed that this example can be modernized to obtain a closed manifold which does not belong to class W. To this end, it suffices to glue up the boundary of the above- mentioned manifold with a full torus so as to obtain a homological (but not a standard) sphere. Thus, *manifolds are explicitly pointed out which are not constant-energy surfaces of integrable systems.*

If we assume that the integral f on Q has more than one minimal and maximal circles, the assertion concerning infinity of the group $H_1(Q, \mathbf{Z})$ becomes invalid. We will give an example: lens spaces obtained by gluing two full tori along the boundary torus. On such spaces one may set a function with exactly two critical circles, minimal and maximal.

1.11 Proof of Claim 2.1.4

We will prove the universal estimate (obtained by Zieschang and the present author) from below on the number m' of full tori in topological decomposition of an isoenergy surface Q^3 into the sum of elementary bricks.

For the detailed proof of Claim 2.1.4, see [289] and [298]. For brevity we will prove only a simpler form of the inequality in Claim 2.1.4, namely,

$$m' = m + s + 2r \geqslant \frac{1}{3}(\varepsilon - 5\beta + 4), q' \geqslant m' - 2, q \geqslant m + r - 2$$

where the numbers m, q, s, r and m', q' give the number of corresponding elementary bricks in the Hamiltonian and topological decompositions of the isoenergy surface:

$$Q = m\ \mathrm{I}\ + p\ \mathrm{II}\ + q\ \mathrm{III}\ + s\ \mathrm{IV}\ + r\ \mathrm{V}\ = m'\ \mathrm{I}\ = p'\ \mathrm{II}\ + q'\ \mathrm{III}\ .$$

We will further examine only the topological decomposition of Q, that is, decomposition into the sum of m' full tori, p' cylinders, and q' trousers.

Unite some trousers, full tori, and cylinders into maximal blocks S_i which are *Seifert manifolds*. Some full tori may remain, namely those, some of whose meridians are fibres of neighbouring trousers. Such full tori will be called *trivializing*. Gluing different blocks S_i first with cylinders, we obtain a "tree". The rest of the cylinders determine handles. Let a block S_i have $m(i)$ singular fibres with invariants $(a_{i1}, b_{i1}), \ldots, (a_{im(i)}, b_{im(i)})$. Here one may assume that $a_{ij} > 1, 0 < b_{ij} < a_{ij}$ and the GCD $(a_{ij}, b_{ij}) = 1$ (that is, the numbers are mutually simple, GCD is the greatest common divisor). Let $k(i)$ be the number of trivializing full tori. Let $n(i)$ be the number of the components of the boundary ∂S_i corresponding to the handles or cylinders in the tree. Let $g(i)$ be the genus of the basis surface of the Seifert space S_i. Recall that S_i is orientable; this affects the relation (2a) below.

CLAIM 2.1.5: (see [276]) The fundamental group $\pi_1(S_i)$ has the following corepresentation:

(1) Generators:
   (a) h_i (corresponds to a fibre),
   (b) $\{r_{ij}|1\leqslant j\leqslant m(i)\}\cup\{x_{ij}|1\leqslant j\leqslant k(i)\}$ (corresponds to components ∂S_i),
   (c) $\{s_{ij}|1\leqslant j\leqslant m(i)\}$ (corresponds to singular fibres of S_i),
   (d$^+$) $\{t_{ij},u_{ij}|1\leqslant j\leqslant g(i)\}$ (corresponds to handles of the base if it is oriented),
   (d)$^-$) $\{v_{ij}|1\leqslant j\leqslant g(i)\}$ (if the base is a nonoriented surface).
(2) The defining relations:
   (a) $[r_{ij},h_i]=1$, $[x_{ij},h_i]=1$, $[s_{ij},h_i]=1$, $[t_{ij},h_i]=1$, $[u_{ij},h_i]=1$, $v_{ij}h_iv_{ij}^{-1}h_i=1$, where j runs through all appropriate numbers (see above),
   (b) $s_{ij}^{a_{ij}}h_i^{b_{ij}}=1$, $1\leqslant j\leqslant m(i)$,
   (c) $h_i^{e_i}\prod_{j=1}^{n(i)}r_{ij}\prod_{j=1}^{k(i)}x_{ij}\prod_{j=1}^{m(i)}s_{ij}\cdot K=1$, where $K=\prod_{j=1}^{g(i)}[t_{ij},u_{ij}]$ if the base is oriented and $K=\prod_{j=1}^{g(i)}v_{ij}^2$ if the base is nonoriented. If $n(i)+k(i)>0$, then we may assume that $e_i=0$.

Gluing cylinders induces the correspondence Φ among all boundary components which are not glued up with trivializing full tori. Here $\Phi(i,j)\neq(i,j)$ and $\Phi^2=\mathrm{id}$. We obtain the following isomorphisms:

$$\psi_{ij}:\langle r_{ij},h_i\rangle\to\langle r_{\phi(i,j)},h_{\phi_1(i,j)}\rangle,$$

where $\phi_1(i,j)$ is the first term of the pair $\phi(i,j)$. Here $\psi_{\phi(i,j)}=\psi_{ij}^{-1}$. Let

$$\psi_{ij}(r_{ij})=r_{\phi(i,j)}^{\alpha(i,j)}\cdot h_{\phi_1(i,j)}^{\beta(i,j)},\;\psi_{ij}(h_i)=r_{\phi(i,j)}^{\gamma(i,j)}\times h_{\phi_1(i,j)}^{\delta(i,j)}.$$

Here the determinant has the form $\alpha(i,j)\delta(i,j)-\beta(i,j)\gamma(i,j)=\pm1$. Now gluings determine the new generators:
(3) $\{w_{ij}|1\leqslant i\leqslant p,1\leqslant j\leqslant n(i)\}$ and the relations:
(4) (a) $w_{ij}^{-1}r_{ij}w_{ij}=\psi_{ij}(r_{ij})=r_{\phi(i,j)}^{\alpha(i,j)}h_{\phi_1(i,j)}^{\beta(i,j)}$,
   (b) $w_{ij}^{-1}h_iw_{ij}=\psi_{ij}(h_i)=r_{\phi(i,j)}^{\gamma(i,j)}h_{\phi_1(i,j)}^{\delta(i,j)}$,
   (c) $w_{\phi(i,j)}=w_{ij}^{-1}$ for $1\leqslant i\leqslant p,1\leqslant j\leqslant n(i)$.

It is clear that each pair $(i,j),\Phi(i,j)$ will require only one pair of relations [4 (a), (b)]. A new generator is required for gluing only if gluing gives rise to a handle. If (i,j) corresponds to a cylinder included into a tree, we have
   (d) $w_{ij}=1$.

Finally, gluing trivializing full tori will add the relations
(5) $w_{ij}=1$, if $k(i)\geqslant 1$.

THEOREM 2.1.5 (FOMENKO, ZIESCHANG). *The fundamental group of the three-dimensional manifold Q has the corepresentation with the generators (1 a–d), (3) and with the defining relations (2 a–c), 4 a–d), (5).*

THEOREM 2.1.6 (FOMENKO, ZIESCHANG). *If all the blocks S_i are "sufficiently complicated" (see [276], [184]) (for instance, if $k(i)+m(i)+n(i)+g(i)>3$) and if there are no trivializing full tori, then the fundamental group $\pi_1(Q)$ determines the manifold Q uniquely up to homeomorphism, and the numbers determining the corepresentation of the group are (almost) invariants of this group.*

To calculate the homology group $H_1(Q, \mathbb{Z})$, we write the group (see above) in an additive form, that is, commute the fundamental group. Let

$$A_{i1} = \langle\langle s_{ij}, h_i | (2b) \rangle\rangle$$

be an Abelian group with the given generators and relations, where j runs through $1, 2, \ldots, m(i)$. Then we have

(6) $\beta(A_{i1}) = 1, \varepsilon(A_{i1}) \leqslant m(i) - 1$ if $m(i) > 0$, and $\beta(A_{i1}) = 0$ if $m(i) = 0$. By $A_{i2} = \langle\langle A_{i1}, (1, b, d), (3) \rangle\rangle$ we denote the Abelian group obtained from A_{i1} by adding the generators $(1b, d), (3)$ for a fixed i. Then we have

(7) $\beta(A_{i2}) = 1 + k(i) + 2n(i) + \omega_i g(i)$ and $\varepsilon(A_{i2}) = \varepsilon(A_{i1})$, where $\omega_i = 2$, if the base is oriented and $\omega_i = 1$ in the contrary case. Let $S_1, \ldots, S_{p_+}$ have an oriented basis surface and $S_{p_++1}, \ldots, S_p$ have a nonoriented base. Let

$$A_{-3} = \langle\langle \text{ all the generators } | (2a - c), (4c, d) \rangle\rangle.$$

If $N = \frac{1}{2} \sum_{i=1}^{p} n(i)$, then

(8a)
$$\beta(A_3) = \sum_{i=1}^{p} \beta(A_{i2}) - N - (q' - 1)$$
$$= \sum_{i=1}^{p} k(i) + \frac{3}{2} \sum_{i=1}^{p} n(i) + 2 \sum_{i=1}^{p_+} g(i) = \sum_{i=p_++1}^{p} g(i) + 1,$$

(8b)
$$\varepsilon(A_3) \leqslant \sum_{i=1}^{p} \varepsilon(A_{i2}) \leqslant \sum_{i=1}^{p} m(i) - p_1,$$

where p_1 is the number of such i with the property: $m(i) > 0$. Let $A_4 \langle\langle A_3, (4a) \rangle\rangle$ and $A_5 = \langle\langle A_4, (4b) \rangle\rangle$. Then we have

(9) $\beta(A_4) = \beta(A_3) - N$, $\varepsilon(A_4) = \varepsilon(A_3)$,

(10)(a) $\beta(A_4) - N \leqslant \beta(A_5) \leqslant \beta(A_4)$,

(b) $\varepsilon(A_4) - N \leqslant \varepsilon(A_5) \leqslant \varepsilon(A_4) + N$,

(c) $-N \leqslant \rho(A_5) - \rho(A_4) \leqslant -(p - 1)$, where $\rho(A) = \text{rank } A$.

Let $A_6 = \langle\langle A_5, (5), (2a) \rangle\rangle$. Then among the relations $(2a)$ of interest are only the relations

$$v_{ij} h_i v_{ij}^{-1} h_i = 1.$$

Let p_0 denote the number of such i for which $k(i) > 0$ and p_{0+} denote the number of such $i \leqslant p_+$ with this property. Then we have: q

(11)(a) $-(p - p_+) - p_{0+} \leqslant \beta(A_6) - \beta(A_5) \leqslant 0$,

(b) $-(p - p_+) - p_{0+} \leqslant \varepsilon(A_6) - \varepsilon(A_5) \leqslant (p - p_+) + p_{0+}$,

(c) $-(p - p_+) - p_{0+} \leqslant \rho(A_6) - \rho(A_5) \leqslant 0$.

Addition of the last relations $(2c)$ gives $H_1(Q) = \langle\langle A_6, (2c) \rangle\rangle$ and the conditions:

(12)(a) $-p \leqslant \beta(Q) - \beta(A_6) \leqslant -p_0$,

(b) $-p + p_0 \leqslant \varepsilon(Q) - \varepsilon(A_6)) \leqslant p - p_0$,

(c) $-p \leqslant \rho(Q) - \rho(A_6) \leqslant 0$.
All the inequalities (6–12) give:
(13)
$$\varepsilon(Q) - \beta(Q) \leqslant \sum_{i=1}^{p} m(i) - p_1 - \sum_{i=1}^{p} k(i) - 2\sum_{i=1}^{p_+} g(i)$$
$$- \sum_{i=p_++1}^{p} g(i) + 3p - 2p_+ - p_0 + 2p_{0+}$$
$$\leqslant \sum_{i=1}^{p} m(i) - p_1 + 2p - p_+.$$

On the other hand, there holds the following assertion:
(14)
$$3q' = 2n + m', \qquad q' \geqslant p, \qquad n \geqslant N,$$
$$m' \geqslant \sum_{i=1}^{p} m(i) + \sum_{i=1}^{p} k(i), \qquad \beta(Q) \geqslant \tilde{\beta} = n - (q' - 1)$$

because $\tilde{\beta}$ is equal to the number of handles. From this we have:
(15) $q' = 2\tilde{\beta} + m' - 2$.
Using (13), (14), (15), we have:
(16)
$$m' \geqslant \sum_{i=1}^{p} m(i) \geqslant \varepsilon(Q) - \beta(Q) - 2p + p_+ \geqslant \varepsilon(Q) - \beta(Q) - 2q' + p_+$$
$$= \varepsilon(Q) - \beta(Q) - 4\tilde{\beta} - 2m' + 4 + p_+$$

and $m' \geqslant \varepsilon(Q) - \beta(Q) - 2\tilde{\beta} - m' + 2$ if $p_+ = p$. This implies the proof of the theorem.

From Theorem 2.1.6 it follows that all "sufficiently large" manifolds Q^3 are classified by their fundamental groups. There remains the question: how many "small" manifolds of the type Q^3 are there? The problem (yet unsolved) consists in complete classification of these manifolds.

Below we will give a complete description of manifolds X^{n+1} which generalize three-dimensional isoenergy surfaces and describe bifurcations of Liouville tori of general position in the neighbourhood of the critical momentum mapping points of an integrable Hamiltonian system. In other words, X^{n+1} is a pre-image of a regular curve (under momentum mapping) which intersects transversally the bifurcation diagram. These manifolds are of the form

$$X^{n+1} = m'(D^2 \times T^{n-1}) + p'(T^n \times D^1) + q'(N^2 \times T^{n-1})$$

where T^r is an (r)-dimensional torus. The problem is to describe the fundamental group of such manifolds. Here one may expect the results close to the theorems

§2 Multidimensional Integrable Systems. Classification of the Surgery on Liouville Tori in the Neighbourhood of Bifurcation Diagrams

presented above. The theory of multidimensional Seifert manifolds developed by Neuman, Raymond, Zimmermann, Zeischang and others will possibly prove useful in solving this problem.

2.1 Bifurcation Diagram of Momentum Mapping for an Integrable System. The Surgery of General Position

In this section we *classify the surgery (modifications) of general position of Liouville tori, which occur at the moment when a torus "intersects" the critical level of the energy integral*. Such surgery turns out to split into compositions of certain canonical modifications of four types, the latter being described explicitly and having a simple geometrical nature. In particular, we develop some ideas expressed by Smale in his remarkable work "Topology and Mechanics" (see [121]).

Suppose $v = \operatorname{sgrad} H$ is a smooth system on a smooth symplectic manifold M^{2n}. Suppose the system v is integrable, that is, there exist n independent (almost everywhere) smooth integrals $f_1, \ldots, f_n$ which are in involution. We will assume that $f_1 = H$. Let $F: M^{2n} \to \mathbb{R}^n$ be *the momentum mapping* corresponding to these integrals, that is, $F(x) = (f_1(x), \ldots, f_n(x))$. Recall that the point $x \in M$ is called *regular for the mapping F* if rank $dF(x) = n$, that is, the mapping $dF(x): T_x M \to \mathbb{R}^n = T_{F(x)}\mathbb{R}^n$ is an epimorphism. Otherwise the point x is called *critical*, and its image $F(x)$ is called *the critical value*. Let $N \subset M$ be the set of all critical points of the mapping. Clearly, the set is closed. Let $\Sigma = F(N)$ be the set of all critical values. This set is called *a bifurcation diagram*. Since the mapping F is smooth, it follows that $\dim \Sigma \leqslant n - 1$. If the point $x \in \mathbb{R}^n$ is not a critical value, that is, $x \in \mathbb{R}^n \setminus \Sigma$ then its complete pre-image $B_a = F^{-1}(a) \subset M$ (i.e. a nonsingular fibre) does not contain critical points of the mapping F. By virtue of the Liouville theorem, each of its compact connectedness components is diffeomorphic to the Liouville torus T^n. Now we assume for simplicity the whole fibre B_a to be compact (Fig. 39).

The corresponding assertions for noncompact fibres B_a, that is, for "cylinders" will easily follow from the results obtained below. If $a \in \Sigma$, then the compact level surface (i.e. the fibre) B_a is *singular (critical)* and $\dim B_a \leqslant n$.

When the point a in $\mathbb{R}^n$ becomes deformed, its pre-image, i.e. the fibre B_a becomes somewhat deformed too. The fibre B_a is transformed through diffeomorphisms, that is, does not undergo qualitative topological surgery until the point a, moving along $\mathbb{R}^n$, meets the bifurcation diagram Σ. In particular, any two fibres B_a and B_b, such that the points a and b may be joined with a smooth curve $\gamma \subset \mathbb{R}^n \setminus \Sigma$ (that is, a curve that contains none of the critical values of momentum mapping), are diffeomorphic, they consist of one and the same number of Liouville tori.

If at some point c a continuous curve γ meets a bifurcation diagram Σ, then the fibres B_a and B_b may be distinct. If, when moving, the point a punctures Σ, then the fibre B_a undergoes, generally, a qualitative topological modification, i.e. surgery (Fig. 40).

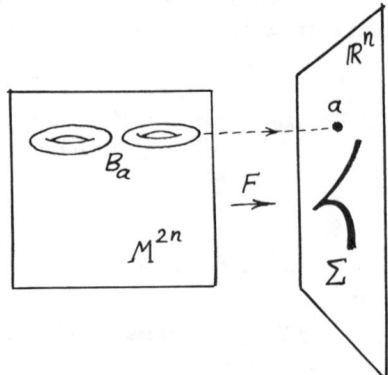

Figure 39

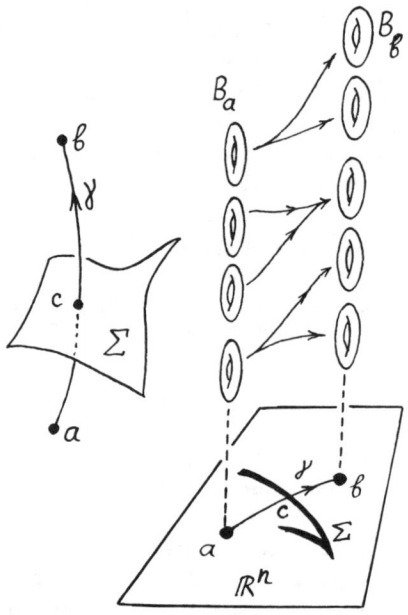

Figure 40

The general problem is *to describe topological surgery on Liouville tori occurring at the moment when the point a intersects the bifurcation diagram Σ.*

The study of the indicated surgery makes it possible to describe the mechanism of modifications of the motion of an integrable Hamiltonian system depending on fixed values of the integrals. Recall that integral trajectories of an integrable system determine (in appropriate action-angle coordinates) rectilinear Liouville tori windings conditionally periodic motion. The transformation of these motions in passing from one common level of integrals to another is set by the surgery on Liouville tori when the point $a \in \mathbb{R}^n$ changes.

As shown below, the modifications (surgery) of general positions can be classified, and they are of a rather simple form. It is clear that there exist the following two cases: (1)$\dim \Sigma < n - 1 \langle n - 1$, (2)$\dim \Sigma = n - 1$. In case 1 the diagram Σ does not separate $\mathbb{R}^n$, that is, any two points $a, b \in \mathbb{R}^n$ are joined by a smooth curve $\gamma \subset \mathbb{R}^n \setminus \Sigma$. Compact nonsingular fibres are therefore mutually diffeomorphic, in particular, they consist of one and the same number of Liouville tori. Substantially more complicated is case 2. Here the diagram Σ, generally, separates $\mathbb{R}^n$ into several open non-intersecting domains. Each of them has, generally speaking, its own topology of the non-singular fibre. This topology changes from domain to domain.

Thus, let $\dim \Sigma = n - 1$. Examine a point c on Σ and investigate the surgery on Liouville tori when a smooth curve γ (the trace of the motion of the point a) punctures the diagram Σ at the point c. It suffices to examine only a small neighbourhood $U = U(c)$ of the point c in $\mathbb{R}^n$.

We will investigate the case of "general position", that is, when the path γ transversally punctures Σ at a nonzero velocity at the point c which lies on an $(n - 1)$-dimensional smooth stratum of the diagram Σ. In other words, we will assume $U \cap \Sigma$ to be a smooth $(n - 1)$-dimensional submanifold in $\mathbb{R}^n$. In the case of general position, one may assume the set $N \cap F^{-1}(U)$ of critical points to be a union of a finite number of smooth submanifolds in M stratified by the rank of dF. This means that N may be represented as a union of non-intersecting smooth submanifolds N_i on each of which the rank of dF is exactly equal to i (some of these submanifolds may, of course, be empty).

The notion of *general position* may be also specified as follows. Since in the neighbourhood of the point c the set Σ is considered to be an $(n - 1)$-dimensional submanifold, one may assume that in the neighbourhood of a certain connected component B_c^o of a singular fibre B_c *the last* $n - 1$ *integrals* $f_2, \ldots, f_n$ *are independent*, and the first integral $f_1 = H$ (energy) becomes dependent on these integrals on the submanifold on critical points $T = N \cap B_c^o$. Indeed, let us restrict the mapping F to the submanifold $N \cap F^{-1}(U)$, which, according to the requirement of general position, is a union of a finite number of smooth submanifolds. Since the restriction of F to each stratum, including the maximal one, $N' \cap F^{-1}(U)$, is a smooth mapping of a smooth submanifold, it follows that $dF(x) : T_x N' \to T_{F(x)} \Sigma$ is an epimorphism and $\operatorname{rank} dF(x) \geqslant n - 1$ because $\dim U \cap \Sigma = n - 1$. At the same time, since $x \in N$ is a critical point, it follows that $\operatorname{rank} dF(x) \leqslant n - 1$. Consequently, $\operatorname{rank} dF(x) = n - 1$. One may therefore assume (changing the basis in the set of intregrals in case of necessity) that $f_2, \ldots f_n$ are independent of B_c^o. Hence, the integral f_1 becomes dependent on these integrals on the set $T = N \cap B_c^o$.

Now we will consider several types of $(n + 1)$-dimensional manifolds whose boundaries are tori.

(1) Examine a "full torus" $D^2 \times T^{n-1}$, whose "axis" is a torus T^{n-1}, embedded in $\mathbb{R}^{n+1}$ in a standard way. Its boundary is a torus T^n. We will call $D^2 \times T^{n-1}$ a *dissipative full torus*. The account for this terminology, which takes root in

mechanics, is given hereafter.

(2) The direct product $T^n \times D^1$ will be called *a cylinder*. Its boundary consists of two tori T^n.

(3) Let N^2 be a disk with two holes. The direct product $N^2 \times T^{n-1}$ will be called *a toric oriented saddle*. Its boundary consists of three tori T^n.

(4) Over the torus T^{n-1} we examine all nonequivalent fibre bundles with a segment $D^1 = [-1, +1]$ as fibre. They are classified by the elements α of the homology group $H_1(T^{n-1}, \mathbb{Z}_2) = \mathbb{Z}_2 \oplus \cdots \oplus \mathbb{Z}_2 (n-1$ times). The space of the fibre bundle corresponding to the element α will be denoted by Y_α^n. It is clear that Y_α^n is an n-dimensional smooth manifold whose boundary consists of two tori T^{n-1} if $\alpha = 0$ and of one torus if $\alpha \neq 0$. We single out *a zero cross-section* and denote it by T^{n-1}. This cross-section is homeomorphic to the base, that is, to the torus. We have the fibre bundle

$$Y_\alpha^n \xrightarrow{D^1} T^{n-1}.$$

Now consider as new fibre bundle (associated with the first one) of the form

$$A_\alpha^{n+1} \xrightarrow{N^2} T^{n-1}$$

with a torus T^{n-1} as the base, and with a fibre N^2. This fibre bundle is defined as follows. A disk with two holes is homotopy equivalent to a figure eight. On N^2 consider a line segment $D^1 = [-1, +1]$ which passes through the disk centre and joins (after its continuation) the centres of the two removed disks (holes). With each fibre bundle

$$Y_{\_a}{}^n \xrightarrow{D^1} T^{n-1}$$

one may associate the fibre bundle

$$A_\alpha^{n+1} \xrightarrow{N^2} T^{n-1}$$

by changing the fibre D^1 by the fibre N^2. Its particular case is the direct product $N^2 \times T^{n-1}$, that is, a manifold of type 3. It is obtained only if $\alpha = 0$. If $\alpha \neq 0$, then the corresponding fibre bundle A_α^{n+1} is nontrivial. Recall that in case $\alpha \neq 0$ the boundary of an n-dimensional manifold Y_α^n is one torus T^{n-1}. If $\alpha \neq 0$, then $(n+1)$-dimensional manifolds A_α^{n+1} will be called *nonoriented toric saddles*. The boundary of the manifold A_α^{n+1} consists of two tori T^n if $\alpha \neq 0$.

It is readily seen that for $\alpha \neq 0$ all manifolds A_α^{n+1} are diffeomorphic to one another, and therefore we will write them as follows:

$$A_\alpha^{n+2} = N^2 \times T^{n-1}$$

(a skew-product). This follows from the fact that any non-self-intersecting trajectory on a torus may be taken as one of the basis generators on the torus. Thus, as in the four-dimensional case, we obtain only two topologically different manifolds of type 4, namely,

$$N^2 \times T^{n-1}, \quad N^2 \times T^{n-1}.$$

Manifolds of type 4 admit a vivid description. For $n = 2$ we obtain the manifold A^3 which we have already dealt with in §1. In the multidimensional case we should take a dissipative full torus (with a torus as the boundary) and drill there a "thin" full torus $D^2 \times T^{n-1}$ which goes ("winds") twice round some of the axes of the large full torus. We obtain one of the manifolds A_α. In the multidimensional (as distinguished from the three-dimensional) case, this drilling is not unique because the axis of the dissipative full torus is the torus T^{n-1}, and on this torus there are $n - 1$ independent cycles-circles (generators) along which the thin full torus can be wound twice.

(5) Let $p : T^n \to K^n$ be a two-sheeted covering over a nonorientable manifold K^n. All such coverings p may be classified. By K_p^{n+1} we will denote the cylinder of the mapping p. It is clear that $\dim K_p^{n+1} = n + 1$ and $\partial K_p^{n+1} = T^n$.

LEMMA 2.2.1. *The manifolds A_α^{n+1} and K_p^{n+1} are represented as the gluing of manifolds of the first three types, that is, from the topological point of view only the manifolds*

$$T^n \times D^1, \qquad N^2 \times T^{n-1}, \qquad D^2 \times T^{n-1}$$

are independent "elementary bricks."

The proof of this Lemma is analogous to that in four-dimensions and therefore we invite the reader to fill in the details. Now we will describe five types of surgery on the torus T^n (Fig. 41).

(1) A torus is set as the boundary of *a dissipative full torus* $D^2 \times T^{n-1}$ and then it is contracted to its axis, that is, to the torus T^{n-1}. This operation will be called *limit degeneracy*. This type of surgery will be denoted by $T^n \to T^{n-1} \to 0$.

(2) Two tori T_1^n and T_2^n, which make up the boundary of the *cylinder $T^n \times D^1$*, move along this cylinder towards each other and *in the middle of the cylinder flow into one torus* T^n. The notation: $2T^n \to T^n \to 0$.

(3) A torus T^n, which is the lower boundary of *the oriented toric saddle $N^2 \times T^{n-1}$* ascends, and according to the topology of the manifold $N^2 \times T^{n-1}$ (see above and §1) *splits into two tori* T_1^n and T_2^n. The notation $T^n \to 2T^n$.

(4) A torus T^n, which is one of the boundaries of the manifold A_α^{n+1}, where $\alpha \neq 0$ (for instance, the boundary of the interior thin full torus, Fig. 24), goes "up" the manifold A_α, and in the middle of it is modified into one torus—the upper boundary of the manifold A_α. The notation: $T^n \xrightarrow{\alpha} T^n$. All such modifications are parametrized by nonzero elements

$$\alpha \in \mathbb{Z}_2^{n-1} = H_1(T^{n-1}, \mathbb{Z}_2).$$

(5) Realize a torus T^n as the boundary of a manifold K_p^{n+1}. Deforming the torus inside K_p^{n+1} along the projection p, we finally doubly cover the manifold K^n with the torus T^n. After this the torus "disappears." The notation: $T^n \to K^n \to 0$.

Now let us formulate the final definition of the surgery on general position of the Liouville torus. Fix the values of the last $n-1$ integrals $f_2, \ldots, f_n$ and examine the obtained $(n+1)$-dimensional level surface X^{n+1}. Restricting the first integral (energy) $f_1 = H$ to this surface, we obtain a smooth function f on the manifold X^{n+1}.

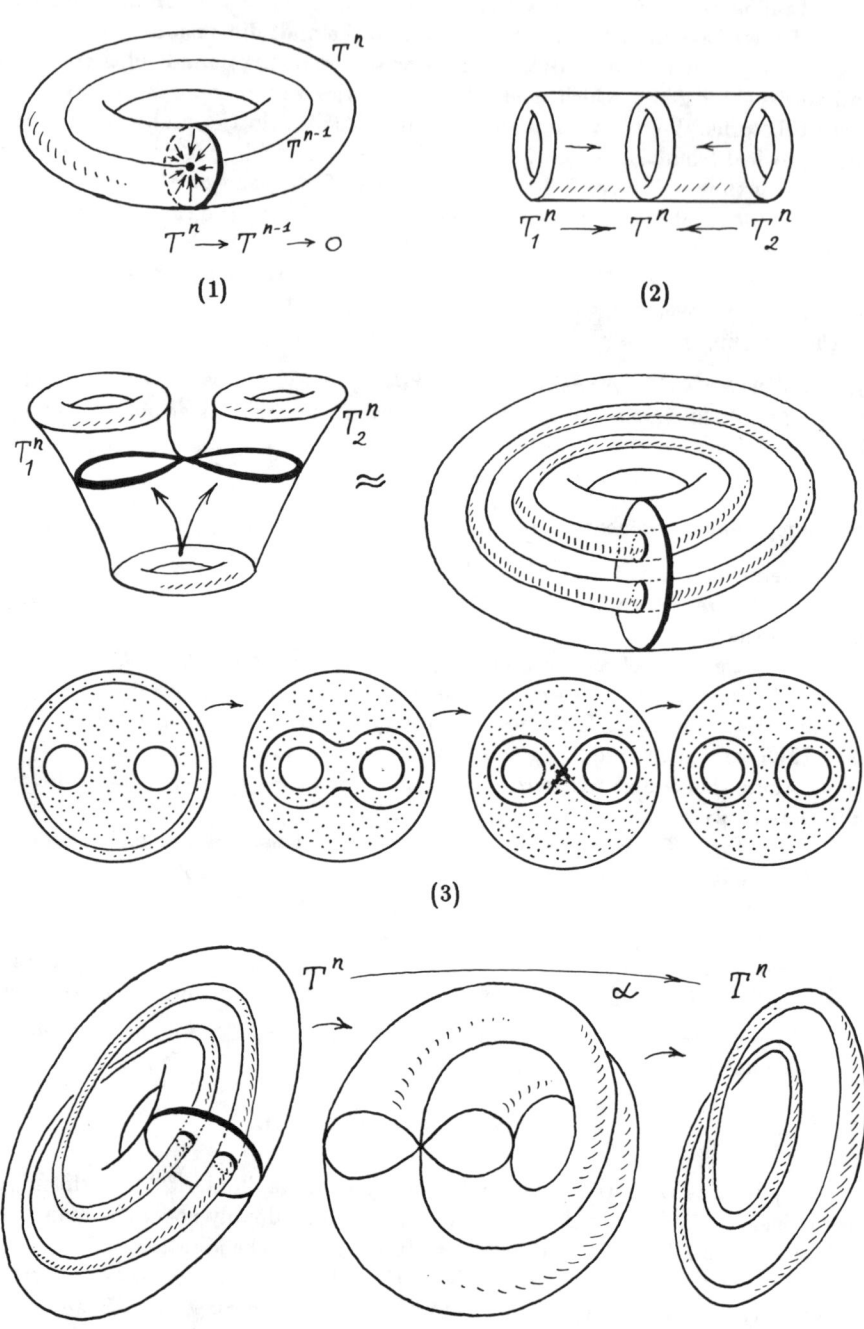

Figure 41

DEFINITION 2.2.1: We will say that the surgery on Liouville tori which form a nonsingular fibre B_a is *the surgery on general position* if in the neighbourhood of a modified torus T^n the surface X^{n+1} is a compact and nonsingular submanifold, and the restriction of the energy $f_1 = H$ to X^{n+1} is (in the same neighbourhood) a Bott function (in the sense of §1).

2.2 The Classification Theorem for Liouville Torus Surgery

THEOREM 2.2.1 (FOMENKO. THE THEOREM ON CLASSIFICATION OF LIOUVILLE TORUS SURGERY). *(1) if* $\dim \Sigma < n - 1$, *then all non-singular fibres* B_a *are diffeomorphic to one another. (2) Let* $\dim \Sigma = n - 1$. *Let a nondegenerate Liouville torus* T^n, *carried away by variation of the value of the energy integral* $f_1 = H$, *move along a common nonsingular level surface* X^{n+1} *of the last integrals* $f_2, \ldots, f_n$. *This is equivalent to the fact that a point* $a = F(T^n) \in \mathbf{R}^n$ *moves along a smooth line segment* γ *towards a bifurcation diagram* Σ. *Let at a certain time moment the torus* T^n *undergo a topological surgery, i.e. get onto the critical energy level. This happens if and only if the torus* T^n *meets on its way the critical points* N *of momentum mapping* $F : M^{2n} \to \mathbf{R}^n$ *(that is, at a point* c *the path* γ *punctures transversally an* $(n-1)$-*dimensional stratum of the diagram* Σ). *Suppose that this is a surgery on general position. Then all possible types of Liouville torus surgery are exhausted by compositions of the above-mentioned types of canonical surgery 1,2,3,4,5. In reality, independent from the topological point of view are only the first three of them, while types 4 and 5 are their compositions. The surgery of the first three types will be called topological, and that of types 1,2,3,4,5 Hamiltonian. In case 1 (the surgery* $T^n \to T^{n-1} \to 0$), *as the energy* H *increases, the torus* T^n *at first transforms into a degenerate torus* T^{n-1}, *and then disappears altogether from the constant-energy surface* $H = $ *const (limit degeneracy). In this case the torus* T^n *"does not pierce" the critical energy level, "is decelerated" and disappears. In case 2 (the modification* $2T^n \to T^n \to 0$), *as the energy* H *increases, the two* T^n_1 *and* T^n_2 *at first flow into one torus* T^n *and then disappear from the surface* $H = $ *const. Here the torus* T^n *"does not pierce" the critical energy level either. The torus* T^n, *may, however, be regarded here as reflected from the critical energy level. In case 3 (the modification* $T^n \to 2T^n$), *as the energy* H *increases, the torus* T^n *"pierces" the critical energy level and splits into two tori* T^n_1 *and* T^n_2 *on the surface* $H = $ *const. In case 4 (the modification* $T^n \xrightarrow{\alpha} T^n$), *as the energy* H *increases, the torus* T^n *also "pierces" the critical energy level and again transforms into the torus* T^n *(a nontrivial modification of double winding, see above). In case 5 (the modification* $T^n \to K^n \to 0$) *the torus* T^n *doubly covers a nonorientable manifold* K^n, *after which it disappears from the surface* $H = $ *const. Changing the direction of motion of the Liouville torus,. we arrive at five inverse processes: (1) formation of a torus* T^n *from a torus* T^{n-1}, *(2) trivial formation of two tori* T^n_1 *and* T^n_2 *from one torus* T^n, *(3) nontrivial confluence of two tori* T^n_1 *and* T^n_2 *into one torus* T^n, *(4) non-trivial modification of a torus* T^n *into a torus* T^n *(Fig. 41): (5)* $K^n \to T^n$.

Some of the above-said types of surgery have already been discovered in concrete important mechanical systems. See the papers by Kharlamov and Pogosyan [177] and [185]. For instance, this is the surgery on tori in the case of Kovalevskaya and in the case of Goryachev–Chaplygin [177]. The surgery on two- dimensional tori suggested in [177] and denoted there as

$$\emptyset \to S^1 \to T^2, \quad T^2 \to R \to 2T^2, \quad 2T^2 \to Q \to 2T^2, \quad T^2 \to P \to T^2$$

may be shown to be particular cases of the surgery which we described in Theorem 2.2.1. The first of these modifications is induced by a dissipative full torus and belongs to type 1. The second is induced by an oriented saddle and belongs to type 3. The third is induced by *two* oriented saddle, that is, splits into a composition $2T^2 \to T^2 \to 2T^2$ of two modifications of type 3. The fourth is induced by a nonoriented saddle A^3 and belongs to type 4.

As in the four-dimensional case, one could distinguish between orientable and nonorientable Hamiltonians H. We call a Hamiltonian *orientable* if all of its critical submanifolds (on X^{n+1}) are orientable, that is, there is not a single critical manifold K^n. Otherwise a Hamiltonian is called *nonorientable*.

But if surfaces X^{n+1} are considered up to two-sheeted coverings, the Hamiltonian H may always be regarded as orientable.

PROPOSITION 2.2.1. *If* $(U(X^{n+1}); \operatorname{sgrad} H; f_2, \ldots, f_n)$ *is an integrable Hamiltonian system with a Bott nonorientable Hamiltonian H on a surface X^{n+1}, then it may always be doubly covered by a Hamiltonian system*

$$(\widetilde{U}(\widetilde{X}^{n+1}); \operatorname{sgrad} \widetilde{H}; \widetilde{f}_2, \ldots, \widetilde{f}_n)$$

with an orientable Hamiltonian $\widetilde{H}$ on a covering $\widetilde{X}^{n+1}$. Here $\widetilde{U}(\widetilde{X}^{n+1})$ is a two-sheeted covering of the neighbourhood $U(X^{n+1})$ of the manifold X^{n+1}.

The proof is completely analogous to the proof of Claim 2.1.1.

Let v be an integrable system on M^{2n}. Fix the values of all last integrals $f_2, \ldots, f_n$ and suppose that the obtained $(n+1)$-dimensional surface X^{n+1} is compact and nonsingular, that is, the integrals $f_2, \ldots, f_n$ are independent on X^{n+1}. The surface X^{n+1} is an invariant submanifold of the system v. Varying the value of the energy H, we displace the Liouville torus T^n along the surface X^{n+1}. Sometimes there arise limit degeneracies, that is, a torus T^n is contracted to the torus T^{n-1} (see the surgery of type 1).

Kozlov has noted to the author that it is of importance that *limit degeneracies actually arise in concrete mechanical systems with dissipation*.

For $n = 1$, the modifications $T^1 \to 2T^1$ and $T^1 \to 0$ can be observed in the problem on the motion of a heavy point in a "two-hump" well. Due to the small energy dissipation, the motion of a point in phase space proceeds along one-dimensional tori (that is, circles) which slightly evolve and finally, meeting with a critical energy level, undergo surgery. If *small friction* is introduced into an integrable system, then to a first approximation one may think that the energy dissiplation is modelled by a decrease inthe energy value $f_1 = H$ and induces, therefore, a slow evolution (drift) of Liouville tori along the level surface X^{n+1}. Examine the motion of a ball in a well under the action of gravity [Fig. 42(2)]. This motion (in the absence of frictional force) may be completely described. If we introduce a small friction, we may assume the motion to be integrable, as before, on each sufficiently small time interval. But as the time interval increases, the friction becomes progressively pronounced. As a result, the ball will rise at increasingly small height [Fig. 42 (2)]. Finally, at a certain time moment the descending level will touch the saddle [Fig. 42 (2)] and the motion will change: the ball will find itself either in the left or in the right well. This is just what is called the surgery on a Liouville torus at the moment of intersecting the critical energy level. It is clear that the character of such surgery is completely determined by the topology of the level surface X^{n+1}.

The answer to the question, how the level surfaces X^{n+1} are organized is given by the following theorem.

THEOREM 2.2.2 (FOMENKO). *Let M^{2n} be a smooth symplectic manifold and let a system $v = \operatorname{sgrad} H$ be integrated by smooth independent commuting integrals $H = f_1, f_2, \ldots, f_n$. Let X^{n+1} be any fixed compact nonsingular common level surface of the last $n - 1$ integrals $f_2, \ldots, f_n$. Let the restriction of H on X^{n+1} be a Bott function. Then the surface X^{n+1} has the following form:*

$$X^{n+1} = m(D^2 \times T^{n-1}) + p(T^n \times D^1) + q(N^2 \times T^{n-1}) + sA_\alpha^{n+1} + rK_p^{n+1},$$

that is, it results from gluing the boundary tori (through some diffeomorphisms) of the following "elementary bricks": m dissipative full tori, p cylinders, q toric oriented saddles, s toric nonoriented saddles and r manifolds K^{n+1}. The number m of dissipative full tori is exactly equal to the number of such limit degeneracies of the system v on the surface X^{n+1} on which the energy H reaches its local minimum or maximum.

2.3 *Toric Handles. A Separatrix Diagram Is Always Glued to a Nonsingular Liouville Torus T^n Along a Nontrivial $(n-1)$-Dimensional Cycle T^{n-1}*

Now we proceed to the proof of Theorems 2.2.1 and 2.2.2. To this end, we have to formulate a new multidimensional Morse-type theory of integrable Hamiltonian systems. We will mainly follow the scheme of §1. We will not repeat the arguments already presented in §1, but only add some new points which should be proved specially because of system multidimensionality.

Let us introduce the following notation. Let $c \in \Sigma$ and a small neighbourhood $U(c) \cap \Sigma$ of the point c on Σ be a smooth $(n-1)$-dimensional submanifold. Let γ be a smooth path transversally puncturing Σ at the point c and joining two noncritical values a and b located on different sides of the hypersurface Σ. Consider a connected component B_c^0 of a singular fibre $B_c = F^{-1}(c)$. Let $T = B_c^0 \cap N$ and let X_0^{n+1} be a connected component of the fibre $F^{-1}U(c)$ lying between two close nonsingular fibres B_a and B_b. Since a and b are noncritical values, it follows that B_a and B_b are unions of Liouville tori. Let $B_a^0 = X_0^{n+1} \cap B_a$ and $B_b^0 = X_0^{n+1} \cap B_b$, that is, $\partial X_0^{n+1} = B_a^0 \cup B_b^0$. Denote by f the restriction of the first integral f_1 to the surface $X_0^{n+1} = X_0$ [Fig. 42(1)].

LEMMA 2.2.1. *A point $x \in X_0$ is critical for the function f if and only if the first integral f_1 in it depends (on M) on the last $n-1$ integrals $f_2, \ldots, f_n$.*

PROOF: Since X_0 is the common level surface of last integrals, their gradients form a basis in a plane normal to X_0 in M. The dependence of the function f_1 on the

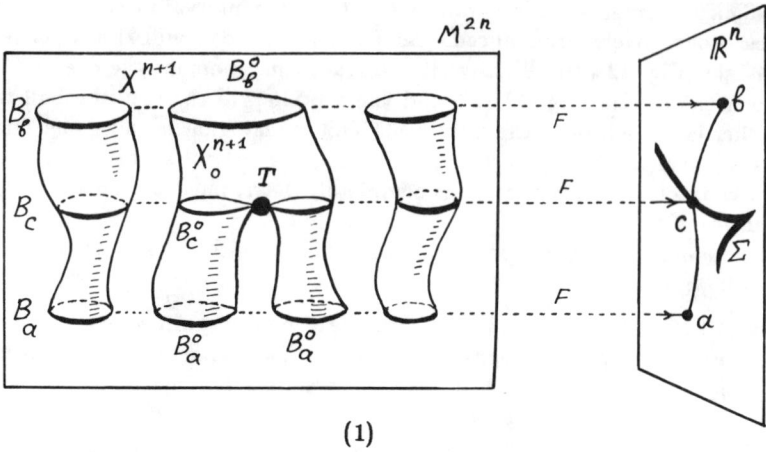

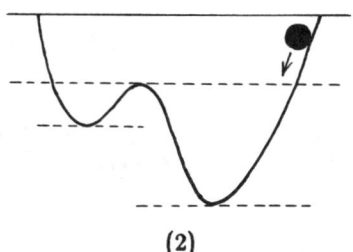

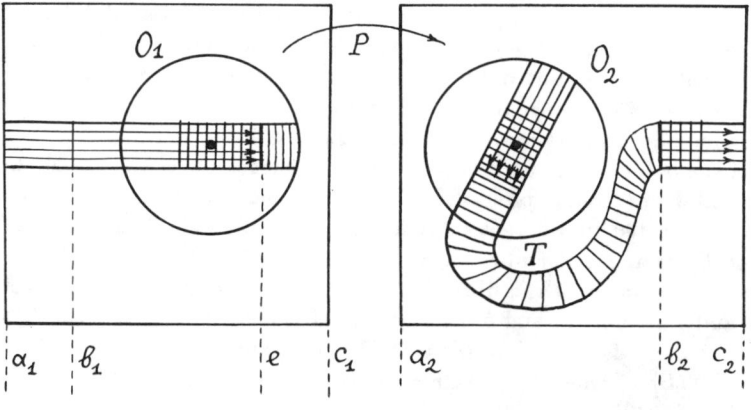

Figure 42

functions $f_2, \ldots, f_n$ is equivalent to the fact that at this point the vector grad f_1 is a linear combination of the gradients grad $f_i, 2 \leqslant i \leqslant n$. It is clear that grad f on X_0 is obtained by orthogonal projection of grad f_1 onto X_0. The lemma follows.

LEMMA 2.2.2. *The set T of critical points of the function f on X_0 is a disconnected union of a certain number of n-dimensional tori T^n, $(n-1)$-dimensional tori T^{n-1}, and n-dimensional nonorientable manifolds K_p^n.*

PROOF: If $T = B_c^0$, then T is a common (singular) level surface of all n integrals $f_1, \ldots, f_n$. Level surfaces R, close to this surface, are nonsingular compact Liouville tori. It is clear that R is the boundary of a tubular neighbourhood V^{n+1} of the submanifold T in X_0. If $\dim T = k$, then $R \approx T^n$ is fibred over T with a fibre S^{n-k}. This may happen only if $n - k$ is equal to zero or to unity, that is, if $\dim T = n$ or $\dim T = n - 1$. If T_0^n is a connected component of T, then T_0^n is equal to T^n or to K_p^n. If $B_c^0 \neq T$, the consideration becomes more complicated. Clearly, in this case $\dim T < \dim B_c^0 = n$, that is, $\dim T \leqslant n - 1$. From the conditions imposed on the integrals of the system it follows that on T the integrals $f_2, \ldots, f_n$ are independent (they are independent on the entire B_c^0). Consequently, on T_0 there exists $n - 1$ independent commuting vector fields sgrad $f_i, 2 \leqslant i \leqslant n$. As is known, from this it immediately follows that $T_0 \approx T^{n-1}$, and the result follows.

If a critical torus has the dimension n, then it is either the set of the local minimum or of the local maximum of the energy H. In this case, either two close nonsingular Liouville tori flow into one torus T^n or the torus T^n splits into two tori T^n. Let $P_-^n = P_-^n(T^{n-1})$ and $P_+^n = P_+^n(T^{n-1})$ be, respectively, in and out separatrix diagrams of the critical submanifold T^{n-1}.

DEFINITION 2.2.2: *The direct product $T^k \times D^\lambda \times D^{n+1-k-\lambda}$ will be called a toric handle of index λ and of degree of degeneracy k.* The part of a handle boundary: $(T^k \times S^{\lambda-1}) \times D^{n+1-k-\lambda}$ will be called the *foot of a handle*. The product $T^k \times S^{\lambda-1}$ will be called the *foot axis*.

Let us define the *operation of gluing a toric handle* to the boundary V^n of an $(n+1)$-dimensional manifold W^{n+1}. Let the boundary contain an embedded submanifold $T^k \times S^{\lambda-1}$. Suppose that its tubular neighbourhood is homeomorphic to the direct product $(T^k \times S^{\lambda-1}) \times D^{n+1-k-\lambda}$. One may remove this tubular neighbourhood whose boundary is homeomorphic to $T^k \times S^{\lambda-1} \times S^{n-k-\lambda}$. On the other hand, the boundary of the foot of the toric handle is also homeomorphic to the product $T^k \times S^{\lambda-1} \times S^{n-k-\lambda}$. Identifying this boundary with the boundary of the removed neighbourhood, we obtain a new $(n+1)$-dimensional manifold. Its boundary will be called a *toric surgery on the boundary V^n*. For our further purposes we may assume the smooth path $\gamma = \gamma(t) \subset \mathbb{R}^n$ to be modelled by a segment on the real axis $\mathbb{R}^1$ on which there lie three points: $a < c < b$, where c is the critical value, a and b are close to c. Let us put $C_a = F^{-1}(t \leqslant a), C_b = F^{-1}(t \leqslant b)$; then $C_a \subset C_b$. To say it differently, we may assume that $f : X_0^{n+1} \to \mathbb{R}^1$ and $C_a = (f \leqslant a), C_b = (f \leqslant b), B_c^0 = f^{-1}(c)$.

LEMMA 2.2.3. *Suppose that on a singular fibre B_c^0 there lies exactly one critical (saddle) torus T^{n-1}. (1) Let $P_-^n(T^{n-1})$ be orientable. Then C_b is obtained from C_a by gluing to the boundary B_a a toric handle of index 1 and of degree of degeneracy $n-1, C_b$ being homotopy equivalent to C_a, to which the manifold $T^{n-1} \times D^1$ is* glued along two non-intersecting tori $T_{1,a}^{n-1}$ and $T_{2,a}^{n-1}$. (2) Let $P_-^n(T^{n-1})$ be nonorientable. Then the set C_b is homotopy equivalent to the set C_a, to the boundary B_a^0 of which an n-dimensional manifold Y_α, which has a boundary T^{n-1} and is the fibration $Y_\alpha^n \xrightarrow{D^1} T^{n-1}$ corresponding to the nonzero element

$$\alpha = \mathbb{Z}_2^{n-1} = H_1(T^{n-1}, \mathbb{Z}_2),$$

is glued along the torus T_a^{n-1}.

The Lemma is proved by the scheme of the proof of Lemma 2.1.4 (see §1) but with some specifications due to system multidimensionality. These technical details are omitted.

LEMMA 2.2.4. *Let a torus T_a^{n-1} embedded in some nonsingular Liouville torus $T_a^n \subset B_a^0$ be either one of the feet of a toric handle (of index 1 and of degree of degeneracy $n-1$) or the boundary of a manifold Y_α^n (in case $P_-(T^{n-1})$ is nonorientable). Then this torus T_a^{n-1} always realized one of the generators in the homology group*

$$H_{n-1}(T_a^n, \mathbb{Z}) = \mathbb{Z}^{n-1}.$$

If both feet of the toric handle are glued to one and the same Liouville torus T_a^n, then the corresponding axes of these feet, that is, the tori $T_{1,a}^{n-1}$ and $T_{2,a}^{n-1}$ do not intersect, realize one and the same generator of the homology group

$$H_{n-1}(T_a^n, \mathbb{Z})$$

and are, therefore, isotopic within the torus T_a^n.

PROOF: Examine a critical saddle torus T^{n-1}. Lemma 2.2.2 implies that this torus is the orbit of action of an Abelian subgroup $\mathbb{R}^{n-1}$ embedded into a group $\mathbb{R}^n$ generated by fields sgrad f_i, $1 \leqslant i \leqslant n$. The basis in the subgroup $\mathbb{R}^{n-1}$ is formed by fields sgrad f_i, $2 \leqslant i \leqslant n$. Fix this subgroup. Since the action of the group $\mathbb{R}^n$ (and $\mathbb{R}^{n-1}$) is defined on the entire M^{2n}, we may examine the orbits of the group $\mathbb{R}^{n-1}$ close to the orbit T^{n-1}. Consider a nonsingular Liouville torus T_a^n which is rather close to the fibre B_c^0 and on which the separatrix diagram P_-^n cuts out some torus T_a^{n-1}. This torus is not, of course, the orbit of action of the group $\mathbb{R}^{n-1}$ on the torus T_a^n. But, as we will show the torus T_a^{n-1}, may be approximated by a certain orbit of action of the group $\mathbb{R}^{n-1}$. To this end, consider the element

$$\alpha \in H_{n-1}(T^{n-1}, \mathbb{Z}_2).$$

From Lemma 2.2.3 we know that the torus T_a^{n-1} is one of the components of the boundary of the manifold Y_α^n glued to the torus T_a^n. If $\alpha \neq 0$, then $\partial Y_\alpha^n = T_a^{n-1}$; if $\alpha = 0$, then

$$\partial Y_\alpha^n = \partial(T^{n-1} \times D^1) = T_{1,a}^{n-1} \cup T_{2,a}^{n-1} \quad \text{and} \quad T_{1,a}^{n-1} = T_a^{n-1}.$$

Setting the element α determines a certain number k of generators in the critical torus T^{n-1}, going round which a normal segment of the separatrix diagram P_-

changes its orientation. Let us single out these generators. In the orientable case $k = 0$ because $\alpha = 0$. Since the torus T^{n-1} is the orbit of action of the group $\mathbf{R}^{n-1}$, then changing the generators in the group $\mathbf{R}^{n-1}$ (if necessary), one may always assume that in the nonorientable case $(k \geqslant 1)$ among the fields sgrad $f_i, 2 \leqslant i \leqslant n$, there exists exactly k fields sgrad $f_2, \ldots,$ sgrad f_{k+1} such that a single detour along the orbits of the point $x \in T^{n-1}$, which are generated by the corresponding one-dimensional subgroups $\mathbf{R}_2^1, \ldots, \mathbf{R}_{k+1}^1$, changes orientation of the normal segment of the separatrix diagram. First consider the orientable case, where $k = 0$. Then in the subgroup $\mathbf{R}^{n-1}$ one may single out an $(n-1)$-dimensional parallelepiped Π, i.e. a fundamental domain of action of the group $\mathbf{R}^{n-1}$ on the torus T^{n-1}, this parallelepiped Π covers the whole torus, that is, the torus T^{n-1} is obtained through identification of opposite sides of this parallelepiped. Since the parallelepiped Π consists of transformations on M, we may consider the orbit of this parallelepiped when it is acting upon a certain point

$$h \in T_{1,a}^{n-1} \subset T_a^n.$$

This orbit will not, of course, be a closed $(n-1)$-dimensional torus in T_a^n. But since the point $h \in T_{1,a}^{n-1}$ is close to the point $x \in T^{n-1}$, one may assume that the orbit is close to the point $x \in T^{n-1}$, one may assume that the orbit $\Pi(h)$ is an "almost-torus", that is, each of the generators of the parallelepiped Π becomes a segment whose ends are close on the torus T_a^n (i.e. we obtain an "almost- circle"). We will choose on the torus T^n some coordinates $\varphi_1, \ldots, \varphi_n$, accord to the Liouville theorem. We make use of the fact that points of the parallelepiped Π are represented by symplectic transformations. Then in these coordinates the "almost-torus" $\Pi(h)$ is *a linear, completely geodesic submanifold*, maybe, with a non-empty boundary. Representing the torus T_a^n (in these coordinates) in the form of a standard cube with opposite sides identified, we obtain in it a plane Π', whose intersections with the opposite sides are $(n-2)$-dimensional subspaces which appear to be close after identification of the sides. It is clear that the plane Π' may be slightly turned so that it becomes (after factorization of the cube) a certain $(n-1)$-dimensional linear, completely geodesic torus T_*^{n-1} in the torus T_a^n. It is clear that the torus T_*^{n-1} is close to the "almost-torus" $\Pi(h)$ and at the same time close to the torus $T_{1,a}^{n-1}$. This implies that these tori are isotopic.

Thus, we have proved that there exists a small isotopy of the torus $T_{1,a}^{n-1}$ in the torus T_a^n, which carries it into a linear torus. But in this case, the torus $T_{1,a}^{n-1}$ realizes a generator in the group $H_{n-1}(T_a^n, \mathbf{Z})$, as required. Thus, this implies the lemma in the orientable case. Note that stirring the plane Π slightly, we have obtained a new plane Π_* whose generators may already include the generator sgrad f_1 which was excluded from the plane Π. It is clear that

$$T_*^{n-1} = \Pi_*(h).$$

Now consider the nonorientable case.

The reasoning will be more delicate. The point is that in this case the parallelepiped Π itself is insufficient. Indeed, the definition of a nonorientable separatrix diagram implies that the orbits

$$(\Pi \cap \mathbf{R}_2^1)h, \ldots, (\Pi \cap \mathbf{R}_{k+1}^1)h$$

of generators $R_2^1, \ldots, R_{k+1}^1$ (corresponding to the fields sgrad $f_1, 2 \leqslant i \leqslant k+1$) are not "almost closed" trajectories on the torus T^n. We will denote the corresponding edges of the parallelepiped Π as Π_i, that is,

$$\Pi_i = \Pi \cap R_i^1, 2 \leqslant i \leqslant k+1.$$

When Π_i acts on the point h, the point has time only to make half the revolution on the torus T_a^n. In order that the point could make almost the whole revolution, one should once again act on it by an edge of the parallelepiped Π_i. In other words, for the point h to make almost the whole revolution on the torus T_a^n, it should be subjected to the action of $2\Pi_i$, that is, the corresponding side of the parallelepiped Π should be doubled., So, we arrive at the following scheme.

One should double all the sides of the parallelepiped $\Pi_2, \ldots, \Pi_{k+1}$. As a result, one obtains a new parallelepiped $\widetilde{\Pi}$ extended in k directions. Now act with this parallelepiped $\widetilde{\Pi}$ upon the point h. As a result, obtain a certain orbit $\widetilde{\Pi}(h)$. It is clear that now this orbit is represented (in action-angle variables on a Liouville torus) by a linear plane which is "almost-closed" after factorization of the cube to the torus. Further arguments repeat those of the orientable case. This implies the lemma.

The other constructions are made by analogy with the scheme of §1, and we omit them here. Theorems 2.2.1 and 2.2.2 are proved.

There arises a natural question:how is the topological surgery of Liouville tori organized when the path γ punctures the bifurcation diagram Σ in its *singular* points which do not lie on the interior of its $(n-1)$-dimensional strata? To put it differently, what is the surgery of special types (not on general position)?

A similar Morse-type theory holds for Hamiltonian systems admitting noncommutative integration which we described in [188], [143]. All the assertions proved in the present chapter hold true, with the only exception that n-dimensional Liouville tori are everywhere replaced by r-dimensional tori, where $r \leqslant n$.

2.4 Any Composition of Elementary Bifurcations (of Three Types) of Liouville Tori Is Realized for a Certain Integrable System On an Appropriate Symplectic Manifold

Above we have displayed three elementary Liouville torus modifications (see the classification theorem). It remains unclear whether any of their compositions is realized in a certain symplectic manifold on which an integrable system is given. It turns out that such a theorem does exist, that is, one can construct a symplectic manifold M^{2n} and such an appropriate set of commuting functions that with an appropriate choice of the path γ in R^n the pre-image of this way, that is, a manifold X^{n+1} is generated by a beforehand given composition of modifications of types I, II, III.

THEOREM 2.2.3 (BRAILOV, FOMENKO). *Let X^{n+1} be a smooth compact closed orientable manifold obtained by gluing an arbitrary number of elementary manifolds of types I, II, III (that is, full tori, cylinders, and trousers) through any diffeomorphisms of their boundary tori T^n. Then there always exists a smooth compact symplectic manifold M^{2n} with a boundary diffeomorphic to a disconnected union* of a certain number of manifolds $S^{n-1} \times T^n$ and such a complete involutive set of smooth functions $f_1, \ldots, f_n$ on M^{2n} that

$$X^{n+1} = \{x \in M^{2n}; \ f_2(x) = \cdots = f_n(x) = 0\}$$

REMARK: There are obviously some topological obstacles to realization of the manifold X^{n+1} in a compact *closed* symplectic manifold M^{2n}.

From Theorem 2.2.3 we immediately obtain an important corollary, namely, the proof of Claim 2.1.2 concerning the possibility to realize any three-dimensional manifold of class (Q) (that is, gluing an arbitrary number of elementary manifolds of types I, II, III) in the form of an isoenergy surface of some integrable Hamiltonian system on an appropriate symplectic manifold M^4. In other words, from Theorem 2.2.3 there immediately follows the equality:$(Q) = (H)$.

PROOF OF THEOREM 2.2.3: The required manifold M^{2n} is constructed of symplectic manifolds corresponding to those "elementary bricks" of which the manifold X^{n+1} is glued. At the first stage of the proof we therefore construct examples of integrable systems in which the surgery of types I, II, III is realized.

Let ω_n be a symplectic structure on a torus

$$T^{2n} = \{(\varphi_1, \ldots, \varphi_{2n}) : \text{mod } 2\pi\}, \omega_n = \sum_{i=1}^{n} d\varphi_{2i-1} \wedge d\varphi_{2i}.$$

On the torus T^2 one can obviously construct smooth functions $f^{[s]}(s = 1, 2, 3)$ and submanifolds $M_s \subset T^2$, so that N_1 is a point of nondegenerate maximum for $f^{[1]}$, N_2 is a nondegenerate saddle point for $f^{[2]}$, N_3 is a nondegenerate critical circle for $f^{[3]}$. One may take, for instance

$$f^{[1]} = f^{[2]} = \cos(\varphi_1 + 2\varphi_2),$$
$$N_1 = \{(0,0)\}, \qquad N_2 = \{(\pi, 0)\},$$
$$f^{[3]} = \cos(\varphi_1), \qquad N_3 = \{(0, \varphi_2) \mod 2\pi\}.$$

If $c_s = f^{[s]}(N_s)$ are critical values, then

$$c_1 = c_3 = 1, \qquad c_2 = -1.$$

EXAMPLE $s(s = 1, 2, 3)$: Let the manifold $M^{2n} = T^{2n}$, and let ω_n be the symplectic structure on this manifold. For $i = 2, \ldots, n$ we have $f_i(\varphi_1, \ldots, \varphi_{2n}) = \sin(\varphi_{2i+1})$. For $i = 1$ we have

$$f_1(\varphi_1, \ldots, \varphi_{2n}) = f^{[s]}(\varphi_1, \varphi_2).$$

Then the Hamiltonian system $v = \text{sgrad} f_1$ is completely integrable on M^{2n} with a complete involutive set of integrals f_i and $N = N_s \times T^{2n-2}$ is a nondegenerate critical submanifold of the function

$$f_1|_{X^{n+1}}, \text{ where } X^{n+1} = \{f_2 = \cdots = f_n = 0\}.$$

The critical value $f_1(N) = c_s$.

Let the manifold $X^{n+1} = \cup_{j=1}^m X_j$ be glued of "bricks" X_j and let $s(j)$ be the type of X_j. If a manifold X_j of type II is glued with various manifolds X_{j_1} and X_{j_2} by diffeomorphisms d_1 and d_2, then from the decomposition of X^{n+1} one can remove X_j by gluing X_{j_1} and X_{j_2} via diffeomorphism $d_2 \cdot d_1^{-1}$. The described process will be called contraction of the decomposition of the manifold X^{n+1} into "elementary bricks." The inverse process, under which in the decomposition of X^{n+1} there appears a new additional "brick" of type II, will be called it an insertion. Neither of these processes changes the manifold X^{n+1} up to a diffeomorphism. Next, it is obvious that by an appropriate succession of contractions and insertions, from the initial decomposition of the manifold X^{n+1} one can obtain a decomposition for which the following conditions will be satisfied: (1) for any "brick" X_j of type II, the numbers of the "bricks" glued to X_j cannot simultaneously be either all greater or all less then j. So, let

$$X^{n+1} = \cup_{j=1}^m X_j$$

be a decomposition satisfying these two conditions. For each $j = 1,\ldots,m$, the copies of the objects f_i, N, M^{2n} from example $s = s(j)$ will be denoted respectively by f_{ij}, N_j, M_j^{2n}. Using an appropriate linear change

$$f_{1,j} \to f'_{1,j} = a_j f_{1,j} + b_j$$

one can succeed in making the momentum mapping

$$F_j(x) = (f_{1j}(x), \ldots, f_{nj}(x))$$

satisfy the following conditions: (1) for each $j = 1,\ldots,m$, a point $y \in \mathbb{R}^n$, when moving along the first coordinate axis from 0 to $m+1$, intersects the bifurcation set Σ_j of the mapping F_j only once, at the point

$$\xi_j = (j, 0, \ldots, 0);$$

(2) if $s(j) = $ I and if a "brick" X_j is glued with a "brick" X_{j_1}, then providing the above-mentioned motion takes place, at the point $y = \xi_j$ a torus $F_j^{-1}(y)$ is "produced" in case $j_1 > j$ and is "annihilated" in case $j_1 < j$; (3) if $s(j) = $ II and X_j is glued with X_{j_1} and X_{j_2}, then at the point $y = \xi_j$, two tori $F_j^{-1}(y)$ are "produced" in case $j_1, j_2 > j$, and two tori are "annihilated" in case $j_1, j_2 < j$; (4) if $s(j) = $ III and if X_j is glued with $X_{j_1}, X_{j_2}, X_{j_3}$, then at the point $y = \xi_j$ one torus is modified into two tori in case $j_1 < j < j_2, j_3$ and two tori are modified into one in case $j_1, j_2 < j < j_3$.

For any $r > 0, a < b$ we will define a cylinder

$$C_r(a,b) = \{y \in \mathbb{R}^n : \sum_{j=2}^n y_i^2 < r^2, \; a < y_1 < b\}.$$

It is readily seen that

$$\Sigma_j \cup C_{\frac{1}{2}}(0, m+1) = \Pi_j \cup C_{\frac{1}{2}}(0, m+1),$$

where
$$\Pi_j = \{y \in \mathbf{R}^n : y_1 = j\}$$
is a hyperplane. A manifold M^{2n} is obtained from the manifolds
$$M'_j = M_j \setminus F_j^{-1}\left(C_{1/2}(0, j - \frac{1}{2}) \cup C_{1/2}(j + \frac{1}{3}, m+1)\right)$$
by gluing the tubes that join these manifolds. The possibility of gluing the tubes follows from the lemma.

LEMMA 2.2.5. *Let the numbers*
$$a_1 < b_1 < c_1 < a_2 < b_2 < c_2, \qquad r > 0$$
be given. Suppose, for each $j = 1, 2$ there exists a symplectic manifold M_j^{2n} and on this manifold there exists such a complete involutive set of smooth functions $f_{ij}(i = 1, \ldots, n)$ that the momentum mapping F_j is fibration with a fibre T^n over the cylinder $C_r(a_j, c_j)$. Then for a sufficiently small $\varepsilon > 0$ there exists a symplectic manifold Z which contains as open submanifolds the pre-images
$$V_1 = F_1^{-1}(C_\varepsilon(a_1, b_1)), \qquad V_2 = F_2^{-1}(C_\varepsilon(b_2, c_2))$$
and such a complete involutive set of smooth functions f_i on v that the corresponding momentum mapping F is fibration with a fibre T^n over $C_\varepsilon(a_1, c_2)$ and $F|_{V_j} = F_j|_{V_j}$ for $j = 1, 2$.

PROOF: Let $b_1 < d_1 < c_1$ and
$$a_2 < d_2 < c_2, \qquad D_j = (d_j, 0, \ldots, 0), \qquad j = 1, 2$$
Let us choose in the neighbourhood $\bar{O}_j$ of the torus $T_j^n = F_j^{-1}(D_j)$ some variables I_{ij}, φ_{ij} of the action-angle type. Diminishing $\bar{O}_j$ if necessary, we will assume the variables I_{ij} to set the fibration $I_j : \bar{O}_j \to B_\delta$ over a small ball
$$B_\delta = \{y \in \mathbf{R}^n : |y| < \delta\}.$$
Thus the variables I_{ij}, φ_{ij} set a diffeomorphism
$$I_j \times \varphi_j : \bar{O}_j \to B_\delta \times T^n$$
Suppose that
$$\bar{P} = (I_2 \times \varphi_2)^{-1}(I_1 \times \varphi_1).$$
It is obvious that $\bar{P} : \bar{O}_1 \to \bar{O}_2$ is a symplectic diffeomorphism. Since the variables I_{ij} are expressed through f_{ij} and are independent, there exists a diffeomorphism
$$J_j : B_\delta \to O_j = F_j(\bar{O}_j)$$

such that $F_j = J_j \cdot I_j$. Let, finally, $P = J_2 \cdot J_1^{-1}$. In the end, we have a commutative diagram:

$$
\begin{array}{ccc}
& B_\delta \times T^n & \\
I_1 \times \varphi_1 \nearrow & \xrightarrow{\bar{P}} & \nwarrow I_2 \times \varphi_2 \\
\bar{O}_1 & & \bar{O}_2 \\
& I_1 \qquad I_2 & \\
F_1 & B_\delta & F_2 \\
& J_1 \searrow \qquad \swarrow J_2 & \\
O_1 & & O_2 \\
& P &
\end{array}
$$

This diagram may be constructed in more than one way because the introduction of action-angle type variables depends on the choice of the basis of cycles on the torus in the neighbourhood of which these variables are introduced. Now let us discuss the character of this ambiguity. Let

$$(h_i), \qquad (h_i') = \Lambda(h_i)$$

be two distinct bases in the group

$$H_1(F_2^{-1}(D_2), \mathbb{Z}) \approx \mathbb{Z}^n.$$

Then $h_i' = \sum_{k=1}^n \Lambda_{ik} h_k$ for an appropriate matrix $\Lambda \in GL(n, \mathbb{Z})$. Since action type variables I_i are specified by way of integrating a fixed 1-form over the cycles h_i, it follows that the change $h_i \to h_i'$ leads to a linear change

$$I_i \to I_i' = \sum_{k=1}^n \Lambda_{ik} I_k.$$

A change of angular variables is made by means of the inverse matrix

$$\varphi_i \to \varphi_i' = \sum_{k=1}^n (\Lambda^{-1})_{ik} \varphi_k.$$

Making, if necessary, the change $h_1 \to h_1' = -h_1$, we may assume that the Jacobian of the diffeomorphism P is greater than zero.

Let $C_\varepsilon(a_1, e)$ be such a cylinder that its right end-face lies within O_1 and the point D_1 within $C_\varepsilon(a, e)$ (Fig. 42(3)). Let us join the cylinder $C_\varepsilon(b_2, c_2)$ by a tube T with the image of the cylinder $C_\varepsilon(a_1, e)$ via the diffeomorphism P in such a way that the tube T be a smooth continuation of both the cylinders and have open non-empty intersections with them. Let

$$A_1 = C_\varepsilon(a_1, e), \qquad A_2 = T \cup C_\varepsilon(b_2, c_2), \qquad Z_j = F_j^{-1}(A_j), \qquad j = 1, 2$$

The diffeomorphism $\bar{P}$ is symplectic, and therefore, gluing Z_1 and Z_2 by means of $\bar{P}$ we obtain a symplectic manifold

$$Z = Z_1 \cup_{\bar{P}} Z_2.$$

For $a = A_1 \cup_P A_2$, with an account of positiveness of the Jacobian of the diffeomorphism P, there exists a diffeomorphism

$$L : A \to C_\epsilon(a_1, c_2) \text{ identical on } C_\epsilon(a_1, b_1) \text{ and } C_\epsilon(b_2, c_2).$$

The mapping $F(x) = L(F_j(x)), x \in Z_j$, is the required fibration.

Let us go back to the proof of the theorem. For each $j = 1, \ldots, m$ the restriction of the momentum mapping F_j to M'_j will be denoted as F'_j. For any $a < b$, let $[a, b]_1$ be a segment of the first coordinate axis in $\mathbf{R}^n$ from a to b. One can readily see that the pre-image

$$(F'_j)^{-1}([0, m+1]_1) = (F'_j)^{-1} \times ([j - \tfrac{1}{3}, j + \tfrac{1}{3}]_1)$$

is diffeomorphic to the "brick" X_j. Next, these manifolds are identified. Each torus T_{k_1}, which restricts X_j, is obviously a connected component of the set

$$(F'_j)^{-1}(j - \tfrac{1}{3}, 0, \ldots, 0) \cup (F'_j)^{-1}(j + \tfrac{1}{3}, \ldots, 0).$$

for the "bricks" $X_{j_1}, X_{j_2}, j_1 < j_2$, glued by the tori $T_{k_1} \subset \partial X_{j_1}, T_{k_2} \subset X_{j_2}$, we set

$$\alpha = (X_{j_1}, T_{k_1}, X_{j_2}, T_{k_2}).$$

Using Lemma 2.2.5 for any such set α one can construct a symplectic manifold Z_α which has non-empty (of dimension $2n$) intersections of V_{k_1} and V_{k_2} with M_{j_1} and M_{j_2}, respectively, and also has the momentum mapping $F_\alpha : Z_\alpha \to \mathbf{R}^n$ which extends the mappings $F_{j_1}|_{V_{k_1}}$ and $F_{j_2}|_{V_{k_2}}$. In more detail: let V_{k_1} be a connected component of the pre-image

$$(F'_{j_1})^{-1}\left(\bar{C}_\epsilon(j_1 + \tfrac{1}{5}, j_1 + \tfrac{1}{3})\right)$$

which contains a torus

$$T_{k_1} \subset F_{j_1}^{-1}(j_1 + \tfrac{1}{3}, 0, \ldots 0);$$

let V_{k_2} be a connected component of the pre-image

$$F_{j_1}^{-1}(\bar{C}_\epsilon(j_2 - \tfrac{1}{3}, j_2 - \tfrac{1}{5})),$$

which contains a torus

$$T_{k_2} \subset F_{j_2}^{-1}(j_2 - \tfrac{1}{3}, 0, \ldots, 0).$$

The bar denotes here a topological closure in $\mathbf{R}^n$. From Lemma 2.2.5 it follows that there exists such a symplectic manifold Z_α and such a momentum mapping

$$F_\alpha : Z_\alpha \to \bar{C}_\epsilon(j_1 + \tfrac{1}{5}, j_2 - \tfrac{1}{5})$$

that for each $t = 1, 2$ the intersection
$$M'_{j_t} \cap Z_\alpha = V_{k_t}, \quad F_\alpha|_{V_{k_t}} = F'_{j_t}|_{V_{k_t}}$$
and the mapping F_α is a fibration with a fibre T^n. Let
$$F''_\alpha : M_{j_1} \cup Z_\alpha \cup M'_{j_2} \to \mathbf{R}^n$$
be a mapping coinciding with F'_{j_1} on M'_{j_1}, with F_α on Z_α, and with F'_{j_2} on M'_{j_2}. Then the pre-image
$$(F''_\alpha)^{-1}([0, m+1]_1)$$
is diffeomorphic to the gluing $X_{j_1} \cup_\Theta X_{j_2}$ of the "bricks" X_{j_i} and X_{j_2} via diffeomorphism of boundary tori $\theta : T_{k_1} \to T_{k_2}$. It is possible, however, that the diffeomorphism θ is not isotopic to the diffeomorphism θ_α, by means of which the tori T_{k_1} and T_{k_2} are identified in the manifold X^{n+1}. In this case, when constructing the manifold Z_α, one should choose the basis cycles
$$h_i \text{ in } H_1(T_{k_2}, \mathbf{Z}), \quad h_i \to h'_i = \sum_{t=1}^{n} \Lambda_{it} h_t, \quad \Lambda \in \mathrm{SL}(n, \mathbf{Z})$$
in such a way that the diffeomorphism $\theta' = \Lambda^{-1} \cdot \theta$ become isotopic to θ_α. Thus we may assume each manifold
$$(F''_\alpha)^{-1} \times ([0, m+1]_1)$$
to be diffeomorphic to the union of X_{j_1} and X_{j_2} in X^{n+1}. For the manifold
$$M' = (\cup_j M'_j) \cup (\cup_\alpha Z_\alpha)$$
and for the mapping F which coincides by definition with F'_j on M'_j and with F_α on Z_α, we therefore have
$$F^{-1}([0, m+1]_1) \approx X^{n+1}.$$
Unfortunately, M' is a "manifold with a piecewise smooth boundary", and for this reason it does not meet the requirements of the theorem. We may smoothly "cut" the angles of the manifold M' and obtain a manifold $M \subset M'$ which differs arbitrarily little from M' and has smooth boundary ∂M. The set W, which we have cut out, should be taken in the form
$$W = M' \setminus M = F^{-1}(Y), \quad Y \subset F(M').$$
Then the boundary ∂M trivially fibres over $F(\partial M)$ with a fibre T^n. The image $F((\partial M)_\alpha)$ of each connectedness component $(\partial M)_\alpha$ of the boundary ∂M is a connected sum of two $(n-1)$-dimensional spheres and is therefore diffeomorphic to S^{n-1}. Consequently, the boundary ∂M is diffeomorphic to the disconnected sum of r copies of the manifold $S^{n-1} \times T^n$, where r is the number of pairs of glued tori in a given decomposition of the manifold X^{n+1}.

2.5 Classification of Nonorientable Critical Submanifolds of Bott Integrals

We have discovered nonorientable manifolds K_p^n, which are minima or maxima of a Bott integral f on a surface X^{n+1}. In the present subsection we give a complete classification of such manifolds.

For each p we denote by G_p (for $p = 0, 1$) a group of transformations of a torus $T^n = \mathbf{R}^n/\mathbf{Z}^n$ generated by the involution

$$R_p(a) = \begin{cases} (-a_1, a_2 + \frac{1}{2}, a_3, \ldots, a_n), & p = 0 \\ (a_2, a_1, a_3 + \frac{1}{2}, a_4, \ldots, a_n), & p = 1, \end{cases}$$

where

$$a = (a_1, \ldots, a_n) \in \mathbf{R}^n/\mathbf{Z}^n.$$

It is assumed here that $n \geqslant 2$ for $p = 0$ and $n \geqslant 3$ for $p = 1$. The group G_p acts on the torus T^n without motionless points, and consequently the factor set

$$K_p^n = T^n/G_p$$

is a smooth manifold. The transformation R_p changes orientation, and therefore the manifold K_p^n is nonorientable. The manifolds K_0^n and K_1^n are not homeomorphic because they have different homology groups, namely,

$$H_1(K_0^n, \mathbf{Z}) = \mathbf{Z}^{n-1} \oplus \mathbf{Z}_2 \text{ and } H_1(K_1^n, \mathbf{Z}) = \mathbf{Z}^{n-1}.$$

From the definition of K_p^n, it immediately follows that

$$K_0^n = K_0^2 \times T^{n-2}, \qquad K_1^n = K_1^3 \times T^{n-3}.$$

Here K_0^2 is a usual Klein bottle and K_1^3 is its natural generalization.

THEOREM 2.2.4 (BRAILOV, FOMENKO). *Let $f_1, \ldots f_n$ be a complete involutive set of smooth functions on a symplectic manifold M^{2n} and let $F : M^{2n} \to \mathbf{R}^n$ be the corresponding momentum mapping. Let*

$$X^{n+1} = \{f_i(x) = \xi_i, i = 1, \ldots, n-1\}$$

be an integral submanifold of the momentum mapping. Suppose that the restriction $f_n|_{X^{n+1}}$ is a Bott function and that

$$X_0 = \{f_i(x) = \xi_i, \ i = 1, \ldots, n\}$$

is a nonorientable manifold of the minimum function $f_n|_{X^{n+1}}$. Then X_0 is diffeomorphic either to $K_0^n (n \geqslant 2)$ or to $K_1^n (n \geqslant 3)$.

PROOF: On a manifold X_0, vector fields $V_i = \operatorname{sgrad} f_i (I = 1, \ldots, n-1)$ are defined globally and a vector field

$$V_n = \operatorname{sgrad} \sqrt{f_n - \xi_n}$$

is defined locally. These fields commute with one another and at each point $x \in X_0$ are linearly independent. Let W be a tubular neighbourhood of X_0 in X^{n+1} of the form
$$W = \{f_i(x) = \xi_i, i = 1, \ldots, n-1, f_n(x) \leqslant \xi_n + \varepsilon\}.$$
Let $\pi : \partial W \to X_0$ be a corresponding two-sheeted covering. It is readily seen that on ∂W there exists a globally defined vector field V'_n such that $\pi_* V'_n = V_n$. The vector fields $V_i, i = 1, \ldots, n-1$, also rise in ∂W, $V_i = \pi_* V'_i$. All vector fields $V'_i (i = 1, \ldots n)$ commute as before and are independent at each point $x \in \partial W$. The transformation of the manifold ∂W which consists in the shift of the field V'_i along the integral curve within the time t will be denoted by g^t_i. For $a = (a_1, \ldots, a_n) \in \mathbf{R}^n$ the composition $g_1^{a_1} \cdots g_n^{a_n}$ will be denoted by g^a. Thus, the group $\mathbf{R}^n$ acts on ∂W by the formula $x \to g^a(x)$. Nonorientability of X_0 implies that the manifold ∂W is connected. Therefore, for any point $x_0 \in X_0$ its $\mathbf{R}^n$-orbit coincides with ∂W. Let $L = L_{x_0}$ be a stationary subgroup of the point x_0 in $\mathbf{R}^n$. Then the mapping
$$a \to g^a(x_0), a \in \mathbf{R}^n/L$$
is a diffeomorphism of $\mathbf{R}^n/L$ on ∂W. For any point $x \in \partial W$, the components a_i of a vector $a \in \mathbf{R}^n$, such that $g^a(x_0) = x$, will be called (multivalued) coordinates of the point x. Let σ be an involution of ∂W, under which the sheets change places. Then $\sigma_* V'_n = -V'_n$. Consequently,
$$\sigma(g^a(x)) = g^{\sigma_0(a)}(\sigma(x)), \text{ where } \sigma_0(a) = (-a_1, a_2, \ldots a_n).$$
Fix the point $x_0 \in \partial W$ and the corresponding (multivalued) coordinates a_i. In these coordinates the transformation σ is given by the formula
$$\sigma(a) = (q_1 - a_1, q_2 + a_2, \ldots, q_n + a_n),$$
where q_1 are coordinates of the point $\sigma(x_0)$. Changing, if necessary, the point x_0 by the point with coordinates $(\frac{1}{2}q_1, 0, \ldots, 0)$, we assume $q_1 = 0$. Let $T_q(a) = a + q$ be translation by the vector q modulo L. Clearly, $\sigma = \sigma_0 \cdot T_q$ and $\sigma_0 \cdot T_q = t_q \cdot \sigma_0$. Since $\sigma_0 = \sigma T_{-q}$ is a well defined transformation of the manifold ∂W and $\sigma_0(x_0) = x_0$, it follows that $\sigma_0(L) = L$. Since
$$T_{2q} = \sigma_0^2 T_{2q} = \sigma^2 = Id|_{\partial W},$$
it follows that $2q \in L$. Let us define the lattices
$$L^{\pm} = \{l \in L : \sigma_0(l) = \pm l\}, L' = L^+ + L^-.$$
Then either $L = L'$ or $L \neq L'$. Now consider the first case. Choose a basis $(l_i)_{i=1}$ in L^- and a basis $(l_i)_{i=2}, \ldots, n$ in L^+. Then $(l_i) = (l_i)_{i=1,2,\ldots,n}$ is the basis in $L' = L$. Hence, in the basis (l_i) the linear operator σ_0 is given by the matrix $\mathrm{diag}(-1, 1, \ldots, 1)$. We have seen above that $q_1 = 0$ and $2q \in L$. Therefore $2q \in L^+$ and we may assume that the basis (l_i) is chosen in such a way that $l_2 = 2q$. In the coordinates associated with the basis
(l_i) we have $\sigma(a_1, \ldots, a_n) = (-a_1, a_2 + \frac{1}{2}, a_3, \ldots, a_n),$

and accordingly X_0 is diffeomorphic to K_0^n

Let $L' \neq L$. For any $a \in \mathbf{R}^n/L$ define $a^{\pm} = \frac{1}{2}(a \pm \sigma_0(a))$. Let $l_1 \in L \setminus L'$. Suppose that $L_1^- \in l^-$. Since $l_1^+ = l_1 - l_1^-$, it follows that $l_1^+ \in L^+$. Consequently, $l_1 = l_1^+ + l_1^- \in L'$, a contradiction. That is why $l_1^- \notin L^-$, while it is obvious that $2l_1^- \in L^-$. We will show that l_1 induces L modulo L'. Indeed, let $l \in L \setminus L'$. Then $2l^- \in L^-$ and $l^- \in L^-$. Therefore, $l_1^- - l^- \in L^-$ and $(l_1 - l)^+ \in L^+$. That is why, $l_1 - l \in L'$, as required. Let l^* be a generating element in L^-. Then for $l_1 \in L \setminus L'$ and for an appropriate integer m we have $2l_1^- = ml^*$. The number m is odd because $l_1^- \notin L^-$. Make a substitution $l_1 \to l_1' = l_1 - nl^*$, where $2n + 1 = m$. Then $2l_1^- = ml^* - 2nl^* = l^*$. Thus, $l_1 \in L \setminus L'$ may be so chosen that $2l_1^-$ is a generating element in L^-. Let $(l_i)_{i=2,\ldots,n}$ be a basis in L^+. Prove that $(l_i) = (l_i)_{i=1,2,\ldots n}$ is a basis in L. Let $l \in L$ be an arbitrary element. It is necessary to find integers $m_i, i = 1, \ldots, n$ such that $l = \Sigma_i m_i l_i$. Since l_1 generates L modulo L', we may assume that $l \in L'$. But in this case $l = l^+ + l^-$, where $l^\pm \in L^\pm$. And since $l^- = 2ml_1^-$, it follows that $l = \sum_i m_i l_i$ for appropriate numbers m_i. In the basis (l_i) the linear operator σ_0 is given by the matrix:

$$ (s_{ij}) = \begin{pmatrix} s_{11} & 0 & \cdots & & 0 \\ s_{21} & 1 & & & \\ & & 1 & & 0 \\ & & & \ddots & \\ s_{n1} & 0 & & & 1 \end{pmatrix}. $$

Since $\det(s_{ij}) = -1$, it follows that $s_{11} = -1$. Making elementary basis transportations $(l_i)_{i=2,\ldots,n}$ in L^+, we can arrive at

$$ s_{31} = s_{41} = \cdots = s_{n1} = 0 $$

In the end, σ_0 is given by the matrix

$$ \begin{pmatrix} -1 & 0 & & & \\ & & & & 0 \\ p & 1 & & & \\ & & 1 & & 0 \\ & 0 & & \ddots & \\ & & 0 & & 1 \end{pmatrix} $$

Now in the basis $(l_i)_{i=1,\ldots,n}$ change the first vector $l_1 \to l_1' = l_1 + kl_2$. Since

$$ \begin{pmatrix} 1 & 0 \\ k & 1 \end{pmatrix}, \quad \begin{pmatrix} -1 & 0 \\ p & 1 \end{pmatrix}, \quad \begin{pmatrix} 1 & 0 \\ -k & 0 \end{pmatrix} = \begin{pmatrix} 1 & 0 \\ p - 2k & 1 \end{pmatrix}, $$

it follows that in the new basis the operator σ_0 is given by the same matrix but with the substitution $p \to p' = p - 2k$. That is why we may assume that $p = 0$ or $p = 1$. The first case corresponds to $L = L'$ and has already been analyzed. It remains to examine the case $p = 1$. The vector $q \in L^+ \cap (\frac{1}{2}L \setminus L)$ of the coordinates of the point $\sigma(x_0)$ is defined modulo L. We therefore assume that $q = \sum_{i=2}^{n} q_i l_i$, where $q_i = 0, \frac{1}{2}$. If $q_2 = \frac{1}{2}$, then replace the reference point x_0 by the point with coordinates $(\frac{1}{2}, 0, \ldots, 0)$. Since $\sigma_0(l_1) = -l_1 + l_2$, it follows that the translation of q in the new coordinates will give $q + \frac{1}{2}l_2 \equiv 0$ (mod L). We may therefore assume that $q_2 = 0$. Next, making elementary transformations of the vectors $l_i, i = 3, \ldots n$, we can arrive at $q \equiv \frac{1}{2}l_3 (\bmod L)$. Finally, making the change $l_2 \to l'_2 = -l_1 + l_2$, we obtain that σ is given in coordinates associated with the basis (l_i) by the formula

$$\sigma(a_1, \ldots, a_n) = (a_2, a_1, a_3 + \frac{1}{2}, a_4, \ldots, a_n).$$

Consequently, the manifold X_0 is diffeomorphic to K_1^n, which implies the theorem.

In conclusion we prove the tubular neighbourhood W of the minimum X_0 in X^{n+1} is cut into "elementary bricks" of types I and III described above. Since the manifold X^{n+1} is orientable and $X_0 \approx K_p^n$, it follows that W is diffeomorphic to the factor manifold W_p^{n+1} of the manifold $T^n \times [-1, 1]$ by the involution R_p^W defined by the formula

$$R_p^W(t, a) = (-t, R_p(a)), \quad t \in [-1, 1], \quad a \in T^n.$$

Because

$$W_0^{n+1} = W_0^3 \times T^{n-2}, \quad W_1^{n+1} = W_1^4 \times T^{n-3},$$

then, in effect, it suffices to consider W_0^3 and W_1^4. Both these manifolds can be obtained by gluing "trousers" with two full tori. In the case W_0^3, both the identification mappings $T^2 \to T^2$ are given by one and the same matrix $\begin{pmatrix} 2 & -1 \\ 1 & 0 \end{pmatrix}$. In the case W_1^4, the identification mappings $T^3 \to T^3$ are given by the matrices

$$\begin{pmatrix} 2 & 0 & -1 \\ 0 & 1 & 0 \\ 1 & 0 & 0 \end{pmatrix}, \begin{pmatrix} 2 & 0 & 1 \\ 1 & 0 & 0 \\ 0 & 1 & 0 \end{pmatrix}.$$

§3 The Properties of Decomposition of Constant-Energy Surfaces of Integrable Systems into the Sum of Simplest Manifolds

3.1 *A Fundamental Decomposition $Q = m\,\mathrm{I} + p\,\mathrm{II} + q\,\mathrm{III} + s\,\mathrm{IV} + r\,\mathrm{V}$ and the Structure of Singular Fibres*

We have already proved that each nonsingular three-dimensional constant-energy surface Q^3 of an integrable system can be represented in the form of a union $Q = m\mathrm{I}+p\mathrm{II}+q\mathrm{III}+s\mathrm{IV}+r\mathrm{V}$, where m, p, q, s, r are non-negative integers and I, II, III, IV, V denote elementary manifolds of the five simplest types described in §1,2. Thus, each manifold Q is naturally associated with five integers. If such a decomposition is generated by a Bott integral f, i.e. if it is Hamiltonian, then the number m is equal to the number of stable periodic solutions of a given Hamiltonian system. If one forgets about the existence of a second integral, then the above-mentioned decomposition can be simplified and considered separately from the point of view of the theory of three-dimensional manifolds. In this case, the numbers m', p', q' are determined by a given manifold Q, generally, not uniquely. For instance, Fig. 34 (f) presents two graphs $\Gamma(Q)$ which set manifolds Q admitting different representation in the form of gluing simplest manifolds.

§3 The Properties of Decomposition of Constant-Energy Surfaces

PROPOSITION 2.3.1. *Each of the elementary manifolds I, II, III, IV, V is actually encountered in decompositions of the form* $Q = m\,I + p\,II + q\,III + s\,IV + r\,V$ *for constant-energy surfaces of concrete mechanical Hamiltonian integrable systems.*

The next question concerns already the properties of the numbers m, p, q, s, r, namely, the question is whether or not any three-dimensional manifold of the form $Q = mI + pII + qIII + sIV + rV$ where m, p, q, s, r are five arbitrary numbers and the gluings of boundary tori are also arbitrary) can be realized in the form of a constant-energy surface Q of an integrable system. The answer is affirmative. See above the theorem of Brailov and Fomenko (Claim 2.1.2).

A similar question arises, of course, in the multidimensional case and is also answered in the affirmative. We have seen that any surgery on Liouville tori of general position splits into a composition of five canonical simple modifications of types I, II, III, IV, V.

PROPOSITION 2.3.2. *In the four-dimensional case, each of the indicated five canonical types of surgery is actually realized in concrete mechanical integrable systems.*

This assertion is apparently valid in the multidimensional case as well. All these types of surgery are possibly realized in integrable systems of equations of motion of a multidimensional rigid body [143].

The results obtained above make it possible to *describe visually the structure of singular fibres*, that is, singular level surfaces of a second integral f (in the four-dimensional case) and singular fibres of the momentum mapping F (in an arbitrary multidimensional case). For the sake of definiteness we will dwell on the four-dimensional case.

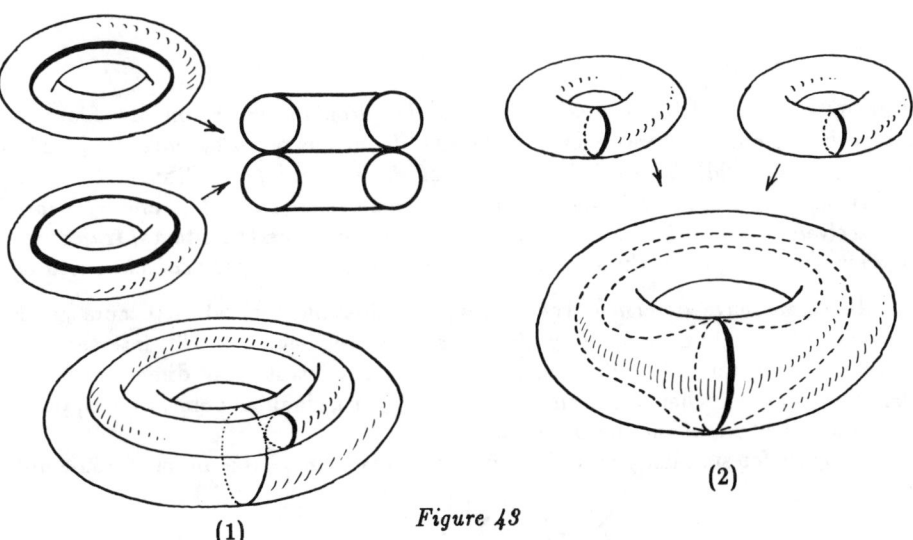

Figure 43

As proved above, each singular fibre containing a saddle critical circle is obtained as follows. *One takes two nonsingular Liouville tori, sets a cycle on each of them* (that is, draws noncontractible circles), *after which glues both tori by these cycles*. In other words, two tori must be tangent to each other in such a way that the two circles drawn on them coincide (Fig. 43). Figure 43 illustrates the simplest cases of tangency of two tori which can be pictured in $\mathbf{R}^3$. An analogous picture is obtained in the multidimensional case (see §2). One should only replace here a circle S^1 by a torus T^{n-1}. Two Liouville tori T_1^n and T_2^n must be tangent to each other in such a manner that the two $(n-1)$-dimensional tori "drawn" on them coincide, i.e. flow into one torus T^{n-1}. As a result, one obtains a singular fibre of the momentum mapping $F : M^{2n} \to \mathbf{R}^n$.

There arises the following question: how are integral trajectories of a system organized on singular fibres (singular level surfaces)? According to Lemma 2.2.2 (see §2, Ch. 2), the saddle critical set of an integral f on a surface X^{n+1} consists of $(n-1)$-dimensional tori (in the compact case). It can be proved that integral trajectories of a system on a singular level either lie on these critical tori or asymptotically wind on these tori (Fig. 44). Thus, we have obtained a complete description of the behaviour of the trajectories of a system on singular level surfaces of integrals.

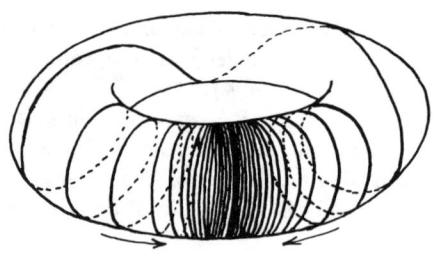

Figure 44

THEOREM 2.3.1. *Let $v = \operatorname{sgrad} H$ be a Hamiltonian system on M^4 Liouville-integrable on one constant-energy surface Q^3 by means of a Bott integral f. Then each singular saddle level surface $f^{-1}(c)$ of the integral f is obtained by gluing two two-dimensional tori T_1^2 and T_2^2 by nontrivial cycles (circles) γ_1 and γ_2 located respectively in the tori T_1^2 and T_2^2. On such a singular fibre the integral trajectories of a system asymptotically wind around this nontrivial cycle (or coincide with it).*

In §2 we have described the surgery on Liouville tori which occurs at the moment when a point a, moving along a smooth segment γ in $\mathbf{R}^n$, intersects the bifurcation diagram $\sum$ at its nonsingular point on a stratum of dimension $n-1$. What will happen when the path γ punctures $\sum$ at a singular point, that is, passes through a stratum of smaller dimension?

Suppose for simplicity that the bifurcation diagram $\sum$ is a union of non-empty strata

$$\sum\nolimits^0, \sum\nolimits^1, \sum\nolimits^2, \ldots, \sum\nolimits^{n-1},$$

each of which is characterized by a corresponding integral. Then the operation of "puncturing" the diagram $\sum$ at some of its singular points can be decomposed into a composition of successive similar operations shown in Fig. 45. The path γ first meets transversally with an $(n-1)$-dimensional stratum $\sum^{n-1}$, then moves along this stratum until it meets transversally with the next, $(n-2)$-dimensional stratum $\sum^{n-1}$, then it moves along the stratum $\sum^{n-2}$ until it meets with the stratum $\sum^{n-3}$, etc. Each time the appearing Liouville torus surgery may be assumed to belong to one of the types I, II, III, IV, V (with a decreasing dimension). The point is that the above-mentioned assumptions one may believe that on the stratum $\sum^i$ exactly $n-i+1$ integrals are dependent. The dependent integrals can be eliminated to reduce the problem to a smaller dimension. In the general case the problem is, of course, much more complicated.

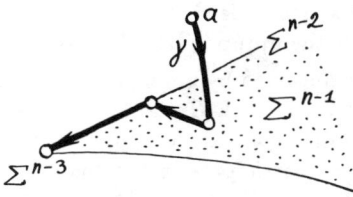

Figure 45

3.2 Homological Properties of Constant-Energy Surfaces

We know already that a three-dimensional manifold Q, which is a nonsingular level surface of an integral H and has an oriented integral f, is glued of manifolds of four types: 1) full tori $S^1 \times D^2$, 2) cylinders $T^2 \times D^1$, 3) oriented saddles $(S^2 \backslash 3D^2) \times S^1$, 4) nonoriented saddles A^3. Thus, the manifold Q is described exactly if we know how many copies of elementary manifolds of each type it contains and, which is of particular importance, by means of what gluings of boundaries it is obtained, in other words, what diffeomorphisms of boundary tori we employ. Gluing elementary manifolds we must obtain a connected closed (that is, without boundary) manifold. This means that all boundary tori must be glued. We put a question: what connected closed oriented three-dimensional manifolds are obtained through such gluings? As proved above, the class of such manifolds Q^3 does not coincide with the class of all manifolds M^3. Moreover, manifolds of type Q^3 form a "gaunt" subset in the set of all manifolds M^3. Following the advise of the author, Mamedov has undertaken a further investigation of this class. We will present some of his results here.

It is of practical interest to answer the following question: *how many full tori (ordinary and n-dimensional $D^2 \times T^{n-1}$) participate in gluing the manifold Q^n?* Here

$$Q^n = (f_2 = c_2, \ldots, f_{n-1} = c_{n-1}) \subset M^{2n-2}.$$

Recall that in the case Q^3 the number m of full tori $S^1 \times D^2$ is associated with stable periodic solutions of a system on Q^3 and is, therefore, of special interest. First we describe two special cases.

CASE 1: A manifold Q^3 is glued of two full tori. For this gluing it is necessary to set a diffeomorphism between two tori, each of which is a boundary of a corresponding full torus. Two manifolds Q thus obtained do not differ topologically, if the corresponding diffeomorphisms participating in their construction are homotopic in the set of continuous mappings of torus into torus. Recall that the classes of homotopic continuous mappings of torus into torus, each of which contains a diffeomorphism, are in one-to-one correspondence with matrices

$$A = \begin{pmatrix} a_1 & a_2 \\ a_3 & a_4 \end{pmatrix},$$

where $a_i \in \mathbb{Z}$ and $\det A = \pm 1$. In this case we can obtain: a) a three-dimensional sphere, b) a manifold $S^1 \times S^2$, c) lens spaces. Pay attention to the fact that in cases a) and b) the rank of the group $H_1(Q, \mathbb{Z})$ is equal to zero.

CASE 2: A manifold Q is obtained by gluing two or several cylinders. In this case, one can obtain a manifold Q in which $H_1(Q, \mathbb{Z}) = \mathbb{Z}$ (for an example see §1 above).

The following important fact has been observed. If *some other elementary manifolds*, besides cylinders, participate in the construction of a manifold Q, then this manifold can be glued without any cylinders. Indeed, a manifold Q is obtained from $Q \setminus$ (cylinder) by adding a cylinder through gluing the cylinder boundary (which is a union of two tori) to two tori from the set $Q \setminus$ (cylinder). It is obvious that a cylinder $T^2 \times D^1$ can be deformed into a torus (contracted to a torus). This means that instead of "joining" two tori of the set $Q \setminus$ (cylinder) by means of a cylinder, one can simply glue these two tori by a diffeomorphism (Fig. 46).

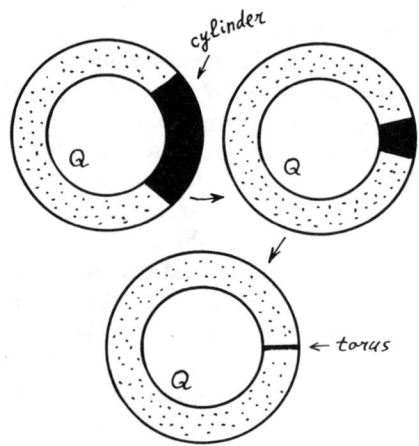

Figure 46

Having performed this operation on all the cylinders, we obtain a manifold glued without the help of cylinders. If a manifold Q is glued *of cylinders only*, then the described reduction process is, of course, impracticable.

§3 The Properties of Decomposition of Constant-Energy Surfaces

THEOREM 2.3.2. *If a three-dimensional connected manifold Q is glued of q oriented saddles, m full tori, and an arbitrary number s of nonoriented saddles A^3 and if it is known that $\operatorname{rank} H_1(Q, \mathbb{Z}) \leqslant k$, then there holds the inequality $m \geqslant q + 2 - 2k$.*

The theorem is true for $q = 0$ as well.

COROLLARY 2.3.1. *If the rank of $H_1(Q, \mathbb{Z})$ is equal to zero, then in the composition of Q there are exactly $q + 2$ full tori.*

COROLLARY 2.3.2. *If the rank of $H_1(Q, \mathbb{Z})$ is equal to zero, then $m \geqslant 2$.*

These and the following assertions proposed by Mamedov are based on purely topological considerations.

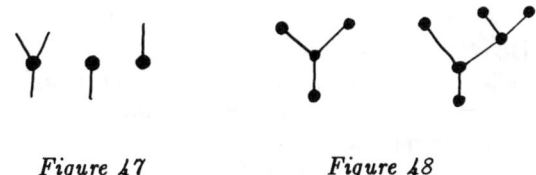

Figure 47   Figure 48

Recall that for each manifold Q one can construct the graph $\Gamma(Q)$ of this manifold. Each oriented saddle will be pictured in the form of a "tripod" (trefoil) (Fig. 47) and a full torus in the form of a black circle with one edge going from it. The gluing places in the graph Γ are not shown. Examples of the graphs are given in Fig. 48. We are interested only in connected manifolds and, therefore, in connected graphs. So, the number of black circles (vertices of the graph) is equal to m, the number of tripods to q.

The ends of a tripod are tori. In the graphs illustrating manifolds Q there cannot be free ends because we consider closed manifolds. The graphs glued of tripods (with not yet glued up ends) may contain closed subgraphs or closed curves, or they may contain not a single closed curve. The latter type of graphs will be called *branching graphs*.

LEMMA 3.2.1. *The number of ends of a connected graph glued of q tripods does not exceed $q + 2$, the number of ends being equal to $q + 2$ if and only if the graph is branching.*

This lemma is easily proved using induction by q.

So, if m is the number of black circles with which (full tori) the ends of the graph are glued up), then $m \leqslant q + 2$. Now suppose that a non-branching graph is given. This means that it contains a certain closed curve. Let us cut the graph in any gluing place lying on this curve. If after this the graph has not become branching, it contains another closed curve. Further cuts will make the graph branching. Each cut 1) leaves the graph connected, 2) does not change the number of tripods 3) increases the number of free ends of the graph by two. If we have made $\tilde{k}$ cuts, then $q + 2 = m + 2\tilde{k}$, where m is the number of ends of the initial graph and q is the number of tripods (that is, oriented saddles). The number of ends of the final graph is equal to $q + 2$.

PROOF OF THEOREM 2.3.2: For simplicity we first consider the orientable case, that is, the case where the manifold Q does not involve nonoriented saddles. We will prove that rank $H_1(Q, \mathbb{Z}) \geqslant \tilde{k}$, where $\tilde{k}$ is the number of cuts necessary for transforming the initial graph into a branching one. Examine any cut. A small neighbourhood of this cut is a cylinder $T^2 \times D^1$. The manifold Q may be assumed to be obtained as a union of two of its submanifolds 1) $\tilde{k}$ cylinders, 2) a connected manifold, which is a complement of the interior of the first submanifold.

If a space Q is a union of two of its subspaces Y_1 and Y_2, $Y = Y_1 \cap Y_2$ being their intersection, then under insignificant restrictions there holds the following exact sequence of groups and homomorphisms called *Mayer–Vietoris sequence*:

$$\ldots H_{i+1}(Q) \to H_i(Y) \xrightarrow{j_*} H_i(Y_1) \oplus H_i(Y_2) \to H_i(Q) \to \ldots,$$

where $j_*(x) = (j_{1*}(x), -j_{2*}(x))$, and $j_1 : Y \to Y_1$, $j_2 : Y \to Y_2$ are embeddings.

If we denote a union of $\tilde{k}$ cylinders as Y_1 and the complement of their interior in Q as Y_2, then $Y = Y_1 \cap Y_2$ is a disconnected union of $2\tilde{k}$ tori. Examine the end of the Mayer–Vietoris sequence:

$$\cdots \to H_1(Q) \to H_0(Y) \to H_0(Y_1) \oplus H_0(Y_2) \to H_0(Q) \to 0.$$

In our case it has the following form (where $a \cdot \mathbb{Z} \approx \mathbb{Z}^a = \mathbb{Z} \oplus \cdots \oplus \mathbb{Z}$):

$$\cdots \to H_1(Q) \to 2\tilde{k}\mathbb{Z} \to (\tilde{k}+1)\mathbb{Z} \to \mathbb{Z} \to 0.$$

From the accuracy of the sequence it easily follows that rank $H_1(Q, \mathbb{Z}) \geqslant \tilde{k}$, and since by the condition $k \geqslant$ rank $H_1(Q, \mathbb{Z})$, then we obtain $k \geqslant \tilde{k} = \frac{1}{2}(q + 2 - m)$, that is, $m \geqslant q + 2 - 2k$.

It is clear that proceeding to the general case (that is, gluing nonoriented saddles) does not affect the result. The point is that as before 1) each manifold has a graph, 2) the number of cuts necessary for the graph to become branching is equal to $\tilde{k} = \frac{1}{2}(q + 2 - m)$, 3) in the place of each cut one may cut out a cylinder (whose boundary consists of two tori T^2. This implies that rank $H_1(Q, \mathbb{Z}) \geqslant \tilde{k}$, that is, $2k \geqslant 2\tilde{k} = q + 2 - m$, where $k =$ rank $H_1(Q, \mathbb{Z})$. From this we find that $m \geqslant q + 2 - 2k$. This completes the proof of the theorem.

Corollary 2.3.2 is obvious. Taking into account that $m \leqslant q + 2$, we obtain Corollary 2.3.1.

The proof was, in fact, carried out for $q \geqslant 1$. But if saddles are not used in the construction of a manifold, then only two cases are possible. 1) Two full tori are joined by means of a certain number of cylinders. This case is, in fact, reduced (see above) to gluing two full tori. 2) A manifold is glued of cylinders only. We may assume the number of cylinders to be equal to two.

In case 1, the condition of Theorem 2.3.2 is trivially fulfilled. In case 2, this condition can be easily proved. Namely, we will prove that a manifold glued of two cylinders has rank $H_1(Q, \mathbb{Z}) \geqslant 1$. Indeed, Q is a union of cylinders Y_1 and Y_2, $Y = Y_1 \cap Y_2$ being two tori. From the Mayer–Vietoris sequence we have:

§3 The Properties of Decomposition of Constant-Energy Surfaces 133

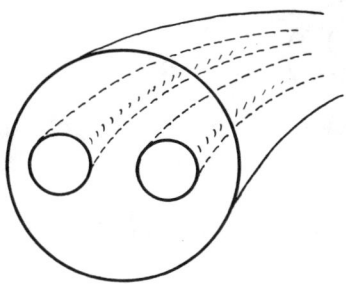

Figure 49

$\cdots \to H_1(Q, \mathbb{Z}) \to \mathbb{Z} \oplus \mathbb{Z} \to \mathbb{Z} \oplus \mathbb{Z} \to 0$. This implies that rank $H_1(Q, \mathbb{Z}) \geq 1$. Thus, the assertion of Theorem 2.3.2 is true for $q = 0$ as well.

We have proved in §1 that of two cylinders a manifold Q can be glued for which $H_1(Q, \mathbb{Z}) = \mathbb{Z}$.

As has already been mentioned, *an arbitrary three-dimensional manifold cannot be glued of the four types of elementary manifolds which we have discovered.* In this connection of special interest is the following result (Mamedov).

THEOREM 2.3.3. *Gluing oriented saddles and full tori one can obtain a three-dimensional oriented manifold with any possible (for oriented three-dimensional manifolds) groups of integer-valued homologies.*

This result follows from the Seifert theory of manifolds; but we present here an elementary proof. Note that the homology groups of compact manifolds are always finitely generated and Abelian. It is well known that any three-dimensional oriented manifold W is glued of two balls with g handles. For such manifolds, $H_2(W, \mathbb{Z})$ is a free group (Abelian) and rank $H_2(W, \mathbb{Z}) = $ rank $H_1(W, \mathbb{Z})$. For oriented connected closed three-dimensional manifolds we have: $H_3(W, \mathbb{Z}) = \mathbb{Z}$. This implies that to prove the theorem it suffices that oriented saddles and full tori be glued to form a three-dimensional manifold with any Abelian finitely generated group H_1.

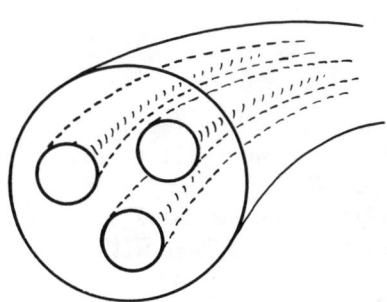

Figure 50

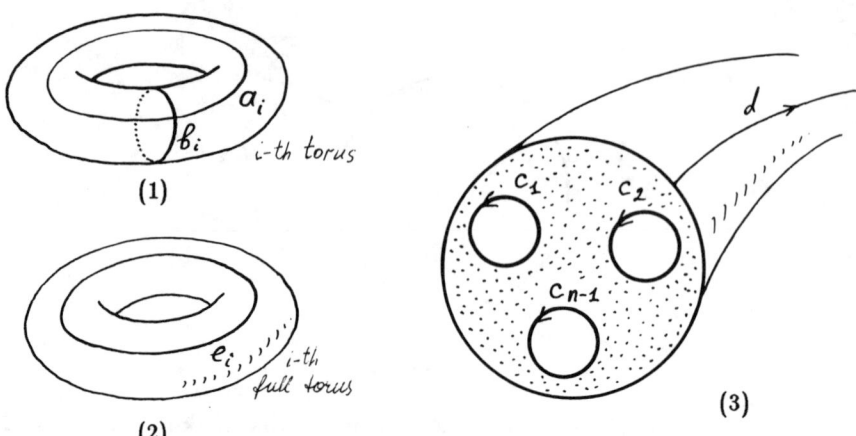

Figure 51

It is clear that an oriented saddle is a full torus of which two parallel full tori are cut out (Fig. 49). Gluing-in a new oriented saddle instead of one of these full tori, we obtain a full torus with three cut out full tori (Fig. 50). Thus, of $q-2$ full tori one can glue a full torus, of which $q-1$ parallel full tori are cut out, that is, one can glue $(S^2\backslash(q\cdot D^2))\times S^1$. We will prove that $q-1$ full tori can be glued-in instead of those which have been cut out, and one full torus can be glued form the outside in such a manner that any Abelian finitely generated group is obtained as $H_1(Q,\mathbb{Z})$. Consider the Mayer–Vietoris sequence

$$0 \to H_3(Q) \to H_2(Y) \to H_2(Y_1) \oplus H_2(Y_2) \to H_2(Q)$$
$$\to H_1(Y) \to H_1(Y_1) \oplus H_1(Y_2) \to H_1(Q) \to q\mathbb{Z} \to (q+1)\mathbb{Z} \to \mathbb{Z} \to 0.$$

Here Y_1 consists of q full tori, Y_2 is a full torus with $q-1$ cut out tori, that is, $(S^2\backslash qD^2)\times S^1$ and $Y = Y_1 \cap Y_2$ are q two-dimensional tori. Since $H_3(Q,\mathbb{Z}) = \mathbb{Z}$ and $H_i(Y)$, $H_i(Y_k)$ for $i=1,2$ and $k=1,2$ are free groups, it is readily seen that the fragment

$$0 \to H_2(Q) \to H_1(Y) \xrightarrow{f} H_1(Y_1) \oplus H_1(Y_2) \to H_1(Q) \to 0,$$

where

$$H_1(Y) = H_1(Y,\mathbb{Z}) = 2q\mathbb{Z}, \qquad H_1(Y_1) = q\cdot \mathbb{Z}, \qquad H_1(Y_2) = q\mathbb{Z}$$

can be singled out of the Mayer–Vietoris sequence. Thus, $f : 2q\mathbb{Z} \to 2q\mathbb{Z}$. Gluing-in full tori in a different manner, we obtain different homomorphisms f. Obviously, $2q\mathbb{Z}/\operatorname{Im} f = H_1(Q)$ and $H_2(Q) = \operatorname{Ker} f$. To examine the homomorphisms f, we should represent it in the matrix form. To this end, we should set the bases (generating cycles) in the groups $H_1(Y)$, $H_1(T_1)$, $H_1(Y_2)$. Figure 51 presents the corresponding generators. 1) For the torus see Fig. 51(1), 2) for the full torus see Fig 51(2), for the full torus with cut out full tori see Fig. 51(3). If for the ith full torus the gluing matrix is of the form $\begin{pmatrix}\alpha_i & \gamma_i \\ \beta_i & \delta_i\end{pmatrix}$, it follows that a_i transforms to $\alpha_i c_i + \beta_i d$ and b_i transforms into $\gamma_i c_i + \delta_i d$, $1 \leqslant i \leqslant q-1$.

§3   *The Properties of Decomposition of Constant-Energy Surfaces*   135

A special case is a full torus glued from the outside to a full torus with cut out full tori. Here a_q transforms into $\alpha_q(c_1 + c_2 + \cdots + c_{q-1}) + \beta_i \cdot d$ and $b_q \to \gamma_q \cdot (c_1 + c_2 + \cdots + c_{q-1}) + \delta_i \cdot d$. Obviously, the embedding $Y \to Y_1$ acts upon the generators $a_i \to 1 \cdot e_i$, $b_i \to 0 \cdot e_i$, $1 \leqslant i \leqslant q$, in the following way. Order the bases $H_1(Y)$ and $H_1(Y_1) \oplus H_1(Y_2)$: for $H_1(Y) - (a_1, \ldots, a_q, b_1, \ldots, b_q)$, for $H_1(Y_1) \oplus H_1(Y_2) - (e_1, \ldots, e_q, c_1, \ldots, c_{q-1}, d)$.

From this it follows that the matrix of the homomorphism f in these bases is of the form:

All empty cells are occupied by zeros

$$\begin{pmatrix} 1_1 & & & & & & & & & & \\ & \ddots & & & & & & & & & \\ & & \ddots & & & & & & & & \\ & & & \ddots & & & & & & & \\ & & & & 1 & & & & & & \\ \alpha_1 & & & & & \alpha_q & \gamma_1 & & & & \gamma_q \\ & \alpha_2 & & & & \alpha_q & & \gamma_2 & & & \gamma_q \\ & & \ddots & & & \vdots & & & \ddots & & \vdots \\ & & & \alpha_{q-1} & & \alpha_q & & & & \gamma_{q-1} & \gamma_q \\ \beta_1 & \beta_2 & \cdots & \cdots & \cdots & \beta_q & \delta_1 & \delta_2 & \cdots & \delta_{q-1} & \delta_q \end{pmatrix}$$

Strictly speaking, in the lower part of the matrix, the minus sign should stand before the numbers, but assuming the gluing matrices to be of the form $\begin{pmatrix} -\alpha_i & -\gamma_i \\ -\beta_i & -\delta_i \end{pmatrix}$, one may remove the sign. The numbers $\alpha_i, \beta_i, \gamma_i, \delta_i$ depend on how we glue the tori. Let

$$\begin{pmatrix} \alpha_q & \gamma_q \\ \beta_q & \delta_q \end{pmatrix} = \begin{pmatrix} 1 & 0 \\ 0 & 1 \end{pmatrix}$$

The matrix of the homomorphism f is of the form

$$\begin{pmatrix} 1 & & & & & & & & & \\ & \ddots & & & & & & & & \\ & & \ddots & & & & & & & \\ & & & 1 & & & & & & \\ \alpha_1 & & & & 1 & & \gamma_1 & & & \\ & & & & & \ddots & & & & \\ & & \alpha_{q-1} & & & 1 & & & \gamma_{q-1} & \\ \beta_1 & \cdots & \beta_{q-1} & & 0 & & \delta_1 & \cdots & \delta_{q-1} & 1 \end{pmatrix}$$

The group into which f maps will be denoted by N. We have: $f : 2q\mathbb{Z} \to 2q\mathbb{Z} = N$. We have already set a basis in N. Let us numerate it with the same letter—$(e_1, \ldots, e_q, \ldots, e_{2q})$. If we change the basis in the group N, the matrix of the homomorphism will change too. Suppose that the gluing has already been made. Note that if for some i there holds the equality $\gamma_i = 0$, then from the integer-valuedness and unimodularity condition on the matrix $\begin{pmatrix} \alpha_i & \gamma_i \\ \beta_i & \delta_i \end{pmatrix}$ it follows $\delta_i = \pm 1$.

Now suppose that for some i, for instance, for $i = 1$, we have $\gamma_1 \neq 0$. Make the following change of the basis in the group N:

$$\tilde{e}_1 = e_1$$
$$\vdots$$
$$\tilde{e}_q = e_q$$
$$\tilde{e}_{q+1} = e_{q+1} + \frac{\delta_1}{\gamma_1} \times e_{2q}$$
$$\tilde{e}_{q+2} = e_{q+2}$$
$$\vdots$$
$$\tilde{e}_{2q} = e_{2q}$$

In the new basis $\tilde{e}_1, \ldots, \tilde{e}_{2q}$, the matrix of the homomorphism f is of the form

$$\begin{pmatrix} 1 & & & & & & & & & & \\ & \ddots & & & & & & & & & \\ & & \ddots & & & & & & & & \\ & & & \ddots & & & & & & & \\ & & & & 1 & & & & & & \\ \alpha_1 & & & & 1 & \gamma_1 & & & & & \\ & \ddots & & & \vdots & & \gamma_2 & & & & \\ & & \ddots & & \vdots & & & \ddots & & & \\ & & & \alpha_{q-1} & 1 & & & & \gamma_{q-1} & & \\ t & \beta_2 & \ldots & \beta_{q-1} & 0 & 0 & \delta_2 & & \delta_{q-1} & 1 \end{pmatrix}$$

where $t = -\alpha_1 \frac{\delta_1}{\gamma_1} + \beta_1$. We see that the matrix has changed as follows: 1) the number δ_1 is now equal to zero, 2) the number β_1 is replaced by the number $t = \beta_1 - \alpha_1 \frac{\delta_1}{\gamma_1}$. Make a similar change for all i such that $\gamma_i \neq 0$. For convenience of notation we assume that $\gamma_i \neq 0$ for $1 \leqslant i \leqslant k$ and $\gamma_i = 0$ for $k+1 \leqslant i \leqslant q-1$. After the change of the basis in the group N, the matrix of the homomorphism f

§3 *The Properties of Decomposition of Constant-Energy Surfaces*

takes the form:

$$\begin{pmatrix} 1 & & & & & & & & & & & \\ & \ddots & & & & & & & & & & \\ & & \ddots & & & & & & & & & \\ & & & \ddots & & & & & & & & \\ & & & & \ddots & & & & & & & \\ & & & & & \ddots & & & & & & \\ & & & & & & 1 & & & & & \\ \alpha_1 & & & & & & 1 & \gamma_1 & & & & \\ & \alpha_2 & & & & & & \vdots & \ddots & & & \\ & & \ddots & & & & & \vdots & & \ddots & & \\ & & & \alpha_k & & & & \vdots & & & \gamma_k & \\ & & & & \pm 1 & & & \vdots & & & & 0 \\ & & & & & \pm 1 & 1 & & & & & \\ & & & & & & & \ddots & & & & 0 \\ & & & & & & & & & & & \ddots \\ t_1 & \ldots & t_k & \beta_{k+1} & \ldots & \beta_{q-1} & 0 & \ldots \ldots \ldots & \pm 1 & \pm 1 & 1 \end{pmatrix}$$

Denote the new basis, as before, by $e_1, \ldots, e_{2q}$ and make the last change:

$$\tilde{e}_1 = e_1 + \alpha_1 e_{q+1} + t_1 e_{2q}$$
$$\tilde{e}_2 = e_2 + \alpha_2 e_{q+2} + t_1 e_{2q}$$
$$\vdots$$
$$\tilde{e}_k = e_k + \alpha_k e_{q+k} + t_k e_{2q}$$
$$\tilde{e}_{k+1} = e_{k+1} \pm 1 \cdot e_{k+q+1} + \beta_{k+1} e_{2q}$$
$$\vdots$$
$$\tilde{e}_{q-1} = e_{q-1} \pm 1 \cdot e_{q-1} + \beta_{q-1} e_{2q}$$
$$\tilde{e}_q = e_q + e_{q+1} + \cdots + e_{2q-1}$$
$$\tilde{e}_{q+1} = e_{q+1}$$
$$\vdots$$
$$\tilde{e}_{2q} = e_{2q}$$

In the new basis the homomorphism f has the matrix:

$$\begin{pmatrix} 1 & & & & & & & & \\ & \ddots & & & & & & & \\ & & \ddots & & & & & & \\ & & & \ddots & & & & & \\ & & & & 1 & & & & \\ & & & & & \gamma_1 & & & \\ & & & & & & \ddots & & \\ & & & & & & & \gamma_k & \\ & & & & & & \pm 1 & \pm 1 & \ldots & 1 \end{pmatrix}$$

It is now evident that

$$N/\operatorname{Im} f \approx \mathbb{Z}_{\gamma_1} \oplus \cdots \oplus \mathbb{Z}_{\gamma_k} \oplus \underbrace{\mathbb{Z} \oplus \cdots \oplus \mathbb{Z}}_{q-k+1 \text{ times}},$$

where $\mathbb{Z}_{\gamma_i} = 0$ provided that $\gamma_i = 1$. Taking a sufficiently large q and gluing matrices with appropriate numbers γ_i, one obtains, as $N/\operatorname{Im} f$, an *arbitrary Abelian finitely generated group*, which proves Theorem 2.3.3.

In particular, this implies that any finitely generated Abelian group may be a one-dimensional homology group of a connected compact oriented closed manifold.

It is known that three-dimensional manifolds may have similar homology groups, but not be diffeomorphic.

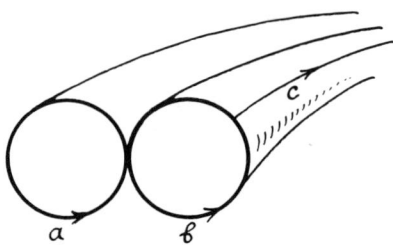

Figure 52

THEOREM 2.3.4. *Gluing three full tori and one oriented saddle gives a homological, but not a homotopical sphere.*

PROOF: An oriented saddle is a full torus minus two full tori. It has a homotopical type of the complex obtained by gluing two parallel tori (Fig. 52). The generators of the fundamental group are presented in the same figure. This group has three generators a, b, c, the generator c commuting with a and b. Now glue-in two full

§3   The Properties of Decomposition of Constant-Energy Surfaces

tori instead of those which were cut out, and from the outside glue one torus with the following gluing matrices:

$$\begin{pmatrix} 2 & -1 \\ -5 & 3 \end{pmatrix}, \quad \begin{pmatrix} 1 & 1 \\ 4 & 5 \end{pmatrix}, \quad \begin{pmatrix} 1 & -1 \\ -8 & 7 \end{pmatrix}$$

The following fragment may be singled out of the Mayer–Vietoris sequence:

$$0 \to H_2(Q) \to H_1(\text{of three tori}) \xrightarrow{f} H_1(\text{of three full tori}) \oplus H_1(\text{of oriented saddle}) \to H_1(Q) \to 0.$$

The homomorphism f has the following matrix:

$$A = \begin{pmatrix} 1 & & & & & \\ & 1 & & & & \\ & & 1 & & & \\ 5 & & 8 & -3 & & -7 \\ & -4 & & & -5 & -7 \\ -2 & -1 & -1 & 1 & -1 & -1 \end{pmatrix}$$

Here $\det A = 1$, that is, f is an isomorphism. Clearly, $H_1(Q, \mathbb{Z}) = H_2(Q, \mathbb{Z}) = 0$. Thus, a glued manifold is a homological sphere. Now calculate the fundamental group. Each glued-in full torus annihilates that element of the fundamental group into which the meridian of the full torus is mapped. Thus, upon gluing with the indicated matrices in the fundamental group there appear three relations: $c^{-1}a^3 = 1$, $cb^5 = 1$, and $c^{-1}(ab)^7 = 1$. We have obtained a group with three generators, one of which (c) commutes with the other two (a and b) and with the three relations. This group turns out to be nontrivial. Add another relation: $c = 1$. This may only diminish our group. The new group has two generators, a and b, and three relations, $a^3 = 1$, $b^5 = 1$, and $(ab)^7 = 1$. Prove that this group is nontrivial. In the permutation group S_7, consider the subgroup generated by the elements

$$\tilde{a} = \begin{pmatrix} 1 & 2 & 3 & 4 & 5 & 6 & 7 \\ 2 & 3 & 1 & 4 & 5 & 6 & 7 \end{pmatrix}$$

and

$$\tilde{b} = \begin{pmatrix} 1 & 2 & 3 & 4 & 5 & 6 & 7 \\ 1 & 2 & 4 & 5 & 6 & 7 & 3 \end{pmatrix}$$

(multiplication from right to left),

$$\tilde{a}\tilde{b} = \begin{pmatrix} 1 & 2 & 3 & 4 & 5 & 6 & 7 \\ 2 & 3 & 4 & 5 & 6 & 7 & 1 \end{pmatrix}.$$

From the group with generators a and b there exists an obvious epimorphism onto the subgroup in S_7 generated by $\tilde{a}$ and $\tilde{b}$ (that is, $a \to \tilde{a}$, $b \to \tilde{b}$). But the subgroup generated by a and b is nontrivial, and therefore the fundamental group of the manifold Q is also nontrivial. This proves Theorem 2.3.4.

We now proceed to the n-dimensional case ($n \geq 4$). Manifolds Q^n of dimension n will be glued of four types of manifolds: 1) full tori $T^{n-2} \times D^2$, 2) oriented saddles $(S^2 \backslash 3D^2) \times T^{n-2}$, 3) cylinders $D^1 \times T^{n-1}$, 4) nonoriented saddles A_α^n.

For now let there be no nonoriented saddles in the composition of Q. We will distinguish between two cases depending on whether or not oriented saddles are used in gluing. If not, then two versions are possible: 1) A manifold is glued of cylinders only. As in the three-dimensional case, we may assume the number of cylinders to be equal to two. As in the three-dimensional case, it can be proved that for such manifolds rank $H_1(Q, \mathbb{Z}) \geq 1$. We can also give an example of a manifold glued of two cylinders, for which $H_1(Q, \mathbb{Z}) = \mathbb{Z}$.

2) A manifold is glued of two full tori and several cylinders. Cylinders may be removed. We obtain a manifold glued of two full tori. As distinct from the three-dimensional case, one cannot glue a sphere of two n-dimensional (dissipative) full tori.

THEOREM 2.3.5. *A manifold Q such that $H_1(Q, \mathbb{Z}) = 0$ cannot be glued of n-dimensional oriented saddles, nonoriented saddles, cylinders, and full tori.*

PROOF: Let at first there be no nonoriented saddles. It has already been mentioned that if a manifold Q is glued of cylinders only, then rank $H_1(Q, \mathbb{Z}) \geq 1$. Consider the gluing of two full tori, which we denote as Y_1 and Y_2. The intersection $Y = Y_1 \cap Y_2$ is a torus T^{n-1}. It can be easily calculated that $H_1(Y, \mathbb{Z}) = (n-1) \cdot \mathbb{Z}$, $H_1(Y_1, \mathbb{Z}) = H_1(Y_2, \mathbb{Z}) = (n-2)\mathbb{Z}$. Consider the end of a Mayer–Vietoris sequence:

$$\cdots \to H_1(Y) \to H_1(Y_1) \oplus H_1(Y_2) \to$$
$$H_1(Q) \to H_0(Y) \to H_0(Y_1) \oplus H_0(Y_2) \to H_0(Q) \to 0.$$

In this case, it is of the form:

$$\cdots \to (n-1)\mathbb{Z} \to 2(n-2)\mathbb{Z} \to H_1(Q) \to \mathbb{Z} \to \mathbb{Z} \oplus \mathbb{Z} \to \mathbb{Z} \to 0.$$

Taking into account that $n \geq 4$, we obtain: rank $H_1(Q, \mathbb{Z}) \geq n - 3 \geq 1$. Now proceed to the general case. As has already been mentioned, cylinders may be disregarded.

As in the three-dimensional case, the manifold Q has the graph $\Gamma(Q)$, and analogously if it contains at least one closed curve, rank $H_1(Q, \mathbb{Z}) \geq 1$. There remains to consider manifolds with branching graphs, that is, manifolds glued of q oriented saddles and $q + 2$ full tori.

We will prove that if Q is glued of q oriented saddles in such a way that we obtain a branching graph, then rank $H_1(Q, \mathbb{Z}) = n - q + 1$. This is proved by induction by q. For $q = 1$ we have: $H_1($ an oriented saddle; $\mathbb{Z}) = H_1((S^1 \vee S^1) \times T^{n-2}) = n \cdot \mathbb{Z}$. Let this hold true for $q = k$. Any manifold with a branching graph, which is glued of $k + 1$ oriented saddles, is obtained from a branching manifold glued of k oriented saddles by gluing an additional saddle.

Let Y_1 be an oriented saddle, Y_2 a manifold with a branching graph glued of k saddles. Then $Y = Y_1 \cap Y_2$ is an $(n-1)$-dimensional torus. The embedding

§3   *The Properties of Decomposition of Constant-Energy Surfaces*   141

$i_1 : Y \to Y_1$ induces a monomorphism $i_{1*} : H_1(Y) \to H_1(Y_1)$. Consider the following Mayer–Vietoris sequence:

$$\cdots \to H_1(Y) \xrightarrow{f} H_1(Y_1) \oplus H_1(Y_2) \to H_1(Q) \to \mathbb{Z} \to \mathbb{Z} \oplus \mathbb{Z} \to \mathbb{Z} \to 0.$$

From this sequence we may single out a piece: $H(Y) \xrightarrow{f} H_1(Y_1) \oplus H_1(Y_2) \to H_1(Q) \to 0$. By definition, $f = (i_{1*}, -i_{2*})$, where $i_2 : Y \to Y_2$ is also an embedding. From injectiveness of i_{1*} there follows injectiveness of f. Taking into account that $\operatorname{rank} H_1(Y) = n - 1$, $\operatorname{rank} H_1(Y_1) = n$, $\operatorname{rank} H_1(Y_2) = n - k - 1$, one can readily see that $\operatorname{rank} H_1(Q) \doteq n + k$, which concludes the step of induction.

Now we will prove that by gluing $q + 2$ full tori to a branching manifold glued of q saddles it is impossible to obtain a manifold Q with $H_1(Q, \mathbb{Z}) = 0$. Let Y_1 consist of $q+2$ full tori and Y_2 be a branching manifold of q oriented saddles. Then $Y = Y_1 \cap Y_2$ is a disconnected union of $q+2$ tori of dimension $n - 1$. We have:

$$H_1(Y_1, \mathbb{Z}) = (q+2)(n-2)(n-2)\mathbb{Z},$$
$$\operatorname{rank}(H_1(Y_2, \mathbb{Z})) = n + q - 1,$$
$$H_1(Y, \mathbb{Z}) = (q+2)(n-1)\mathbb{Z}.$$

Consider the fragment of a Mayer–Vietoris sequence:

$$H_1(Y) \xrightarrow{f} H_1(Y_1) \oplus H_1(Y_2) \to H_1(Q) \to \cdots.$$

In order that the group $H_1(Q)$ be trivial, it is necessary that $\operatorname{rank} H_1(Y)$ be not smaller than $\operatorname{rank} H_1(Y_1) + \operatorname{rank} H_1(Y_2)$, but $\operatorname{rank} H_1(Y) = (q+2)(n-1)$, whereas $\operatorname{rank} H_1(Y_1) + \operatorname{rank} H_1(Y_2) = (q+2)(n-2) + (n+k-1) = (q+2)(n-1) + (n-3)$. It is clear that this may not hold for $n \geqslant 4$. It can be easily shown that gluing nonoriented saddles does not affect the result. This proves the theorem.

This implies, in particular, that using the four above-mentioned types of elementary manifolds, it is impossible to glue any n-dimensional oriented connected closed compact manifold.

In conclusion we present a theorem proved by analogy with Theorem 2.3.2.

THEOREM 2.3.6. *If an n-dimensional connected manifold Q is glued of q oriented saddles, m full tori, and an arbitrary number of nonoriented saddles, and if $\operatorname{rank} H_1(Q, \mathbb{Z}) \leqslant k$, then $m \geqslant q + 2 - 2k$.*

In this case there are no corollaries analogous to those to Theorem 2.3.2, because $\operatorname{rank} H_1(Q, \mathbb{Z}) \geqslant 1$ (which follows from Theorem 2.3.5).

We will give an example of such a manifold Q^{n+1} glued of two cylinders $Y_1 = Y_2 = T^n \times D^1$ that $H_1(Q, \mathbb{Z})_+ = \mathbb{Z}$. By definition, $Y = Y_1 \cap Y_2$. We will list the gluing matrices f, where $f : H - 1(Y) \to H_1(Y_1) \oplus H_1(Y_2)$.

For $n = 2k+1$

$$\begin{pmatrix} 1 & & & & & 1 & & & & \\ & \ddots & & & & & \ddots & & & \\ & & \ddots & & & & & \ddots & & \\ & & & \ddots & & & & & 1 & \\ & & & & 1 & & & & & 1 \\ 1 & & & & & & & & & \\ & \ddots & & & & & & A & & \\ & & \ddots & & & & & & & \\ & & & \ddots & & & & & & \\ & & & & 1 & & & & & \end{pmatrix}$$

For $n = 2k$

$$\begin{pmatrix} 1 & & & & & 1 & & & & & \\ & \ddots & & & & & \ddots & & & & \\ & & \ddots & & & & & \ddots & & & \\ & & & \ddots & & & & & \ddots & & \\ & & & & 1 & & & & & 1 & \\ 1 & & & & & 1 & & & & & \\ & \ddots & & & & & & & 1 & 1 & \\ & & \ddots & & & & & & 1 & 0 & \\ & & & \ddots & & & & & & & \ddots \\ & & & & 1 & 1 & & & & & 0 \end{pmatrix}$$

where

$$A = \begin{pmatrix} 2 & & & & & & 1 & \\ & & & & & & & 1 \\ & & & & & 1 & & \\ & & & & 1 & 1 & & \\ & & & \ddots & & & & \\ & & & & & & \ddots & \\ 1 & 1 & & & & & & 1 \end{pmatrix}$$

CHAPTER 3

Some General Principles of Integration of Hamiltonian Systems of Differential Equations

§1 Noncommutative Integration Method

1.1 *Maximal Linear Commutative Subalgebras in the Algebra of Functions on Sympletic Manifolds*

In Ch. 1, we got acquainted with the Liouville theorem which makes it possible to describe the behaviour of integral trajectories of systems possessing a complete set of integrals in involution. In this chapter, we deal with modern methods of integration, a particular case of which is the method of integration by means of the Liouville theorem. We develop here, in particular, the fundamental investigations of E. Cartan, Marsden, Weinstein, Moser, Bernat, Conze, Duflo, and Vergne.

A linear (infinite-dimensional) space of all smooth functions on a sympletic manifold M^{2n} will be denoted by $C^\infty(M)$. As we already know from Ch. 1, this space naturally transforms into an infinite-dimensional Lie algebra with respect to the Poisson bracket $\{f, g\}$, where $f, g \in C^\infty(M)$. Different subalgebras (both finite- and infinite-dimensional) in the Lie algebra $C^\infty(M)$ are of great interest from the point of view of Hamiltonian mechanics. By using this approach, the Liouville theorem is elucidated as follows. The Hamiltonian H of a system $v = \operatorname{sgrad} H$ can be represented as a vector (that is, a function) in the Lie algebra $C^\infty(M)$. If the system v is *completely Liouville-integrable*, then H is included in a *commutative* subalgebra of functions $G(H)$, the dimension of which is exactly equal to n (half the dimension of the manifold) and in which one can choose an *additive basis* $f_1, \ldots, f_n$, all of whose functions are functionally independent (almost everywhere) on M^{2n}. We may assume that $f_1 = H$ (Fig 53).

Of course, such commutative subalgebra $G(H)$ may not exist. Then the system $v = \operatorname{sgrad} H$ is not integrable in the sense of Liouville. As shown in Ch. 5, most of the systems are not integrable in this sense. That is why one rarely finds a sufficiently large commutative subspace generated by the function H.

If $g \in G(H)$ is any other function in the commutative subalgebra $G(H)$, then the system $\operatorname{sgrad} g$ is also integrable on M^{2n} with the same set of first integrals as

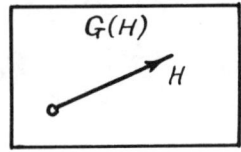

Figure 53

in the system sgrad H. Therefore, the discovery on M^{2n} of a commutative subalgebra $G \subset C^\infty(M)$ of dimension n, which has an additive basis of n functionally independent functions, immediately gives us a whole family of integrable systems.

DEFINITION 3.1.1: We will say that on a symplectic manifold M^{2n} a *maximal linear commutative subalgebra of functions* G_0 is given (in the Lie algebra $C^\infty(M)$ with respect to the Poisson bracket) if $\dim G_0 = n$ and if in G_0 one can choose an additive basis consisting of n functions $f_1, \ldots, f_n$ functionally independent on M^{2n} (almost everywhere). Such an algebra of functions will be sometimes called a *complete involutive (commutative) set of functions*.

In what follows, the term "almost everywhere" is omitted, as it is obviously implied. If g is an arbitrary element (function) of G_0, then $g = c_1 f_1 + \ldots c_n f_n$, where $c_1 \ldots, c_n$ are some constant (real) numbers. Consequently, any element $g \in G_0$ is represented as a *linear combination with constant coefficients* of independent basis functions $f_1, \ldots, f_n$. In Definition 3.1.1 we have therefore introduced the term "linear algebra." The term "maximal" in Definition 3.1.1 also has a clear meaning explained by the lemma that follows.

LEMMA 3.1.1. *If T is a commutative linear subalgebra (not necessarily maximal) in $C^\infty(M)$, whose additive basis forms functions $f_1, \ldots, f_n$ independent on M^{2n}, then always $r \leqslant n$.*

The proof immediately follows from Lemma 1.2.2 (§2, Ch. 1). Indeed, the linear subspace L generated by linearly independent vectors sgrad $f_1, \ldots,$ sgrad f_r has the dimension r and is an isotropic plane in a tangent space to M^{2n} (at the point of general position). Since the dimension of isotropic planes does not exceed n, the assertion of the lemma follows.

Thus, the maximal linear commutative subalgebra G_0 actually *has a maximal possible dimension* among all commutative subalgebras whose *additive* basis consists of independent functions. If we reject the latter condition, then in $C^\infty(M)$ there exist many other commutative subalgebras of dimension greater than n. But in these subalgebras one cannot choose an additive independent (in the functional sense) basis. The class of such commutative subalgebras involves those of infinite dimension. To construct such subalgebras, it suffices for instance to consider the linear subspace generated by the following infinite sequence of independent (in the additive sense, but dependent in the functional sense) functions $f, f^2, f^3, \ldots, f^i, \ldots$.

From now on, the terminology will not be as confusing, and by speaking of a maximal commutative subalgebra, we will automatically imply that we deal with the subalgebra which possesses an additive functionally independent basis. In this sense, the theory of infinite-dimensional Lie in algebras is substantially more complicated than the theory of finite-dimensional algebras, because the concept of Cartan subalgebra requires a special discussion in this case.

Thus, an important problem of Hamiltonian geometry is formulated as follows.

A symplectic manifold M^{2n} with a two-form ω is given. *Does there exist at least one maximal linear commutative subalgebra G_0 in the Lie algebra of functions $C^\infty(M)$ (with respect to the Poisson bracket) on this manifold?* If such subalgebras do exist, one should find out *how many there are*, by how many parameters they are given, *what the algorithms of their construction are, how these subalgebras (and their amount) are associated with the metric and topological properties of the manifold*, and whether or not there exist *topological obstacles* for the existence of such subalgebras on the manifold M^{2n}.

If this problem is solved (in a certain sense), the next question is: What concrete and interesting (from the mechanical, physical point of view) Hamiltonians exist in such maximal commutative subalgebras? All such Hamiltonians give rise to integrable Hamiltonian systems.

Let a sympletic manifold M^{2n} be fixed. It is not clear a priori whether on this manifold there exists *at least one* maximal linear commutative subalgebra of functions, that is, whether there exists atleast one integrable system on M^{2n}. We shall specify the formulation.

If M^{2n} is a smooth manifold, then one can always find *at least one* maximal linear commutative subalgebra G_0. It is constructed in a very simple way. It turns out that on $C^\infty(M)$ a closed 2n-dimensional ball D^{2n}, in which the canonical sympletic coordinates $p_1, q_1, \ldots, p_n, q_n$ are given, one can always construct a set of n independent smooth functions $f_1, \ldots, f_n$ which are in involution and vanish on the ball boundary. For this consider the functions

$$g_i = p_i^2 + q_i^2, \qquad 1 \leqslant i \leqslant n,$$

and the function

$$\alpha = \sum_{i=1}^n g_i = \sum p_i^2 + q_i^2.$$

Let $\rho(r)$, where r is the distance to the centre of the ball D^{2n}, be such a smooth function, equal to unity within the ball D_1^{2n}, r be the distance to the centre of the ball D^{2n}, strictly embedded in the ball D^{2n}, which is equal to zero on the boundary of the ball D^{2n} and monotonously decreases from unity to zero within the ring $D^{2n} \setminus D_1^{2n}$. Suppose $\varphi(p,q) = \rho(\alpha(p,q))$ take the functions φg_i, $1 \leqslant i \leqslant n$. It is easy to calculate that the functions φg_i are in involution, independent on the interior of the ball D^{2n}, and equal to zero on the ball boundary. Covering the manifold with nonintersecting balls, so that the complement of their union has a measure zero, and "sewing" the functions constructed above, we obtain just the maximal linear commutative subalgebra on M.

At the same time, it is clear that this set of functions is pathological and is of no interest whatsoever from the point of view of applications. The class of functions among which one seeks maximal commutative subalgebras should therefore be narrowed. It is natural to examine the following classes.

1) *Bott* functions on smooth manifolds.
2) *Real analytic* functions on manifolds of the same class.
3) *Algebraic* functions on algebraic manifolds.
4) It is reasonable to analyze the *complex-analytic* case.

It is not yet clear whether on any of the symplectic manifolds of the four classes listed above there exists a corresponding maximal linear commutative (MLC) subalgebra of functions.

It turns out that a Hamiltonian system can be integrated not only in case its Hamiltonian is included in the MLC subalgebra, but also in case it is included in a *noncommutative* Lie algebra G of functions on M which has the property $\dim G + \operatorname{rank} G = \dim M^{2n}$. This question is considered in the next subsection.

1.2 A Hamiltonian System Is Integrable if Its Hamiltonian Is Included in a Sufficiently Large Lie Algebra of Functions

The highly important role of commutative maximal linear subalgebras is well known. The discovery of MLC subalgebras is therefore an extremely urgent problem. But, as is seen in the recent studies, in many important cases, complete involutive sets of functions (MLC algebras) are so successively "masked" in the space of all smooth functions that one should make great effort to discover them. So, we are facing the problem of *effective identification of MLC subalgebras*. We have revealed that in some cases the existence of an MLC subalgebra of functions (integrals) may be derived from the fact of the existence of a *noncommutative* algebra of functions *(which are not, generally, integrals of the system)*. Naturally, this sometimes *simplifies the search for completely integrable systems*. We will now proceed to this *method of noncommutative integration*.

Let G be a finite-dimensional Lie algebra, G^* a dual space, $\xi \in G^*$ a certain covector. Consider in the algebra G a linear subspace $H_\xi = \operatorname{Ann} \xi$ consisting of all vectors X such that $\operatorname{ad}_X^* \xi = 0$. The subspace H is called an *annihilator of the covector ξ*. We will say that $\xi \in G^*$ is a *covector of general position* if the dimension of its annihilator is *the smallest*. The dimension of the annihilator of a covector in general position will be called *the index* $\operatorname{ind} G$ of the Lie algebra G. If G is a semisimple Lie algebra, then its index coincides with its *rank*, that is, with the dimension of the Cartan subalgebra (the maximal commutative subalgebra in G).

We will formulate the theorem proved by Fomenko and Mishchenko (see [90], [93], [143]), which generalizes the Liouville theorem and the theorems of E. Cartan, Marsden, and Weinstein.

DEFINITION 3.1.2: Let G be a finite-dimensional subalgebra in a Lie algebra $C^\infty(M)$ (with respect to the Poisson bracket). The subalgebra G will be called a *maximal linear (ML) subalgebra* on a symplectic manifold M^{2n} if $\dim G + \operatorname{ind} G = \dim M^{2n}$ and if we can choose in G an additive basis consisting of functionally independent (almost everywhere) functions on the manifold M.

It is clear that if G is an MLC subalgebra in $C^\infty(M)$, then it is an ML subalgebra. Generally, the converse is false, first of all because a maximal linear subalgebra may be noncommutative. We will soon give such examples (many of them).

Let G be a certain subalgebra (not necessarily maximal) in which one can choose an additive basis of the functionally independent functions $f_1, \ldots, f_k$. Then

§1 Noncommutative Integration Method

$\dim G = k$. Consider the momentum mapping $F: M^{2n} \to G^*$, where

$$F(x) = (f_1(x), \ldots, f_k(x)) = \xi \in G^*.$$

Here G^* is a k-dimensional linear space. Each point ξ of this space determines a certain common level surface M_ξ of the functions $f_1, \ldots, f_k$, that is, $M_\xi = F^{-1}(\xi)$, $M_\xi \subset M^{2n}$.

Fix a nonsingular level surface M_ξ by setting a certain covector $\xi \in G^*$. Then ξ also generates the subalgebra $H_\xi = \operatorname{Ann} \xi$, that is, its own annihilator. Thus, each covector ξ is in one-to-one correspondence with two objects which are of importance to us:

a) the *level surface* $M_\xi = F^{-1}(\xi)$,
b) the *annihilator* $\operatorname{Ann} \xi$.

If $\xi = \sum_{i=1}^{n} \xi_i e_i$, where the covectors $e_1, \ldots, e_k$ are dual to the basis vectors $f_1, \ldots, f_k$, then the surface M_ξ is given in M by the system of equations: $f_1(x) = \xi_1, \ldots, f_k(x) = \xi_k$. The annihilator $\operatorname{Ann} \xi$ lies in the original algebra of functions G.

As the covector $\xi \in G^*$ changes, its annihilator also changes somewhat in the Lie algebra G. The same change can be modulated by changing the common level surface M_ξ of the basic functions $f_1, \ldots, f_k$.

The level surface M_ξ will be called *nonsingular* if all the functions $f_1, \ldots, f_k$ are independent at each point of this surface. Then $\dim M_\xi = 2n - k$.

THEOREM 3.1.1. *Let G be a maximal linear finite-dimensional algebra of functions on M^{2n}, let $\xi \in G^*$ be a covector of general position and M_ξ the corresponding common nonsingular level surface of independent functions $f_1, \ldots, f_k$ which form an additive basis in G. If the surface M_ξ is compact and connected, then it is diffeomorphic to an r-dimensional torus T^r, where $r = \operatorname{ind} G$ (that is, $r = 2n - k$). On the torus T^r and on all level surfaces which are close to this torus and diffeomorphic to it, one can introduce regular coordinates $\varphi_1, \ldots, \varphi_r$, such that in these coordinates all vector fields of the form $\operatorname{sgrad} h$, where $h \in \operatorname{Ann} \xi$, set a conditionally periodic motion along the torus T^r (that is, they are fields with constant components).*

In this theorem, we did not imply any concrete Hamiltonian system, but described the properties of the whole class of fields of the form $\operatorname{sgrad} h$ generated by the annihilator of the covector of general position. *A particular case of Theorem 3.1.1 is, of course, the classical Liouville theorem.* Indeed, if the maximal linear subalgebra of functions G is commutative then its index $r = \operatorname{ind} G$ is equal to its dimension k and, theoreofore, the maximality condition becomes $k + k = \dim M = 2n$, that is $k = n$. In this case, all the tori T^r from Theorem 3.1.1 are ordinary n-dimensional Liouville tori.

Now we shall look at Theorem 3.1.1 from the point of view of the possibility of integrating concrete systems $v = \operatorname{sgrad} H$. By fixing the Hamiltonian H, we set a certain element of a Lie algebra $C^\infty(M)$. Consider in this algebra various maximal linear subalgebras G (if they do exist) and try to find at least one such subalgebra that the Hamiltonian H get into G.

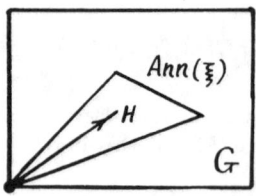

Figure 54

DEFINITION 3.1.3: We shall say that the Hamiltonian H admits a *correct embedding* in the maximal linear algebra of functions G if $H \in G$ and also if there exists a covector $\xi \in G^*$ of general position, such that $H \in \text{Ann}\,\xi \subset G$ (Fig. 54).

If the Hamiltonian H admits a correct embedding in a certain maximal linear subalgebra G, then defined is a certain (at least one) common nonsingular level surface $F^{-1}(\xi) = M_\xi \subset M$ which corresponds to the covector of general position $\xi \in G^*$.

THEOREM 3.1.2. Let $v = \text{sgrad}\,H$ be a Hamiltonian system on M^{2n}, where the Hamiltonian H admits a correct embedding in a certain maximal linear subalgebra of functions G on M. Then there exists at least one covector $\xi \in G^*$ of general position, such that $H \in \text{Ann}\,\xi$ and the common nonsingular level surface M corresponding to this covector is invariant with respect to the field v. If the surface M_ξ is compact and connected, then it is diffeomorphic to the torus T^r, where $r = \dim M - \dim G$, and on T^r there exist regular coordinates $\varphi_1, \ldots, \varphi_r$, with respect to which the field v determines a conditionally periodic motion on the torus T^r (that is, the integral trajectories of the field v set a rectilinear torus winding).

DEFINITION 3.1.4: The Hamiltonian systems admitting integration in the sense of Theorem 3.1.2 will be called *completely integrable in the noncommutative sense (noncommutatively integrable).*

Theorem 3.1.2 is an obvious corollary to Theorem 3.1.1 We should emphasize an important fact: *not nearly all* functions g from the maximal linear subalgebra G are *usual integrals* of the field $v = \text{sgrad}\,H$. Integrals (in the usual sense) are only the functions h from the annihilator $\text{Ann}\,\xi \subset G$ of the covector ξ. If the algebra G is noncommutative then $\{g, H\} \neq 0$ for $g \notin \text{Ann}\,\xi$.

How should Theorems 3.1.1 and 3.1.2 be applied in practice? One should first find the largest number of independent functions commuting with H. Next, one should try *to include* this family A into a certain larger *(in general, already noncommutative)* Lie algebra G, so that A may be represented in the form of the *annihilator of a covector of general position*. If the algebra G turns out *maximal linear*, one can integrate the original system in the same terms (tori, conditonally periodic motion, etc.) as in the classical Liouville theorem. Such situations are frequently encountered in concrete problems of Hamiltonian geometry (see Ch. 4). For the proof of Theorems 3.1.1 and 3.1.2, see our papers [90], [93], [143]. It employs, in particular, the reduction of Marsden and Weinstein [78], the results of Duflo and Vergne [40], Bernat, Conze, and Vergne [199].

Before proving the theorems, we will comment on their assertions.

1) In noncommutative Lie algebras G, one has the inequality: $k = \dim G > \operatorname{ind} G = r$. For instance, this takes place in a semisimple case, where $r \approx \sqrt{k}$, that is, r is substantially smaller than k. Since $r + k = 2n$, the r-dimensional invariant tori T^r, along which integral tranjectories of the system move, are "few-dimensional," that is, their dimension is less (sometimes substantially) than half the dimension of the symplectic manifold M^{2n}. This means that a Hamiltonian system, which admits noncommutative integration, is *degenerate* in the sense that its integral trajectories lie not on n-dimensional Liouville tori, but on tori *of smaller dimension*.

2) By virtue of remark 1, it will be natural to call the Hamiltonian systems admitting noncommutative integration *systems with degeneracies*. Therefore, on one hand, they are more complicated than systems integrable in the usual commutative sense. On the other hand, *the collection of maximal linear subalgebras is much larger than the collection of maximal linear commutative subalgebras*. In geometry, one often encounters Hamiltonian systems whose Hamiltonian is correctly included in the maximal linear algebra of functions (and, as we already know, these systems admit noncommutative integration). For this reason, *our method now allows us to integrate a substantial collection of concrete systems, the usual commutative integration of which is difficult.* The experience of recent years shows that to discover a noncommutative ML algebra is often much simpler than to discover an MLC algebra of integrals.

3) Of interest is the case where a Hamiltonian H is correctly included into a maximal linear algebra of functions G and gets into the centre of the algebra G. In this case, all the functions from the algebra G commute with the Hamiltonian H, that is, are *usual integrals. The integrals themselves do not have to commute with one another*, that is, on the whole, the algebra G may be noncommutative. It is natural to call such an algebra G *a noncommutative algebra of integrals* of a given system. By virtue of Theorems 3.1.1 and 3.1.2, the trajectories of such a system set a conditionally periodic motion on tori T^r, where $r = \operatorname{ind} G$.

1.3 Proof of the Theorem

Let the Hamiltonian of a system $v = \operatorname{sgrad} H$ be correctly included into a certain finite-dimensional maximal linear algebra G of functions on M^{2n}. We denote by $\mathfrak{G}$ the corresponding simply connected Lie group which acts on M by symplectic transformations. Let $f_1, \ldots, f_k$ be an additive basis in G, which consists of independent functions, and let $F : M \to G^*$ be momentum mapping. By $M_\xi = F^{-1}(\xi)$, where $\xi \in G^*$, we denote the common level surface of the functions $f_1, \ldots, f_k$, which is therefore set by the system of equations:

$$f_i(x) = \xi_i, 1 \leqslant i \leqslant n, \qquad \xi = (\xi_1, \ldots, \xi_n).$$

The operation in the Lie algebra G is given by the Poisson bracket: $f, g \to \{f, g\}$. If $\xi \in G^*$ and $g \in G$, then the value of the covector (linear functional) ξ on the element g is denoted as $\langle \xi, g \rangle = \xi(g)$. Then

$$\langle \xi, \{g, f\} \rangle = \langle \operatorname{ad}_g^* \xi, f \rangle.$$

It is convenient to denote the covector $\mathrm{ad}_g^* \xi$ by $\{\xi, g\}$. One should bear in mind that this is not a usual Poisson bracket, because here

$$\xi \in G^*, \qquad g \in G, \qquad \{\xi, g\} \in G^*.$$

In this notation, we have:

$$\langle \xi, \{g, f\}\rangle = \langle \{\xi, g\}, f\rangle.$$

Clearly,

$$H_\xi = \mathrm{Ann}\, \xi = (g \in G : \{\xi, g\} = 0).$$

LEMMA 3.1.1. *Let an element f of the algebra G lie in the annihilator H_ξ and $x \in M_\xi$, where M_ξ is a nonsingular level surface. Then $\mathrm{sgrad}\, f(x) \in T_x M_\xi$. This means that the skew gradients of the functions from the annihilator of the covector ξ are tangent to the surface M_ξ.*

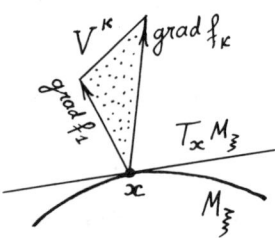

Figure 55

PROOF: Consider the additive generators $f_1, \ldots, f_k$ in G and the gradients $\mathrm{grad}\, f_1, \ldots, \mathrm{grad}\, f_k$. Since M_ξ is a nonsingular surface, it follows that all the gradients are independent at all points of the surface, and therefore they are transversal to M_ξ. Thus, a k-dimensional plane V spanned by $\mathrm{grad}\, f_i$ intersects $T_x M_\xi$ only at a zero point, and $V \oplus T_x M_\xi = T_x M$ (Fig 55). We will assume that a Riemannian metric is given on M. Then the vectors $\mathrm{grad}\, f_i$ are orthogonal to M_ξ. To prove the relation $\mathrm{sgrad}\, f(x) \in T_x M_\xi$, it suffices to check that $(\mathrm{sgrad}\, f)g = 0$ for any function $g \in G$, i.e., that the derivative along $\mathrm{sgrad}\, f$ of any function g from G constant on M_ξ is equal to zero. Indeed,

$$(\mathrm{sgrad}\, f)g = \langle \mathrm{sgrad}\, f, \mathrm{sgrad}\, g\rangle,$$

where $\langle,\rangle$ is the Riemann metric on M. From equality to zero of the scalar products of $\mathrm{sgrad}\, f$ to the plane V, that is,

$$\mathrm{sgrad}\, f \in T_x M_\xi.$$

Thus,

$$\{f, g\} = (\mathrm{sgrad}\, f)g.$$

Next,

$$\{f, g\}(x) = \langle \xi, \{f, g\}\rangle = \langle \{\xi, f\}, g\rangle = 0$$

because $\{\xi, f\} = 0$, $f \in \mathrm{Ann}\, \xi$, as required.

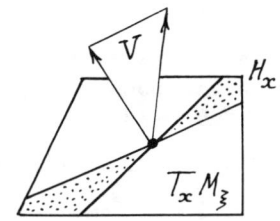

Figure 56

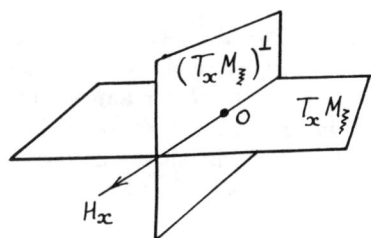

Figure 57

LEMMA 3.1.2. *We have the inequality*

$$T_x M_\xi \cap (\operatorname{sgrad} f; f \in G) = H_x = (\operatorname{sgrad} h; h \in H_\xi)$$

(Fig. 56).

PROOF: We have already proved that

$$(\operatorname{sgrad} h; h \in H_\xi) \subset T_x M_\xi \cap (\operatorname{sgrad} f; f \in G).$$

We shall now prove the inverse inclusion. Let

$$X \in T_x M_\xi \quad \text{and} \quad X = \operatorname{sgrad} f,$$

where $f \in G$. It is required to prove that $f \in H_\xi$. Examine

$$\langle \{\xi, f\}, g \rangle = \langle \xi, \{f, g\} \rangle = 0,$$

for any $g \in G$, because $\{f, g\}(x) = 0$. The latter inequality follows from the fact that all the functions $g \in G$ are constant on the leven surface M_ξ. Thus, $\langle \{\xi, f\}, g \rangle = 0$ for any $g \in G$. This means that $\{\xi, f\} = 0$, that is, $f \in H_\xi$. (see Fig. 57). This proves the lemma.

Consider a group $\mathfrak{H}_\xi$ with a Lie algebra H_ξ, that is, $\mathfrak{H}_\xi \subset \mathfrak{G}$.

LEMMA 3.1.3. *The level surface M_ξ is invariant under the action of the group $\mathfrak{H}_\xi$ on the manifold M.*

Examine the form ω, and let $\tilde{\omega}$ be its restriction to the surface M_ξ. The action of the group $\mathfrak{H}_\xi$ on M_ξ induces at each point $x \in M_\xi$ a plane $H_x \subset T_x M_\xi$ generated by the vectors sgrad f, where $f \in H$ (see Fig. 56).

LEMMA 3.1.4. *The kernel of the form $\tilde{\omega}$ coincides with the plane H_x.*

PROOF: We will prove that $H_x \subset \operatorname{Ker} \tilde{\omega}$. Let $X = \operatorname{sgrad} h$, where $h \in H_\xi, X \in H_x$. It should be proved that X lies in the kernel of the form, i.e., that $\omega(X, Y) = 0$, for any vector Y from the plane $T_x M_\xi$. Indeed,

$$\omega(X, Y) = \omega(\operatorname{sgrad} h, Y) = Y(h) = 0$$

because the vector Y is tangent to the level surface, and being an element of the algebra G, the function h is constant on the level surface. Let us prove the inverse, i.e., that $H_x \supset \operatorname{Ker} \tilde{\omega}$. Let $\omega(X, Y) = 0$ for any vector $Y \in T_x M_\xi$. It is required to represent the vector X in the form $X = \operatorname{sgrad} h$ for a certain function $h \in H_\xi$. Consider the form ω as a skew-symmetric scalar product on $T_x M$ and denote by $(T_x M_\xi)^\perp$ an orthogonal complement of the plane $T_x M_\xi$ determined by the form ω. Since the form ω is nondegenerate, then

$$\dim(T_x M_\xi)^\perp = \dim M - \dim T_x M_\xi = \dim V = k.$$

It is clear that $\operatorname{Ker} \tilde{\omega} = T_x M_\xi \cap (T_x M_\xi)^\perp$. We will prove that sgrad f, $f \in G) = (T_x M_\xi)^\perp$. Indeed, let $Y \in T_x M_\xi$. Then $\omega(\operatorname{sgrad} f, Y) = Yf = 0$, because $f = \operatorname{const}$ on M_ξ. Thus, $(T_x M_\xi)^\perp \supset (\operatorname{sgrad} f, f \in G)$. Next, $\dim(\operatorname{sgrad} f, f \in G) = k = \dim G$. This follows from the fact that the linear subspace generated by the gradients grad $f, f \in G$ is of dimension k (see the definiton of G). Since the skew-symmetric scalar product is nondegenerate, it follows that the linear subspace generated by the skew gradients is also of dimension k. Finally, it has been proved that $\dim(T_x M_\xi)^\perp = k$. That is why (sgrad f, $f \in G) = (T_x M_\xi)^\perp$. (Fig. 57). This implies the lemma.

Let us summarize the facts obtained above. The main objects are:
a) the *level surface* M_ξ, where $\dim M_\xi = 2n - k$;
b) the *orbit* $\mathfrak{G}(x)$ of the point x, $\dim \mathfrak{G}(x) = k$;
c) the *orbit* $\mathfrak{H}_\xi(x)$ of the point x under the action of the subgroup $\mathfrak{H}_\xi = \exp H_\xi$.

Clearly, $T_x \mathfrak{G}(x) = (\operatorname{sgrad} f, f \in G)$, $T_x \mathfrak{H}(x) = H_x$. This implies that $\mathfrak{G}(x) \cap M_\xi = \mathfrak{H}_\xi(x)$ (Fig. 58). Note that the dimension of the orbit $\mathfrak{H}_\xi(x)$ is equal to the dimension of the orbit $\mathfrak{H}_\xi$ and equal to r.

Consider the action of the group $\mathfrak{G}$ on M and suppose that in a small neighbourhood of the surface M_ξ this action has one and the same type of stationary subgroups, that is, all orbits of the group $\mathfrak{G}$ close to the orbit $\mathfrak{G}(x)$ are diffeomorphic to it. Consider the projection $p : M \to M/\mathfrak{G}$ of the manifold M onto the orbit space $M/\mathfrak{G} = N$. This space may be not a smooth manifold and may have singularities. But in view of the above remark, in a small neighbourhood of the point $p\mathfrak{G}(x) \in N$, the space N is a smooth manifold of dimension $2n - k$.

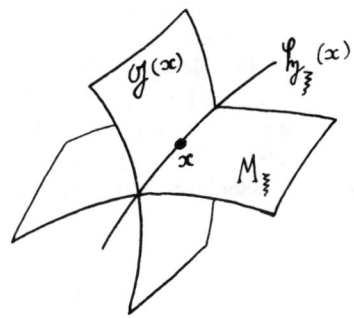

Figure 58

In reality, if the group $\mathfrak{G}$ is, for instance, compact then the set of orbits of general position (of maximal dimension) diffeomorphic to one another is an everywhere open dense subset in M. Therefore, N is a manifold everywhere except for the set of points of measure zero. The manifold N is not necessarily symplectic. For instance, it may be of odd dimension.

Being restricted to the surface M_ξ, the projection p projects this surface onto the space $Q_\xi = M_\xi/\mathfrak{H}_\xi$. That is why N is fibred into surface Q_ξ (Figure 59). This conclusion is based on Lemma 3.1.3.

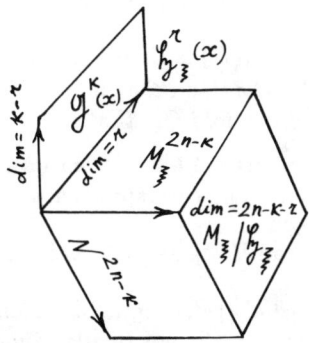

Figure 59

LEMMA 3.1.5. *The manifold Q_ξ are symplectic manifolds with a nondegenerate closed form ρ, which is a projection of the form $\tilde{\omega}$ onto M_ξ, under the mapping $p: M_\xi \to Q_\xi$. In this case $p^*\rho = \tilde{\omega}$.*

The proof follows from Lemma 3.1.4 because the kernel of the form $\tilde{\omega}$ on $T_x M_\xi$ coincides with the plane H_ξ.

Let us go back to studying Hamiltonian systems on M. Let sgrad $H = v$ be a system with the algebra of integrals G, that is, $\{H, G\} = 0$. Since H commutes

with all elements from G, the function H is invariant with respect to the group $\mathfrak{G}$. Indeed, $(\operatorname{sgrad} f)H = \{f, H\} = 0$, $f \in G$. In particular, acting on M, the subgroup $\mathfrak{H}_\xi$ also carries the function H into itself. Thus, the natural projection of the field sgrad H onto space N is defined. The field sgrad H is tangent to the surface M_ξ and is also projected into a certain field $E(H)$ on the factor Q_ξ because the field sgrad H is invariant with respect to the group $\mathfrak{H}_\xi$. The space N is therefore fibred into simplectic manifolds Q_ξ, and the vector field $E(H)$ tangent to all surfaces Q_ξ is defined on N (Fig. 60). Thus, we have associated with the triplet $(M, \operatorname{sgrad} H, \omega)$ a new triplet
$(Q_\xi,\ E(H), \rho)$.

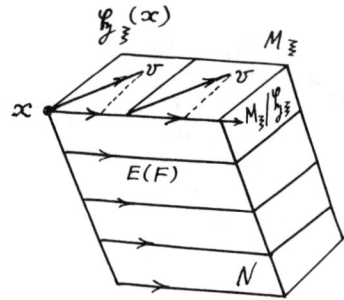

Figure 60

LEMMA 3.1.6. *The vector field $E(H)$ is Hamiltonian with respect to the symplectic form ρ on the manifold Q_ξ for the Hamiltonian function $\widetilde{H}$ equal to the projection of the function $H|_{M_\xi}$ onto the manifold Q_ξ, that is, $E(H) = \operatorname{sgrad}_\rho(p_* H)$.*

The proof follows from Lemma 3.1.5 and from invariance of the Hamiltonian H under the action of the group. The correspondence constructed above

$$(M, \operatorname{sgrad} H, \omega) \to (Q_\xi,\ E(H), \rho),$$

is called the *reduction of the initial system* sgrad H. Thus, on the manifold Q_ξ, we have obtained a *new Hamiltonian system* of smaller dimension $2n - k - r$. In this case, $\dim Q_\xi < \dim M_\xi = 2n - k$. The reduced system on Q_ξ may turn out to be simpler than the initial system on M.

Suppose that one succeeded in integrating the reduced system. This enables the number of integrals of the initial system sgrad H on M to be increased by "raising" these new integrals from the manifold N to the manifold M.

LEMMA 3.1.7. *Let G be a finite-dimensional algebra of integrals of the field sgrad H on M, which satisfies the conditions listed above. Let $E(F)$ be a reduced system on the manifold $N = \cup Q_\xi$, which is Hamiltonian on each submanifold Q_ξ. Let G' be a linear space of functions on N, such that their restrictions to submanifolds Q_ξ form a finite-dimensional algebra of integrals of the flow $E(H)$.*

Then the space of functions $G \oplus G''$, where $G'' = p^*G'$, that is, $G'' = (gp, g \in G', p: M \to N)$ is the Lie algebra of integrals of the system sgrad H, and $\{G, G''\} = 0$.

PROOF: Let g be a certain function on the space N. Then under the mapping $p: M \to N$ the preimage gp of this function is a function on M, which is invariant under the action of the group $\mathfrak{G}$ on M. This means that the function gp is in involution with the whole initial algebra of integrals G. Thus, any new function g, which is an integral for a reduced flow $E(H)$ on N, gives an additional integral gp of the initial field sgrad H on M. That these additional integrals are independent of the functions of the algebra G follows from the fact that their gradients are different from zero in the direction of submanifolds Q_ξ which lie (locally) in the level surface M_ξ, whereas the gradients of the functions from G are orthogonal to M_ξ. This completes the proof.

PROOFS OF THEOREMS 3.1.1 AND 3.1.2:

Let v be one of the systems mentioned above, that is, either $v = $ sgrad H, $\{H, G\} = 0$, or $v = $ sgrad h, where $h \in $ Ann$_\xi$. Examine the reduction described above. Since now the additional condition dim $G + $ ind $G = $ dim M, that is, $k + r = 2n$ is fulfilled, the dimension of the surface M_ξ is equal to r. The dimension of the orbit $\mathfrak{H}_\xi(x)$ contained in M_ξ is also equal to r (Fig. 59). This implies that $M_\xi = \mathfrak{H}_\xi(x)$, that is, in the assumptions of Theorem 3.1.2, the level surface M_ξ is the orbit of the point x under the action of $\mathfrak{H}_\xi$. In particular, dim $Q_\xi = 2n - k - r = 0$. In this case, the structure of the reduced system is therefore particularly simple. Since Q_ξ is a point, the flow $E(H)$ is zero (Fig.61). Here the space N has the dimension n. Since M_ξ is the level surface of integrals G for a flow v, this flow is tangent to M_ξ, that is, M_ξ is an r-dimensional submanifold invariant with respect to all fields of the form sgrad h, $h \in $ Ann$_\xi$, and sgrad H, $\{H, G\} = 0$. It remains to prove that the level surface is an r-dimensional torus, provided that M_ξ is compact and connected.

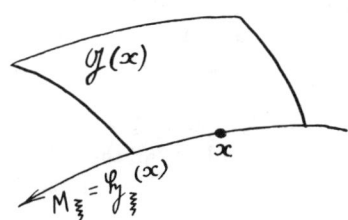

Figure 61

LEMMA 3.1.8 [40], [199]. *Let $\xi \in G^*$ be a covector of general position. Then its annihilator* Ann$_\xi$ *is commutative, in particular, the subgroup* $\mathfrak{H}_\xi$ *is commutative.*

PROOF: Consider a coadjoint action of the group $\mathfrak{G}$ on the coalgebra G^*. We denote the orbit passing through the point $\xi \in G^*$ by $O^*(\xi)$. Since dim $H_\xi = r$ and dim $G^* = k$, it follows that dim $O^*(\xi) = k - r$. Since ξ is a covector of

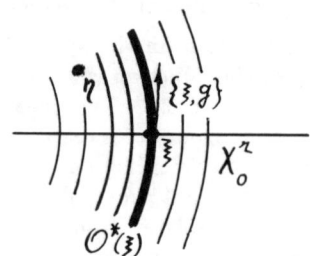

Figure 62

general position, the orbits close to the orbit $O^*(\xi)$ are diffeomorphic to it. One may assume that a sufficiently small neighbourhood U of the point ξ is fibred into homeomorphic fibres (Fig. 62). A local cross-section of the fibration of U into orbits of action of the group $\mathfrak{G}$ will be denoted by X_0. Then U can be represented as the direct product of the base X_0 by a fibre–a part of the orbit (Fig 62). Let $h(\eta)$ be a smooth function on U constant on the orbits. We assert that $\{\xi, dh(\xi)\} = O$, where dh (the differential of h) is interpreted as an element of a dual space G^{**}, that is, $dh(\xi) \in G$. In other words, we assert that $dh(\xi) \in \operatorname{Ann} \xi$. One should make sure that $\{\xi, dh(\xi)\}(g) = O$ for any $g \in G$. Thus we have:

$$\{\xi, dh(\xi)\}(g) = \langle \{\xi, dh(\xi)\}, g \rangle = \langle \xi, [dh(\xi), g] \rangle = -\langle \{\xi, g\}, \quad dh(\xi) \rangle = O,$$

because the covector $\{\xi, g\} = \operatorname{ad}_g^* \xi$ lies in the tangent plane $T_\xi O^*$ at the point ξ, and the function h is constant on the orbits, in particular, on the orbit O^* (Fig. 62).

On a cross-section X_0, which is a smooth surface of dimension r transversally intersecting the orbits close to the orbit $O^*(\xi)$, consider a set of r independent functions $h_1, \ldots, h_r$ and continue them up to the smooth functions on the entire neighbourhood U, by way of continuing their cross-sections X_0, using the values constant along the orbits O^*.

We obtain $\{\xi, dh_i(\xi)\} = 0$, $1 \leqslant i \leqslant r$. Thus, $dh_i(\xi) \in \operatorname{Ann} \xi$. Since the functions h_i were chosen as independent, all the differentials $dh_i(\xi)$ are independent in $\operatorname{Ann} \xi$ and their number is equal to r, that is, it exactly coincides with the dimension of $\operatorname{Ann} \xi$.

Thus, the differentials $dh_i(\xi)$ form a basis in $\operatorname{Ann} \xi$, and to prove commutativity of the annihilator it suffices to prove that pairwise commutators of these differentials are equal to zero, i.e., that $0 = [dh_i(\xi), dh_j(\xi)]$. Since $\{\xi, dh_i(\xi)\} = 0$, we have the inequality

$$b(\xi) = \{\xi, dh_i(\xi)\} \quad (dh_j(\xi)) = 0.$$

Hence,

$$b(\xi) = \langle \{\xi, dh_i(\xi)\}, \quad dh_j(\xi) \rangle = \langle \xi, [dh_i(\xi), \quad dh_j(\xi)] \rangle = 0.$$

Consider an arbitrary direction of η in the neighbourhood U and differentiate the function $b(\xi) = 0$ along this direction. We obtain:

$$0 = \frac{d}{d\eta}b(\xi) = \frac{d}{d\eta}\langle\xi[dh_i(\xi), dh_j(\xi)]\rangle$$

$$= \langle\eta, [dh_i(\xi), dh_j(\xi)]\rangle + \langle\xi, [\frac{d}{d\eta}dh_i(\xi), dh_j(\xi)]\rangle +$$

$$\langle\xi, [dh_i(\xi), \frac{d}{d\eta}dh_j(\xi)]\rangle = \langle\eta, [dh_i(\xi), dh_j(\xi)]\rangle -$$

$$\langle\{\xi, dh_j(\xi)\}, \frac{d}{d\eta}dh_i(\xi)\rangle - \langle\{\xi, dh_i(\xi)\}, \frac{d}{d\eta}dh_j(\xi)\rangle =$$

$$\langle\eta, [dh_i(\xi), dh_j(\xi)]\rangle = 0,$$

because $\{\xi, dh_j(\xi)\} = \{\xi, dh_i(\xi)\} = 0$. Since $\eta \in U$ is arbitrary, the equality $\langle\eta, [dh_i(\xi), dh_j(\xi)]\rangle = 0$ implies that $[dh_i(\xi), dh_j(\xi)] = 0$. The assertion follows.

Let us go back to the proof of the theorem. We have proved that M is the orbit of the group $\mathfrak{H}_\xi$, and since $\dim M_\xi = \dim \mathfrak{H}_\xi$, it follows that M_ξ is the factor group of the group $\mathfrak{H}_\xi$ by the discrete lattice r. As the group $\mathfrak{H}_\xi$ is commutative, then if compact and connected, M_ξ appears as an n-dimensional torus. The other assertions are proved in the same way as the Liouville theorem.

§2 The General Properties of Invariant Submanifolds on Hamiltonian Systems

2.1 *Reduction of a System on One Isolated Level Surface*

Consider the results obtained in §1 from a more general point of view. Let us single out in the pure form the most general elements of the constructions described above. Recall that a smooth function f on a manifold M is called the *integral* of a vector field v if it is constant along all integral trajectories of the field. In other words, an arbitrary function f in the direction of a field v is equal to zero, that is, $v(f) = 0$. In this case, integral trajectories of the field lie on the levels $M_c = (f = c)$ of the function f.

A smooth function f is called a *particular integral* of a field v if one (!) of its level surfaces is invariant with respect to the field v (in particular, with respect to the flow induced by the field). If this level has the form $M_0 = (f = 0)$, then the indicated condition will be written as $v = (f)|_{(f=0)} = 0$. In the case of the general position, M is an invariant submanifold of codimension 1. More generally, a submanifold L of codimension k in M given by the equations $L = (f_1 = 0, \ldots, f_k = 0)$ is called *invariant* (with respect to the field v) if $v(f_i)|_L = 0$. In particular, if several functionally independent first integrals $f_1, \ldots, f_k$ of the system v are given, then each submanifold of the form

$$M_{c_1}, \ldots, c_k = (f_1 = c_1, \ldots, f_k = c_k)$$

will be invariant.

PROPOSITION 3.2.1. *Let $v = \operatorname{sgrad} H$ be a Hamiltonian field on a symplectic manifold M^{2n} and f be a smooth function commuting (in the sense of the usual Poisson bracket) with the Hamiltonian H of this field, that is, $\{H, f\} = 0$ on M. Then the function f is an integral of the field v. Next, a submanifold $L^{2n-k} = (f_1 = 0, \ldots, f_k = 0)$ is invariant with respect to the field v under the condition that $\{H, f_i\}|_L = 0$.*

REMARK: We may say that the functions f_i must commute with the function H on the surface L only (and generally, they do not already commute) outside this surface.

PROOF: Calculating the derivative of the function f along the field v, we obtain

$$v(f) = \omega(v, \operatorname{sgrad} f) = \omega(\operatorname{sgrad} H, \operatorname{sgrad} f) = \{H, f\} = 0$$

either on the entire manifold M or only on the surface L, as required.

The Liouville theorem asserts that each invariant compact submanifold L^n in M^{2n} (for an integrable system) is a torus. Nekhoroshev and Kozlov have noticed that one may single out the case where a single isolated invariant submanifold $L = (f_1 = \ldots f_{n+k} = 0)$ in M^{2n} is a torus. To this end, one should require that on the submanifold L we have the relations:

$$\{f_i, f_l\} = 0, \quad i = 1, n-k, \; l = 1, \ldots, n+k$$

(that is, the vector fields $\operatorname{sgrad} f_i$, $1 \leqslant i \leqslant n-k$, are tangent to the surface L), and also that

$$d\{f_i, f_j\} = 0 \quad \text{for} \quad i, j = 1, 2, \ldots, n-k,$$

these identities being fulfilled at each point of the submanifold L (but, in general, not outside it). This means that the restrictions of the fields $\operatorname{sgrad} f_i$ to the surface L commute (on L but not outside L).

Now we will describe a simple and useful construction enabling a Hamiltonian system, which, in general, has a noncommutative symmetry group (see §1 above) or, more generally, an invariant submanifold, to be transformed into a Hamiltonian system on a sympelctic manifold of smaller dimension. This procedure is called reduction of a Hamiltonian system on one surface and, as has been noted by Tatarinov, in fact, goes back to Cartan. The modern version of this procedure, although expressed in different terminology, is presented in the paper by Losco [200].

Let a Hamiltonian system $v = \operatorname{sgrad}$ be given on a sympectic manifold (M^{2n}, ω) and let $L = (f_1 = \cdots = f_k = 0)$ be an invariant manifold of this system, that is $\{f_i, H\}|_L = 0$.

PROPOSITION 3.2.2. *Let G_L be a set of all smooth functions F, such that the submanifold L is invariant for all corresponding Hamiltonian system $\operatorname{sgrad} F$. Then G_L is the Lie algebra with respect to the Poisson bracket.*

PROOF: It is obvious that G_L is a vector space. Let $F_1, F_2 \in G_L$. Then by virtue of the Poisson-Jacobi identity we have:

$$\{f_i, \{F_1, F_2\}\} + \{F_1, \{F_2, f_i\}\} + \{F_2, \{f_i, F_1\}\} = 0.$$

§2 The General Properties of Invariant Submanifolds

If $f|_L = 0$, then $f = \sum_{\alpha=1}^k g_\alpha f_\alpha$. In particular,

$$\{f_i, F_1\} = \sum \varphi_\alpha^{1i} f_\alpha, \qquad \{f_i, F_2\} = \sum \varphi_\alpha^{2i} f_\alpha.$$

Substituting these expressions into the Poisson-Jacobi identity, we obtain:

$$\{f_i, \{F_1, F_2\}\} = -\{\{f_i, F_2\}, F_1\} + \{\{f_i, F_1\}, F_2\}$$
$$= -\{\sum \varphi_\alpha^{2i} f_\alpha, F_1\} + \{\sum \varphi_\alpha^{1i} f_\alpha, F_2\} = -\sum \varphi_\alpha^{2i} \{f_\alpha, F_1\}$$
$$+ \sum \varphi_\alpha^{1i} \{f_\alpha, F_2\} - \sum f_\alpha \{\varphi_\alpha^{2i}, F_1\} + \sum f_\alpha \{\varphi_\alpha^{1i}, F_2\}.$$

The whole of this expression vanishes on L, as required.

Let Y be a vector field on a manifold P and let λ be some exterior 2-form on P. We denote by $i_Y(\lambda)$ the 1-form obtained by substituting one argument, the field Y, into the form λ. If

$$\lambda = \sum \lambda_{ij} dx_i \wedge dx_j,$$

then

$$i_Y(\lambda) = \sum_j (\sum_i \lambda_{ij} dx_i(Y)) dx_j = \sum_j (\sum_i \lambda_{ij} Y^i) dx_j.$$

PROPOSITION 3.2.3. *Consider a restriction ω_L of the canonical structure ω to be a submanifold L, and in each tangent space $T_z L \subset T_z M$, single out a plane*

$$\prod_z (L) = \text{Ker}_z \omega_L = Y \in T_z L : i_Y \omega_L = 0).$$

In $T_z M$, this is an intersection of $T_z L$ with its skew-orthogonal complement. Suppose that $\dim \prod_z = \text{const}$. Then: 1) The distribution $\prod_z$ is completely integrable by the Frobenius theorem. We denote by $z_1 \sim z_2$ the equivalence relation, by virtue of which z_1 and z_2 belong to one and the same integral manifold of this distribution (maximally continued and, generally, immersed). Suppose that the factor manifold $N = L/\sim$ is well defined and denote the corresponding projection by $\pi : L \to N$.

2) Suppose $\tilde{\omega}(\tilde{X}_1, \tilde{X}_2) = \omega_L(X_1, X_2) = \omega(X_1, X_2)$, where $\tilde{X}_i = \pi_* X_i, X_i \in T_z L$. Then $\tilde{\omega}$ is a canonical structure well defined on N.

It is clear that this proposition is similar to the assertions proved above for the family invariant submanifolds.

PROOF: Note, first, that ω_L is the so-called *precanonical structure* on L (the nondegeracy condition is lifted, but the rank is constant, and the closedness of the form $d(\omega_L) = (d\omega)_L = 0$ is retained). In his paper [200], Losco made use of the fact that the Darboux theorem remains valid for precanonical structures. But we shall take a different approach, the one suggested by Tatarinov. Note that if a vector field Y (on L) is such that $Y(z) \in \prod_z$, then

$$\mathcal{L}_Y(\omega_L) = i_Y(d\omega_L) + d(i_Y \omega_L) = 0,$$

where $\mathcal{L}$ is a Lie derivative. The distribution $\prod_z$ is the so-called *characteristic distribution* for

$$\omega_L : \prod_z = \text{Ker}\,\omega_L \cap \text{Ker}\,d\omega_L (= \text{Ker}\,\omega_L).$$

It is integrable by the Frobenius theorem. Namely, if $Y_1(z)$, $Y_2(z) \in \prod_z$ then

$$i_{[Y_1, Y_2]}\omega_L = [\mathcal{L}_{Y_1}, i_{Y_2}]\omega_L = \mathcal{L}_{Y_1}(i_{Y_2}\omega_L) - i_{Y_2}(\mathcal{L}_{Y_1}\omega_L) = 0,$$

because

$$i_{Y_2}\omega_L = 0, \qquad \mathcal{L}_{Y_1}\omega_L = 0.$$

As far as the correctness of the definition of $\tilde{\omega}$ is concerned, it is implied by the following remarks (always $Y \in \prod_z$). First,

$$\omega(X_1 + Y, X_2) = \omega(X_1, X_2) + \omega(Y, X_2) = \omega(X_1, X_2),$$

that is, the value of $\tilde{\omega}$ does not depend on the choice of X_1, X_2 at a given point $X \in L$. Second, if $X_i(z)$, $Y(z)$ are vector fields then

$$[Y, X_i] = 0 \Longrightarrow Y(\omega_L(X_1, X_2)) = (\mathcal{L}_Y \omega_L)(X_1, X_2) = 0,$$

according to the definition of $\mathcal{L}$, that is, any shift of X_1, X_2 along fibres does not change their skew-scalar product.

The form $\tilde{\omega}$ is nondegenerate (which is obvious) and closed because

$$\omega = \pi^* \tilde{\omega} \Longrightarrow d\omega_L = \pi^* d\tilde{\omega} = 0.$$

PROPOSITION 3.2.4. *Let the assumptions of Proposition 3.2.3 be fulfilled and, besides, let the submanifold L be invariant with respect to the field $\operatorname{sgrad} F$. Then:*
1) The function F is constant on integral manifolds of the distribution $\prod_z$, so that the function $\tilde{F} : N \to \mathbb{R}$, such that $F|_L = \tilde{F} \cdot \pi$, is well defined.
2) In the structure $\tilde{\omega}$, we have the identity: $\widetilde{\operatorname{sgrad} \tilde{F}} = \operatorname{sgrad} F$.

PROOF: Since $\operatorname{sgrad} F \in T_z L$, for any $Y \in \prod_z$, we have: $Y(F) = \omega(Y, \operatorname{sgrad} F) = 0$. Next,

$$\tilde{\omega}(\tilde{Y}, \widetilde{\operatorname{sgrad}\tilde{F}}) = \tilde{Y}(\tilde{F}) = Y(F) = \omega(Y, \operatorname{sgrad} F) = \tilde{\omega}(\tilde{Y}, \widetilde{\operatorname{sgrad} F}),$$

as required.

2.2 Further Generalizations of the Noncommutative Integration Method

The *noncommutative integration method* described in §1 was proposed by Fomenko and Mishchenko and then developed by Brailov and extended to the case of a larger collection of subalgebras in the Lie algebra $C^\infty(M)$. In this subsection, we briefly present the result obtained by Brailov.

Recall the definition of the index of a finite-dimensional Lie algebra G. Let G^* be a dual space. Then

$$(1) \qquad \operatorname{ind} G = \inf_{\xi \in G^*} \dim H_\xi,$$

where

$$H_\xi = (g \in G : \operatorname{ad}_g^* \xi = 0).$$

§2. The General Properties of Invariant Submanifolds

In this subsection, we shall deal with (possibly, infinite-dimensional) Lie algebras of functions on a manifold M with the Poisson bracket $\{\,,\,\}$; we shall define in terms of the Lie algebra of (first) integrals the *differential index of such Lie algebras* and shall give the *integrability condition in quadratures* of dynamic systems on M Hamiltonian with respect to the Poisson bracket $\{\,,\,\}$.

Let M be a symplectic manifold. It is also convenient to consider the case where $M = G^*$, that is, where such a manifold is not symplectic in the usual sense. The *Poisson bracket* on G^* is given by the following formula. Let f, g be smooth functions on G^*. Then:

$$\{g, f\}(\xi) = \langle \xi, [dg(\xi), df(\xi)]\rangle.$$

Let $\mathcal{F} \subset C^\infty(M)$ be a certain, possibly infinite-dimensional subalgebra in the Lie algebra $C^\infty(M)$. The maximum number of linearly independent differentials among $df \in T_x^* M$, where $f \in \mathcal{F}$, is called the *differential dimension of the algebra at a point* $x \in M$. The maximum of differential dimensions of $\mathcal{F}$ at a point x, for all $x \in M$ is called the *differential dimension of the algebra* $\mathcal{F}$. Those points x at which this maximum is attained will be called *nonsingular* points. The differential dimension will be denoted by $d.\dim \mathcal{F}$. For any point x, define

$$\mathcal{F}_x = (f \in \mathcal{F} : \{f, g\}(x) = 0, \forall g \in \mathcal{F}).$$

It can be easily seen that $\mathcal{F}_x$ is a subalgebra in $\mathcal{F}$. Let $x \in M$ be a nonsingular point.

Define the *finite-dimensional Lie algebra* $K_x \mathcal{F} = (df \in T_x^* M : f \in \mathcal{F}_x)$, in which the *commutation operation* is given by the formula

(2) $$[df, dg] = d\{f, g\}; \qquad f, g \in \mathcal{F}_x$$

We shall prove correctness of this definition. Indeed, let $df = 0$. The local system of coordinates $y_1, \ldots, y_m$ on M in the neighbourhood of the point x can always be chosen that any function $g \in \mathcal{F}$ is expressed through $y_1, \ldots, y_k$, $k = d.\dim \mathcal{F} \leqslant m$. We have

$$d\{f, g\} = d\left(\sum_{i=1}^m \frac{\partial f}{\partial y_i}\{y_i, g\}\right) = \sum_{i=1}^k \{y_i, g\} d\frac{\partial f}{\partial y_i} + \sum_{i=k+1}^m \{y_i, g\} d\frac{\partial f}{\partial y_i}.$$

The second sum is equal to zero because f does not depend on y_i for $i \geqslant k+1$. The first sum is equal to zero because $g \in \mathcal{F}_x$. Thus, the commutator $[\xi, \eta]$ of the elements ξ, η does not depend on the arbitrariness of the choice of f, g such that $\xi = df, \eta = dg$. The Jacobi identity:

$$I = [\xi, [\eta, \varsigma]] + [\eta, [\varsigma, \xi]] + [\varsigma, [\xi, \eta]] = 0,$$

for $\xi, \eta, \varsigma \in K_x \mathcal{F}$ is verified as follows. Let $\xi = df$, $\eta = dg$, $\varsigma = dh$, where $f, \gamma, h \in \mathcal{F}_x$. Then:

$$I = [df, [dg, dh]] + [dg, [dh, df]] + [dh, [df, dg]]$$
$$= [df, d\{g, h\}] + [dg, d\{h, f\}] + [dh, d\{f, g\}]$$
$$= d\{f, \{g, h\}\} + d\{g, \{h, f\}\} + d\{h, \{f, g\}\} = 0,$$

by virtue of the Jacobi identity for the Posson bracket. Thus, $K_x \mathcal{F}$ is the finite-dimensional Lie algebra for each nonsingular point $x \in M$. The Lie algebra $K_x \mathcal{F}$ is a differential analogue of the Lie algebra H_x. The *differential index of the Lie algebra $\mathcal{F}$ of functions* on a manifold M with a Poisson bracket is determined by the formula

(3)
$$d.\operatorname{ind} \mathcal{F} = \inf_{x \in M} \dim K_x \mathcal{F}$$

If, for instance, $M = G^*$, $x = \xi$, $\mathcal{F} = G \subset C^\infty(G^*)$, then the differential index (3) coincides with the usual index (1) and $K_x G = H_x$. If $M \subset G^*$ is the orbit of the coadjoint representation and G is interpreted as the algebra of functions on M, then $d.\dim G = \dim M$, $d.\operatorname{ind} G = 0$. If for any smooth function f on M the derivative $(\operatorname{sgrad} H)f$ along the vector field $\operatorname{sgrad} H$ is expressed by the Poisson bracket $(\operatorname{sgrad} H)f = \{H, f\}$ then the field $\operatorname{sgrad} H$ is called Hamiltonian. Suppose that a Hamiltonian system possesses a number of known integrals $f_1, \ldots, f_k$. Calculating pairwise Posson brackets, we obtain certain, possibly new, integrals of the system (see Ch. 1). Adding new integrals to the initial set $f_1, \ldots, f_k$, one can obtain an extended set of integrals $f_1, \ldots, f_m$, $m > k$. Next, the procedure of calculating the Poisson brackets is repeated with the extended set of integrals.

After a certain finite number of extensions of the initial set of integrals, one arrives at a set of integrals $f_1, \ldots, f_m$, such that for any i, $j = 1, 2, \ldots, m$ the Poisson bracket $\{f_i, f_j\}$ depends functionally on $f_1, \ldots, f_m$. This property of the set $F = (f_1, \ldots, f_m)$ can be written as follows:

(4)
$$\{f_i, f_j\} \in \mathcal{F}(F), \qquad \forall f_i, f_j \in F$$

From this point on, $\mathcal{F}(F)$ is a set of smooth functions on M functionally dependent on $f_i \in F$. An arbitrary set F of smooth functions on M will be called *closed (with respect to the Poisson bracket)* if the condition (4) is fulfilled for it. Thus, a search for new integrals of a Hamiltonian equation inevitably leads to closed sets of integrals.

Let F be a closed set of functions. For the set F, one can define the *Lie algebra* $K_x F$, the *dimension*, and the *index*, as it was done above for the Lie algebras of functions. Since the set F is *neither a Lie algebra* nor even a vector space, one may omit the adjective "differential" and only imply it in defining the index and the dimension of F. By definition, we set

$$\dim F = d.\dim \mathcal{F}(F), \qquad \operatorname{ind} F = d.\operatorname{ind} \mathcal{F}(F), \qquad K_x F = K_x \mathcal{F}(F).$$

Let $F = (f_1, \ldots, f_m)$ be a closed set consisting of independent functions and $m = \dim F$. Let $g_{ij} = \{f_i, f_j\}$ (where $i, j = 1, \ldots, m$) be pairwise Poisson brackets. It is readily seen that the Lie algebra $K_x F$ admits the following effective description:

$$K_x F = \{\xi = \sum_{i=1}^{m} \xi_i df_i \in T_x^* M : \sum_{i=1}^{m} \xi_i g_{ij}(x) = 0, \forall j = 1, \ldots, m\},$$

$$[\mathcal{E}, \eta] = \sum_{i,\, j=1}^{m} \xi_i \eta_j dg_{ij}(x), \qquad \forall \xi, \quad \eta \in K_x F.$$

§2 The General Properties of Invariant Submanifolds

Consider the construction of the basis of the Lie algebra $K_x F$ which depends smoothly on the point $x \in M$. Solving the system

$$(5) \qquad \sum_{i=1}^{m} \xi_i g_{ij}(x) = 0, \qquad j = 1, \ldots, m$$

with respect to ξ_i, we find the rational functions $\xi_i^j(g_{11}, \ldots, g_{mm}), i = 1, \ldots, m;$ $j = 1, \ldots, k; k = \operatorname{ind} F$, such that the differentials

$$\xi^j = \sum_{i=1}^{m} \xi_i^j(g_{11}, \ldots, g_{mm}) df_i \in T_x^* M$$

form the basis of the Lie algebra $K_x F$ for each nonsingular point $x \in U$. Here and below $U \subset M$ is a domain in which the denominators of the rational functions ξ_i^j do not vanish. Associated with the differentials ξ^j are the vector fields

$$Y^j = \sum_{i=1}^{m} \xi_i^j(g_{11}, \ldots, g_{mm}) \operatorname{sgrad} f; \ j = 1, \ldots, k.$$

Note that, in the general case, the vector fields Y^j are not Hamiltonian, because the differentials ξ^j are not exact. A common level surface $M_z = (x \in M; f_i(x) = f_i(z),$ for all $i = 1, \ldots, m)$ is called nonsingular if each point $x \in M_z$ is a nonsingular point for the algebra $\mathcal{F}(F)$.

The next theorem is closely connected without generalizations of the Liouville theorem on completely integrable systems, see, for instance [143].

THEOREM 3.2.1 (SEE [20]). *Let M be a symplectic manifold and let $F = (H = f_1, f_2, \ldots, f_m)$ be a set of independent known functions closed with respect to the Poisson bracket.*

a) The Hamiltonian equation $\overset{\circ}{x} = \operatorname{sgrad} H$ with the initial data $x(0) = z$ is integrated in quadratures, provided that M_z is a nonsingular isotropic surface, $dH \in K_z F$, and $\mathbb{R}dH$ is a one-dimensional ideal in a solvable Lie algebra $K_z F$.

b) The vector fields $Y^1, \ldots, Y^k$, defined in the domain U, pairwise commute, and $K_z F$ is a commutative Lie algebra for a point z of general position in M.

c) If $F = (H = f_1, f_2, \ldots, f_m)$ is a set of integrals of the Hamiltonian system $\overset{\circ}{x} = \operatorname{sgrad} H$, and we have $M_z \subset U$, for a point z of general position in M, and if M_z is a nonsingular isotropic surface, then the equation $\overset{\circ}{x} = \operatorname{sgrad} H$ with initial data of general position is integrated in quadratures.

We shall say that a Hamiltonian system $\overset{\circ}{x} = \operatorname{sgrad} H$ is *completely integrable* if for this system we have the conditions of Theorem 3.2.1, c).

REMARK 1: In the neighbourhood of a compact level surface M_z of a completely integrable system, there always exist generalized (in the sense of [99]) action-angle variables. This follows readily from Nekhoroshev's theorem [99].

REMARK 2: To check the isotropy of a level surface, one can employ Cartan's formula (see [49]):

(6) $$\dim(T_z \cap T_z^{\perp}) = m - \operatorname{rank}(g_{ij}(z)),$$

where $T_z^{\perp}$ is a skew-orthogonal complement determined by the symplectic form.

As shown below, the complete integrability of a Hamiltonian system makes it possible not only to integrate this system in quadratures (an example of integration of an interesting system by means of integrals which form no Lie algebra but have Poisson brackets expressed through these very integrals is given in [201]), but also to organize the isotropic tori M_z into Langrangian tori (Liouville tori) of dimension $\frac{1}{2} \dim M$.

PROOF OF THEOREM 3.2.1:

a) The linear subspace generated by Hamiltonian vector fields sgrad f_i (where $i = 1, \ldots, m$) on a level surface M_z will be denoted by $X(M_z)$. The linear subspace in $X(M_z)$ consisting of vector fields tangent to M_z will be denoted by $K(M_z)$. The space $K(M_z)$ is the Lie algebra with respect to the operation of vector field commutation. The commutator [sgrad f, sgrad h] of any Hamiltonian vector fields sgrad f, sgrad h is known to coincide with the Hamiltonian field sgrad$\{f, h\}$. This implies that the mapping $\rho : K_z F \to K(M_z)$ defined according to the formula

$$\rho(\xi) = \sum_{i=1}^{m} \xi_i \operatorname{sgrad} f_i,$$

where

$$\xi = \sum_{i=1}^{m} \xi_i df_i \in K_z F,$$

is an isomorphism of Lie algebras. Let $D_1, \ldots, D_k$ be linear generators of the Lie algebra $K(M_z)$. We shall assume that $D_1 = \operatorname{sgrad} H$. The vector fields $D_1, \ldots, D_k$ form a solvable Lie algebra and $[D_i, D_1] = c_i D_1$ for each $i = 1, \ldots, k$. Here c_i are appropriate constants independent of the point $x \in M_z$. The vector fields $D_1, \ldots, D_k$ are linearly independent and cover the level surface M_z due to its isotropy. From what has been said it follows that the differential equation $\frac{dx}{dt} = D_1(x \in M_z)$ is integrated in quadratures by virtue of the well-known theorem of Sophus Lie.

b) For each surface M_z lying in the domain U, the vector fields $Y^1, \ldots, Y^k$ form the basis of the Lie algebra $K(M_z)$. Since the Lie algebras $K_z F$ and $K(M_z)$ are isomorphic, it follows that, to prove the assertion b), it suffices to prove that the vector fields Y^j commute. We have

(7) $$[Y^p, Y^q] = \sum_{i,j} \xi_i^p \xi_j^q [\operatorname{sgrad} f_i, \operatorname{sgrad} f_j]$$
$$+ \sum_{i,j} \xi_i^p \{f_i, \xi_j^q\} \operatorname{sgrad} f_j$$
$$- \sum_{i,j} \xi_j^q \{f_j, \xi_i^p\} \operatorname{sgrad} f_i.$$

Next,
$$\xi_i^p\{f_i, \xi_j^q\} = \sum_s \xi_i^p\{f_i, f_s\}\partial_s \xi_j^q = 0,$$

where $\partial_s = \partial/\partial f_s$. Similarly, $\xi_j^q\{f_j, \xi_i^p\} = 0$. Consider the sum

(8)
$$\sum_{ij} \xi_i^p \xi_j^q g_{ij} = 0.$$

All the functions ξ_i^p, ξ_j^q, and g_{ij} contained in this sum are functions of $f_1, \ldots, f_m$. Differentiating (8) with respect to f_s, we find

(9)
$$\sum_{ij} \xi_i^p \xi_j^q \partial_s g_{ij} = 0.$$

From (9) it follows that

$$\sum_{ij} \xi_i^p \xi_j^q [\operatorname{sgrad} f_i, \operatorname{sgrad} f_j] = \sum_{ij} \xi_i^p \xi_j^q \left(\sum_s (\partial_s g_{ij}) \operatorname{sgrad} f_s\right) = 0.$$

Thus, we have proved that all three sums on the right-hand side of equality (7) are equal to zero. Conseqently, $[Y^p, Y^q] = 0$.

c) From assertion b) it follows that $K_x F$ is a commutative Lie algebra, and therefore assertion c) follows from assertion a) proved above. This concludes the proof.

§3 Systems Completely Integrable in the Noncommutative Sense Are Often Completely Liouville-Integrable in the Conventional Sense

3.1 The Formulation of the General Equivalence Hypothesis and Its Validity for Compact Manifolds

HYPHOTHESIS. *Let a Hamiltonian system $v = \operatorname{sgrad} H$ admitting a noncommutative integration on M^{2n} be given. This means that the Hamiltonian H is correctly included in a certain maximal linear Lie algebra G of functions (that is, $\dim G + \operatorname{ind} G = \dim M$). Then, obviously, the same system is completely Liouville-integrable in the conventional sense as well. This means that there exists another maximal linear commutative Lie algebra G_0 of functions on M (such that $\dim G_0 = \frac{1}{2}\dim M$, i.e., induced by n independent involutive integrals of the system) containing the Hamiltonian H.*

It is natural to assume that the integrals of the new commutative algebra G_0 are functionally expressed through the former functions of the algebra G. In particular, the integrals from G_0 must, evidently, belong to the same functional class as the functions of the initial Lie algebra G.

This hypothesis, expressed by Fomenko and Mischenko in [90], [93] and discussed in [143], is actually valid for the case of compact symplectic manifolds, which we have also proved in [93].

THEOREM 3.3.1. *Let M^{2n} be a compact symplectic manifold and let $v = $ sgrad H be a Hamiltonian system completely integrable in the noncommutative sense on M (or in the neighbourhood of one fixed level surface of functions generating the maximal linear algebra G; see above). Then the same system is completely Liouville-integrable in the usual sense, the commuting integrals of the new algebra G_0 (that is, of the maximal linear commutative subalgebra on M) being functionally (and by explicit formulae) expressed by the former functions of the Lie algebra G.*

This theorem immediatley offers new opportunities for seeking (Liouville)-integrable systems. We have revealed that such systems may "have the appearance" of systems admitting noncommutative integration. This has substantially widened the collection of signs for identifying and concretely indicating integrable systems on symplectic manifolds. As has already been mentioned, in many concrete cases it is simpler to discover a maximal linear noncommutative algebra G of functions than to find immediately a maximal commutative set of integrals of the algebra G_0. Most often (as shown below) the completely Liouville-integrable systems "are disguised" as noncommutatively integrable systems in the case of geodesic flows of left-invariant metrics on some Lie groups and symmetric spaces.

The proof of Theorem 3.3.1 is given below. It is instructive to realize its geometrical meaning. If an initial Hamiltonian system v possesses a noncommutative symmetry group, that is, if it admits a maximal linear noncommutative algebra G, then by virtue of Theorem 3.1.1 (see §1) the trajectories of the system define the conditionally periodic motions on r-dimensional tori T^r (where $r = \dim M - \dim G$). As has already been said, these tori are "few-dimensional," that is, $r < n = \frac{1}{2} \dim M$. If, in addition, the system v admits the maximal commutative algebra G_0 of integrals, then the same integral trajectories may also lie on Liouville tori of dimension n. Consequently, in this case the few-dimensional tori can be organized into "large" Liouville tori T^n.

Since the tori T^n are *fibred* into invariant tori T^r, it follows that the trajectories of the system v *do not set everywhere dense (irrational) winding of the tori* T^n (Fig. 63). In this sense, the system v is degenerate.

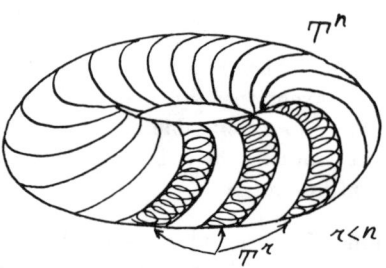

Figure 63

It is exactly this circumstance that accounts for the "experimental" fact that for many systems of this kind it is easier first to find a noncommutative maximal linear algebra of functions (which correctly contains the Hamiltonian of the system) and then, by our general Theorem 3.1.1 (in the compact case), to find the "genuine" maximal commutative algebra of integrals.

However, this scheme works only on *compact* symplectic manifolds. The point is that, in the case of noncompact manifolds, the analogue of Theorem 3.1.1 has not yet been fully proved. It is therefore not excluded a priori that there may exist Hamiltonian systems completely integrable in the noncommutative sense, but nonintegrable in the classical commutative sense. This question is still open.

Summary

Thus, a *wide class of noncommutatively integrable Hamiltonian systems* is naturally under consideration. It is not yet clear whether it coincides with the class of ordinary completely Liouville-integrable systems. Furthermore, the overlapping of the class of noncommutatively integrable systems and the class of systems integrable in the sense of Liouville consists of interesting "systems with degeneracies." Integral trajectories of such systems on Liouville tori do not determine the irrational widing of these tori. The trajectories lie on *few dimensional tori* which regularly fibre large Liouville tori. Thus, establishing the fact of noncommutative integrability of a system provides substantially more information on the behaviour of its integral trajectories than is derived from the usual Liouville integrability. Additional information on the behaviour of the trajectories of a system is provided by considering the fibration of Liouville tori into few-dimensional tori (there may be many such nonequivalent fibre bundles). The data on this fibration can be obtained from the formulae that express the relation between commuting and noncommuting integrals of a system. In this field, there are many unsolved problems. For instance, it is of interest to extablish the geometrical meaning of the analytic formulae which transform a noncommutative algebra of integrals into a commutative one. As shown below, these formulae describe the momentum mapping and the commutative algebra of integrals which often exists on the noncommutative maximal linear algebra of functions. Changing these algebras, we change the fibrations of Liouville tori into smaller (few-dimensional) tori.

3.2 *The Properties of Momentum Mapping of a System Integrable in the Noncommutative Sense.*

Let G be a finite-dimensional Lie algebra and G^* a dual space fibred into orbits O^* of the coadjoint representation Ad^* of the group $\mathfrak{G}$ which corresponds to the Lie algebra G, that is, $\mathfrak{G} = \exp G$. Then the form of Kirillov and the corresponding Poisson bracket are defined on each orbit O^* (see Ch. 1). But, for our immediate purposes, another operation , $\{\,,\}'$, is more convenient, which will be called a *degenerate Poisson bracket* on G^*. The point is that separate Poisson brackets given on orbits O^* may be glued into a single bracket $\{\,,\}'$. We shall do this as follows. Let $f, g \in C^\infty(G^*)$. Then by definition, we set

$$\{f, g\}'(x) = \{f|_{O^*(x)}, g|_{O^*(x)}\}_{O^*(x)}(x).$$

Since G^* fibres into orbits O^*, this definition is correct. This bracket can be explicitly calculated in terms of the structural tensor c^i_{jp} which sets the Lie algebra G. It turns out that

$$\{f, g\}' = \sum c^p_{ij} x_p \frac{\partial f}{\partial x^i} \frac{\partial g}{\partial x^j},$$

where $e_1, \ldots, e_k$ is the basis of the Lie algebra $G, e^1, \ldots, e^k$ and a conjugate basis in G^*, and the coordinates corresponding to these bases are denoted respectively as $x^1, \ldots, x^k$ and $x_1, \ldots, x_k; c_{ij}^p$ is the structure tensor of the algebra G in the basis e_i, that is,

$$[e_i, e_j] = \sum_p c_{ij}^p e_p.$$

Although degenerate, this bracket satisfies the Jacobi identity.

Not to complicate the notation, we will denote the Poisson bracket constructed above by $\{\,,\}$. This is quite natural because restricting this bracket to a separate orbit O^*, we obtain the *former nondegenerate Poisson bracket*.

Thus, we can already speak of functions which are in involution on the entire coalgebra G^* and not only on the orbits O^*. We shall now specify the class of Lie algebras important for our further purposes.

DEFINITION 3.3.1: We shall say that a Lie algebra G is *integrable* if on G^* there exists a linear subspace $V \subset C^\infty(G)^*$ in which one can single out an additive basis of functionally independent functions $g_1, \ldots, g_q$ (where $q = \dim V$), such that they are in involution on G^* with respect to the Poisson bracket, and

$$q = \dim V = \frac{1}{2}(\operatorname{ind} G + \dim G) = \frac{1}{2}(r + k).$$

Therefore, the restrictions of the functions $g_1, \ldots, g_q$ to the orbits O^* of general position in G^* set on these orbits the maximal linear commutative Lie algebra of functions.

Recall what the latter condition means. The restriction of the functions $g_1, \ldots, g_q$ of the algebra V to an orbit O^* yields a set of functions $h_1, \ldots, h_s$, where

$$s = \frac{1}{2}\dim O^* = \frac{1}{2}(\dim G^* - r) = \frac{1}{2}(k - r), \qquad r = \operatorname{ind} G.$$

The result which follows was proved by Fomenko and Mischenko in [90] and [93] (see also [143]).

THEOREM 3.3.2. *Let M^{2n} be a symplectic manifold (compact or noncompact) and let a maximal linear Lie subalgebra G of smooth functions on M^{2n} be given (that is, $\dim G + \operatorname{ind} G = \dim M$). If the Lie algebra G is integrable (see Definition 3.3.1), then on M^{2n} there always exists a maximal linear commutative algebra G_0 of functions (functionally dependent on the functions of the subalgebra G), that is, such an algebra that $\dim G = \frac{1}{2}\dim M^{2n}$.*

Thus, this theorem asserts, in particular, that a system v, which is completely integrable on M^{2n} in the noncommutative sense by means of the integrable Lie algebra G of functions, automatically turns out to be integrable also in the classical, commutative (Liouville) sense.

We will proceed to the proof of Theorem 3.3.2. At the same time, we investigate some important properties of momentum mapping, which are also of interest independent of Theorem 3.3.2. Let a maximal linear Lie algebra G of smooth functions be given on M^{2n}. Consider the corresponding momentum mapping

$$F : M \to G^*,$$

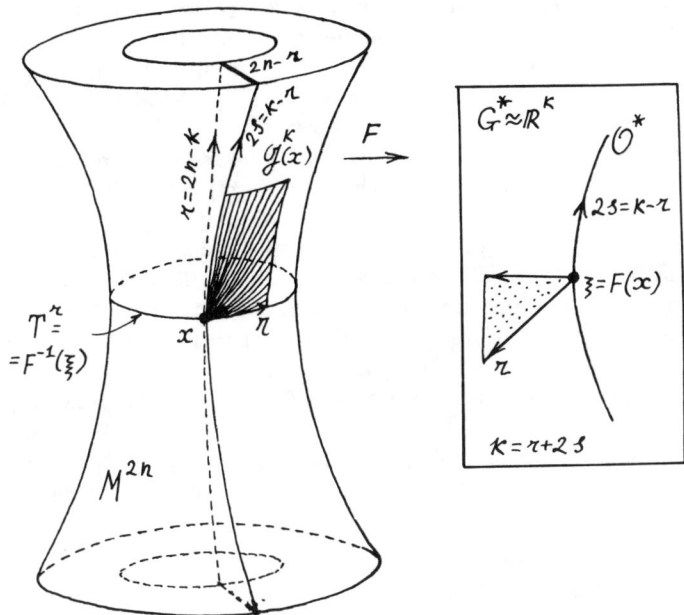

Figure 64

that is,
$$F(x) = (f_1(x), \ldots, f_k(x)),$$
where $f_1, \ldots, f_k$ is an additive independent basis in G. In other words, $F(x) = \xi \in G^*$, where $\xi(f) = f(x)$ for $f \in G$. Since, according to Theorem 3.1.1 (see Ch.3, §1), the compact connected common level surfaces $M_\xi = F^{-1}(\xi)$ of all the functions of the Lie algebra G are diffeomorphic to r-dimensional tori T^r (where $r = \text{ind}\, G$), only these tori T^r are nonsingular fibres of general position of the momentum mapping F.

For convenience of references, we give here a list of notations used hereafter.

1) The algebra $G : k = \dim G$, $f_1, \ldots, f_k$ are functionally independent additive generations (functions) of the algebra G,
$$r = \text{ind}\, G, \quad s = \frac{1}{2}\dim O^*,$$
where O^* is an orbit of general position in G^*, $s = \frac{1}{2}(k-r)$, $2n = k + r$, that is, $\dim M = \dim G + \text{ind}\, G$.

2) The algebra $V : q = \dim V$, $g_1, \ldots, g_q$ are functionally independent additive generators in the commutative algebra V, $\text{ind}\, V = \dim V = q$, $q = \frac{1}{2}(k+r) = \frac{1}{2}(\dim G + \text{ind}\, G)$; $q = s + r = \frac{1}{2}\dim O^* + \text{ind}\, G$. It is clear that $q < k$ (in the case of noncommutative algebra G), because $q = \frac{1}{2}(k+r)$ and $r < k$.

3) The algebra $G_0 : n = \dim G_0$, $\text{ind}\, G_0 = \dim G_0 = \frac{1}{2}\dim M^{2n}$.

Figure 64 provides a visual representation of all surfaces and dimensions participating in the consideration. The mapping F fully covers the entire open neighbourhood of the point $\xi \in G^*$. Let the preimage $F^{-1}(\xi)$ contain a compact connected component. Then this component is homeomorphic to the torus T^r (Fig. 64). From a point x, there originates an orbit $\mathfrak{G}(x)$ of the group $\mathfrak{G}$ (symplectically acting on M) which is locally homeomorphic to the neighbourhood of the unit element of the group $\mathfrak{G}$. A part of this orbit is a torus T^r. Along the normal to the torus, there remain $2n-k$ independent directions. Under momentum mapping, the corresponding "normal" submanifold (in the orbit $\mathfrak{G}(x)$) is (locally) homeomorphically carried onto the orbit $O^* \subset G^*$. The number of independent directions on the normal to the orbit O^* of general position in G^* is equal to r, that is, to ind G. Under the mapping F, the torus T^r transforms into a point ξ, and the area of dimension $2n-r$, normal to this torus, transforms (by virtue of the condition $2n-r = k$) into an open neighbourhood (of dimension k) of the point ξ in G^*. If $\alpha(\xi)$ is a certain function on G^*, then, by carrying it onto a manifold M by means of momentum mapping, we obtain a smooth function $\alpha \cdot F$ on M^{2n}. This function is obviously constant on each torus $T^r \subset M^{2n}$. If two functions α and β are independent (at a point of general position) on G^*, then their preimages, i.e., the functions αF and βF are also independent on M. The following property of momentum mapping is very important.

LEMMA 3.3.1. *The momentum mapping $F : M \to G^*$ is $\mathfrak{G}$ invariant, where the group $\mathfrak{G} = \exp G$ acts on G^* in a coadjoint manner, and on the manifold M it acts as a group of symplectic transformations generated by the Lie algebra G of vector fields sgrad f, $f \in G$.*

The proof follows immediately from the definitons of the action of the group.

LEMMA 3.3.2. *If functions α and β on G^* were in involution with respect to the canonical Poisson bracket, that is, $\{\alpha, \beta\}_{G^*} = 0$ (see the definition of the bracket above), then their preimages are in involution with respect to the initial Poisson bracket induced on M by the 2-form ω, that is, $\{\alpha F, \beta F\}_M = 0$.*

PROOF: This assertion follows from $\mathfrak{G}$-invariance of the momentum mapping and from the definition of the Poisson bracket. We shall dwell on this fact in more detail. In fact, for any functions α and β on G^*, we have the identity

$$\{\alpha F, \beta F\}_M = \{\alpha, \beta\}_{G^*}.$$

Let $f_1, \ldots, f_k$ be a linear basis in the Lie algebra G. Consider the functions $a(f_1, \ldots, f_k)$ and $b(f_1, \ldots, f_k)$ of the initial functions $f_1, \ldots, f_k$. Then the Poisson bracket of the functions a and b is set as follows:

$$\{a, b\}_M = \sum_{ij} \frac{\partial a}{\partial f_i} \frac{\partial b}{\partial f_j} \{f_i, f_j\}_M.$$

At the same time, the functions a and b may be regarded from a different point of view. Each function a of k variables $f_1, \ldots, f_k$ defines the function a^* on a dual space G^* by the formula:

$$a^*(\xi) = a(\langle \xi, f_1 \rangle, \ldots, \langle \xi, f_k \rangle) = a(\xi(f_1), \ldots, \xi(f_k)).$$

Next, the Poisson bracket of the functions $a^*(\xi)$ and $b^*(\xi)$ on G^* has the form:

$$\{a^*, g^*\}_{G^*}(\xi) = \sum_{ij} \frac{\partial a}{\partial f_i}(\langle\xi, f_1\rangle, \ldots, \langle\xi, f_k\rangle)$$

$$\frac{\partial b}{\partial f_j}(\langle\xi, f_1\rangle, \ldots, \langle\xi, f_k\rangle)\{\langle\xi, f_i\rangle, \langle\xi, f_j\rangle\}_{G^*}$$

$$= \sum_{ij} \frac{\partial a}{\partial f_i}(\langle\xi, f_1\rangle, \ldots, \langle\xi, f_k\rangle)$$

$$\frac{\partial b}{\partial f_j}(\langle\xi, f_1\rangle, \ldots, \langle\xi, f_k\rangle)\langle x_i, \{f_i, f_j\}_M\rangle.$$

Consequently, the natural mapping $a \to a^*$ *carries a Poisson bracket again into a Poisson bracket* and therefore determines the isomorphism of two infinite-dimensional Lie algebras. Consider the inverse mapping $\rho : a^* \to a$. Through this mapping the function $a^*(\xi)$ on G^* is asociated with the function $a(f_1, \ldots, f_k)$, that is, the function $a(f_1, \ldots, f_k)(x)$ on M^{2n}. Obviously, $\rho(a^*) = a^* F$. Thus, we have proved that the momentum mapping F induces the counter mapping $C^\infty(G^*) \to C^\infty(M)$ which carries the bracket $\{\,,\,\}_{G^*}$ into the bracket $\{\,,\,\}_M$. It is apparent that this implies the assertion of Lemma 3.3.2, as required.

Now we are in a position to conclude the *proof of Theorem 3.3.2*. By the assumption of the theorem, in the Lie algebra $C^\infty(G^*)$ a commutative subalgebra V is singled out which meets the requirements of Definition 3.3.1. In this subalgebra, single out an additive basis $g_1, \ldots, g_q$ (the functions on G^*) and consider their preimages, the functions $g_1 F, \ldots, g_q F$ on M. According to Lemmas 3.3.1 and 3.3.2, all of them are independent on M (at points of general position) and are in involution. The total number of these functions

$$q = \frac{1}{2}(k + r) = \frac{2n}{2} = n.$$

Consequently, we may take these functions as additive generators of the commutative algebra G_0, which is obviously the maximal linear commutative Lie algebra of functions on M. We have reached the desired conclusion.

3.3 *Theorem on the Existence of Maximal Linear Commutative Algebras of Functions on Orbits in Semisimple and Reductive Lie Algebras*

According to Theorem 3.3.2 of the preceding subsection, in order to prove the commutative integrability of systems that admit noncommutative integration, one should have the largest possible collection of integrable Lie algebras G, that is, algebras possessing a set of functions V on G^*, which satisfy the conditions of Definition 3.3.1 (see §3.2).

We will begin with an important case of semisimple Lie algebras G. Here G is canonically identified with the dual space G^* by means of a nondegenerate $\mathfrak{G}$-invariant Killing scalar product. The ring (algebra) of invariants of the algebra G is denoted by IG. Recall that a function $f(X)$ on G is called an *invariant of the algebra G* if $f(X) = f(\text{Ad}_g X)$ for any $g \in \mathfrak{G}$. In the case of matrix Lie algebras, this condition is equivalent to the following: $f(X) = f(gXg^{-1})$. In the semisimple case, all algebras of invariants are found in an explicit form, and their properties are thoroughly investigated (see, for instance, [202], [203]). We are interested only in polynomial invariants of semisimple algebras, and therefore we assume hereafter that IG denotes the ring of invariant polynomials on G. The generators in the ring IG are explicitly calculated and well known.

We have proved the following result in [89] and in [91] (for a detailed presentation, see [92], [93], [94], [143]).

THEOREM 3.3.3. *Let G be one of the following Lie algebras: a) a semisimple complex Lie algebra, b) a compact real form G_c of a semisimple complex Lie algebra, c) a normal compact subalgebra G_n in a compact Lie algebra. Then each of these Lie algebras in integrable. Reductive Lie algebras are also integrable. Maximal linear commutative algebras of functions are constructed explicitly and are polynomial algebras.*

This means that on each such Lie algebra $G \approx G^*$, there always exists a set $g_1, \ldots, g_q$ of independent functions which are in involution on G^*, and $q = \frac{1}{2}(\dim G + \operatorname{ind} G)$. In our case, $\operatorname{ind} G = \operatorname{rank} G$, that is, $\operatorname{ind} G$ is equal to the dimension of the Cartan subalgebra in G. In particular, restrictions of the functions $g_1, \ldots, g_q$ to the orbit of general position in G determine the maximal commutative algebra of functions on the orbit.

In Theorem 3.3.3, the functions $g_1, \ldots, g_q$ are explicitly indicated, and therefore the algorithm of seeking the functions g_i is absolutely *effective*. Namely, the following assertions are true.

THEOREM 3.3.4. *Let G be one of the Lie algebras listed in Theorem 3.3.3. Let IG be a ring of invariant polynomials on the algebra G. Consider in G an element a of general position, and let $f(X)$ be an arbitrary invariant polynomial on G. Then the family of all functions of the form $f(X + \lambda a)$, where λ is an arbitrary number, yields a set V, which transforms the algebra into an integrable algebra (see Definition 3.3.1, §3.2). In particular, restrictions of all functions of the form $f(X + \lambda a)$ to orbits of general position in G yield the maximal linear commutative algebras of fuctions on the orbits.*

The assertion of Theorem 3.3.4 can be reformulated more visually. Let $P_1(X), \ldots, P_g(X)$ be invariant homogeneous polynomials, which are generators in the ring IG. Consider the shift of the argument $X \to X + \lambda a$, that is, calculate the functions $P_i(X + \lambda a)$. Expanding the polynomials obtained in power series of the parameter λ, we obtain:

$$P_i(X + \lambda a) = \sum_j Q_{ij}(X, a)\lambda^j.$$

Here the polynomials $Q_{ij}(X, a)$ depend on X and on a, and a may be viewed as a parameter. Fixing the element a (of general position), we obtain the set of functions $Q_{ij}(X, a)$ on orbits of the algebra G. They turn out to form a commutative Lie algebra of dimension $\frac{1}{2} \dim O$, where O is an orbit of general position. It should be emphasized that the set of polynomials $Q_{ij}(X, a)$ is, generally, excessive. Being restricted to the orbits O, some of the indicated polynomials become dependent. But nevertheless, one can choose from them exactly $\frac{1}{2} \dim O$ independent polynomials, commuting with respect to the canonical Poisson bracket.

Brailov has strengthened Theorem 3.3.4 in his paper [20]. Namely, he proved that a maximal linear commutative algebra of functions can be found not only on orbits of general position in G (see Theorem 3.3.4), but also on singular orbits, that is, on orbits of smaller dimension. We will formulate this result. Let G be a semisimple Lie algebra. An element $X \in G$ is called *semisimple* if the endomorphism ad_x is semisimple. An orbit O in G is called semisimple if each of its elements is semisimple (that is, the orbit is generated by a semisimple element). Semisimple orbits may be either in general position or singular. *If a Lie algebra G is compact, then all its orbits ar semisimple* (because in this case each orbit necessarily intersects the Cartan subalgebra).

THEOREM 3.3.5 (BRAILOV). *Let O be an arbitrary semisimple orbit (of general position or singular) in a semisimple Lie algebra G. Let a be a regular semisimple element from G (that is, an element of general position). Let $P_1(X), \ldots, P_g(X)$ be independent invariant polynomials generating the algebra of invariants IG. Then from the restrictions of all functions of the form $P_i(X + \lambda_a)$ (where $\lambda \neq 0$) to any semisimple orbit (including a singular one), one can always choose independent functions in involution equal to $\frac{1}{2} \dim 0$, i.e., these restrictuions generate the maximal linear commutative algebra of functions on the orbit.*

3.4 *Proof of the Hypothesis for the Case of Compact Manifolds*

Now we are in a position to prove our Theorem 3.3.1.

CLAIM 3.3.1 (SEE [6]). *Let M be a compact symplectic manifold on which a finite-dimensional Lie algebra G of functions (with respect to the Poisson bracket) is given that effectively acts on M (that is, each nonzero element $g \in G$ is represented by a nonzero vector field sgrad g on M). Then the algebra G is reductive.*

In the assumption of Theorem 3.3.1, a finite-dimensional maximal linear Lie algebra G is given on the manifold. If M is compact, then by virtue of Claim 3.3.1 this algebra is reductive. Due to Theorem 3.3.3, this algebra is integrable, and due to Theorem 3.3.2, on M there exists a maximal linear commutative algebra G_0 of functions funcitonally dependent on the functions of the algebra G. This concludes the proof of Theorem 3.3.1.

3.5 *Momentum Mapping of Systems Integrable in the Noncommutative Sense by Means of an Excessive Set of Integrals*

We present the extension (proposed by Brailov) of the theorems of the preceding subsections to the case where an algebra of functions G given on M is excessive in the sense that the dimension of the subspace swept out at the point $x \in M$ by the vectors sgrad G, where $g \in G$, is smaller than the dimension of the algebra G.

Suppose that the functions $f_1, \ldots, f_k$ are integrals of a Hamiltonian system sgrad H (where $x \in M$) which provide its complete noncommutative integrability (see above). It follows from the definition of complete integrability that the property of the functions $f_1, \ldots, f_k$ to provide complete integrability is an intrinsic property of the set $f_1, \ldots, f_k$, that is to say, it does not depend on which particular Hamiltonian system of the form sgrad H the integrals are. This intrinsic property of the set of functions $f_1, \ldots, f_k$ admits the following redefiniton.

PROPOSITION 3.3.1. *Let G be a finite- or infinite-dimensional subalgebra in the Lie algebra $C^\infty(M)$; $f_1, \ldots, f_k \in G$ are such functions that any function $f \in G$ depends on $f_1, \ldots, f_k$. Then the following two conditons are equivalent.*

a) Any nonsingular surface $M_z = (x \in M : f_i(x) = f_i(z))$ is an isotropic submanifold in M.

b) $\dim M = d.\dim G + d.\operatorname{ind} G$.

PROOF: Suppose that condition b) is fulfilled. Let z be a nonsingular point. Choose m funcions from the set $f_1, \ldots, f_k$, where $m = d.\dim G$. The differentials of these funcions are independent. For simplicity we assume the functions $f_1, \ldots, f_m$ to be independent. By applying the Cartan formula, to $f_1, \ldots, f_m$, we find that
$$\dim T_z \cap T_z^\perp = d.\dim G - \operatorname{rank}(\{f_i, f_j\}(z)).$$
Consequently, on one hand,
$$d.\dim G - \operatorname{rank}(\{f_i, f_j\}(z)) \leqslant \dim T_z = \dim M - d.\dim G,$$
and on the other hand,
$$d.\dim G - \operatorname{rank}(\{f_i, f_j\}(z)) \geqslant d.\operatorname{ind} G.$$
Since $\dim M - d.\dim G = d.\operatorname{ind} G$, it follows that both these inequalities are, in fact, equalities. In particular, $\dim T_z \cap T_z^\perp = \dim T_z$. Hence, b) implies a). Inversely, suppose that condition a is fulfilled. It follows from the definition of $d.\operatorname{ind} G$ that there exists a nonsingular point $z \in M$ such that
$$d.\dim G - \operatorname{rank}(\{f_i, f_j\}(z)) = d.\operatorname{ind} G.$$
From the Cartan formula for the point z, it follows that
$$T_z \cap T_z^\perp = T_z = \dim M - d.\dim G = d.\operatorname{ind} G.$$
Therefore, $\dim M = d.\dim G + d.\operatorname{ind} G$. This proves the proposition.

Let $f_1, \ldots, f_m$ be smooth functions on M generating a set closed with respect to the Poisson bracket, and

(1) $$\{f_i, f_j\}(x) = \sum_{k=1}^{m} c_{ij}^k f_k(x),$$

where the numbers c_{ij}^k are independent of the point x. The linear subspace G generated by the functions f_i in the space $C^\infty(M)$ is a Lie algebra finite-dimensional with respect to the Poisson bracket. Suppose that G is a reductive Lie algebra and $\langle \, , \, \rangle$ is a fixed bilinear symmetric nondegenerate form on G invariant under all inner automorphisms. The dual space G^* is identified with G by $\langle \, , \, \rangle$.

Recall the definition of the momentum mapping with an account of identification of G and G^*. The momentum mapping $F : M \to G$ brings a point $X \in M$ into correspondence with an element $F(x) \in G$, such that $\langle F(x), f_i \rangle = f_i(x)$ for each $i = 1, \ldots, m$. The element ξ is called *semisimple* if the operator ad_ξ is diagonalizable or becomes such after complexification.

THEOREM 3.3.6. *Let $G \subset C^\infty(M)$ be a reductive subalgebra in a Lie algebra of smooth functions on a symplectic manifold M. Let $F : M \to G$ be the corresponding momentum mapping. Let $I_1, \ldots, I_r$ be a set of homogeneous generators of the algebra of invariants of the Lie Algebra G, and let a be an element of G. The functional coefficient λ^j in the polynomial $P_i(\lambda, \xi) = I_i(\xi + \lambda a)$ is denoted by $I_{i,a}^j(\xi)$, $\xi \in G$. Suppose that*

$$\dim M = d.\dim G + d.\operatorname{ind} G$$

and that for an element x of general position in M the element $F(x)$ is semisimple in G. Then all the functions $I_{i,a}^j \cdot F$ commute pairwise and one can choose among them $\frac{1}{2} \dim M$ independent functions, for the element a of general position G. In particular, if a Hamiltonian system sgrad H (where $x \in M$) can be completely integrated (in the noncommutative sense) by a set of integrals $f_1, \ldots, f_m$, such that $G = \oplus_{i=1}^m \mathbf{R} f_i$, then the funcions $I_{i,a}^j(\xi)$ form the complete commutative set of integrals of this Hamiltonian system, that is, their linear combinations generated the maximal linear commutative subalgebra in $C^\infty(M)$.

Our Theorem 3.3.1 (see [90]) asserts that on a symplectic manifold M a Hamiltonian system sgrad H possessing a set of independent integrals $f_1, \ldots, f_m$, such that

$$G = \oplus_{i=1}^m \mathbf{R} f_i,$$

is a reductive Lie algebra, and

(2) $$\dim M = \dim G + \operatorname{ind} G$$

is completely Liouville-integrable. Theorem 3.3.6 develops this theorem, since in Theorem 3.3.6 the complete integrability is asserted without assuming integrals to be independent.

It follows from Proposition 3.3.1 that the definition of complete noncommutative integrability given above may be formulated in the form close to formula (2). Namely, a Hamiltonian system on a symplectic manifold M with the Lie algebra G of integrals is completely integrable in the noncommutative sense if

(3) $$\dim M = d.\dim G + d.\operatorname{ind} G.$$

LEMMA 3.3.3. *Let G be a complex or real reductive Lie algebra, $a \in G$ a semisimple element, G^a the centralizer of the element a, ZG^a its centre, and let $I_1, \ldots, I_r$ be generators of the algebra of invariants of the Lie algebra G. Then the differentials $dI_1(a), \ldots, dI_r(a)$ generate ZG^a.*

PROOF: It suffices to prove the Lemma for a semisimple Lie algebra G. First consider the case of a complex-valued field $\mathbf{C}$. Let K be the Cartan subalgebra. As is known, the mapping of the restriction, $j : IG \to SK$, where IG is the algebra of invariants and K the algebra of polynomial functions on K, is a monomorphism and $j(IG) = SK^W$. Here SK^W is a set of polynomial funcions on K, invariant with respect to the Weyl group W. Let b belong to ZG^a, W^a be the stabilizer of a in W, and W^b the stabilizer of b. Then $W^a \subset W^b$. Let $(a_1, \ldots, a_s)$ be the orbit of a with respect to W. Let g be a polynomial function on K such that $dg(a) = b$ and $dg(a_i) = 0$ for $a_i \neq a$. Define the function $\tilde{g} = \frac{s}{|W|} \sum_{v \in W} g \cdot v$, where $|W|$ is the number of elements in the Weyl group. For $\tilde{g}$ we have $d\tilde{g}(a) = b$ and $\tilde{g} \in \mathrm{SK}^W$. Therefore, $I = j^{-1}(\tilde{g})$ is an invariant of G, such that $dI(a) = b$. Since $I_1, \ldots, I_r$ are generators of the algebra of invariants, it follows that $I = P(I_1, \ldots, I_r)$ for an appropriate polynomial P. Hence,

$$b = \sum_{i=1}^{r} \frac{\partial P}{\partial I_i} dI_i(a).$$

This proves the lemma for the case of the field $\mathbf{C}$. The real case follows from the complex one.

PROOF OF THEOREM 3.3.6: Let G be the Lie algebra of the Lie group $\mathfrak{G}$, x an element of general position in M, O the orbit of a (co)adjoint action of $\mathfrak{G}$ on G passing through $F(x)$. Let k be the number of independent functions among $f_1, \ldots, f_m$, $k = d.\dim G$. There exists a small disk D^s of dimension $s = k - \dim O = d.\mathrm{ind}\, G$, which transversally intersects the orbit O at the point

$$F(x), \qquad T_{F(x)} O \oplus T_{F(x)} D^s = T_{F(x)} F(M).$$

By virtue of Lemma 3.3.1, among the invariants $I_1, \ldots, I_r$, there exist s invariants whose restrictions to the disk D^s are independent. By virtue of Theorem 3.3.5 (§3.3), from the functions $I_{i,a}^j$ one can choose $\frac{1}{2} \dim O$ functions in such a way that, being restricted to the orbit O, they remain independent. Therefore, on M there exist $s + \frac{1}{2} \dim O$, independent functions of the form $I_{i,a}^j \cdot F$. Since

$$\dim M = d.\dim G + d.\mathrm{ind}\, G$$

and

$$d.\dim G = d.\mathrm{ind}\, G + \dim O,$$

it follows that

$$s + \frac{1}{2} \dim O = d.\mathrm{ind}\, G + \frac{1}{2} \dim O$$
$$= \frac{1}{2}(2 d.\mathrm{ind}\, G + \dim O) = \frac{1}{2}(d.\mathrm{ind}\, G + d.\dim G) = \frac{1}{2} \dim M.$$

Since the functions $I_{i,a}^j \cdot F$ are obviously integrals, the Hamiltonian system sgrad H is completely Liouville-integrable, which implies the theorem.

3.6 Sufficient Conditions for Compactness of the Lie Algebra of Integrals of a Hamiltonian System

A Lie algebra G is called *compact* if there exists a positive definite scalar product $\langle\, ,\, \rangle$ on G invariant under all inner automorphisms. Let a reductive Lie algebra G be a Lie algebra of functions with respect to the Poisson bracket on a symplectic manifold (M, ω). The *semisimplicity condition for the image of the momentum mapping* $F : M \to G$ is automatically fulfilled by virtue of Theorem 3.3.6 for compact Lie algebras G. In this connection, the sufficient compactness conditions for the Lie algebra G of integrals of a Hamiltonian system sgrad H are of interest. As follows from [6], such a sufficient condition is compactness of the manifold M. In this case, an invariant scalar product

$$\langle f, g \rangle = \int_M fg\omega^n$$

is specified for any functions $f, g \in G$. Below, we adduce a weaker condition which provides compactness of the Lie algebra of integrals (Lemma 3.3.4) and, using Theorem 3.3.6, formulate the following theorem (Brailov).

THEOREM 3.3.7. *Let a Hamiltonian system* sgrad H *on* M *be completely integrable (in the noncommutative sense) by means of a set of integrals* $f_1, \ldots, f_m$ *which form the Lie algebra* $G = \oplus_{i=1}^m Rf_i$. *If any isoenergy surface* $M_h = \{x \in M : H(x) = h\} i = 1$ *is compact, then the Hamiltonian system* sgrad H *is completely Liouville-integrable by means of a complete set of commuting integrals which polynomially depend on* $f_1, \ldots, f_m$.

Theorem 3.3.7 develops our initial theorem (see [93]) on complete integrability of the Hamiltonian system H provided that M is compact and completely integrable by means of a set of independent integrals $f_1, \ldots, f_m$, forming a Lie algebra G such that $\dim M = \dim G + \operatorname{ind} G$.

PROOF OF THEOREM 3.3.7: Let G be the Lie algebra of integrals of a Hamiltonian system sgrad H each of whose isoenergy surfaces is compact. Then G turns out to be a compact Lie algebra. Although the proof of this assertion easily follows from the results of §3.4, we will present the reasoning necessary for our further purposes.

Let g be an arbitrary element in G and sgrad g the corresponding Hamiltonian vector field. The vector field sgrad g is tangent to M_h for any h. Since the manifold M_h is compact, all integral trajectories of the field sgrad g are infinitely continued. For this reason, defined is an action of a connected and simply connected Lie group $\mathfrak{G}$ (whose Lie algebra is G) on M that the element $\exp(tg)$ of the group $\mathfrak{G}$ carries each point $\gamma(O)$ of the manifold M, lying on an integral trajectory $\gamma : \mathbb{R} \to M$ of the vector field sgrad g into a corresponding point $\gamma(t)$.

The momentum mapping $F : M \to G^*$ carries the above-said action of the group $\mathfrak{G}$ on M into a coadjoint action on G^*, and therefore the image $F(M_h)$ of the surface M_h is invariant with respect to Ad^*.

Let $g \in G$ be a nilpotent element and $\xi_0 \in F(M_h)$. Consider the mapping $\xi : \mathbb{R} \to G^*$ defined by the formula $(\operatorname{Ad}_{\exp(tg)})^* \xi_0 = \xi(t)$. The invarance of $F(M_h)$ implies that $\xi(t) = F(M_h)$. The nilpotence of g implies that ξ is a polynomial mapping. The compactness of M_h implies the compactness of $F(M_h)$. Since the mapping $\xi : \mathbb{R} \to F(M_h)$ is polynomial and the set $F(M_h)$ is compact, it follows that $\xi(t) = \xi_0$ for any t. Accordingly, $\operatorname{ad}_g^*(\xi_0) = 0$ for each $\xi_0 \in F(M_h)$ and for any nilpotent $g \in G$. Hence, for a nilpotent $g \in G$ and for any $f \in G$, $x \in M$ we have: $F(x)[g, h] = \{g, f\}(x) = 0$. Each nilpoint element g of the Lie algebra G is therefore contained in its centre which will be denoted by Z. From this, there readily follows compactness of the Lie algebra G. Indeed, let R be a solvable radical of the algebra G. Then the set $[R, R]$ consists of nilpotent elements of the Lie algebra G. Conseqently, the whole radical consists of nilpotent elements. Therefore, $[R, R] \subset Z = R$. Thus G, is a reductive Lie algebra. Let $S = [G, G]$ be a semisimple ideal in G. Since $S \cap Z = 0$, it follows that S contains no nonzero nilpotent elements and is, accordingly, a semisimple compact Lie algebra. Consequently, G is a compact Lie algebra, and the assertion of Theorem 3.3.7 follows.

§4 Liouville Integrability on Complex Symplectic Manifolds

4.1 Different Notions of Complex Integrability and Their Interrelation

Fomenko has formulated [143] the following general problem: how can one algorithmically find a maximal linear commutative algebra of functions on a symplectic manifold and establish how many parameters describe the set of all such algebras? Above we have discussed the real version of this problem, now we shall briefly treat the compact version.

The specific feature of the compelx analytic case is that on complex manifolds there are few holomorphic functions. For instance, if a manifold is compact then, by the well-known Liouville theorem, there is not a single non-constant holomorphic function on this manifold. Therefore there is no point in literally transferring the standard definition of the completely integrable symplectic structure from the smooth case. We shall further analyze several distinct notions of Liouville integrability. This question has been examined by Markushevich, and the results are presented below.

A symplectic manifold will be understood as a *complex manifold* M on which there exists a closed holomorphic 2-form $\omega \in \Gamma(M, \Omega_M^2)$ nondegerate at all points of M. The nondegeneracy condition automatically implies that the dimension of M equals $2n$, i.e., is even (we mean the complex dimension) and that the canonical fibre bundle of holomorphic differential forms whose highest degree equals $2n$ is trivial.

Indeed, the form $\omega^n = \omega \wedge \cdots \wedge \omega$ (n times) is its basis. The latter condition will be denoted as $K_M = 0$, where K_M is the so-called *canonical class of divisors* on M, and if this condition is fulfilled, we say that the canonical class M is trivial. The pair (M, ω), in which the form ω is specified up to proportionality, will be called a *symplectic structure*.

Note that if on an even-dimensional complex manifold M there exists a holomorphic (or, more generally, meromorphic) 2-form, which has the maximal rank at the general point, then its restriction to an appropriate subset $U \subset M$ open in Zariski topology sets on U is a symplecftic structure, and thus U is a symplectic manifold.

Next, if there exists an analytic subset Y in M, such that codim $Y \geqslant 2$, and if $U = M \backslash Y$ is a symplectic manifold, then the manifold M itself is also symplectic. This follows from the Riemann continuation theorem for holomorphic forms.

In the present subsection, the manifold M is assumed to be *compact*.

DEFINITION 3.4.1: The symplectic structure (M, ω) is called *completely integrable* if there exists a proper surjective morphism of smooth complex manifolds $f : M \to N$ whose general fibre is a disconnected union of several completely isotropic n-dimensional tori (in particular, a general fibre may appear to be a torus).

This means that there exists a proper closed analytic subset $Z \subset N$ containing all critical values of f, such that for all $y \subset N\setminus Z$, we have

$$f^{-1}(y) = \cup_{i=1}^{r} T_i(y),$$

where $T_i(y)$ are n-dimensional complex tori whose tangent spaces are maximal isotropic subspaces with respect to the form ω at all points, and r is a certain natural number independent of the point $y \subset N\setminus Z$.

DEFINITION 3.4.2: The symplectic structure (M, ω) is called *integrable in the weak sense* if there exists a surjective morphism $f : M \to Y$ of normal complex spaces whose general fibre is a completely isotropic connected submanifold in M. The symplectic structure is called *integrable without degeneracy* if the base Y of the morphism f is a complex manifold and all the fibres of f are n-dimensional connected submanifolds in M.

COMMENT: It is relevant to somewhat specify our terminology. The fibre of the morphims $f : M \to Y$ over the point $y \in M$ is understood as a fibre in the sense of the theory of analytic spaces. Let $a_y \subset O_Y$ be the ideal of the point y in the structural sheaf 0_Y of germs of holomorphic functions on Y, and let $f^{-1}O_Y \to O_X$ a natural embedding of sheafs corresponding to the morphism f. Then a_y may be regarded as a subsheaf of $f^{-1}O_y$ moduli in O_X. The ideal of the fibre $f^{-1}(y)$ is defined by the formula $I_y = a_y O_X$, and then

$$f^{-1}(y) = V(I_y) := (\text{Supp}(O_X/I_y), \, O_X/I_y)$$

is a ringed space with a structural sheaf O_X/I_y. This definition implies, for instance, that if the fibre $f^{-1}(y)$ over any nonsingular point y is an n-dimensional complex manifold, then y is a noncritical value of the mapping f.

The definition of integrability adduced above may be equally formulated in the *algebraic category*. Speaking of integrability in the algebraic sense, we shall always imply that all manifolds M, N, Y and the morphisms $f : M \to N$ or $f : M \to Y$ are projective.

DEFINITION 3.4.3: The symplectic structure (M, ω) is called *meromorphically* (respectively, *additively*) *integrable* if on M there exists n independent meromorphic (respectively, additive) functions $h_1, \ldots, h_n$ which are in involution with respect to the Poisson bracket and possesses the following property: for a certain set of complex numbers $(c_1, \ldots, c_n) \in \mathbb{C}^n$, the closure of the level set $\{h_1 = c_1, \ldots, h_n = c_n\}$ does not intersect the divisor of the poles $h_1, \ldots, h_n$.

Recall that *additive* is the term to specify, possibly, a *multivalued* analytic function on M whose branches differ by a constant, all branches having only pole-type singularities. Such functions are defined like the Picard integrals of type I and II on M, see [225], Ch. II.

The Poisson bracket is defined in the standard way: $\{\varphi, \psi\} = \omega(d\varphi, d\psi)$. Let us now compare different types of integrability.

PROPOSITION 3.4.1 (1). *Complete integrability always implies integrability in a weak sense.* (2) *If the symplectic structure (M, ω) is integrable in a weak sense, then the general fibre of the integrating morphism $f : M \to Y$ is an n-dimensional* complex torus (here, as above, $n = \frac{1}{2}\dim M = \dim Y$). In particular, if the base Y of an integrating family has no singularities, then the symplectic structure (M,ω) is completely integrable. (3) If the symplectic structure (M,ω) is integrataed meromorphically, then in M there exists such a subset open in Zariski topology that the symplectic structure $(U, w|_U)$ is completely integrable. (4) Additive integrability does not, generally, imply integrability in any other sense. (5) Meromorphic integrability follows from integrability in the weak sense if and only if, for a certain integrating morphism $f: M \to Y$, the base Y is a quasi-algebraic manifold.

Recall that the *algebraic dimension* $\text{alg.dim}\, M = tr.\deg_{\mathbb{C}} \mathbb{C}(M)$, that is, the degree of transcendence of the field of meromprphic functions on M over the complex-valued field $\mathbb{C}$ is determined for a complex manifold. The manifold M is called *quasi-algebraic* if $\text{alg.dim}\, M = \dim M$.

PROOF OF PROPOSITION 3.4.1: Item (1) follows from Stein's factorization existence theorem. To prove the second assertion, choose a nonsingular point y, which is a noncritical value of the integrating morphims $f: M \to Y$, and an arbitrary set of local coordinates $z_1, \ldots, z_n$ on Y in the neighbourhood of the point y. Like in §44 of the book [226], we check that the vector fields $X_1, \ldots, X_n$ on M, $X_i = \text{sgrad}\, z_i$ form the basis of the tangential distribution of fibres and thus cause the action of the cotangent bundle T^*Y on the foliation f, which is transitive along the fibres and is defined in the neighbourhood of the point y. In particular, all leaves of f in the neighbourhood of y are tori.

The assertion of item (3) follows from the holomorphic version of the Liouville theorem on integrable systems, see §49 of the book [3]. The proof of the holomorphic version is quite similar to the standard one.

An irreducible even-dimensional complex torus (see Example 1 below) may serve as an example of a manifold integrated additively, but not integrated in the weak sense. However, any symplectic form on a torus in some linear coordinates is written in the canonical form as

$$\omega = dz_1 \wedge dz_{n+1} + \cdots + dz_n \wedge dz_{2n}$$

and is therefore additively integrated; the integrals are, for instance, the functions $z_1, \ldots, z_n$. With respect to item (5), we should note that in the case $\text{alg.dim}\, Y = n$ one can choose preimages of n independent mermorphic functions on Y as independent meromorphic integrals on M. The involutivity of this set of functions follows from the arguments which we have used in the proof of item (2). Conversely, if $\text{alg.dim}\, Y < n$, then the set of n independent meromorphic functions does not exist on Y at all.

An example of a manifold which is completely integrable (and even integrable without degeneracies) is given by the product of two nonalgebraic complex tori of equal dimension with a symplectic structure, in which multipliers are isotropic. But this manifold is not meromorphically integrable..

Thus, we see that in the complex-analytic case, the existence of n independent meromorphic integrals in involution on a symplectic manifold of dimension $2n$ meets with obstacles of two kinds:

1) obstacles to the foliation into isotropic tori which, as shown below, are of analytic or semi-analytic character and are hidden in the structure of the group of analytic cycles of half dimension;

2) if foliation into tori does exist, then the obstacle is the defects between the dimension and the algebraic dimension of the base of this foliation. In the algebraic category, the obstacles of the second kind do not arise, which is asserted in the propositon that follows.

PROPOSITION 3.4.2. *In the algebraic category, the complete integrability of the symplectic structure (M, ω) implies the meromorphic integrability, that is, on M there exist n meromorphic integrals in involution, which are independent at the general point.*

4.2 Integrability on Complex Tori

On an even-dimensional complex torus there are many symplectic structures. All of them are uniquely determined by their values at a certain point. It is therefore reasonable first to investigate separately the question of the existence of foliation into tori of smaller dimensions and then to find out with respect to which symplectic structure this foliation is isotropic. It turns out that *on a "general" complex torus there are no foliations into tori of smaller dimensions*, and if they do exist, they are locally trivial.

THEOREM 3.4.1 (MARKUSHEVICH). *Let M be a complex torus of dimension m; let p be an integer, and $1 \leqslant p < m$. Then:*

(1) If $T \subset M$ is a p-dimensional subtorus, then T is embedded in M linearly and is a shift of the subgroup in M.

(2) If $f : M \to N$ is a proper surjective morphism whose general fibre is a p-dimensional torus, then almost all fibres of f are shifts of the p-dimensional subgroup $T \subset M$, and there exists a bi-meromorphic mapping $\alpha : N \to M/T$ such that $f = \alpha^{-1}\pi$, where $\pi : M \to M/T$ is a canonical projection. If the base N of the family $f : M \to N$ is a (nonsingular) complex manifold, then α is a regular isomorphism of N and M/T.

(3) If $f : M \to N$ is a proper surjective morphism whose general fibre is a disconnnected union of several p-dimensional tori, then each of these tori is a shift of the torus-subgroup T, and there exists a meromorphic mapping $\beta : M/T \to N$ such that $f = \beta\pi$, where $\pi : M \to M/T$ is a canonical projection.

PROOF: Let $i : T \to M$ be an embedding of tori, $\tilde{\imath}$ an induced mapping of universal coverings, and $\tilde{\imath} : \mathbb{C}^p \to \mathbb{C}^m$. Then the Jacobian matrix of $\tilde{\imath}$ is a periodic holomorphic function on $\mathbb{C}^n$ with a complete lattice of periods of rank $2p$ and is therefore constant. Consequently, $\tilde{\imath}$ is an affine mapping, and we have come to the desired conclusion.

(2) The mapping α can be defined by setting its graph: $\Gamma_\alpha = (f, \mathrm{id})(\Gamma_\pi) \subset N \times M/T$, where $\Gamma_\pi \subset M \times M/T$ is a graph of π. The regularity of α in the case of a smooth N follows from the Weyl theorem stating that any meromorphic mapping from a smooth manifold into a torus is a morphism (see, for instance, Lang, *Abelian Varieties*. New York, Wiley Interscience, 1962).

(3) Apply Stein's factoriation and item (2).

PROPOSITION 3.4.3. *Let M be a manifold with a trivial canonical class and let $f : M \to N$ be a morphism of manifolds whose general fibre is a torus. Then f cannot have simple multiple fibres.*

PROPOSITION 3.4.4. *For a "general" complex torus M of dimension m there exists not a single proper subtorus $T \subset M$.*

PROOF: The torus M is uniquely defined by setting an integer-valued lattice $H_{\mathbb{Z}} \subset \mathbb{R}^{2m} = H_{\mathbb{R}}$ of rank $2m$ and an operator of complex structure on $H_{\mathbb{R}}$: $J : H_{\mathbb{R}} \to H_{\mathbb{R}}$, $J^2 = -1$. The exact meaning of the assertion of Proposition 3.4.4 is as follows. Let P be a smooth manifold of all complex structures $J \subset \operatorname{End} H_{\mathbb{R}}$ (it is readily seen that they form a manifold). Then there exists an everywhere dense set $P' \subset P$, such that for any complex structure $J \in P'$, a complex torus $M_J = (H_{\mathbb{R}}/H_{\mathbb{Z}}, J)$ does not have proper subtori.

The proof is obvious: P' consists of complex structures J which have no proper invariant subspaces in $H_{\mathbb{Q}} = H_{\mathbb{Z}} \otimes \mathbb{Q}$.

COROLLARY 3.4.1. *On a "general" even-dimensional complex torus, there is not a single completely integrable simplectic structure.*

4.3 Integrability on $K3$-Type Surfaces

After a two-dimensional complex torus, the next simplest example of a compact complex symplectic manifold is a $K3$ type surface.

DEFINITION 3.4.4: A *$K3$-type surface* is a simply connected compact complex surface with a trivial canonical class.

Recall the main facts regarding $K3$ surfaces. All $K3$ surfaces are known to be diffeomorphic to each other, [232], and Kählerian (the recent result obtained by Siu). The manifold of moduli of $K3$-type surfaces is a connected compact twenty-dimensional complex manifold. $K3$-type surfaces have the following discrete invariants:

$$c_1(S) = 0, \ c_2(S) = \chi(S) = 24, \quad b_1 = b_3 = 0, \ b_2 = 22,$$
$$h^{2,0} = h^{0,2} = 1, \quad h^{1,1} = 20.$$

Besides, the integer-valued homologies of the $K3$ surface S are known to have no torsion. The signature of the form of the intersection on a 22-dimensional integer-valued lattice $H_2(S, \mathbb{Z})$ is equal to $(3, 19)$ (that is, 3 positive and 19 negative squares).

Choosing an arbitrary basis $\Gamma_1, \ldots, \Gamma_{22}$ of the homology lattice of a $K3$ surface S_0, we can define the map of periods on the set of all $K3$ surfaces (or on the manifold of their moduli). This is done in the following way.

Since all $K3$ surfaces are differentiably equivalent, the classes of isomorphism of $K3$ surfaces correspond to different complex structures on S_0. For each such complex structure, corresponding to a $K3$ surface S, there exists a 2-form ω_S on S_0. *The map of periods is thus*

$$F : \{\text{The set of all } K3 \text{ surfaces}\} \to \mathbb{C}P^{21},$$
$$F(S) = (\int_{\Gamma_1} \omega_S, \ldots, \int_{G_2 2} \omega_S).$$

The condition $\omega_S \wedge \omega_S = 0$ implies that the image of F lies in a twenty-dimensional quadric Q. The image of F is known to fill up the whole quadric Q and the class of isomorphism S to be uniquely restored from the point $F(S) \in Q$. The latter assertion is known as the Torelli theorem for Kählerian $K3$ manifolds, see [233].

The Lefschetz theorem on (I,I)-classes (see [234]) states that the topological cohomology class $\lambda \in H^2(S, \mathbb{Z})$ is realized as the 1-st Chern class of a holomorphic linear bundle on S if and only if $\lambda \in H^2(S,\mathbb{Z}) \cap H^{1,1}(S, \mathbb{C})$. But if $\lambda \in H^{1,1}(S, \mathbb{C})$, then denoting the dual homology class by the same letter, we have $\int_\lambda \omega_S = 0$. Thus, the class $\lambda \in H^2(S_0, \mathbb{Z})$ is analytic on the $K3$ surface S if and only if the point $F(S)$ lies in a 19-dimensional hyperflat cross-section of the quadric Q defined by the linear functional $\int_\lambda$.

Let λ be an analytic class on $S, \lambda = c_1(L)$ for a certain linear bundle L on S and let $\lambda^2 = 0$. From the Riemann–Roch theorem, we immediately obtain that $h^0(L) \geq 2$ or $h^0(L^{-1}) \geq 2$. We may assume that $h^0(L) \geq 2$. Then from the condition $\lambda^2 = 0$, it follows that the linear system $|L|$ maps S onto $\mathbb{C}P^1$ with an elliptic curve as typical fibre.

A $K3$-type surface, such that there exists a morphism $\pi : S \to \mathbb{C}P^1$ with an elliptic curve as the typical fibre, is called a surface with an elliptic bundle. Items (1) and (2) of the theorem which follows are proved above; the other items are more or less obvious.

THEOREM 3.4.2 (MARKUSHEVICH). *Let S be a $K3$-type surface, and let ω_S be a symplectic form. The following assertions hold true:*

(1) An elliptic bundle exists on S if and only if there exists a cohomology class

$$\lambda \in H^2(S, \mathbb{Z}) \cap H^{1,1}(S, \mathbb{C}) \subset H^2(S, \mathbb{C})$$

with the self-intersection squared λ^2 equal to zero.

(2) For the general point of the quadric of periods Q the corresponding $K3$ surface has no elliptic bundle. But there exists a countable everywhere dense in Q union T of hyperflat cross-sections of Q, to the points of which there correspond $K3$ surfaces with an elliptic bundle:

$$T = \cup_{\lambda^2 = 0} Q \cap H_\lambda, \qquad H_\lambda = \{a_1 X_1 + \cdots + a_{22} X_{22} = 0\},$$

where

$$\lambda = a_1 \Gamma_1 + \cdots + a_{22} \Gamma_{22} \in H_2(S, \mathbb{Z}).$$

(3) The symplectic structure (S, ω_S) is Liouville-integrable if and only if there exists an elliptic bundle on S.

(4) Complete meromorphic integrability of the symplectic structure on a $K3$ surface and integrability in the weak sense are equivalent.

(5) On an algebraic $K3$ surface, there always holds an additive integrability; for a general algebraic $K3$ surface, integrability in any other sense does not exist.

4.4 Integrability on Beauville Manifolds

The Beauville manifold $S^{[2]}$ constructed by means of the $K3$ surface S is the first nontrivial example of an irreducible four-dimensional symplectic manifold.

DEFINITION 3.4.5: Let S be a $K3$-type surface. The Douady space parametrizing all zero-dimensional analytic cycles $Z \subset S$ of length r, lgth $Z = r$, is called the *Beauville manifold* constructed by means of the $K3$ surface S. The Beauville manifold is denoted by $S^{[r]}$ and is a $2r$-dimensional symplectic manifold [231].

We may present the following effective and *vivid description of the Beauville manifold* $S^{[2]}$. Let $S \times S$ be the direct produce of a $K3$ surface S by itself, the group $\mathbb{Z}_2$ acting on this product by permutation of multipliers. Then the symplectic square $S^{(2)} = S \times S/\mathbb{Z}_2$ of the surface S has an analytic type singularity along the diagonal

$$\Delta' \subset S^{(2)} : (S^{(2)}, \Delta') \sim (\mathbb{C}^2 \times Q, \mathbb{C}^2 \times \{0\}),$$

where Q is a nondegenerate quadratic cone in $\mathbb{C}^3$. The inflation of $S^{(2)}$ centered at Δ' is already a nonsingular manifold and is isomorphic to the Beauville manifold

$$S^{[2]} : S^{[2]} = B_{\Delta'}(S^{(2)}).$$

The symplectic form φ on $S^{[2]}$ is induced by a $\mathbb{Z}_2$-invariant 2-form $pr_1^* \omega_S + pr_2^* \omega_S$ on $S \times S$. Consider the mapping

$$S \underset{pr_2}{\overset{pr_1}{\rightleftarrows}} S \times S \xrightarrow{\varepsilon} S^{(2)} \xleftarrow{\sigma'} S^{[2]}.$$

With the help of these mappings, we may define the mapping

$$i : H^2(S, \mathbb{C}) \to H^2(S^{[2]}, \mathbb{C})$$

by the formulae $i(a) = \sigma'^* \varepsilon_*(pr_1^* a + pr_2^* a)$. In the paper [231], it is proved that i is an embedding and a morphism of the Hodge structures and that

$$H^2(S^{[2]}, \mathbb{C}) = i H^2(S, \mathbb{C}) + \mathbb{C}[E],$$

where $[E]$ is the class of exceptional divisor $(\sigma')^{-1}(L') = E$. Let now $M = S^{[2]}$. On two-dimensional cohomologies, one can define the quadratic form

$$q(\alpha) = \int_M \alpha^2 \varphi \bar{\varphi} - \int_M \alpha \varphi \bar{\varphi} 2 \cdot \int_M \alpha \varphi^2 \bar{\varphi}.$$

In the paper [231], it is proved that $q|i H^2(S, \mathbb{C})$ coincides with the form of intersection on S and that $iH^2(S, \mathbb{C}) \perp [E]$ with respect to the form q. Suppose that S is an algebraic $K3$ surface and that $f : M \to N$ is an integrating morphism. Choose on N an abundant divisor. Its inverse image on M will be denoted by D. Then $D^3 = 0$, $\varphi | D^2 = 0$. Assume in addition that $D^2 E = 0$. Then obviously, $\varphi | DE = 0$. Thus, in this assumption, we obtain from the de Rham theorem: $q([E]) = 0$, $q([D], [E]) = 0$. Since the mapping i corresponds to the Hodge structures and to the integer-valued structure, therefore $[D] = i(a)$, where a is the class of divisors on S determining the structure of the elliptic bundle. If $D^2 E \neq 0$, then more delicate geometric arguments ar needed, but the existence of an elliptic bundle on S can be proved in this case as well.

THEOREM 3.4.3. *Let $M = S^{[2]}$ be an algebraic Beauville manifold. For the symplectic structure on M be Liouville-integrable, it is necessary and sufficient that on S there exist a bundle of elliptic curves. Furthermore, if the symplectic structure on $S^{[2]}$ is integrable, then there exists an integrating morphism $\rho: S^{[2]} \to \mathbb{C}P^2$, the set of degeneracies Z of which contains a smooth curve C of second degree in $\mathbb{C}P^2$, and a certain number of straight lines L_j tangent to this curve. A degenerate fibre over a general point has the form $2F_1 + F_2$, where F_1 and F_2 are transversally intersecting surfaces with an elliptic base; the type of degeneracies over L_j is determined by the type of degeneracies of the elliptic bundle on S.*

COROLLARY 3.4.2. *For the general algebraic K3 surface S, the symplectic structure $S^{[2]}$ is not Liouville-integrable.*

Remark 1. Algebraic K3 surfaces fill a countable number of 19-dimensional families in a 20-dimensional manifold of moduli of Kählerian K3 surfaces. The term "general algebraic K3 surface" describes each of these families. In each of these 19-dimensional families, there exists an everywhere dense countable union of 18-dimensional hyperflat cross-sections, whose terms admit an elliptic bundle.

Remark 2. The space of moduli of Beauville manifolds $S^{[2]}$ is larger than that of K3-type surfaces. Thus, the general deformation of $S^{[2]}$ is not represented in the form $X^{[2]}$ for any surface X. Obviously, in the set of all deformations of manifolds $S^{[2]}$, there exists a countable number of divisors, to the points of which there correspond integrable symplectic structures, but it is not at all clear how this should be proved.

Remark 3. The assumption of fibre isotropy in Theorem 3.4.3 is not needed. It suffices to assume that the general fibre of f is a disconnected union of two-dimensional tori. This alone is sufficient for S to have an elliptic bundle.

Remark 4. In the case of a K3 surface, any divisor A for which $A^2 = 0$ gives a bundle of elliptic curves. As is seen in the example which follows, on the manifold $S^{[2]}$, not all sliding divisors D with a zero cube ($D^3 = 0$) give a family of Abelian surfaces. Let A be an Abelian surface and $K = A/(x \to -x)$ its Kummer surface. Then the minimal desingularization of K is a K3 surface. We shall denote it by $\tilde{K}$ (see, for instance, Ch. 6 in the book [234]). It can be easily verified that the image of a general shift of the diagonal

$$\Delta_z \subset A \times A, \ \Delta_z = \{(x, x+z) \in A \times A | x \in A\}$$

under the mappings

$$A \times A \to K \times K \dashrightarrow K \times \tilde{K} \to K^{(2)} \dashrightarrow K^{[2]}$$

in $K^{[2]}$ is a K3 surface isomorphic to $\tilde{K}$. Thus, we have constructed a two-dimensional family of K3 surfaces in $\tilde{K}^{[2]}$ without general points. The corresponding mapping $\rho: \tilde{K}^{[2]} \to \mathbb{C}P^2$ is given by the linear system of the divisor D with a zero cube.

4.5 Symplectic Structures Integrated Without Degeneracies

We shall show that a *sufficiently nontrivial symplectic structure does not admit an integrating family without degeneracies*. We can impart an exact meaning to this assertion and prove it only for Kählerian symplectic manifolds of dimension not exceeding four (that is, 2 or 4). But this statement has evidently a more general character and is valid in a much more general situation.

DEFINITION 3.4.6: A symplectic structure (M,ω) will be called *isotrivial* if there exists a finite nonbranching covering $\pi : \widetilde{M} \to M$, such that $\widetilde{M}$ splits into the direct product of an n-dimensional complex torus, completely isotropic with respect to the form

$$\widetilde{\omega} = \pi^*\omega,$$

by a certain n-dimensional complex manifold

$$N : \widetilde{M} = T^n \times N, \ \widetilde{\omega}|T^n \times \{y\} = 0, \qquad \forall y \in N.$$

A symplectic structure will be called *locally trivial* if it is integrable and if there exists a locally trivial integrating family.

We shall similarly define local triviality and isotriviality for an arbitrary morphism (a morphism is isotrivial if, after a finite-sheeted nonbranching change of the base, it becomes a projection onto a multiplier of the direct product).

THEOREM 3.4.4. *Let M be a Kählerian manifold of dimension $m = 2n \leqslant 4$ with a symplectic form ω. Suppose that the symplectic structure (M,ω) is integrable without degeneracies. Then it is locally trivial. If M is a projective manifold and if the symplectic structure (M,ω) is integrable without degeneracies in the algebraic category, then it is also isotrivial.*

This theorem is immediate from a more general assertion.

THEOREM 3.4.5. *Let M be a compact Kählerian manifold of dimension $m \leqslant 4$ and $c_1(M) = 0$. Suppose that a morphism of complex manifolds $f : M \to N$ is given and that all fibres of this morphism are p-dimensional complex tori, where*

$$p = m - n, \qquad n = \dim N, \qquad 1 \leqslant p \leqslant 3.$$

Then the morphism f is locally trivial. If M, N, f are projective, then f is isotrivial.

CHAPTER 4

INTEGRATION OF CONCRETE HAMILTONIAN SYSTEMS IN GEOMETRY AND MECHANICS. METHODS AND APPLICATIONS

§1 Lie Algebras and Mechanics

1.1 Embeddings of Dynamic Systems into Lie Algebras

In Ch. 1 we have already encountered an important fact: some equations describing the motion of mechanical systems may be interpreted as Hamiltonian systems (vector fields) on suitable Lie algebras. In this chapter we give a more thorough analysis to this effect which turns out to enable many interesting and important mechanical systems and their multidimensional analogues to be integrated. There are now many available methods of integration at our disposal, and therefore we are not able to give a more or less detailed account of all of them. So we refer the reader to the list of references at the end of the book. We proceed as follows: we concentrate attention only on several essential mechanisms which control integrability or nonintegrability of Hamiltonian systems.

In mechanics, many classical equations are written as systems of differential equations on Euclidean space. As we have seen, for instance, in Ch. 1, the equations of motion of a three-dimensional rigid body fixed in the centre of mass are written in $\mathbb{R}^3(x, y, z)$ as follows:

$$\dot{x} = \frac{\lambda_1 - \lambda_2}{\lambda_1 + \lambda_2} yz, \qquad \dot{y} = \frac{\lambda_3 - \lambda_1}{\lambda_3 + \lambda_1} xz, \qquad \dot{z} = \frac{\lambda_2 - \lambda_3}{\lambda_2 + \lambda_3} xy.$$

As has already been noted (Ch. 1), this system is closely connected with the three-dimensional Lie algebra so(3) of the rotation group SO(3). The point is that the space $\mathbb{R}^3(x, y, z)$ can be naturally identified with the space so(3) of skew-symmetric real matrices $X = (x_{ij})$. To this end, it suffices to put $x_{12} = x, x_{13} = y, x_{23} = z$, that is, to associate with the vector with coordinates (x, y, z) the matrix

$$X = \begin{pmatrix} 0 & x & y \\ -x & 0 & z \\ -y & -z & 0 \end{pmatrix}.$$

Next, in the Lie algebra so(3), we consider the (co)adjoint representation (action) of the group SO(3). It is readily seen that the orbits of this action are standard two-dimensional spheres centered at the origin.

Identifying $\mathbf{R}^3$ with so(3) we interpret the Euler equations (see above) as a certain vector field $\tilde{v}$ on the Lie algebra so(3). Now make a simple transformation induced by the following change of coordinates in $\mathbf{R}^3$:

$$x \to \frac{x}{\lambda_1 + \lambda_2}, \qquad y \to \frac{y}{\lambda_1 + \lambda_3}, \qquad z \to \frac{z}{\lambda_2 + \lambda_3}$$

(similarity transformation). The field $\tilde{v}$ will rotate and become a new field v.

CLAIM 4.1.1: *A field v described by the Euler equations of motion of a rigid body fixed in the centre of mass is tangent to the orbits 0 of the adjoint representation in the Lie algebra so(3) and is Hamiltonian on these orbits (which are homeomorphic to the spheres S^2).*

The proof is reduced to a direct calculation, and we leave it to the reader. Thus, on embedding the equations of motion of a rigid body into a suitable Lie algebra we find out that they have become a Hamiltonian system on the orbits of the action of the group SO(3). This fact is a particular case of the general mechanism which makes it possible to integrate many mechanical systems similar to the one just described.

Now we are in a position to formulate the *general idea of algebraization of mechanical systems* which permits integration of several multidimensional Euler equations.

DEFINITION 4.1.1: *We say that a dynamic system v on a manifold M admits embedding into a Lie algebra (or admits algebraization) if the manifold M can be identified with a certain submanifold $N \subset G^*$ in the dual space G^* of a certain Lie algebra G (for a certain Lie group $\mathfrak{G}$) so that there hold the conditions:*

1) The submanifold N is a union of orbits $O^*(\xi)$ of the coadjoint representation of the group $\mathfrak{G}$ on G^*.

2) The vector field v is tangent to the orbits $O^*(\xi)$.

3) The vector field v on N turns out to be Hamiltonian on the orbits $O^*(\xi)$ with respect to the canonical symplectic structure w and has the form $v = \text{sgrad } H$, where $H \in C^\infty(O^*(\xi))$. If the initial manifold was symplectic, it is natural to require that the symplectic structure on it be induced by the canonical symplectic structure given on G^*.

In other words, a system v *admits algebraization* if it possesses hidden *group symmetries* that manifest themselves under embedding into a suitable Lie algebra.

What does algebraization of any concrete mechanical system result in? It allows us to apply the developed apparatus of the theory of Lie groups and Lie algebras. As is seen from the studies carried out in recent years (see, in particular, [89]–[94], [130]–[149] etc.), this makes it possible to exhibit rather efficiently and in an explicit form the polynomial and rational integrals of many interesting dynamic systems.

First of all this concerns the various multidimensional analogues of the equations of motion of a rigid body: in the absence of gravity, in the presence of gravity, inertial motion in an ideal incompressible liquid, in a magnetized liquid, etc. See below.

We have already seen above that one of the central positions in the modern integration theory is taken by maximal linear commutative (and noncommutative) Lie subalgebras V of functions on the orbits $O^* \subset G^*$ of (co)adjoint representations. Many of them are already available. In the next subsection we give a list of the basic examples. For the moment we restrict ourselves to the question: what is the discovery of such maximal linear algebras of functions from the point of view of applications? The answer is very simple. The point is that any smooth function f from a maximal linear commutative algebra V of functions generates a completely integrable (in the sense of Liouville) Hamiltonian system sgrad f. Varying the function f within the algebra V we change a corresponding system.

The collection of such systems is therefore determined by the dimension of the commutative Lie algebra V of integrals. It appears that one often obtains interesting mechanical systems (and their analogues). To say it differently, the commutative Lie algebras V already constructed often prove to involve interesting mechanical Hamiltonians. We should only be able to find them among the functions of the Lie algebra V.

There exist circumstances that sometimes facilitate the search. In fact the Hamiltonians appearing in mechanics and physics are usually quadratic (bilinear) functions. Therefore one should first examine the linear subspace of quadratic functions contained in a commutative Lie algebra V. As a rule, this examination is rather simple, and we will demonstrate it on concrete examples.

1.2 List of the Discovered Maximal Linear Commutative Algebras of Polynomials on the Orbits of Coadjoint Representations of Lie Groups

We will enumerate several examples of the above-mentioned algebras, namely those which are of interest from the point of view of multidimensional analogues of the equations of motion of the various mechanical systems.

THEOREM 4.1.1 (FOMENKO, MISHCHENKO) (SEE [89], [91], [93]). *Let G be one of the following finite-dimensional Lie algebras: a) a complex semisimple Lie algebra G_s; b) a compact real form G_c of a complex semisimple Lie algebra G_s; c) a normal compact subalgebra G_n in a compact Lie algebra G_c. Then on each orbit of general position O in the algebra G (canonically identified with G^*) there always exists a maximal linear commutative algebra of polynomials (with respect to the canonical Poisson bracket in G^*). These polynomials are calculated in an explicit form as functions of the canonical coordinates in the Lie algebra G.*

It is instructive to emphasize that the generators of the commutative algebras in Theorem 4.1.1 are given in an explicit form. Moreover, they are extremely simple.

PROPOSITION 4.1.1 (SEE [89], [91], [93]). *Let IG be a ring on invariant polynomials on a Lie algebra G, where G is either a complex semisimple Lie algebra, its compact real form, or a normal compact subalgebra. Let a be an element of general position in G. By shifting the argument towards the invariants $f(X)$ of the algebra G, we obtain new polynomials of the form $f(X + \lambda a)$, where $X, a \in G, \lambda \neq 0$. Restricting these functions to orbits of general position, one obtains the maximal commutative linear algebra of polynomials on the orbit. Next, let $G_n \to G_c$ be the* canonical embedding on a normal compact subalgebra into a compact real form. Restricting polynomials of the form $f(X + \lambda a')$, $f \in IG_c$, $a' \in G_c$, defined on the algebra G_c to the orbits of general position in G_n, we also obtain the maximal linear commutative algebra of polynomials on the orbit in G_n (which is, generally speaking, different from the preceding one).

Let us comment upon these theorems. Cases a), b), and c) are similar from the point of view of constructing explicit formulae for the generators of the commutative algebras of functions. The case of a normal compact subalgebra is richer. It is distinguished in that we can also consider the embedding $G_n \subset G_c$ and choose the element a' as the element of general position from the point of view of the larger enveloping algebra G_c (but not the subalgebra G_n).

Above we have dealt with typical orbits, that is, orbits of general position. *In a semisimple case, however, these results also remain valid for singular orbits* (which occupy the set of measure zero in a Lie algebra). This assertion was published for the first time in [39] by Dao Chong Thi, but later on several authors found lacunas in his proof. The accurate and complete final proof was obtained by Brailov.

THEOREM 4.1.2 (DAO CHONG THI, BRAILOV). *Let O be an arbitrary semisimple orbit (either of general position or singular) in a semisimple Lie algebra G (either complex or real). Then on it there always exists a maximal linear commutative algebra of polynomials. As in Proposition 4.1.1, their explicit construction looks as follows. One should examine all polynomials of the form $f(X+\lambda a)$, where $f(X)$ are invariant polynomials on G and a is an arbitrary fixed regular semisimple element in G.*

Since $f(X)$ are polynomials then expanding $f(X + \lambda a)$ in a power series of the parameter λ we obtain $\sum P_i(X, a)\lambda^i$. Then the set of polynomials $P_i(X, a)$ forms the maximal linear commutative algebra of functions on the orbit O.

In the semisimple compact Lie algebra G, each orbit (of general position or singular) necessarily meets with the Cartan subalgebra, and therefore each orbit is semisimple. In an arbitrary semisimple algebra this is not the case. In the complex semisimple (and also in the real) Lie algebra there exist both semisimple and non-semisimple orbits. The latter do not meet with the Cartan subalgebra and fill the set of measure zero in the Lie algebra. As far as these orbits are concerned, the situation is not yet clear, i.e. it is unknown whether or not maximal linear commutative algebras of polynomials exist on such orbits.

In all the enumerated cases, the maximal set of independent functions in involution on orbits was obtained *by restricting the shifts of invariants of a Lie algebra to the orbits*. This mechanism is sufficiently general although, as shown below, it does *not always* yield a sufficient set of functions in involution. This idea (the shift of invariants) was first suggested by Manakov [76] for the Lie algebra $so(n)$. The proof (in algebraic terms) of involutivity and maximality of the obtained set of functions was proposed by Fomenko and Mishchenko who then extended this method already to arbitrary semisimple Lie algebras (see Theorem 4.1.1 of this subsection), [89], [91]–[94], [143], [186], [182].

Some maximal linear commutative algebras of functions on semisimple Lie algebras were later constructed in [16] by Bogoyavlensky.

§1. Lie Algebras and Mechanics

The next class of Lie algebras, in some sense rather close to the class of semisimple algebras, is an extension of semisimple Lie algebras by means of linear representations of minimal dimension. Recall some definitions.

Let ρ be the linear representation of a certain Lie algebra K by means of linear operators on the linear space V of the representation ρ, that is, $\rho: K \to \operatorname{End} V$. Then one can construct a new Lie algebra denoted as $G = K \oplus \rho V$ and formed by the pairs (X, v), where $X \in K, v \in V$, and the element commutation rule is given by the following formula. The linear space of the new algebra G may be interpreted as the direct sum of the linear subspaces K and V, that is, $G = K \oplus V, q = X + v, q \in G$. Then $[q_1, q_2] = [X_1 + v_1, X_2 + v_2] = [X_1, X_2] + (\rho X_1)v_2 - (\rho X_2)v_1$.

This means that the subspace V in G is an Abelian subalgebra, and the subalgebra K acts on V by means of the representation ρ. Among the described extensions there is an important class corresponding to the exact representations of minimal dimension of semisimple Lie algebras, that is, to the standard representations ρ under which the Lie algebras are described by matrices of the least possible dimension. One of such most popular Lie algebras is the Lie algebra

$$e(n) = so(n) \oplus_\rho \mathbf{R}^n.$$

It is the Lie algebra of the group $E(n)$, the group of proper transformations of the Euclidean space $\mathbf{R}^n$. In the matrix form this algebra is given as

$$\left(\begin{array}{c|c} so(n) & \begin{array}{c} b_1 \\ \vdots \\ b_n \end{array} \\ \hline 0 \ldots 0 & 0 \end{array} \right),$$

where $b = (b_1, \ldots, b_n)$ is a vector from $\mathbf{R}^n$. It is clear that

$$E(n) = \left(\begin{array}{c|c} so(n) & \begin{array}{c} a_1 \\ \vdots \\ a_n \end{array} \\ \hline 0 \ldots 0 & 1 \end{array} \right).$$

Here $a = (a_1, \ldots, a_n) \in \mathbf{R}^n$, and if $g \in E(n)$ then $g(x) = Ax + a$, where

$$x \in \mathbf{R}^n, \qquad A \in so(n), \qquad a \text{ is translation}$$

It is natural that the Lie algebra is noncompact and is not semisimple.

THEOREM 4.1.3 (FOMENKO AND TROFIMOV [141], [142], [163], [186]). *On each orbit O^* of general position in the dual space $e(n)^*$ to the Lie algebra $e(n)$ of the group of motions of the space $\mathbf{R}^n$ there always exists a maximal linear commutative algebra of polynomials. It is formed by all polynomials of the form $f(X + \lambda a)$, where $f(X)$ is invariant of the algebra $e(n)$ and $a \in e(n)^*$ is a covector of general position.*

In the case of singular orbits the situation is not yet clear. The commutative algebra of polynomials which we have discovered is in a certain exact sense an analogue of the "integrals of compact series" constructed in Theorem 4.1.1 for compact real forms of semisimple Lie algebras. It has turned out that on the noncompact Lie algebra $e(n)$ there also exist other commutative algebras of polynomials, similar to the "normal series" (see Theorem 4.1.1). This result belongs to Brailov and will be presented in the sequel.

The sets of functions of this series are integrals of the following equation:

$$\dot{x} = \operatorname{ad}^*_{dH(x)}(x), \qquad x \in e(n)^* \tag{1}$$

These sets of functions are therefore completely Liouville integrable. Note that for $n = 3$ and for a positive definite quadratic function H this equation coincides with the familiar *Kirchhoff system of the dynamics of a rigid body in an ideal liquid*. In our case, however, the quadratic function H is nondegenerate but sign indefinite. Determine the standard basis of the Lie algebra $e(n)$, consisting of the elements $\bar{x}_{ij}$ and $\bar{y}_k$, where $\bar{x}_{ij}$ is an infinitesimal rotation in the (i,j)-plane and $\bar{y}_k$ is an infinitesimal motion along the k-th coordinate. The linear coordinate functions on $e(n)^*$ corresponding to the elements $\bar{x}_{ij}$ and $\bar{y}_k$ will be denoted by x_{ij} and y_k. We will require the $(n+1) \times (n+1)$ matrices X and A. Let $a_1, \ldots, a_n$ be arbitrary numbers. Define the matrices:

$$E_{ks} = (\delta_{ik}\delta_{sj}); \qquad Y_{ii} = E_{ii} - \frac{1}{n+1}\sum_{j=1}^{n+1} E_{jj};$$

$$Y_{ij}^{\pm} = E_{ij} \pm E_{ji} (i \neq j); \qquad A_1 = \sum_{i=1}^{n-1} a_i Y_{ii};$$

$$A_{-1} = a_n Y_{n,n+1}^+; \qquad A = A_1 + A_{-1}; X_1 = (x_{ij}),$$

$$x_{i,n+1} = x_{n+1,i} = 0; \qquad X_{-1} = \sum_{i=1}^{n} y_i Y_{i,n+1}^-;$$

$$X = X_1 + X_{-1}.$$

THEOREM 4.1.4 (BRAILOV). *Let $a_1 > \cdots > a_{n-1}, a_n \neq 0, b_1, \ldots, b_n$ be arbitrary numbers. Then equation (1) with the quadratic function*

$$2H = \sum_{i=1}^{n-2}\sum_{j=i+1}^{n-1}\frac{b_i - b_j}{a_i - a_j}x_{ij}^2 - 2\sum_{i=1}^{n-1}\frac{b_i}{a_n}y_i x_{in} = \sum_{i=1}^{n}\frac{a_n b_n - a_i b_i}{a_n^2}y_i^2 + \frac{b_n}{a_n}y_n^2$$

is completely Liouville integrable on the orbit O^* (of general position) of the coadjoint action of the group $E(n)$ in $e(n)^*$. The complete set of commuting integrals are the coefficients h_{ks} (where $k = 2, \ldots, n+1, s = 0, \ldots K$) for $\lambda^2 \mu^{-2}$ in the polynomial $h_k(\lambda, \mu^{-1}) = \Delta^k(X_1 + \lambda A_1 + \mu^{-1}(X_{-1} + \lambda A_{-1}))$, where Δ^k is the sum of all symmetric minors of order k.

We may proceed to prove the theorem only after some necessary preliminaries. Let G be a Lie algebra, θ an involutive automorphism. Examine the decomposition $G = G_1 + G_{-1}$ such that the automorphism $\theta|_{G_i} = i$(where $i = 1, -1$). Establish the correspondence $T_\mu : g_1 + g_{-1} \to g_1 + \mu^{-1}g_{-1}$, where $g_i \in G_i$. The image of the commutator $[\,,\,]$ of the Lie algebra G for T_μ will be denoted by $[\,,\,]_\mu$. It is readily seen that

$$[\,,\,]_\mu = \begin{pmatrix} [\,,\,]_{1,1} & [\,,\,]_{1,-1} \\ [\,,\,]_{-1,1} & \mu^2[\,,\,]_{-1,-1} \end{pmatrix},$$

where $[\,,\,]_{ij}$ is the restriction of $[\,,\,]$ to $G_i \times G_j$. Hence for $\mu \to 0$ the commutator $[\,,\,]_\mu$ has a limit which will be denoted by $[\,,\,]_0$. The Lie algebra corresponding to $[\,,\,]_\mu$ is denoted by $G_{\theta,\mu}$ and in the case of $\mu = 0$ simply by G_θ. By definition, the vector spaces $G, G_\theta, G_{\theta,\mu}$ coincide. Let f be a polynomial on G^*, T_μ^* a linear mapping conjugate to T_μ. Define the function f_θ as the coefficient of the polynomial $f(\mu^{-1}) = f_0 T_\mu^* = f_0 + \mu^{-1} f_1 + \cdots + \mu^{-k} f_k$ for μ^{-k}, $f_\theta = f_k$.

LEMMA 4.1.1. *Let G, G_θ be Lie algebras of Lie groups $\mathfrak{G}$, $\mathfrak{G}_\theta$. For each invariant f of the representation Ad^* of the group $\mathfrak{G}$ the function f_θ is the invariant of the representation Ad^* of the group $\mathfrak{G}_\theta$. For the semisimple Lie algebra G the index is of the form:* $\mathrm{ind}\, G_\theta = \mathrm{ind}\, G$.

PROOF: By definition, $T_\mu : G \to G_\mu$ is isomorphism of Lie algebras for $\mu \neq 0$. Consequently, the function $f \circ T_\mu^*$ is the invariant of the representation Ad^* of the Lie group $\mathfrak{G}_\mu$ which corresponds to the Lie algebra G_μ. By continuity, $\lim_{\mu \to 0} \mu^k f \circ T_\mu^* = f_\theta$ is the invariant of $\mathfrak{G}_\theta$. We will prove the second assertion. Since the index of a Lie algebra does not change under complexification, G may be assumed to be a complex Lie algebra. Let H be the Cartan subalgebra in G such that:

(1) $H \cap G_{-1} = H_{-1}$ is a maximal commutative subspace in G_{-1}, 2) the centralizer K of the space H_{-1} in G_1 is a reductive Lie algebra and $H_1 = H \cap G_1$ is the Cartan subalgebra in K. The existence of such Cartan subalgebras H is proved, for instance, in [204], p. 78. Identify G_θ^* and G_θ by means of the Killing form of the Lie algebra G. Our immediate aim is to find the subalgebra $G_\theta^x = \{y \in G_\theta : \mathrm{ad}_{0,y}^*(x) = 0\}$, where $\mathrm{ad}_{0,y}^*$ is the operator conjugate to the operator $\mathrm{ad}_{0,y}(g) = [y,g]_0$ for the element x of general position in H_{-1}. Since

(2) $$\mathrm{ad}_{0,y}^*(x) = [x_1, y_1] + [x_{-1}, y_{-1}] + [x_{-1}, y_1],$$

where $x = x_1 + x_{-1}$, $y = y_1 + y_{-1}$; $x_i, y_i \in G_i$, then for $y \in G_\theta^x$ we have

(3) $$[x_1, y_1] + [x_{-1}, y_{-1}] = 0$$

(4) $$[x_{-1}, y_1] = 0$$

From (4) we obtain that $y_1 \in K$, whence $[x_1, y_1] = -[x_{-1}, y_{-1}] \in K$. Consequently, $[x_{-1}, [x_{-1}, y_{-1}]] = 0$. Since x_{-1} is a semisimple element in G then $[x_{-1}, y_{-1}] = 0$. Therefore, $y_{-1} \in H_{-1}, [x_1, y_1] = 0, y_1 \in H_1$ and $y \in H$. Hence $G_\theta^x \in H$. The inverse inclusion is verified in a trivial manner. Eventually we have $G_\theta^x \in H$. Thus, $\mathrm{ind}\, G_\theta \leqslant \mathrm{ind}\, G$. The inverse inequality is immediate from the continuous dependence of the commutator $[\,,\,]_\mu$ on μ. This proves the lemma.

PROOF OF THEOREM 4.1.4: The functions h_{ks} will be shown to be integrals. Let θ be such an automorphism of the Lie algebra $\mathrm{sl}(n+1, \mathbb{R})$ that for the matrix

$Z \in \mathrm{sl}(n+1, \mathbb{R})$ we have: $\theta(Z) = C_n Z C_n$, where $C_n = \mathrm{diag}(c_1, \ldots, c_{n+1})$, $c_1 = \cdots = c_n = 1, c_{n+1} = -1$. Lemma 4.1.1 implies that the functions $f_k(Z) = \Delta_\theta^k(Z)$ are invariants of the representation Ad^* of the Lie group $\mathrm{SL}(n+1, \mathbb{R})_\theta$ which corresponds to the Lie algebra $\mathrm{sl}(n+1, \mathbb{R})_\theta$. Determine the matrix $B = B_1 + B_{-1}$, $B_1 = \sum_{i=1}^{n-1} b_i Y_{ii}$, $B_{-1} = b_n Y_{n,n+1}^+$. From this point on the matrix B is interpreted as the element of the Lie algebra $\mathrm{sl}(n+1, \mathbb{R})_\theta$. Using formula (2) we find that for any element $x \in \mathrm{so}(n+1)_\theta^*$ there holds the equality $\mathrm{ad}_{0^*, dH(x)}^*(A) = \mathrm{ad}_{0,B}^*(x)$. From this it follows that equation (1) for any number λ is equivalent to the equation $\frac{d}{dt}(x + \lambda A) = \mathrm{ad}_{0, dH(x) + \lambda B}^*(x + \lambda A)$. Hence we identify the Lie algebras $E(n) = \mathrm{so}(n+1)_\theta$. This implies that the functions $f_k(x + \lambda A)$ are integrals of equation (1).

Let σ be such an involutive automorphism of the algebra $\mathrm{sl}(n+1, \mathbb{R})_\theta$ that $\sigma(Z) = -Z^T$ for any matrix Z, where T stands for transposition. Since the set of σ- invariant matrices in the Lie algebra $\mathrm{sl}(n+1, \mathbb{R})_\theta$ is $\mathrm{so}(n+1)_\theta = E(n)$ and $\sigma^*(A) = -A$, then we obtain involutivity of the integrals h_{ks}. It remains to check completeness of the set of integrals h_{ks}. Let f_{ks} be the coefficient at λ^s in the polynomial

$$f_k(\lambda, Z) = f_k(Z + \lambda A'), Z \in \mathrm{sl}(n+1, \mathbb{R})_\theta^*, A' = \sum_{i=1}^n Y_{i,i+1}^-.$$

Before proving completeness of the set of integrals h_{ks}, we will find the explicitly linear subspace generated by the differentials df_{ks} at a special point $Z' \in \mathrm{sl}(n+1, \mathbb{R})_\theta^*$. In effect, we will prove completeness of the commutative set of functions f_{ks} on $\mathrm{sl}(n+1, \mathbb{R})_\theta^*$ although in the sequel this is not used explicitly. Define the Cartan subalgebra H in $\mathrm{sl}(n+1, \mathbb{R})$, $H = H_1 + H_{-1}$; $H_1 = (\sum_{i=1}^{n-1} x_i Y_{ii}$; $x_i \in \mathbb{R})$, $H_{-1} = (y_n Y_{n,n+1}^+; y_n \in \mathbb{R})$. Restricting the functions f_k to the Cartan subalgebra H, we obtain $f_k|_H = \sigma_{k-2}(x'_1, \ldots, x'_{n-1}) y_n^2, 2 \le k \le n+1$, where σ_{k-2} is an elementary symmetric function of degree $k - 2$, $x'_i = x_i - \frac{1}{n+1} \sum_{j=1}^{n-1} x_j$. Therefore, at the point Z of general position in H the differentials $df_2, \ldots, df_{n+1}$ generate H. Let Z' be a point of general position. For any $k = 2, \ldots, n+1$; $s = 0, \ldots, k-1$ there holds the equality

(5) $$\mathrm{ad}_{0, U_{k,s+1}}^*(Z') + \mathrm{ad}_{0, U_{k,s}}^*(A') = 0,$$

where $U_{ks} = df_{ks}(Z')$ is a differential. From the definition we know that the differentials U_{k0} generate H. From formula (5) we find that U_{k1} generate the space $F_1 + H$ modulo H, where F_1 is the linear subspace generated by the elements:

$$Y_{1,2}^-, Y_{2,3}^-, \ldots, Y_{n-2,n-1}^-, Y_{n-1,n+1}^-, Y_{n,n+1}^-.$$

Next, once again making use of formula (5), we find that the elements $U_{k,2}$ generate the space $F_2 + F_1$ modulo F_1, where F_2 is the linear subspace generated by the elements:

$$Y_{1,3}^+, Y_{2,4}^+, \ldots, Y_{n-3,n-1}^+, Y_{n-2,n+1}^+, Y_{n-1,n+1}^+.$$

Proceeding by induction we find the liner subspace generated by all the differentials $df_{ks}(Z')$ to be of the form

(6) $$V(Z', A') = \mathrm{sl}(n+l, \mathbf{R})_{-1} \oplus \left(\bigoplus_{\substack{i<j \\ 1 \leqslant i,j \leqslant n-1}} \mathbf{R}Y_{ij}^{\pm} \right),$$

In the last sum the sign "+" is taken in the case of even difference $j - i$, the sign "−" in the case of odd difference $j - i$. Then $V(Z', A') = V(x, A)$, where $x = A'$, $A = Z'$. The differentials of the functions h_{ks} are obtained by projection of the differentials of the functions f_{ks} onto $\mathrm{so}(n+1)_\theta$. From this and from formula (6) we find that the dimension of the linear subspace generated by the differentials of the functions h_{ks} at the point $x = A'$ is equal to m^2 in case $n+1 = 2m$ is even and to $m^2 + m$ in case $n+1 = 2m+1$ is odd. On the other hand, $\dim \mathrm{so}(n+1)_\theta + \mathrm{ind}\,\mathrm{so}(n+1)_\theta = \dim \mathrm{so}(n+1) + \mathrm{ind}\,\mathrm{so}(n+1)$, which is equal to $2m^2$ in case $n+1 = 2m$ and to $2(m^2+m)$ in case $n+1 = 2m+1$. This proves completeness of the commutative set of integrals h_{ks} of equation (1).

EXAMPLE: $n = 3$. Equation (1) always has three integrals, $I_1 = H$, $I_2 = y_1^2 + y_2^2 + y_3^2$ and $I_3 = x_{12}y_3 + x_{23}y_1 - x_{13}y_2$. If the function H is given by the formula from Theorem 4.1.4, an additional integral I_4 can be found from the cubic invariant Δ_θ^3 of the algebra $\mathrm{sl}(4, \mathbf{R})_\theta$,

$$I_4 = a_1 y_2^2 + a_2 y_1^2 + (a_1 + a_2) y_3^2 - 2a_3(x_{13}y_1 + x_{23}y_2).$$

REMARK: In the proof of the equality $\mathrm{ind}\,G_\theta + \mathrm{ind}\,G$ for semisimple Lie algebras, we may use the Rais's formula [205] for the index of a semidirect product. But in this case, as before, the Lie algebra G_θ^x is to be found, which makes up the crucial point in the proof of the formula for the index of the Lie algebra G_θ. The theorem follows.

As concerns the Lie algebra $G = K \oplus_\rho V$, where ρ is the representation of minimal dimension and K is a special simple Lie algebra (that is, one of the five algebras G_2, F_4, E_6, E_7, E_8), no completed results have been obtained.

Examine a Lie algebra G which admits the representation in the form $G = K \oplus V$, where K is the subalgebra, V is the commutative ideal. Let $\rho = \mathrm{ad}_K : V \to V$ be the adjoint representation of the algebra K in the space V. The algebra G is, in fact, a semidirect extension $K \oplus_\rho V$ of the Lie algebra K via the representation $\rho = \mathrm{ad}$. Such Lie algebras are called *affine Lie algebras*. Of importance is a particular case where the space V is the linear space of the algebra K, on which K acts in an obvious adjoint way. We treat the second (right) copy of K, on which the first copy of K acts, as an Abelian ideal.

THEOREM 4.1.5 (TROFIMOV FOR PARTICULAR CASES, BRAILOV FOR THE GENERAL CASE). Let $G = K \oplus \rho K$ be an extension of a compact Lie algebra K by means of the adjoint representation of K on an Abelian ideal K, that is, $\rho = \mathrm{ad}_K : K \to K$. Then on each orbit of general position in G^* there always exists a maximal linear commutative algebra of polynomials. Its generators are

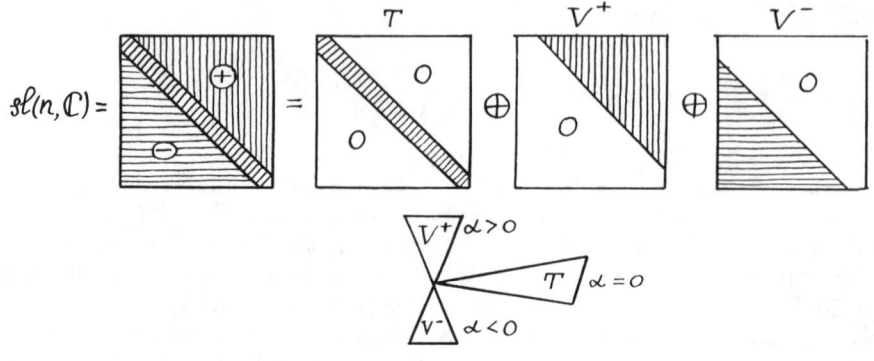

Figure 65

written explicitly in terms of the polynomial generators of the ring IK of invariant polynomials on the algebra K.

We do not dwell on the details of the explcit construction of the generators of a commutative algebra of functions. If ρ is a certain representation of a Lie algebra then $\wedge^k \rho$ and $S^k \rho$ will respectively denote the kth external and the kth symmetric degree of the representation ρ.

THEOREM 4.1.6. (PEVTSOVA [106], [111]). *Let $G = K \oplus_\rho V$ be the semidirect sum of a simple Lie algebra K and an Abelian ideal $V = \mathbf{R}^N$ via the representation $\mu : K \to \mathrm{End}\, \mathbf{R}^N$. Now if:*

a) the Lie algebra $K = \mathrm{gl}(2n)$ and the representation $\mu = \wedge^2 \rho$;

b) the lie algebra $K = \mathrm{sl}(2n)$ and the representation $\mu = S^2 \rho$;

c) the Lie algebra $K = \mathrm{sp}(n)$ and the representation $\mu = \rho + \tau$, where ρ is a representation of minimal dimension and τ is a one-dimensional trivial representation, then on the orbits of general position in G^* there always exists a maximal linear commutative algebra of rational functions. Rationality is understood relative to the canonical coordinates in the Lie algebra G.

Let G be a complex simple Lie algebra, T its Cartan subalgebra, $G = T \oplus \sum_{\alpha \neq 0} G_\alpha$ the Cartan decomposition, E_α the vectors described in Ch. 1.

The root decomposition of the algebra G can now be written in the form $G = T \oplus V^+ \oplus V^-$, where $V^+ = \sum_{\alpha>0} G_\alpha, V^- = \sum_{\alpha<0} G_\alpha$. In the case of $\mathrm{sl}(n, \mathbf{C})$, the subspace V^+ is obviously identified with the subspace of all uppertriangular matrices with zeroes upon the principal diagonal. The subspace V^- is identified with the subspace of all lower-triangular matrices with zeroes upon the principal diagonal. This decomposition is depicted in Fig. 65. Note that the scalar product of the basis vectors (which is generated by the Killing form) is organized as follows $\langle E_\alpha, E_{-\alpha} \rangle = -1, \langle E_\alpha, E_\alpha \rangle = 0$. This implies, in particular, that the vectors E_α are isotropic with respect to the nondegenerate Killing form. This is due to the fact that the Killing form is indefinite on the comples algebra G. In particular, the restriction of this form to the planes V^+ and V^- is identical zero, which follows from the explicit representation of the form.

§ 1 Lie Algebras and Mechanics

LEMMA 4.1.2. *The subspaces V^+ and V^- are nilpotent subalgebras in the Lie algebra G.*

PROOF: Let E_α, E_β belong to V^+. Then we have: $[E_\alpha, E_\beta] = N_{\alpha\beta} E_{\alpha+\beta}$ and since $\alpha > 0, \beta > 0$ then $\alpha + \beta \neq 0$ and $\alpha + \beta > 0$, that is, $E_{\alpha+\beta} \in V^+$ as required. In the case of V^- the reasoning is similar. Nilpotence of both subalgebras immediately follows from the explicit form of the constituent matrices (see above).

LEMMA 4.1.3. *The subspace $T \oplus V^+$ and $T \oplus V^-$ are solvable subalgebras in the Lie algebra G.*

The only difference from Lemma 4.1.2 is that one adds commutators of the form $[H'_\alpha, E_\beta]$, which coincide with $\beta(H'_\alpha) E_\beta$ (Fig. 66).

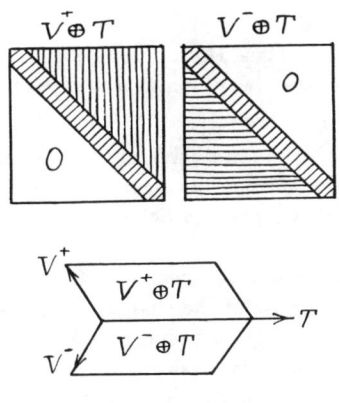

Figure 66

Examine now the subalgebra $T \oplus V^+$. The basis in this subalgebra (over the field of complex numbers) are the vectors $E_\alpha, \alpha > 0$ in G_α and H'_α in T. Consider a real subalgebra $BG = (\oplus_\alpha \mathbf{R} H'\alpha) \oplus (\sum_{\alpha>0} \mathbf{R} E_\alpha)$. Obviously, it is embedded in $T \oplus V^+$. The subalgebra BG is called the real form of the Borel subalgebra (in a semisimple Lie algebra). It is a solvable Lie algebra.

THEOREM 4.1.7 (TROFIMOV [130]–[133]). *Let G be a simple Lie algebra of one of the following types: $so(n), su(n), sp(n), G_2$. Then on each orbit of general position in the real form of the Borel (solvable) subalgebra BG (of the algebra G) there always exists a maximal linear commutative algebra of polynomials. These polynomials are written by explicit formulae.*

A particular case of Borel subalgebras is the Lie algebra T_n of real upper-triangular matrices of order n. The maximal linear commutative algebra of polynomials for this algebra has been discovered by Arkhangelsky [12]. It has become clear that (as in the general case of Borel subalgebras) shifts of invariants do not yield a complete set of independent commuting polynomials. This required the development of a new deeper technique based on the consideration of chains of subalgebras. See [130]–[133]. The following more general assertion is valid.

THEOREM 4.1.8 (LE NGOK TYEUEN [67]). *Let G be an arbitrary subalgebra in a solvable Lie algebra T_n of upper-triangular matrices, which is of the form:*

$$G = L \oplus \sum_{i<j} \mathbf{R} E_{ij} = L \oplus \sum_{\alpha>0} \mathbf{R} E_\alpha,$$

where E_{ij} is an elementary matrix of order n (which has a unity at the intersection point of the ith row and jth column, the rest of the elements being zero), L an arbitrary subspace of the n-dimensional space of diagonal matrices. Then on the dual space G^* there always exists a maximal linear commutative algebra of polynomials.

It is clear that $\sum_{i<j} \mathbf{R} E_{ij} = \sum_{\alpha>0} \mathbf{R} E_\alpha$ is the "real part" of the complex subalgebra V^+ in a complex semisimple algebra. It consists of upper-triangular matrices with zeros upon the diagonal. Le Ngok Tyeuen has strengthened the results obtained by Arkhangelsky and Trofimov. Namely, he has proved the following result.

THEOREM 4.1.9. *Let BG be a Borel subalgebra in a semisimple complex Lie algebra G. Then on BG there always exists a maximal commutative linear algebra of polynomials.*

It is of interest that the commutative algebras of polynomials constructed by Le Ngok Tyeuen differ from the similar algebras constructed by Trofimov.

Theorem 4.1.10, which we present below, and the corollaries to this theorem were obtained by Bolsinov.

Let $\mathfrak{G}$ be a simply-connected Lie group and G its Lie algebra. Let $f_1, \ldots, f_k$ be a complete set of local invariants of the coadjoint representation of the group $\mathfrak{G}$ in a certain neighbourhood of a regular point $x \in G^*$. Considering the shifts of these invariants by the covector of general position $a \in G^*$, we obtain an involutive family of functions. It should be noted that such a construction is local, i.e., it generally does not allow us to obtain functions in involution, which are globally defined on G^*, if the invariants are not globally defined. But this shortcoming can be easily eliminated by replacing the family of shifts by an equivalent set of polynomials. In this case, equivalence implies that the spaces generated by the differentials of the functions from these families at points $y \in G^*$ coincide for almost all y from the general interval on which these functions are defined. A very simple way for such a replacement was proposed by Brailov. For this it suffices to examine series expansions of local invariants at the point $a \in G^*$:

$$g_i(a+x) = p_i^0(x) + p_i^1(x) + p_i^2(x) + \ldots,$$

where p_i^k is a polynomial of degree k homogeneous in x. It can be easily checked that the set composed of polynomials p_i^k is equivalent in the indicated sense to the family of shifts of invariants by the covector $a \in G^*$. Note that this method of constructing functions in involution is constructive. Indeed, the differentials dp_i^k satisfy the Fomenko–Mishchenko recurrence relations(cascade):

(1)
$$\begin{aligned}\{dp_i^1, a\} &= 0 \\ \{dp_i^2, a\} + \{dp_i^1, x\} &= 0 \\ \{dp_i^3, a\} + \{dp_i^2, x\} &= 0 \\ &\ldots\end{aligned}$$

These relations are systems of equations in partial derivatives with constant coefficients at the derivatives and can be successively integrated. Integration is not single-valued at each stage and as a result we will in general not obtain the polynomials p_i^k. Nonetheless, the set of polynomials obtained will be equivalent to the set $\{p_i^k\}$.

Thus, a family of shifts of invariants can be hereafter understood either as a set $\{p_i^k\}$ or as a set resulting from the solution of the relations (1). After this remark, we may formulate the criterion of completeness of the family of shifts of invariants without imposing any additional limitations on the Lie algebra G.

THEOREM 4.1.10 (BOLSINOV). *Let G be an arbitrary finite-dimensional complex Lie algebra, $S = \{y \in G^* | \operatorname{codim} O(y) > \operatorname{ind} G\}$ a set of singular elements from G^*. Shifts of invariants by the covector of general position form a complete involutive set on G^* if and only if $\operatorname{codim} S \geqslant 2$.*

COROLLARY. *Let G by an arbitrary finite-dimensional real Lie algebra, G^C its complexification. The shifts of invariants of the Lie algebra G by the covector of general position form a complete involutive set on G^* if and only if $\operatorname{codim} S^C \geqslant 2$, where S^C is a set of singular elements from $(G^C)^*$.*

This assertion is immediate from the fact that invariants of the Lie algebras G and G^C are, in fact, one and the same functions.

COROLLARY. *Let G be an arbitrary finite-dimensional complex (real) Lie algebra and S a set of singluar elements from G^*, $x \in S$. Let $\operatorname{ind} \operatorname{Ann} x = \operatorname{ind} G$, $\operatorname{codim} S \geqslant 2$ ($\operatorname{codim} S^C \geqslant 2$, where S^C is a set of singular elements from $(G^C)^*$). Then the shifts of invariants of the Lie algebra G by a certain covector of general position $a \in G^*$ form a complete involutive set on a singular orbit $O(x)$.*

This assertion is a generalization of the results deduced by Dao Chong Thi and Brailov for complete involutive sets on singular orbits of semisimple Lie algebras.

Let G be a semisimple symmetrically graduated Lie algebra (complex or real), that is, G is decomposed into the direct sum of two subspaces, K and V, the following relations being fulfilled

$$[K,K] \subset K, \quad [K,V] \subset V, \quad [V,V] \subset K.$$

Consider the so-called contraction of the Lie algebra G, that is, the semi-direct sum of the Lie algebra K and the commutative ideal V via the representation ad. Denote this Lie algebra by $\widetilde{G}$. The dual spaces G^* and $\widetilde{G}^*$ will be identified with the direct sum $K^* \oplus V^*$ and the elements from $K^* \oplus V^*$ will be written in the form $k + v$, where $k \in K^*, v \in V^*$. Let $f_1, \ldots, f_s$ be invariants of the Lie algebra G. Let $a \in V^*$ be a regular element with respect to the representation $\operatorname{ad}^* : K \to \operatorname{End}(V^*)$. Examine the functions $f_i^\lambda(k+v) = f_i(\lambda k + v + \lambda^2 a), \lambda \in \mathbf{R}$. Let $\operatorname{St}(a) \subset K$ be the stationary subalgebra of the element $a \subset V^*$ with respect to the representation $\operatorname{ad}^* : K \to \operatorname{End}(V^*)$. Let $\mathcal{F}$ be a complete involutive set of functions on $\operatorname{St}(a)^*$. The family of functions $\{f_i^\lambda\} \cup \mathcal{F}$ is known (see [177]) to be involutive with respect to the bracket on $\widetilde{G}^*$.

COROLLARY. *The involutive family of functions $\{f_i^\lambda\} \cup \mathcal{F}$ is complete.*

PROOF OF THE THEOREM: Fix an arbitrary element of general position $a \in G^*$. Consider a regular element $x \in G^*$. Let $f_1, \ldots, f_k$ be a complete set of local invariants at the point $x \in G^*$. The completeness means that at the point x the differentials $df_1, \ldots, df_k$ generate $\operatorname{Ann} x$, where $\operatorname{Ann} x = \{\xi \in G | \operatorname{ad}_\xi^* x = 0\}$. The differentials of shifts $f_i^\lambda(x) = f_i(x + \lambda a)$ at the point x are known to generate a subspace $M \subset G$ which can be represented as the sum $M = U_0 + U_1 + U_2 + \ldots$, where $U_0 = \operatorname{Ann} x$ and the subspaces $U_1, U_2, \ldots$ satisfy the following chain of relations:

(2)
$$\{U_1, x\} = \{U_0, a\}$$
$$\{U_2, x\} = \{U_1, a\}$$
$$\{U_k, x\} = \{U_{k-1}, a\}$$
$$\ldots$$

Using this chain one can unambiguously construct the subspace $M \subset G$, and we therefore forget about the origin of the space M and employ only the relations (2).

A family of shifts of invariants forms a complete involutive set of G^* if and only if for almost all $x \in G^*$ a corresponding subspace M is of a dimension equal to $\frac{1}{2}(\dim G + \operatorname{ind} G)$; however, this definition of completeness is not very convenient in this case and we will therefore use another. For this purpose consider the subspace $\widetilde{M} = \{\xi \in G | (x, [\xi, \eta]) = 0 \forall \eta \in M\}$. The subspace $\widetilde{M} \subset G$ can be conditionally called a skew-orthogonal complement of M at the point $x \in G^*$. It is clear that $M \subset \widetilde{M}$. Therefore, if the element $x \in G^*$ is regular, then the condition $\dim M = \frac{1}{2}(\dim G + \operatorname{ind} G)$ is equivalent to $M = \widetilde{M}$. The idea of the proof consists in comparison of the two subspaces, M and $\widetilde{M}$.

LEMMA 4.1.4. *A subspace $\widetilde{M}$ is maximal among all subspaces $L \subset G$ satisfying the condition $\{L, a\} \subset \{L, x\}$.*

PROOF OF THE LEMMA: First show that any subspace $L \subset G$ such that $\{L, a\} \subset \{L, x\}$ is contained in $\widetilde{M}$. Prove by induction that $(x, [L, U_k]) = 0$ for any k. For $k = 0$ we have

$$(x, [L, U_0]) = (\{U_0, x\}, L) = (\{\operatorname{Ann} x, x\}, L) = 0$$

Let $(x, [L, U_{k-1}]) = 0$, then

$$(x, [L, U_k]) = (\{U_k, x\}, L) = (\{U_{k-1}, a\}, L) = -(\{L, a\}, U_{k-1}).$$

But $\{L, a\} \subset \{L, x\}$ and $(\{L, x\}, U_{k-1}) = -(x, [L, U_{k-1}]) = 0$, and therefore $(x, [L, U_k]) = 0$. Thus, $(x, [L, M]) = 0$, that is, $L \subset \widetilde{M}$.

We will now show that any element $\xi_0 \in \widetilde{M}$ is included in a natural manner into a certain subspace $L \subset \widetilde{M}$ with the result that $\{L, a\} \subset \{L, x\}$. Prove that there exists $\xi_1 \in G$ such that $\{\xi_0, a\} = \{\xi_1, x\}$. To this end it is necessary and sufficient that there hold the condition $(\{\xi_0, a\}, \operatorname{Ann} x) = 0$. This condition does hold because

$$(\{\xi_0, a\}, \operatorname{Ann} x) = -(\{\operatorname{Ann} x, a\}, \xi_0) = -(\{U_1, x\}, \xi_0) = -(x, [\xi_0, U_1]) = 0.$$

Next, $\xi_1 \in \widetilde{M}$. Indeed,

$$(x, [\xi_1, U_k]) = -(\{\xi_1, x\}, U_k) = -(\{\xi_0, a\}, U_k)$$
$$= (\{U_k, a\}, \xi_0) = (\{U_{k+1}, x\}, \xi_0) = (x, [\xi_0, U_{k+1}]) = 0$$

for any k. Thus, we can construct a family of vectors $\xi_2, \xi_3, \ldots$ such that $\xi_i \in \widetilde{M}$ and $\{\xi_{i+1}, x\} = \{\xi_i, a\}$ for any $i = 1, 2 \ldots$. Now as the subspace $L \subset \widetilde{M}$ it suffices to take the linear subspace generated by the vectors $\xi_0, \xi_1, \xi_2, \ldots$. Obviously, $\{L, a\} \subset \{L, x\}$. This proves the lemma.

The lemma immediately implies, in particular, that the subspace $\widetilde{M}$ contains annihilators of all elements of the form $a + \lambda x, \lambda \in \mathbb{C}$. On the other hand, from the recurrence relations (2) it can be easily established that the subspace M satisfies the condition $\{M, a + \lambda x\} = \{M, x\}$. Consequently, $\dim M \cap \mathrm{Ann}(a + \lambda x) = \dim M \cap \mathrm{Ann}\, x = \dim \mathrm{Ann}\, x = \mathrm{ind}\, G$. Therefore, if for a certain $\lambda \in \mathbb{C}$ the element $a + \lambda x$ is not regular, that is, $\dim \mathrm{Ann}(a + \lambda x) > \mathrm{ind}\, G$, then $\mathrm{Ann}(a + \lambda x) \not\subset M$ and, accordingly, $M \neq \widetilde{M}$.

Choose an arbitrary algebraic complement of $\mathrm{Ann}\, x$ in $\widetilde{M}$ in a way that $\widetilde{M} = \mathrm{Ann}\, x \oplus K$. Define the operator $C : \widetilde{M} \to \widetilde{M}$ by the relation

$$\{\xi, a\} = \{C(\xi), x\}, \xi \in \widetilde{M}, C(\xi) \in K \in \widetilde{M}.$$

It is clear that if K is fixed, the operator C is well defined.

LEMMA 4.1.5. *The subspace $K \subset \widetilde{M}$ can be so chosen that the subspace $M \subset \widetilde{M}$ be invariant with respect to the operator C and the restriction $C|_M$ be nilpotent.*

The proof of this lemma is exclusively technical, and we leave it to the reader.

Show that if all elements of the form $a + \lambda x$ are regular then $M = \widetilde{M}$. Let $K \subset \widetilde{M}$ be precisely the subspace whose existence is stated in Lemma 4.1.5. Then the operator $C : \widetilde{M} \to \widetilde{M}$ is nilpotent. Indeed, there otherwise exists a nonzero-weight eigenvector $\xi \in \widetilde{M}$, that is, $C(\xi) = \lambda \xi, \lambda \neq 0$ or $\{\xi, a\} = \{\lambda \xi, x\}$. Then $\xi \in \mathrm{Ann}(a - \lambda x)$. It has been exhibited above that $\dim M \cap \mathrm{Ann}(a - \lambda x) = \mathrm{ind}\, G$, but the element $a - \lambda x$ is regular, and therefore $\dim M \cap \mathrm{Ann}(a - \lambda x) = \dim \mathrm{Ann}(a - \lambda x)$, that is, $\mathrm{Ann}(a - \lambda x) \subset M$. Thus, $\xi \in M$, which contradicts the nilpotence of the restriction $C|_M$.

Prove now that $\widetilde{M} = M$. Suppose the contrary, that is, $\widetilde{M} \neq M$. Then (by virtue of nilpotence of the operator C) there exists a number i such that $C^{i-1}(\widetilde{M}) \not\subset M$, $C^i(\widetilde{M}) \subset M$. Let $\eta \in C^{i-1}(\widetilde{M})$ such that $C(\eta) \in M, \eta \notin M$. Then $\{\eta, a\} = \{C(\eta), x\}$. But $\{M, x\} = \{M, a\}$, and therefore these exists $\eta' \in M$ such that $\{C(\eta), x\} = \{\eta', a\}$. Then $\eta - \eta' \in \mathrm{Ann}\, a$. The element $a \in G^*$ is regular, and therefore $\mathrm{Ann}\, a \subset M$ (see above). We have $\eta' \in M, \eta - \eta' \in M$, consequently, $\eta \in M$, is a contradiction.

Thus we have proved that $M = \widetilde{M}$ if and only if elements of the form $a + \lambda x$ are regular for any $\lambda \in \mathbb{C}$. Geometrically, the regularity of elements of the form $a + \lambda x$ implies that the complex affine straight line passing through the regular element $a \in G^*$ with a directing vector $x \in G^*$ does not intersect the set of singular points S. Clearly, this is possible for almost all $x \in G^*$ if and only if $\mathrm{codim}\, S \geqslant 2$. This completes the proof of Bolsinov's theorem.

In the subsections which follow we will briefly touch upon a new operation of "duplication" of the Lie algebras possessing maximal linear commutative algebras of polynomials on orbits. Such a duplication will result in the appearance of new Lie algebras (distinct from the initial ones) possessing the same property, the integrability (in our terminology).

The results of the present subsection are tabulated on the following pages.

Lie Algebras	Complete involutive sets of functions. Orbits of general position.	Complete involutive sets of functions. Semisimple (singular) orbits.	The remaining singular orbits.
1 Arbitrary semisimple complex (next, real semisimple).	Fomenko and Mishchenko, see Theorem 4.1.1 and [89], [91], [93]. "General series." Polynomials.	Dao Chong Thi (proof with lacunas). Brailov, Theorem 4.1.2. "Singular series." Polynomials.	?
2 Compact real form G_c of a semisimple Lie Algebra.	Fomenko and Mishchenko, see Theorem 4.1.1 and [89], [91], [93]. "Compact series." Polynomials.	Dao Chong Thi (proof with lacunas). Brailov, Theorem 4.1.2. "Compact series." Polynomials.	No such orbits.
3 Normal compact subalgebra G_n in the compact real form G_c.	Fomenko and Mishchenko, see Theorem 4.1.1 and [89], [91], [93]. "Normal series." Polynomials.	Brailov, see Theorem 4.1.2, "Normal series." Polynomials.	No such orbits.
4 $E(n) = so(n) \oplus_\rho \mathbb{R}^n$, ρ is representation of minimal dimension.	1) Fomenko and Trofimov, see Theorem 4.1.3 and [141], [142]. This algebra of integrals V_1 is an analogue of the algebra of integrals of the "compact series" for $so(n)$. Polynomials. 2) Brailov, see Theorem 4.1.4. Another algebra of integrals V_2 is an analogue of the "normal series" for $so(n)$. Polynomials.	?	?

	Lie Algebras	Complete involutive sets of functions. Orbits of general position.
5	$su(n) \oplus_\rho \mathbb{C}^n$, where ρ is representation of minimal dimension.	a) Brailov (polynomials), b) Bolsinov (polynomials).
6	$u(n) \oplus_\rho \mathbb{C}^n$, where ρ is representation of minimal dimension.	Brailov (polynomials).
7	$K \oplus_\rho V$, where $\rho = \text{ad}_K : V \to V$; V is ideal, K compact Lie algebra.	Trofimov (in several particular cases), Brailov (in the general case). See above theorem (Polynomials).
8	$gl(2n) \oplus_\mu \mathbb{R}^n$, where $\mu = \wedge^2 \rho$, ρ is representation of minimal dimension, $\wedge^2 \rho$ is the second outer degree of the representation ρ.	Pevtsova [106], [111]. See Theorem 4.1.6. Rational functions.
9	$sl(2n) \oplus_\mu \mathbb{R}^N$, where $\mu = S^2 \rho$, ρ is representation of minimal dimension, $S^2 \rho$ is the second symmetric degree of the representation ρ.	Pevtsova [106], [111]. See Theorem 4.1.6. Rational functions.
10	$sp(2n) \oplus_\mu \mathbb{R}^n$, where $\mu = \rho + \tau$, ρ is representation of minimal dimension, τ is a one-dimensional trivial representation.	Pevtsova [106], [111]. See Theorem 4.1.6. Rational functions.
11	Real forms BG of Borel subalgebras of simple Lie algebras: $B(so(n))$, $B(su(n))$, $B(sp(n))$, BG_2, BF_4, BE_6, BE_7, BE_8.	Trofimov, see Theorem 4.1.6 [130], [133]. For the particular case of the algebra of upper-triangular matrices—earlier Arkhangelsky, [12]. For BE_8 Syarov. Polynomials.
12	$L \oplus \sum_{i<j} \mathbb{R} E_{ij} = L \oplus \sum_{\alpha>0} \mathbb{R} E_\alpha$ a solvable subalgebra where L is a subalgebra in diagonal matrices. $\sum_{\alpha>0} \mathbb{R} E_\alpha$ is the algebra of upper-triangular nilpotent matrices.	Le Ngok Tyeuen [67]. See Theorem 4.1.8. Polynomials.

13	BG, a Borel subalgebra in a semisimple complex Lie algebra G.	Le Ngok Tyeuen. See Theorem 4.1.9. Polynomials. This set differs from the ones constricted by Trofimov (item 11).
14	Series of Lie algebras $h(G)$ obtained by "duplication" of Lie algebras G which already possess on the orbits the maximal linear commutative algebras of polynomials (rational functions). Then (under an appropriate choice of the functor h) the algebras $h(G)$ also have on their orbits maximal linear commutative algebras of polynomials (rational functions).	Trofimov [135],[138]. Brailov [19]. Le Ngok Tyeuen. Polynomials. The Lie algebra $h(G)$ is of the form $G \otimes A$, where A is a Frobenius algebra.
15	Ratiu. On a singular orbit in $(so(n) \times S)^*$, where S are symmetric matrices.	Ratiu [118]. In item 9 see particular cases.
16	Lie algebras of small dimension.	Trofimov [252].
17	Semisimple Lie algebras. The new sets of functions (see on the right) differ from the familiar ones, see items 1–3 above.	Bogoyavlensky [15], [16]. The complete commutative sets of functions constructed by him rest upon the new idea of using filtrations in Lie algebras. Polynomials.
18	$K \oplus_\rho V$, where K is an arbitrary simple Lie algebra and ρ is an arbitrary irreducible linear representation, V Abelian ideal, V being the representation space for ρ.	Bolsinov. This result generalizes many of the above-mentioned results. Polynomials. Thus, the Abelian extension of simple Lie algebras via any exact irreducible representation always possesses a complete commutative set of polynomials on orbits of general position.

19	$K \otimes_\rho V$, where K is an arbitrary semisimple Lie algebra (real or complex), ρ an arbitrary linear representation (reducible or irreducible), $\dim V$ being greater than $\dim K$, and in the decomposition of the representation ρ into irreducible components trivial summands are absent.	Brailov, Pevtsova [106], [111]. All the functions of these complete involutive sets are linear.
20	$G \oplus_{ad} G$, where G is a semisimple Lie algebra (complex or real). This algebra is a particular case of Lie algebras of the type $G \otimes A$ (see item 14 above). To this end, one should take $A = \mathbb{R}[x]/(x^2)$ (or in the complex case $A = \mathbb{C}[x]/(x^2)$).	Bolsinov [277]. His complete involutive sets on these algebras differ from those constructed earlier by Trofimov and Brailov (see item 14 above).
21	$G = so(n) \oplus_\rho \mathbb{R}^n$ where ρ is representation of minimal dimension.	Perelomov [107]. Polynomials, integrals of the "n-dimensional Clebsh case." See also Reyman.
22	$G = so(n) \oplus_\rho \mathbb{R}^n$, where ρ is representation of minimal dimension.	Perelomov [281]. Polynoms, analogues of Kovalevskaya's integral in the n-dimensional case. See also Reyman.
23	$G = so(n) \oplus_\rho \mathbb{R}^n$, where ρ is representation of minimal dimension.	Belyayev [13]. Polynomials, integrals of the analogue of the Lagrange case in the n-dimensional case.
24	Lie algebras $so(4)$ and $so(3) \oplus_\rho \mathbb{R}^3$.	See §2.2 (below) containing the corresponding results.
25	G, a nilpotent real Lie algebra.	Vergne (Bull. Soc. Mat. France, 1972. 100, N.3, pp. 301–335). Complete involutive set of polynomials.

26	G, completely solvable real algebraic Lie algebra.	Ginzburg (1980). Complete involutive set of functions, consists of polynomials. This set is invariant under a coadjoint group representation (but a separate polynomial is, of course, not invariant).
27	G, solvable real algebraic Lie algebra, for G there exists a flag of algebraic subalgebras $G = G_n \supset G_{n-1} \supset \cdots \supset G_1 \supset G_O; \dim G_i = i.$	Bolsinov (1985). Complete involutive set of polynomials. These two sets (see the preceding item) are possibly distinct.

§2 Integrable Multidimensional Analogues of Mechanical Systems Whose Quadratic Hamiltonians are Contained in the Discovered Maximal Linear Commutative Algebras of Polynomials on Orbits of Lie Algebras

2.1 The Description of Integrable Quadratic Hamiltonians

The maximal linear commutative algebras of polynomials (and of rational functions) on orbits of Lie algebras enumerated in §1 (and those not included for lack of space) are of interest not only from the point of view of interrelation between the commutative and noncommutative methods of integrating Hamiltonian systems. *These algebras V of polynomials often turn out to contain important quadratic Hamiltonians of analogues of classical mechanical systems*. This means that all such systems are completely Liouville integrable. Moreover, the algebras V often involve as the subspace a whole linear multidimensional space of quadratic Hamiltonians, which offers the opportunity of verifying them and seeking interesting mechanical systems.

Not to burden the presentation, we dwell only on one series of such examples associated with various equations of motion of a rigid body (in the multidimensional situation). To demonstrate our general method, we elucidate more or less comprehensively the procedure of studying the maximal linear commutative algebra of polynomials performed in Theorem 4.1.1 for a complex semisimple Lie algebra G.

Let V be the maximal linear commutative algebra of polynomials, on orbits of a semisimple Lie algebra G, presented in Theorem 4.1.1 and in Proposition 4.1.1 (Sec. 1.2). We will now describe the subspace of quadratic Hamiltonians contained in V.

THEOREM 4.2.1 (SEE [92], [93]). *Let G be a finite-dimensional Lie algebra and let P be a set of functions functionally generated by functions of the form $f(\xi + \lambda a)$, where $f(\xi)$ is the invariant of the algebra G, that is, a function constant on the orbits $O^* \subset G^*$ of the coadjoint representation $\mathrm{Ad}^*_\mathfrak{G}$; let $a \in G^*$ be a fixed covector, λ a real number, and $\lambda \neq 0$. The quadratic polynomial $H(\xi)$ belongs to* the set P if and only if there exists an element (vector) $b \in G$ such that there hold the identities $\{a, dH(\xi)\} + \{\xi, b\}$ (where $\{\xi, b\} = \mathrm{ad}_b^*(\xi)$) and $\{a, b\} = 0$, where $\{,\}$ is the operation in the Lie algebra G (a commutator or a Poisson bracket).

COMMENT: Here $dH(\xi)$ (that is, the differential of the function H) is given as the vector (element) of the Lie algebra G because $H(\xi)$ is a function on the dual space G^* and $G^{**} = G$.

Theorem 4.3.1 yields an important consequence.

THEOREM 4.2.2 (SEE [92], [93]). *Let G be a semisimple complex Lie algebra, $a \in G$ a vector in general position, V a set of functions on the algebra G, which are functionally generated by functions of the form $f(X + \lambda a)$, where $f(X)$ are functions constant on the orbits $O \subset G$ of the coadjoint representation of the group $\mathfrak{G}$. The quadratic function $H(X)$ belongs to the family V if and only if there exists a vector $b \in G$ (not necessarily in general position) such that there hold the following identities:* $[\mathrm{grad}\, H(X), a] = [X, b]; [a, b] = 0$. *The Hamiltonian vector field $\dot X = \mathrm{sgrad}\, H(X)$ corresponding to such function H will be completely Liouville-integrable on all the orbits of general position in the Lie algebra G.*

Here $\mathrm{grad}\, H(X)$ may be treated as the element of the Lie algebra G identified with G^* by means of the Killing form.

PROOF OF THEOREM 4.2.1: The Hamiltonian function $H(\xi)$ must depend functionally on functions of the form $f(\xi + \lambda a)$, where λ are some numbers, a covector, $a \in G^*$, f functions constant on the orbits of the coadjoint representation. Thus, $H(\xi) = F(f_1(\xi + \lambda_1 a), \ldots, f_N(\xi + \lambda_N a))$. We suppose that all the functions $f_1, \ldots, f_N$ and F are smooth functions. Without loss of generality we may assume the functions $f_k, 1 \leqslant k \leqslant N$ to be homogeneous polynomials of certain degrees p_k. Indeed, the constancy condition for the function f on the orbits implies fulfillment of the identities: $\{\xi, df(\xi)\} = 0, \xi \in G^*$. Therefore, the same identities are also fulfilled for the formal Taylor series and, accordingly, for all its separate terms which are already homogeneous polynomials. Hence, the terms of the Taylor series can be taken for new generators in the new representation for the Hamiltonian H. Consider the expansion of the function $f_k(\xi + \lambda a)$ in powers of λ: $f_k^0(\xi) + f_k^1(\xi)\lambda + \cdots + f_k^{p_k}(\xi)\lambda^{p_k}$. The polynomial f_k^j is a homogeneous polynomial of degree $p_k - j$, in particular, $f_k = \mathrm{const}$. Thus, in the new representation for H one may assume without loss of generality that

$$H = F(f_1^{p_1-1}, \ldots, f_N^{p_N-1}; f_1^{p_1-2}, \ldots, f^{p_N-2}),$$

where $f_k^{p_k-1}, 1 \leqslant k \leqslant N$, are polynomials of the first degree, $f_k^{p_k-2}$ of second degree in ξ. Then

$$H = f_1(f_1^{p_1-1}, \ldots, f_N^{p_N-1}) + F_2(f_1^{p_1-2}, \ldots, f_n^{p_N-2}).$$

where $F_1 = F_1(u_1, \ldots, u_N)$ is a quadratic function and $F_2 = F_2(u_1, \ldots, u_N)$ is a linear one. We have

$$dH = \sum_{k=1}^N \frac{\partial F_1}{\partial u_k}(f_1^{p_1-1}, \ldots, f_N^{p_N-1}) df_k^{p_k-1} + \sum_{k-1}^N \frac{\partial F_2}{\partial u_k} df_k^{p_k-2}$$

whence
$$0 = \{(\xi + \lambda a), df_k^0(\xi) + df_k^1(\xi) + \cdots + df_k^{p_k}(\xi)\lambda^{p_k}\},$$
or:
$$\{a, df_k^{p_k}\} = 0, \{\xi, df_k^{p_k}\} + \{a, df_k^{p_k-1}\} = 0,$$
$$\{\xi, df_k^{p_k-1}\} + \{a, df_k^{p_k-2}\} = 0.$$

Since $df_k^{p_k} = 0$, then there remain only two out of the first three identities:
$$\{a, df_k^{p_k-1}\} = 0, \qquad \{\xi, df_k^{p_k-1}\} + \{a, df_k^{p_k-2}\} = 0$$

Calculate $\{a, df_k^{p_k-1}\}$. We obtain

$$\{a, dH(\xi)\} = \sum_{k=1}^{N} \frac{\partial F_1}{\partial u_k}(f_1^{p_1-1}(\xi), \ldots, f_N^{p_N-1}(\xi))\{a, df_k^{p_k-1}\}$$
$$+ \sum_{k=1}^{N} \frac{\partial F_2}{\partial u_k}\{a, df_k^{p_k-2}(\xi)\}$$
$$= \sum_{k=1}^{N} \frac{\partial F_2}{\partial u_k}\{a, df_k^{p_k-2}(\xi)\}$$
$$= -\sum_{k=1}^{N} \frac{\partial F_2}{\partial u_k}\{\xi, df_k^{p_k-1}\} = \{\xi, b\},$$

where $b = -\sum_{k=1}^{N} \frac{\partial F_2}{\partial u_k} df_k^{p_k-1} \in G$. Since $\{a, df_k^{p_k-1}\} = 0$, we obtain that $\{a, b\} = 0$. Inversely, if there exists a vector $b \in G$ satisfying these conditions and if the covector $a \in G^*$ is in general position then the function $H(\xi)$ depends functionally on the functions $f_k^j(\xi)$. Indeed, inasmuch as $b \in \text{Ker}(a) = (X \in G : \{a, X\} = 0)$ and the covector a is in general position then there exists a polynomial f, constant on the orbits, such that $df(a) = b$. Then we put

$$f(\xi + \lambda a) = f^0(\xi) + f^1(\xi)\lambda + \cdots + f^{p-2}(\xi)\lambda^{p-2} + f^{p-1}(\xi)\lambda^{p-1} + f^p(\xi)\lambda^p.$$

For $\xi = 0$ we obtain $df(a) = df^{p-1} = b$. Thus, $\{\xi, b\} = \{a, d\widetilde{H}(\xi)\}$, that is, the function $\widetilde{H} = H + f^{p-2}$ satisfies the relation $\{a, d\widetilde{H}(\xi)\} = \{\xi, b\}$; for $b = 0$: $\{a, d\widetilde{H}(\xi)\} = 0$. This means that $d\widetilde{H}$ is a linear self-conjugate operator $d\widetilde{H}: G^* \to \text{Ker}(a) \subset G$. Fix the basis $b_1, \ldots, b_s \in \text{Ker}(a)$ and the functions $f_1, \ldots, f_s$, constant on the orbits, for which $df_k(a) = b_k, 1 \leqslant k \leqslant s$. Assume that $f_k(a) = 0, 1 \leqslant k \leqslant s$. Put, as before,

$$f_k(\xi + \lambda a) = f_k^0(\xi) + \cdots + f_k^{p_k-2}(\xi)\lambda^{p_k-2} + f_k^{p_k-1}\lambda^{p_k-1}$$

Here $f_k^{p_k} = 0$, that is, $f_k(a) = 0$. Examine the function

$$F(\xi) = \sum \nu_{ij}(df_i^{p_i-1}(\xi)f_j^{p_j-1}(\xi) + f_i^{p_i-1}(\xi)df_j^{p_j-1}(\xi))$$
$$= \sum \nu_{ij}(b_i < \xi, b_j > + < \xi, b_i > b_j) = \sum b_i < \xi, b_j > (\nu_{ij} + \nu_{ji})$$
$$= \sum_i b_i(\sum_j < \xi, b_j > \mu_{ij}); \qquad \mu_{ij} = \mu_{ji} = \nu_{ij} + \nu_{ji}.$$

On the other hand, it is readily seen that for any self-conjugate operator $\varphi : G^* \to \mathrm{Ker}(a) \subset G$ there exist such numbers $\mu_{ij} = \mu_{ji}$ that $\varphi(\xi) = \sum b_i \langle \xi, b_j \rangle \mu_{ij}$. Indeed, let $V \subset G^*$ be a subspace annihilating $\mathrm{Ker}(a)$, that is, $V = \{\xi : \xi\,|_{\mathrm{Ker}(a)} = 0$. Then $\varphi|_V = 0$ because if $\xi \Delta_1 \in G^*, \xi_2 \in V$ then $\langle \xi_1, \varphi(\xi_2) \rangle = \langle \xi_2, \varphi(\xi_1) \rangle = 0$. Consequently, decomposing the value $\varphi(\xi)$ in the basis $b_1, \ldots, b_s \in \mathrm{Ker}(a)$ we are led to $\varphi(\xi) = \sum b_i \varphi_i(\xi)$, where $\varphi_i(\xi)$ are linear functions equal to zero on V. The complete set of such functions is generated by functions of the form $\langle \xi, b_j \rangle$, that is, $\varphi_i(\xi) = \sum \mu_{ij} \langle \xi, b_j \rangle$. The condition $\mu_{ij} = \mu_{ji}$ follows from self-conjugation of the operator φ. Thus, applying the above reasoning to the operator $d\widetilde{H}(\xi)$, we come to

$$\widetilde{H}(\xi) = \sum \nu_{ij} f_i^{p_i-1}(\xi) \cdot f_j^{p_j-1}(\xi)$$

up to a constant, which proves the theorem.

2.2 Cases of Complete Integrability of Equations of Various Motions of a Rigid Body

Let G be a semisimple Lie algebra. In the preceding subsection we discovered the quadratic integrable Hamiltonians $H(X)$, remarkable in many respects, such that $[\mathrm{grad}\, H(X), a] = [X, b]$ and $[a, b] = 0$. Any quadratic function $H(X)$ on G may be represented in the form $H(X) = \langle X, \varphi(X) \rangle$, where $\langle\ ,\ \rangle$ is the Killing form and $\varphi : G \to G$ is a linear operator symmetric (self-conjugate) with respect to the Killing form. It is clear that $\mathrm{grad}\, H(X) = \varphi(X)$. In our case, the operator φ satisfies the following equation: $[\varphi(X), a] = [X, b]$. It does not, of course, fix the operator φ (and therefore the quadratic function H). But since at a given moment all events happen in the semisimple Lie algebra G, all the solutions of the indicated equation can be easily described. It is convenient to exploit the familiar Cartan decomposition of the algebra G. Since we assume a to be an element of general position in the semisimple Lie algebra G, it can always be included into a certain Cartan subalgebra which we will denote by T. Recall that the algebra G is decomposed into the sum of subspaces $G = T \oplus V^+ \oplus V^-$, where T is the Cartan subalgebra, V^+ and V^- are linear subspaces orthogonal to the subalgebra T and spanned by the root subspaces $E_\alpha, \alpha > 0$ and $E_\alpha, \alpha < 0$, respectively. Consider a linear operator $\mathrm{ad}_a : G \to G$, that is, $\mathrm{ad}_a X = [a, X]$. This operator is invertible on the plane $V^+ \oplus V^-$ and is identical zero on the Cartan subalgebra T. We denote by ad_a^{-1} the inverse operator on the plane $V^+ \oplus V^-$ and by $D : T \to T$ the arbitrary linear self-conjugate operator on the Cartan subalgebra. Let b be an arbitrary element from the Cartan subalgebra T. Consider a linear operator $\varphi : G \to G$ which is set in the following way: $\mathrm{ad}_a^{-1} \mathrm{ad}_b : V^+ \oplus V^- \to V^+ \oplus V^-$ and $D : T \to T$. Denote this operator as $\varphi_{abD} : G \to G$. It is well defined because the operator ad_a^{-1} is meaningful on the plane $V^+ \oplus V^-$, and the operator ad_b is defined on the entire algebra G.

PROPOSITION 4.2.1. *All the solutions of the equation $[\varphi(X), a] = [X, b]$ (where $[a, b] = 0$ and the element a is of general position) on a semisimple Lie algebra G are set in an explicit form by the operators $\varphi_{abD} : G \to G$.*

COROLLARY 4.2.1. *Let G be a semisimple complex Lie algebra, a an element of general position, V an involutive (commutative) set of functions on the algebra G*

which are functionally generated by functions of the form $f(X + \lambda a)$, where $f(X)$ are invariants of the algebra G, that is, functions constant on the orbits O of the adjoint representation in G. The quadratic function $H(X)$ which is set uniquely by the linear operator $\varphi(X) = \operatorname{grad} H(X), \varphi : G \to G$, belongs to the family V if and only if the operator φ has the form φ_{abD} for a certain element b such that $[a, b] = 0$ (that is, an element belonging to the same Cartan subalgebra T, to which the element a belongs) and for a certain symmetric linear operator $D : T \to T$.

It will be seen below that it is natural to refer to left-invariant metrics of the form $\langle X, \varphi_{abD}(X)\rangle$ on the cotangent bundle $T^*\mathfrak{G}$ as *rigid-body metrics* (or *metrics of a rigid body*). The operator φ_{abD} is readily written in an explicit form.

Indeed, $\operatorname{ad}_a E_\alpha = \alpha(a) E_\alpha$, that is, $\alpha(a) \neq 0$. Then the operator ad_a is invertible on the plane $V = V^+ \oplus V^-$, namely, $\operatorname{ad}_a^{-1} E_\alpha = \frac{1}{\alpha(a)} E_\alpha$. The linear operator φ_{abD} is of the form: $\varphi_{abD}(X) = \operatorname{ad}_a^{-1} \operatorname{ad}_b X' = D(t)$, where $X = X' + t$ is a unique decomposition of X in V and T and $D : T \to T$ is an arbitrary linear operator symmetric on T. The operator φ_{abD} is parametrized by a, b, D. It is clear that $\varphi_{abD} E_\alpha = \frac{\alpha(b)}{\alpha(a)} E_\alpha$. In the Weyl basis $E_\alpha, E_{-\alpha}, H'_\alpha$ the operator φ is given by the matrix

$$\varphi_{abD} = \begin{pmatrix} \lambda_1 & & & & & & 0 \\ & \ddots & & & & & \\ & & \lambda_q & & & & \\ & & & \lambda_1 & & & \\ & & & & \ddots & & \\ 0 & & & & & \lambda_q & \\ & & & & & & D \end{pmatrix}$$

and $\varphi_{ab} : V \to V$, where $\lambda_\alpha = \alpha(b)/\alpha(a), q = \dim V^\pm =$ (the number of roots $\alpha > 0$).

LEMMA 4.2.1. *The operator φ_{abD} is symmetric with respect to the Killing form for any a, b, D satisfying the above-mentioned conditions.*

PROOF: Denote the Weyl basis in V by (e_i). It suffices to verify that $\langle \varphi e_i, e_j\rangle = \langle e_i, \varphi e_j\rangle$ for any i, j. We may assume that $i \neq j$. Recall that the plane T is orthogonal to the plane V. Since φ maps V and T into themselves and since D is symmetric on T, it suffices to verify the symmetry of φ on V. Since E_α (where $\alpha \neq 0$) are eigenvectors of φ, it follows that

$$\langle \frac{\alpha(b)}{\alpha(a)} E_\alpha, E_\beta\rangle = \langle E_\alpha, \frac{\beta(b)}{\beta(a)} E_\beta\rangle$$

for $\alpha + \beta \neq 0$ because $\langle E_\alpha, E_\beta\rangle = 0$. If $\alpha + \beta = 0$, then

$$\frac{\alpha(b)}{\alpha(a)} = \frac{-\alpha(b)}{-\alpha(a)}.$$

This completes the proof of the Lemma.

In the case of general position, the operator φ on V has q distinct eigenvalues of multiplicity two. The operator $\varphi : V \to V$ is an isomorphism of V into itself.

Recall that V^+ is a nilpotent subalgebra. In our model example, this is a subalgebra of upper-triangular matrices with zeros upon the principal diagonal. Since V^+ is generated by the vectors $E_\alpha, \alpha > 0$, then $\varphi|_{V^+}$ is symmetric with respect to the Killing form. In the case of general position, the eigenvalues of this operator are distinct: $\lambda_1, \ldots, \lambda_q$. This series will be called *a normal nilpotent series*. By construction, to each *complex series* there corresponds one *normal nilpotent series*. The operator $\varphi : G \to G$ also maps the subalgebra $T \oplus V^+$ into itself, φ being an isomorphism of the space $T \oplus V^+$ onto itself. In our model example, $T \oplus V^+$ is a subalgebra of upper-triangular matrices. As above, in the case of general position, all eigenvalues of the operator $\varphi|_{T \oplus V^+}$ are distinct, and the operator is symmetric.

Thus, to each complex series there corresponds a *normal solvable series*. In the Weyl basis the operators $\varphi|_{V^+}$ and $\varphi|_{T \oplus V^+}$ are given by the matrices

$$\varphi|_{V^+} = \begin{pmatrix} \lambda_1 & & 0 \\ & \ddots & \\ 0 & & \lambda_q \end{pmatrix} \qquad \varphi|_{T \oplus V^+} = \left(\begin{array}{ccc|c} \lambda_1 & & 0 & \\ & \ddots & & 0 \\ 0 & & \lambda_q & \\ \hline & 0 & & D \end{array} \right).$$

Similar operators φ with the same properties also turn out to exist on compact Lie algebras. Each semisimple complex Lie algebra possesses a compact form G_c. Above we have constructed the canonical embedding of this compact subalgebra into the complex algebra G. Recall that $G_c = \langle E_\alpha + E_{-\alpha}, i(E_\alpha - E_{-\alpha}), iH'_\alpha \rangle = W^+ \oplus iT_0$. Let a and b belong to T_0 and a be the element of general position. Since $\mathrm{ad}_a E_\alpha = [a, E_\alpha] = i[a', E_\alpha]$, where $a = ia', a' \in T_0$ then $\mathrm{ad}_a E_\alpha = i\alpha(a')E_\alpha$, where $\alpha(a')$ is real. Consequently, $\mathrm{ad}_a(E_\alpha + E_{-\alpha}) = \alpha(a')i(E_\alpha - E_{-\alpha}); \mathrm{ad}_a i(E_\alpha - E_{-\alpha}) = -\alpha(a')(E_\alpha + E_{-\alpha})$. Thus, the operator $\mathrm{ad}_a : W^+ \to W^+$ turns the vector $E_\alpha + E_{-\alpha}$ into the vector proportional to $i(E_\alpha - E_{-\alpha})$ and vice versa. The operator ad_b acts in a similar manner. By virtue of the choice of $a \in iT_0$ the operator ad_a is invertible on W^+. Then the operator $\varphi_{ab} = \mathrm{ad}_a^{-1} \mathrm{ad}_b : W^+ \to W^+$ has all the vectors $E_\alpha + E_{-\alpha}, i(E_\alpha - E_{-\alpha})$ as its eigenvectors with eigenvalues $\alpha(b)/\alpha(a) = \alpha(b')/\alpha(a')$, where $a = ia', b = ib', a', b' \in T_0$. Analogous events also happen on the subspace W^-. The operator $\varphi_{abD} : G_c \to G_c$ may be defined as follows: $\varphi X = \varphi(X' + t) = \varphi_{ab} X' + Dt = \mathrm{ad}_a^{-1} \mathrm{ad}_b X' + Dt$, where $X = X' + t$ is a unique decomposition of X in $G_c = W^+ \oplus iT_0$, $X' \in W^+, t \in iT_0$ and $D : iT_0 \to iT_0$ is a linear operator symmetric on iT_0. In the basis $E_\alpha + E_{-\alpha}, i(E_\alpha - E_{-\alpha}), iH'_\alpha$ the operator φ is given by the matrix:

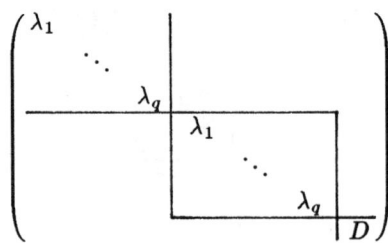

where the numbers $\lambda_\alpha = \alpha(b)/\alpha(a)$ are real, $q = \dim W^+$.

LEMMA 4.2.2. *The operator* $\varphi : G_c \to G_c$ *is symmetric for any* a, b, D *satisfying the above-mentioned limitations.*

The proof is absolutely identical to the proof of Lemma 4.2.1. The only thing to be specially verified is orthogonality of the chosen basis in the plane W^+. This can be checked by simple calculus which we leave to the reader.

In the case of general position, the operator $\varphi_{ab} : W^+ \to W^+$ has q distinct eigenvalues of multiplicity two.

Now construct a similar family of Hamiltonian systems on certain simple compact real Lie algebras which correspond to classical normal compact subalgebras. Above we have presented in each compact form G_c the subalgebra G_n spanned by the vectors $E_\alpha + E_{-\alpha}, \alpha \in \Delta$. Since all of them are eigenvectors for the operators φ of a compact series, it follows that if the operators are restricted to the subalgebra G_n, we obtain a *normal series of operators*. They coincide with the operators

$$\varphi_{ab} : G_n \to G_n, \varphi X = \mathrm{ad}_a^{-1}\mathrm{ad}_b X, X \in G_n; a, b \in iT_0, \alpha(a) \neq 0.$$

In the basis $E_\alpha + E_{-\alpha}$ the operators φ are given by the matrices:

$$\varphi_{ab} = \begin{pmatrix} \lambda_1 & & 0 \\ & \ddots & \\ 0 & & \lambda_q \end{pmatrix}, \qquad q = \dim W^+$$

Note that in this case a and b do not belong to G_n, that is, *to define operators of a normal series, one should have elements of a certain larger Lie algebra.*

This differs the normal series from the complex and compact series for which the elements a and b did belong to the Lie algebra under investigation. Not all compact semisimple Lie algebras can be represented as G_n in a certain compact real form $G_c \subset G$. As shown above, the algebra G_n coincides with the fixed points of the automorphism $\tau : G \to G, \tau X = \overline{X}$, after its restriction to G_c. Let $P \subset G_c$ be a subspace orthogonal to G_n in G_c, on which $\tau = -1$. Then the commutation relations $[G_n, G_n] \subset G_n, [P, P] \subset G_n, [G_n, P] \subset P$.

Thus, we have completely described the subspace of quadratic functionals in the maximal commutative linear algebra V of polynomial on the orbit of general position in a semisimple Lie algebra.

We are going to show now that among the quadratic Hamiltonians which we have discovered there exist extremely interesting ones from the mechanical point of view. First of all we will find out how the operation of calculating the skew gradient of the function $H(X)$ is written *in an explicit form* on the Lie algebra G.

LEMMA 4.2.3. *Given a semisimple Lie algebra G, we have* $\mathrm{sgrad}\, H(X) = [\mathrm{grad}\, H(X), X]$ *for any smooth function $H(X)$ on an orbit O in G.*

PROOF: Let ξ be a vector from $T_X O$, then $\omega(\mathrm{sgrad}\, H, \xi) = H(X) = \langle \mathrm{grad}\, H, \xi \rangle$. By the definition of the form ω we have: $\omega(\mathrm{sgrad}\, H, \xi) = \langle \mathrm{sgrad}\, H, y \rangle$, where $\xi = [X, y]$, whence, $\langle \mathrm{grad}\, H, [X, y] \rangle = \langle \mathrm{sgrad}\, H, y \rangle$, that is, $\langle [\mathrm{grad}\, H, X], y \rangle = \langle \mathrm{sgrad}\, H, y \rangle$. Since this is identically fulfilled for any y, it follows that $\mathrm{sgrad}\, H = [\mathrm{grad}\, H, X]$.

Let H be a quadratic Hamiltonian, then $\operatorname{grad} H(X) = \varphi X$, for which the operator φ is described above in Corollary 4.2.1. We may therefore examine the following Hamiltonian system of equations: $\dot{X} = [\varphi X, X]$, that is, $\dot{X} = \operatorname{sgrad} H(X)$. Such equations are called the *Euler equations*. It turns out that among such systems there exist interesting multidimensional analogues of classical mechanical systems.

We will demonstrate, for instance, that among systems of the form $\dot{X} = \operatorname{sgrad} H(X)$, where $H(X)$ is a quadratic Hamiltonian of the form $H(X) = \langle X, \varphi X \rangle$ ($\varphi = \varphi_{abD}$ is an operator of normal series, see above), there exist classical equations of motion of a multidimensional rigid body with a fixed point. Examine the Lie algebra $\operatorname{so}(n)$ and represent it as the normal form in the Lie algebra $\operatorname{su}(n)$, that is, $G = \operatorname{sl}(n, \mathbb{C}), G_c = \operatorname{su}(n), G_c = \operatorname{so}(n), \sigma X = \bar{X}, n > 1$. The algebra G_n is realized in G_c as a subalgebra of real skew-symmetric matrices. Embed $\operatorname{su}(n)$ into the Lie algebra $\operatorname{u}(n)$ in a standard way and consider two regular elements a, b from the Cartan subalgebra iT_0 in $\operatorname{u}(n)$ (but not in $\operatorname{su}(n)$!). Let

$$a = \begin{pmatrix} ia_1 & & 0 \\ & \ddots & \\ 0 & & ia_n \end{pmatrix}, b = \begin{pmatrix} ib_1 & & 0 \\ & \ddots & \\ 0 & & ib_n \end{pmatrix},$$

where $a_i, b_i \in \mathbb{R}$ and $a_i \neq \pm a_j, b_i \neq \pm b_j$ for $i \neq j$. Then the operator $\varphi_{ab} : G_n \to G_n$ acts as follows: $\varphi_{ab}(E_\alpha + E_{-\alpha}) = \frac{\alpha(b)}{\alpha(a)} \cdot (E_\alpha + E_{-\alpha})$. Since each root α is given by the pair of indices i, j, that is, $\alpha = \alpha_{ij}$ (see above), then each eigenvector $E_\alpha + E_{-\alpha}$ corresponding to the pair i, j is multiplied by the eigenvalue $\lambda_{ij} = \frac{b_i - b_j}{a_i - a_j}$. Thus, under the action of the operator φ, the basis skew-symmetric matrices

$$E_{ij} = T_{ij} - T_{ji} = \begin{pmatrix} \ddots & & 1 \\ & & \\ -1 & & \ddots \end{pmatrix}$$

are multiplied by the numbers λ_{ij}. The Hamiltonian system $\dot{X} = [X, \varphi X]$ has therefore the form

$$\dot{x}_{ij} = \sum_{q=1}^{n} x_{iq} x_{qj} \left(\frac{b_q - b_j}{a_q - a_j} - \frac{b_i - b_q}{a_i - a_q} \right).$$

Let now $a = -ib^2$, that is, $a_p = b_p^2$. From this we have

$$\dot{x}_{ij} = \sum_{q-1}^{n} x_{iq} x_{qj} \left(\frac{1}{a_j + a_q} - \frac{1}{a_i + a_q} \right).$$

Thus, for $a = -ib^2$ we obtain the *system of equations of the dynamics of a rigid body with a fixed point*. Moreover, among the operators φ_{ab} of normal series there exists the classical operator $\psi X = IX + XI$, where I is a real diagonal matrix. Indeed, we set $b = -ia^2$, then $\varphi_{ab} E_{ij} = \frac{b_i - b_j}{a_i - a_j} E_{ij} = (a_i + a_j) E_{ij}$, that is, $\psi - \varphi_{a,-ia}; \varphi_{ab} X = IX + XI$, where $I = -ia$. Thus, we have included the classical Hamiltonian system of equations of motion of a rigid body with a fixed point (without potential) into the multiparameter family of similar Hamiltonian systems naturally defined on simple compact Lie algebras.

We will make some comments. Let $G = so(n)$ be the Lie algebra of the orthogonal group, I a diagonal real matrix

$$I = \begin{pmatrix} \lambda_1 & & 0 \\ & \ddots & \\ 0 & & \lambda_n \end{pmatrix},$$

where $\lambda_i \neq \lambda_j$ for $i \neq j$. Consider on $so(n)$ the operator $\psi X = IX + XI$. The equations $\psi \dot{X} = [X, \psi X]$ are called the *equations of motion of an n-dimensional rigid body*. Write them in an explicit form making use of the standard coordinates in $so(n)$. Represent $so(n)$ as the algebra of the skew-symmetric matrices $X = (x_{ij})$. Then $((\lambda_i + \lambda_j)x_{ij}) = \psi X$. It is clear that $\dot{x}_{ij} = \frac{\lambda_j - \lambda_i}{\lambda_j + \lambda_i} \cdot \sum_{q=1}^{n} x_{iq} x_{qj}$. For $n = 3$ we obtain:

$$\dot{x}_{12} = \frac{\lambda_2 - \lambda_1}{\lambda_2 + \lambda_1} x_{13} x_{32}, \quad \dot{x}_{13} = \frac{\lambda_3 - \lambda_1}{\lambda_3 + \lambda_1} x_{12} x_{23}, \quad \dot{x}_{23} = \frac{\lambda_3 - \lambda_2}{\lambda_3 + \lambda_2} x_{21} x_{13}.$$

Identifying $so(3)$ with $\mathbb{R}^3$ we obtain the *classical equations of the dynamics of a three-dimensional rigid body*. Let the matrix I be such that $\lambda_i + \lambda_j \neq 0$ for all i, j. Then the operator ψ is invertible on $so(n)$, and the inverse operator $\psi^{-1} = \varphi$ is of the form $(\varphi X)_{ij} = \frac{1}{\lambda_i + \lambda_j} x_{ij}$. Perform in $so(n)$ the change of coordinates $Y = \psi X$. Then the equations of motion of a rigid body will become $\dot{Y} = [\psi^{-1} Y, Y]$. Multiplying this equation by -1 and expressing Y in terms of X, we obtain the Euler equation $\dot{X} = [X, \varphi X]$, where $\varphi : so(n) \to so(n)$ is a linear self-conjugate operator. In coordinate notation we have:

$$\dot{x}_{ij} = \sum_{q=1}^{n} x_{iq} x_{qj} \left(\frac{1}{\lambda_j + \lambda_q} - \frac{1}{\lambda_i + \lambda_q} \right).$$

The rigid-body operators φ described above turn out to be particular cases of the general construction proposed by Fomenko in [145], [144], [143] and called *"sectional operators."* In the case of a compact semisimple Lie algebra, a sectional operator becomes an operator of the form φ_{abD}, in particular, an operator described by the classical equations of motion of a rigid body. The sectional operator makes it possible to write on an arbitrary finite-dimensional Lie algebra a certain system of differential equations which possesses some symmetry properties. For instance, for the noncompact Lie algebra $e(n)$ (that is, the Lie algebra of the group $E(n)$ of motions in $\mathbb{R}^n$), this system is transformed into the *equations of inertial motion of a rigid body in an incompressible liquid*.

The properties of this system are studied in the papers by Fomenko and Trofimov, [139], [141], [142], [182], [186]. Later, Trofimov has shown that a particular case of the general construction of sectional operator are the *equations of motion of a rigid body in a magnetized liquid*, [188], [134], [135].

Of importance is the fact that all the above-mentioned systems of equations are flows of the form $\dot{X} = \text{sgrad } H(X)$, where the quadratic Hamiltonian $H(X)$ is included into the maximal linear commutative algebra of integrals on the orbits in a corresponding Lie algebra.

THEOREM 4.2.3 (SEE [89]–[94]). *All the enumerated Hamiltonian "rigid-body" systems of the form $\dot X = [X, \varphi X]$ are completely Liouville-integrable for all the operators φ_{abD} described above (on orbits of general position), the complete set of integrals being indicated explicitly and consisting of polynomials on the Lie algebra.*

Integrability of these equations on singular semisimple orbits in semisimple Lie algebras was later proved by Dao Chong Thi [39] and Brailov [206], [208].

The theorem on complete integrability of multidimensional equations of motion of a rigid body in a liquid (with a nondegenerate and sign indefinite quadratic Hamiltonian) was proved by Fomenko and Trofimov in [139]–[142]. Theorems on complete integrability of other multidimensional analogues of the equations of motion of a rigid body (with a Hamiltonian of fixed sign) were obtained by Perelomov. See, for instance, [107]. See also the papers by Belyayev, Bogoyavlensky, Reyman, Semenov-Tian-Shansky, Ratiu.

2.3 Geometric Properties of Rigid-Body Invariant Metrics on Homogeneous Spaces

Although the geometrical interpretation of the general dynamical problems has been know since Riemann and Poincaré, the interest in the mathematical analysis of dynamical systems on homogeneous nonlinear manifolds has been newly stimulated by the study of separate classes of equations of mathematical physics. In particular, the problem of the geodesics of left invariant (pseudo)Riemannian and Finslerian metrics on Lie groups is urgent as before. The problem of extremals of left-invariant one-dimensional variational problems on Lie groups is entailed by the finite-dimensional approximation of ideal-liquid hydrodynamic equations as well as by the various multidimensional models in mathematical physics.

As demonstrated above, wide classes of completely integrable Hamiltonian systems were found among the dynamical systems on homogeneous spaces (orbits) and Lie groups. These systems appeared to relate to the orbit method in the theory of Lie group representation, to the algebraic geometry of compact Riemann surfaces, to infinite-dimensional graduated Lie algebras, and to the inverse scattering method in the theory of nonlinear differential equations with partial derivatives. See, for instance, the review by Novikov [96].

Two associated directions are distinguished in the study of dynamical systems on homogeneous spaces. One is the development of the methods of constructing a complete set of first integrals and a qualitative analysis of system motion. The other is seeking explicit formulae for Hamiltonian system trajectories and constructing action-angle variables.

The behaviour of extremals of invariant Hamiltonian systems on homogeneous spaces of Lie groups was qualitatively analyzed, in particular, by Meshcheryakov. Such a system is defined as a pair (T^*M, H), where $M = \mathfrak{G}/K$ is a homogeneous space of a finite-dimensional Lie group $\mathfrak{G}$ and $H : T^*M \to \mathbf{R}$ is a function, on the cotangent bundle T^*M of the manifold M, invariant under the induced Hamiltonian action of the group $\mathfrak{G}$ on T^*M.

In the present subsections we briefly describe the results of Meshcheryakov.

THEOREM 4.2.4. *Each extremal of any $\mathfrak{G}$-invariant Hamiltonian system on the* cotangent bundle T^*M of a homogeneous space $M = \mathfrak{G}/K$ is either an embedded circle or an injective immersion of a real-axis interval.

Theorem 4.2.4 is applicable to the case of left-invariant Hamiltonian flows on Lie groups. In particular, Theorem 4.2.4 implies the following assertion. Let $F : T^*M \to G^*$ be momentum mapping.

COROLLARY 4.2.2. *Extremals of left-invariant Hamiltonian flows on $T^*\mathfrak{G}$ with a fixed value of the momentum F do not intersect and cover once the Lie group $\mathfrak{G}$.*

We can point out examples of homogeneous spaces on which the geodesics of an invariant metric are either all closed or all non-closed.

PROPOSITION 4.2.2. *Let $\mathfrak{G}$ by a connected Lie group with a left-invariant Riemannian metric all of whose geodesics are closed. Then $\mathfrak{G}$ is isometric either to the group $SO(3)$ or to the group $SU(2)$ with a biinvariant metric.*

On the other hand, the following result holds true.

THEOREM 4.2.5. *The trajectories of any left-invariant Hamiltonian vector field on the cotangent bundle of a connected and simply connected solvable Lie group, in which a certain degree of each element lies in the image of exponential mapping (of the algebra into the group), are projected onto the group into non-closed curves without self-intersection.*

This approach permits the study of the momentum mapping "rigidity" under pulverizations of left-invariant affine connectednesses on semisimple Lie groups. The results of Meshcheryakov generalize the results of Lacomba [234], Kostant [246], and in the case of the group $SO(3)$, lead to the interpretation of rigid-body motion according to Poinsot, which is well known in classical dynamics.

Meshcheryakov also examines left-invariant quadratic Hamiltonian functions on $T^*\mathfrak{G}$ connected with the construction of the shift of invariants of the coadjoint representation of semisimple Lie groups $\mathfrak{G}$. More precisely, he examines the class of functions $f_\lambda(X) = f(X + \lambda a)$, where $\lambda \in \mathbf{R}, \mathbf{C}$ and f belongs to the ring of invariants IG (see above). The quadratic forms of f on G^*, which functionally depend on functions of the form f_λ are, according to Fomenko and Mishchenko [91], of the following form: $f(X) = \langle X, \varphi_{abD}(X) \rangle$. Here $\langle \ \rangle$ is the Killing form of the algebra G. The definition of the operators φ_{abD} is given above.

Recall that the Poisson bracket on a symplectic manifold $T^*\mathfrak{G}$ induces in the ring of smooth functions on the space G^* the Poisson bracket $\{H, F\}(X) = \langle X, [\operatorname{grad} H(X), \operatorname{grad} H(Y)]\rangle, G \approx G^*$. Let us set in this ring of functions another Poisson bracket, namely, $\{H, F\}_a = \langle a, [\operatorname{grad} H(X), \operatorname{grad} F(X)]\rangle$, in the general case the gradient being taken with respect to the nondegenerate Killing form.

THEOREM 4.2.6. *Let $\varphi : G \to G$ be a linear operator, on a semisimple Lie algebra G, self-conjugate with respect to the Killing form. The Euler equation $\dot{X} = [X, \varphi X]$ is Hamiltonian simultaneously with respect to both Poisson brackets (the element a is a covector of general position) $\{,\}$ and $\{,\}_a$ if and only if $\varphi = \varphi_{abD}$ for a certain element $b \in T$ and for an operator $D : T \to T$.*

Thus, this result of Meshcheryakov complements the characteristic description of the "rigid-body type" operators which we have presented above. The Hamiltonian property of the Euler equations with the operators $\varphi : G \to G$ simultaneously in two Poisson brackets, that is bihamiltonian property, proves to be the characteristic property of the metrics φ_{abD} associated with the construction of the shift of invariants of the coadjoint representation.

Theorem 4.2.6 motivates the study on Lie algebras of pairs of Poisson brackets which appear from two different structures of Lie algebras on one vector space. Two such structures will be called *compatible* if the sum of commutators is again a commutator. The natural class of compatible Poisson brackets is determined by effective symmetric Lie algebras dual in the sense of Cartan. Let the triplet (G, K, σ) be an effective symmetric Lie algebra and $G = K \oplus P$ its decomposition under an involutive automorphism σ. Suppose that the representation ad : $K \to$ End P is irreducible.

PROPOSITION 4.2.3. *The Euler equations defined by the self-conjugate operator $C : G \to G$ such that $C = C_1 \oplus C_2, C_1 : K \to K, C_2 : P \to P$ will be Hamiltonian simultaneously with respect to the Poisson brackets of Lie algebras dual in the sense of Cartan if and only if the operator C_2 is scalar.*

THEOREM 4.2.7. *The functions $f_\lambda(k, s) = f(k + \lambda s), \lambda \in \mathbf{R}, f \in IG$ are first integrals in involution of the Euler equations with the Hamiltonian $H(X) = \langle X, CX \rangle$. Here $X = k + s$ is a decomposition of the vector $X \in G$ which corresponds to the decomposition $G = K \oplus P$.*

Next, it turns out that the Euler equations with the Hamiltonian function H on irreducible effective symmetric Lie algebras possess the following property.

PROPOSITION 4.2.4. *If $(k(t), s(t))$ is a solution of the Euler equations then $(k(t), [k(t), s(t)])$ is also a solution of these equations.*

Compatible Poisson brackets on Lie algebras were analyzed in the paper by Reyman [117], where such brackets appeared from infinite-dimensional graduated Lie algebras and were applied to the study of the various generalizations of Toda chains. In the same paper [117], Reyman pointed out the Hamiltonian property of the Euler equations for the shifts of invariants of semisimple Lie algebras indicated earlier in the paper [247].

Meshcheryakov obtained some results on exact integration of geodesic flows of metrics φ_{abD} on simple Lie groups by means of special functions. The functional nature of the solutions of the equations for geodesics is as follows: they are either quasi-polynomials or rational functions of the restrictions of the theta functions of compact Riemann surfaces to rectilinear windings of Jacobian tori of these surfaces. These methods rest upon the papers by Novikov and Dubrovin [45].

THEOREM 4.2.8. *Let G be a semisimple Lie algebra with a splitting Cartan subalgebra T. The Euler equations $\dot{X} = [X, \varphi X]$ which have T as the algebra of linear first integrals are integrated in elementary functions provided that the initial data are taken in any Borel subalgebra BG containing T. These functions are quasi-polynomials whose degrees do not exceed the least degree of the algebra G.*

Formulae for the solutions are derived by successive integration of a certain system of linear inhomogeneous equations in which the Euler equations are reduced after the choice of the basis from the root vectors. A similar assertion was later announced in the paper [15]. From Theorem 4.2.8 there follows a useful corollary.

THEOREM 4.2.9. *The equations for geodesics of left-invariant metrics on semi-simple linear Lie groups are explicitly integrated by quasi-polynomials provided that the angular velocities of the geodesics lie in Borel subalgebras containing the Cartan subalgebra T if T is the algebra of linear first integrals of the Euler equations.*

Among the obtained geodesics there exist isotropic ones. Their angular velocities belong to the maximal nilpotent subalgebra $N = [BG, BG]$.

It turns out that one may present an exact integration of the equations for geodesics of the metrics φ_{abD} on the group $SL(m, \mathbb{C})$. Metrics of the type φ_{abD} appeared for the first time in the course of construction of nonlinear differential equations integrable by the inverse scattering method. From the paper [38] it readily follows that the Euler equation $\dot{X} = [X, \varphi_{abD}(X)]$ on a classical Lie algebra of series A_{m-1} serves as a commutativity equation for a pair of operators.

$$L_\lambda = \frac{d}{dt} + \varphi_{abD}(X) + \lambda b, \Lambda_\lambda = X + \lambda a : [L_\lambda, \Lambda_\lambda] = 0.$$

The operators act on the space of smooth vector functions, and the Lie algebra A_{m-1} is realized by square zero-trace matrices. The matrices a and b are diagonal with distrinct diagonal elements. According to the finite-zoned integration theory (see [77]), the commutativity equations $[L_\lambda, \Lambda_\lambda] = 0$ are integrated by means of the theta-functions of the Riemann surface of the algebraic curve $Q(W, \lambda) = \det(W - X - \lambda a) = 0$.

The matrix elements of the solutions of the equations for geodesics of left-invariant metrics φ_{abD} on the group $SL(m, \mathbb{C})$ with initial data of general position turn out to be rational functions of the exponents and theta-functions of the Riemann surfaces $\Gamma : Q(W, \lambda) = 0$. More exactly (Meshcheryakov), the solution $g(t)$ has the form $g(t) = \varepsilon(t) r(t) \exp(tD(h))$. The matrices $\varepsilon(t), r(t)$ are expressed here by the following formulae through the data on the Riemann surface Γ:

$$r_{jk}(t) = \exp(t\gamma_k) \frac{\theta(u_\nu(\infty_k) + \kappa_\nu(d_j) + t\eta_\nu)}{\theta(u_\nu(\infty_k) + \kappa_\nu(d_j))};$$

$$u_\nu(P) = \int_{P_0}^{P} \Omega_\nu,$$

$$\varepsilon_{jk}(t) = (-1)^{k+j} \det_{\substack{\alpha \neq j \\ \beta \neq k}}(\psi_\beta(t, \theta_\alpha));$$

$$\exp\left(t \int_{P_0}^{P} \Omega\right) \frac{\theta(u_\nu(\theta_\alpha) + \kappa_\nu(d_\beta) + t\eta_\nu)}{\theta(u_\nu(\theta_\alpha) + \kappa_\nu(d_\beta))} = \psi_\beta(t, \theta_\alpha);$$

$$\gamma_k = \lim_{P \to \infty_k} \left(\int_{P_0}^{P} \Omega + \lambda b_k\right), \kappa_\nu(d_j) = -\sum_{i=1}^{p-1} u_\nu(P_i) - u_\nu(P_{p+j}) - k_\nu$$

The data involve: the basis $A_1, \ldots, A_p, B_1, \ldots, B_p$ in the homology group with the intersection matrix

$$A_i \circ A_j = B_i \circ B_j = 0, \qquad A_i \circ B_j = \delta_{ij}, i, j, = 1, \ldots, p$$

and the basis of holomorphic differentials $\Omega_1, \ldots, \Omega_p$ normalized by the conditions

$$\oint_{A_j} \Omega_k = \delta_{jk};$$

non-special divisors

$$d_j = P_{p+j+1} + \sum_{i=1}^{p-1} P_i, \quad P_j \in \Gamma, j = 1, \ldots, m;$$

p, the genus of the surface Γ; the poles ∞_k and zeros Q_α of the meromorphic function $\lambda = \lambda(P)$; Abelian second-type differential Ω, which has the pole with the leading term $-b_k \lambda^2$ at the point ∞_k and is normalized by the condition $\oint_{A_j} \Omega = 0$, $\oint_{B_j} \Omega = 2\pi i \eta_j$, $i = \sqrt{-1}$; the theta-function

$$\theta(u_1, \ldots, u_p) = \sum_{n \in \mathbb{Z}^p} \exp(\pi i (Bn, n) + 2\pi i (u, n)); \qquad B_{jk} = \oint_{B_j} \Omega_k$$

and the vector of the Riemann constants

$$k = (k_1, \ldots, k_p), \quad k_\nu = -\frac{1}{2} + \frac{B_{\nu\nu}}{2} + \sum_{\nu=j}^{p} \oint_{A_j} \left(\int_{P_0}^{P} \Omega_\nu \right) \Omega_j$$

with a basis point $P_0 \in \Gamma$.

COROLLARY 4.2.3. *The solutions of the Euler equations with the metrics ψ_{abD} on the Lie algebra $\mathrm{sl}(m, \mathbb{C})$ are rational functions of the exponents and restrictions of the theta-functions of the algebraic curves $Q(W, \lambda) = 0$ to rectilinear windings of the Jacobian tori $J(\Gamma)$ of these curves.*

One can construct action-angle variables on orbirts of minimal nonzero dimension in Lie algebras of the classical series A_m and C_m for the Euler equations with an algebra of linear first integrals isomorphic to a Cartan subalgebra. After this, the Euler equations are integrated in elementary functions.

§3 Euler Equations on the Lie Algebra so(4)

Examine a particular case of Theorem 4.2.3 from §2 (Ch. 4). Examine the Euler equations of motion of a multidimensional rigid body realized as Hamiltonian systems on the Lie algebras so(3) and so(4) of small dimension. We have, in fact, considered the case of the algebra so(3) when we demonstrated that the equations integrated by us coincide with the classical Euler equations of motion of a three-dimensional rigid body. The case of the Lie albegra so(4) deserves a more detailed analysis.

As an example we will consider flows of normal series for $G_n = \mathrm{so}(4) \subset G_c = \mathrm{su}(4) \subset G = \mathrm{sl}(4, \mathbb{C})$. The algebra so(4) is realized in the form of skew-symmetric

§3   Euler Equations on the Lie Algebra so(4)

matrices and is spanned by the vectors $E_{ij} = E_\alpha + E_{-\alpha}$ of the standard form. Write $X \in so(4)$ as

$$X = \alpha E_{12} + \beta E_{13} + \gamma E_{14} + \delta E_{23} + \rho E_{24} + \varepsilon E_{34},$$

where all the coefficients are real.

Recall that rank $so(4) = 2$ and that the orbits of general position are four-dimensional manifolds $S^2 \times S^2$. Let $a, b \in T_c \subset su(4)$, then

$$\varphi_{ab}(X) = \alpha \frac{b_1 - b_2}{a_1 - a_2} E_{12} + \beta \frac{b_1 - b_3}{a_1 - a_3} E_{13}$$
$$+ \gamma \frac{b_1 - b_4}{a_1 - a_4} E_{14} + \delta \frac{b_2 - b_3}{a_2 - a_3} E_{23}$$
$$+ \rho \frac{b_2 - b_4}{a_2 - a_4} E_{24} + \varepsilon \frac{b_3 - b_4}{a_3 - a_4} E_{34}.$$

For each pair a, b we obtain a flow $\dot{X}$ on $S^2 \times S^2$. The integrals will be the functions $tr(X + \lambda a)^k, 1 \leqslant k \leqslant 4$, where

$$X + \lambda a = \begin{pmatrix} \lambda a_1 & \alpha & \beta & \gamma \\ -\alpha & \lambda a_2 & \delta & \rho \\ -\beta & -\delta & \lambda a_3 & \varepsilon \\ -\gamma & -\rho & -\varepsilon & \lambda a_4 \end{pmatrix}$$

The calculations give four integrals:

$$h_1 = tr X^2, \quad h_2 = tr X^4, \quad h_3 = tr X^2 a, \quad h_4 = 2tr X^2 a^2 + tr(Xa)^2.$$

The integrals h_1 and h_2 are constant on the orbits (they are invariants) and are of the form

$$h_1 = \alpha^2 + \beta^2 + \gamma^2 + \delta^2 + \rho^2 + \varepsilon^2,$$
$$h_2 = h_1^2 + 4(\beta\delta\gamma\rho - \alpha\delta\gamma\varepsilon + \alpha\rho\beta\varepsilon) - 2(\alpha^2\varepsilon^2 + \beta^2\rho^2 + \gamma^2\delta^2).$$

As a matter of fact, h_2 is the square of the integral q of degree two (after subtraction of the function h_1^2 from h_2), where $q = \alpha\varepsilon - \beta\rho + \gamma\delta$.

Thus, two quadratic integrals h_1 and q are generators of the ring of invariants $I(so(4))$, that is, any polynomial constant on the orbits is expanded in h_1 and q. It can be easily verified that h_1 and q are independent. The equations $h_1 = p, q = t$, where p and t are constants, determine the orbits of general position. These integrals, in particular q, were considered in [65](Langlois).

The integrals h_3 and h_4 are already not constant on the orbits. They have the form:

$$h_3 = \alpha^2(a_1 + a_2) + \beta^2(a_1 + a_3) + \gamma^2(a_1 + a_4)$$
$$+ \delta^2(a_2 + a_3) + \rho^2(a_2 + a_4) + \varepsilon^2(a_3 + a_4);$$
$$h_4 = \alpha^2(a_1^2 + a_1 a_2 + a_2^2) + \beta^2(a_1^2 + a_1 a_3 + a_3^2)$$
$$+ \gamma^2(a_1^2 + a_1 a_4 + a_4^2) + \delta^2(a_2^2 + a_2 a_3 + a_3^2)$$
$$+ \rho^2(a_2^2 + a_2 a_4 + a_4^2) + \varepsilon^2(a_3^2 + a_3 a_4 + a_4^2).$$

It can easily be checked that the integrals h_1, q, h_3, h_4 are functionally independent and the integrals h_3 and h_4 are in involution on the orbits.

Thus, the Euler–Arnold equations $\dot{X} = [X, \varphi X]$ on the Lie algebra so(4) are completely integrated for all the operators $\varphi = \varphi_{abD}$ introduced above.

The Euler equations on the Lie algebra so(4) are of special interest in many respects. A remarkable fact is that for certain metrics on so(4) the corresponding Euler equations have a natural mechanical interpretation. First of all this refers to metrics of the form $\langle X, \varphi X \rangle$ generated by the operators $\varphi = \varphi_{abD}$ studied by Manakov, Fomenko and Mishchenko (see above). Later, Bogoyavlensky, Veselov, and Veselova noted in [209], [15], [210] that Euler equations describe inertial motion of a rigid body with ellipsoidal cavity filled with ideal incompressible liquid which is in homogeneous turbulent motion. (see [211]). The latter system is classical and was investigated by many authors, including Poincaré and Steklov.

Another interpretation of these equations was suggested in [212]: the homogeneous dynamics of magnetic moments for a double-lattice magnetic is described by the Euler equations on so(4). It should be noted that the first (after the important work by Arnold [3]) example of such interpretation of the Euler equations on a Lie algebra was discovered by Novikov [97] for the Lie algebra of the group of motions of the three-dimensional space $e(3)$. The corresponding equations are the Kirchhoff–Thomson equations describing free dynamics of a rigid body in liquid. These equations have been tested for integrability by Kozlov and Onishchenko [167]. The papers by Veselov [209],[210] are devoted to the conditions of the existence of an additional integral for the Euler equations on so(4) and corresponding mechanical systems. In the first part of the present subsection we deal with the results of the latter two papers.

In the second part, we analyze the familiar integrable cases of the Euler equations on so(4). Among them we first distinguish the equations of motion of a "four-dimensional rigid body" which correspond to the metric φ_{abD} of a special form. Quadratic integrals for the equations of motion of an n-dimensional rigid body were found in [84], their involutivity was proved in [43]. In the case $n = 4$, the obtained integrals proved to suffice for complete integrability of these equations. Manakov [76] found the Lax representation for a wider family (including the equations of motion of an n-dimensional rigid body) of metrics on so(n) diagonal in the standard representation. Fomenko and Mishchenko proved independence of a sufficient number of obtained integrals and extended the results of Ref. [76] to the case of arbitrary semisimple Lie algebras. See [91] and also the review in [143].

Another integrable family of metrics on so(4) whose integrals are also quadratic was presented by Steklov in [213] without any interpretation. The latter cases of integrability have been rediscovered many times in recent years, (see [214], [209], [15], [215]). The Lax representation with the spectral parameter for them has been found by Veselov in [212], the explicit formulae in theta-functions for the solutions of these systems, the same as for the systems in [76], have been recently derived by Babenko.

Examples of integrability with an additional integral of degree four for a certain relation among the values of other integrals are given by Bogoyavlensky in [158]. The equations derived are integrated in elliptic functions. In their recent paper, Adler and van Moerbeke [181] present the family of such metrics on so(4) for which the existence of an integral of degree four, without any limitations on the values of the other integrals, is announced.

The integrability conditions for the general Euler equations on so(4). As is well known, the Lie algebra so(4) is not simple and is isomorphic to the direct sum so(3) $\oplus$ so(3). This is precisely the form in which this algebra appears in the description of the mechanical system discussed above. Generally, one copy of so(3) corresponds to a rigid body, and the other to the "quasi-rigid" motion of a liquid [15], [210]. Introduce the following notation: let M_i, l_i (where $i = 1, 2, 3$) be a natural basis in the Lie algebra so(3) $\oplus$ so(3) or the coordinates on the space conjugated to this algebra.

Examine the quadratic form of the variables M_i and l_i corresponding to the left-invariant metric of the general form

(1) $$H = \langle AM, M \rangle + 2\langle BM, l \rangle + \langle Cl, l \rangle,$$

where $\langle \, , \, \rangle$ is the standard scalar product in $\mathbb{R}^3$, the matrices A and C are symmetric, B is arbitrary. On the space congugated to the Lie algebra, the corresponding Lie group acts in a natural way. This action now has the form

(2) $$M \to R_1 M, \quad l \to R_2 l, \quad R_1, R_2 \in \mathrm{SO}(3)$$

Note that although the Lie algebras are isomorphic, the groups SO(4) and SO(3)×SO(3) do not coincide: $(\mathrm{SO}(3) \times \mathrm{SO}(3))/\pm E \approx \mathrm{SO}(4)$, but this is inessential for us.

LEMMA 4.3.1. *By the change* (2) *the form* (1) *can be reduced to the form in which the matrix A is diagonal and B is upper-triangular, that is, $A_{ij} = A_i \delta_{ij}$ and $B_{ij} = 0$ for $i > j$.*

To prove the assertion, we first reduce A to the diagonal form using the change $M \to R_1 M$. The change $l \to R_2 l$ acts upon the obtained matrix B' as follows:

$$B'' = R_2^{-1} B'.$$

The assertion of the lemma is implied by the fact that any matrix can be represented as the product of an orthogonal matrix by an upper-triangular matrix. If the columns of the matric B' are interpreted as a frame in $\mathbb{R}^3$, then the unkown rotation of R_2 turns the frame so that the first vector is collinear to the first basis vector and the second lies in the coordinate plane spanned by the first two basis vectors. This reasoning implies that if the eigenvalues of the matrix A are distinct, then the elements of the upper-triangular matrix B are uniquely determined up to sign.

The Euler equations corresponding to the metric (1) are of the form

(3) $$\begin{cases} \dot{M} = [M, AM + B^t l] \\ \dot{l} = [l, BM + Cl] \end{cases}$$

where $[\, , \,]$ denotes a vector product in $\mathbb{R}^3$. As is known, these are Hamiltonian equations with a Hamiltonian H for the Poisson bracket.

(4) $$\{M_i, M_j\} = \varepsilon_{ijk} M_k, \quad \{l_i, l_j\} = \varepsilon_{ijk} l_k, \quad \{M_i, l_j\} = 0$$

Besides the energy integral $I_1 = H$ Eqs. (3) always possess the integrals $I_2 = M^2, I_3 = l^2$ which generate the annihilator of the Poisson bracket and fix the orbit of the coadjoint representation: $M^2 = m_0^2, l^2 = l_0^2$. On this orbit we have a Hamiltonian system with two degress of freedom for integrability of which (system) an additional, H-independent integral is needed. Consider the question of the existence of such an integral for all sufficiently small m_0 and l_0. More precisely, when does the system (3) have in the neighbourhood of the point $M = l = 0$ an analytic integral $I_4 = F(M, l)$ independent of I_1, I_2, I_3?

First of all, notice that the forms that differ by the change (2) correspond to the equivalent Equations (3) so that the reduction presented in Lemma 4.3.1 may be assumed to be fulfilled. We will restrict ourselves to the case where the eigenvalues of the matrix A are distinct, that is, $A = \mathrm{diag}(A_1, A_2, A_3), A_1 > A_2 > A_3$. The case of coincident eigenvalues is presented in [210]. In what follows we will also require the nondegeneracy condition for the matrix $B : \det B \neq 0$, which we assume to be fulfilled. Along with (3) we consider the system obtained from it in the limit of small l, that is, change l by εl and send ε to zero [167]. We obtain

$$(5) \qquad \dot{M} = [M, AM], \dot{l} = [l, BM]$$

The system (5) always has three integrals: $I_1 = H_0 = \frac{1}{2}\langle AM, M\rangle, I_2 = M^2, I_3 = l^2$, but as distinguished from the initial system (3), it is not already Hamiltonian. Nonetheless, the question of the additional integral is also meaningful for this system (Veselov).

LEMMA 4.3.2. *If in the neighbourhood of the point $M = l = 0$ the system (3) possesses an additional analytic integral $I_4 = F(M, l)$ then the system (5) has a polynomial integral $J_4 = G(M, l)$ independent of the above-mentioned integrals.*

First of all note that by virtue of homogeneity of the system (3) the presence of an integral with the indicated analytic properties entails the existence of a polynomial integral obtained as a component $F_0(M, l)$ of lower degree in M and l in the expansion of $F(M, l)$ into a Taylor series at the point $M = l = 0$. If this component is dependent on the integrals I_1, I_2, I_3, that is, $\Phi(F_0, I_1, I_2, I_3) = 0$, then we may consider the lower component of the function $\widetilde{F} = \Phi(F, I_1, I_2, I_3)$ which is also an integral of the system (3). As a result of such a process, we finally arrive at a polynomial integral independent of I_k because $F(M, l)$ is independent of them. The assertion of the lemma is proved by similar argument with the only difference being that as the unknown integral of the system (5) one should take the component of an integral of the system (3) of a lower degree in l.

Now turn to the system (5). The first three equations

$$(6) \qquad \dot{M} = [M, AM]$$

are ordinary Euler equations of free motion of a rigid body, which are integrated in elliptic functions. The second set of equations

$$(7) \qquad \dot{l} = [l, BM]$$

consists of linear equations with coefficients whose time dependence is determined by the system (6). By virtue of the properties of the general solution of this system, this system is periodic, and if one takes complex values of time, then even doubly periodic. For linear systems with periodic coefficients, the concept of monodromy operator is exceedingly important.

The role of this concept in the problems of the existence of integrals of Hamiltonian systems was demonstrated by Ziglin for Yang–Mills Equations [216], [217]. The arguments presented below essentially use the specificity of the systems (3) and (5) and differ from those exploited in the indicated papers.

Let $\dot{x} = A(t)x$, $x \in \mathbf{R}^n$ be a linear system such that $A(t+T) = A(t)$ is a matrix periodically depending on time, $X(t, t_0)$ is its fundamental solution: $X(t, t_0)|_{t=t_0} = E$. The *monodromy operator (matrix)* $\hat{T}(t_0)$ is defined by the equality: $X(t + T, t_0) = \hat{T}(t_0) X(t, t_0)$. For distinct t_0 the operators $\hat{T}(t_0)$ are mutually conjugate: $\hat{T}(t_1) = C\hat{T}(t_0)C^{-1}$; $C = X(t_1, t_0)$.

In our case, as easily follows from the form of the system, its monodromy operator belongs to the group SO(3), i.e., it is a rotation around a certain axis by an angle φ.

In the assumption that the matrix B is nondegenerate we may show that the rotation angle φ is not constant but depends on the solution of Equation (6). Note that by virtue of conjugation of the monodromy operators, φ is a function of the entire solution but not of a point on it. Hence, for the everywhere dense set of trajectories (6) the angle φ acquires values in commensurable with π.

Suppose now that our system has an additional integral $J_4 = G(M, l)$. Being constant on the solutions $G(M, l) = G(M, \hat{T}l) = G(M, \hat{T}^2 l) = \ldots$, that is, in the case of irrational ratio φ/π, the integral $G(M, l)$ acquires identical values on the dense subset and, therefore, on the entire circle $l^2 = \text{const}$, $\langle n(M), l \rangle = \text{const}$, where $n(M)$ is the *eigenvector of the monodromy operator* and at the same time is the *vector of the rotation axis*.

From this it follows that the vector $n(M)$ may be chosen to be polynomially dependent on M. Indeed, its direction can be found as the vector product $[\xi_1, \xi_2]$, where $\xi_i = [\frac{\partial G}{\partial l}, l_i]$ for certain fixed l_1 and l_2, (Fig. 67).

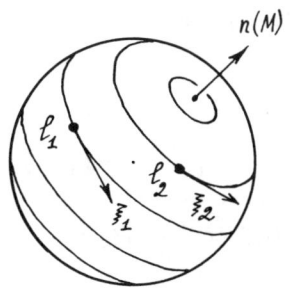

Figure 67

Using the equation for the monodromy matrix one can easily show that this vector satisfies the equation: $\dot{n} = [n, BM]$. Let us summarize our arguments.

LEMMA 4.3.3. *If the system (5) has an additional integral then the corresponding system (7) has the solution* $l = n(M)$ *polynomially depending on* M. *The vector* $n(M)$ *is the eigenvector of the monodromy operator* $\widehat{T}(M)$.

We now consider the question of the existence of such a solution for the system (7). Therefore we make use of the explicity formulae for the solutions of the Euler equations (6) (see, for instance, [218]):

(8) $\qquad M_1 = C_1 cn\lambda t, \quad M_2 = C_2 sn\lambda t, \quad M_3 = C_3 dn\lambda t$

Here $cn\, t$, $sn\, t$, $dn\, t$ are *standard Jacobian elliptic functions* corresponding to an *elliptic curve* with the parameter

$$k^2 = \frac{A_1 - A_2}{A_2 - A_3} \cdot \frac{2H_0 - M^2 A_3}{M^2 A_1 - 2H_0},$$

C_i and λ are constants also expressed in terms of the integrals of the problem, H_0 and M^2. For the formulae see ref. [218].

The above formulae and the properties of Jacobian elliptic functions imply that the coordinates M_i are doubly periodic functions of complex time t which have in the rectangle of periods poles of order one at points A, B, C, D with residues

$$\alpha_i = \frac{\pm\sqrt{\Delta_i}}{\sqrt{\Delta_1 \Delta_2 \Delta_3}}, \quad \text{where} \quad \Delta_1 = A_2 - A_3, \quad \Delta_2 = A_3 - A_1, \quad \Delta_3 = A_1 - A_2$$

for an appropriate choice of signs (Fig. 68).

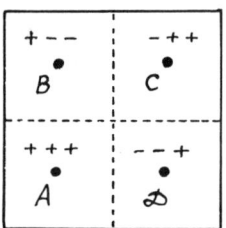

Figure 68

Being polynomial functions of M, the coordinates of the vector $n(M)$ are also elliptic functions with the same poles. Write their decomposition in the neighbourhood of one of the poles: $n(M(t)) = \frac{a_k}{t^k} + \ldots (\vec{a}_k \neq 0, \ k > 0)$. By virtue of the system (7), we have

$$-k\frac{a_k}{t^{k+1}} + \cdots = \left[\frac{a_k}{t^k} + \cdots, \frac{B\alpha}{t} + \cdots\right], \quad \alpha = (\alpha_1, \alpha_2, \alpha_3),$$

whence $[a_k, B\alpha] + ka_k = 0$. The latter equality is a linear homogeneous system whose nonzero solution exists only when the corresponding determinant vanishes.

A simple calculation leads us to the equality $k(k^2 + |B\alpha|^2) = 0$, or since $k > 0$, then

(9) $$k^2 + |B\alpha|^2 = 0.$$

Considering now similar equalities in the remaining three poles, that is, changing signs at the coordinates α, we can show that all the elements of the matrix B which do not stand on the diagonal must necessarily be zero: $B_{ij} = 0$ for $i \neq j$. There remains one requirement

(10) $$b_1^2\alpha_1^2 + b_2^2\alpha_2^2 + b_3^2\alpha_3^2 + k^2 = 0 \qquad b_i = B_{ii}$$
$$\text{or } \Delta_1 b_1^2 + \Delta_2 b_2^2 + \Delta_3 b_3^2 + k^2\Delta_1\Delta_2\Delta_3 = 0, \qquad k \in N.$$

These calculations resemble the reasoning of Kovalevskaya, who discovered her case of integrability of the motion of a heavy top proceeding from the meromorphy of the general solution of the system. Such an approach for the Euler equations on so(4) was proposed by Adler and van Moerbeke [181]. They considered the problem in which the Euler equations on so(4) have an additional algebraic integral such that the common level surface of all the integrals of the problem, when extended to the complex region, is an Abelian manifold.

The dynamics on this manifold is assumed to be linearized, the initial coordinates to be Abelian functions and, therefore, to depend meromorphically on complex time. All these properties fulfilled, the system is called *algebraically completely integrable*.

The main result of the paper [181] is as follows. Among the Euler equations corresponding to the metrics diagonal in the standard representation of so(4), under some nondegeneracy assumptions, algebraically integrable are only the metrics generated by the operator φ_{abD} described in detail above.

This result was later extended by Haine (see [219]) to the case of so(n) for an arbitrary n. In this formulation of the problem, the described approach does not require that the general solution be single-valued and no limitations are imposed in advance on the level surface of integrals. Now summarize the results.

THEOREM 4.3.1 (VESELOV [209]). *The Euler equations on the Lie algebra so(4) with a Hamiltonian of the general form (1) which corresponds to a nondegenerate matrix B and to a symmetric matrix A with distinct eigenvalues may have an additional analytic integral in the neighbourhood of $M = l = 0$ only provided that the matrices A and B may be simultaneously diagonalized by the change (2). The corresponding diagonal elements $B_{ii} = b_i$ must satisfy the relation (10) for a certain natural k. If for all that the eigenvalues of the matrix C are also distinct then as a result of the same change (2) it is also reduced to the diagonal form and its eigenvalues c_i satisfy a similar relation*

(11) $$\delta_1 b_1^2 + \delta_2 b_2^2 + \delta_3 b_3^2 + m^2\delta_1\delta_2\delta_3 = 0, \qquad m \in N,$$
$$\delta_1 = c_2 - c_3, \qquad \delta_2 = c_3 - c_1, \qquad \delta_3 = c_1 - c_2$$

Only the last assertion of the theorem is to be proved. By virtue of the symmetry between M and l the matrices C and B can also be reduced to the diagonal form using the change (2). But the matrix B being already diagonal and besides nondegenerate, only the transformations are admissible $l_i \to \pm l_i$, $M_i \to \pm M_i$ which do not change the matrix C.

It can be shown (see [210]) that under the conditions (10) and (11) the numbers k and m are only odd, the conditions (10) and (11) being sufficient for integrability of linearized systems.

In the paper [209], the theorem formulated above is proposed for $k = m = 1$. The concluding part of the proof makes use of the separatrix splitting method, in line with [167], but as is seen from the further analysis (see [210]), this step in the proof requires additional motivations.

The above-mentioned result implies that under appropriate assumptions of nondegeneracy, among the general metrics (depending on 15 parameters if all the changes (2) are preliminarily made) it is only the 7-parameter families (10) and (11) that can correspond to integrable cases of the Euler equations.

The familiar cases of integrability which are considered below make up 6-parameter families. The obtained conditions are unlikely to be sufficient for integrability although they limit rather strongly the class of corresponding matrices.

We present two corollaries of the theorem. For the details see [210], [212].

COROLLARY 4.3.1. *The equations of free dynamics of a rigid body with ellipsoidal cavity filled with an ideal liquid in homogeneous turbulent motion do not have an additional analaytic integral in the general case.*

COROLLARY 4.3.2. *The Landau–Lifshitz equation for the double-sublattice magnetic (see [212]) is nonintegrable in the general case.*

Integrable cases of the Euler equations on so(4). Consider separately the question of additional quadratic integral. This case is of interest because the whole commutative algebra of integrals consists of polynomials of one and the same degree, and the problem reduces to describing all such four-dimensional algebras. We will restrict our consideration to the case of nondegenerate algebras when for a certain representative the eigenvalues of the matrices A and C are distinct. Besides, we assume for simplicity that all the matrices A, B, and C corresponding to the elements of the algebra can be diagonalized simultaneously. According to the theorem from the preceding section, for this purpose it suffices to require that the matrix B be nondegenerate for one of the representatives mentioned above.

A commutative algebra is generated by the forms M^2, l^2, as well as by certain $H = H_0 + H_1 + H_2$ and $I = I_0 + I_1 + I_2$, where H_k and I_k are terms of degree k in l. We will assume H to be a nondegenerate representative, that is, the eigenvalues of the forms H_0 and H_2 are distinct. Writing the condition $\{H, I\} = 0$ in components, we are led to (cf. [222]) : $\{H_0, I_0\} = 0$, whence $I_0 = \lambda H_0 + \mu M^2$. Now let us turn to another integral

$$J = I - \lambda H - \mu M^2; \qquad J = J_1 + J_2, \qquad J_0 = 0.$$

Precisely in the same manner $\{H_2, J_2\} = 0$ or $J_2 = \alpha H_2 + \beta l^2$. For this there are two possibilities: $\alpha = 0$ and $\alpha \neq 0$.

Case 1. Let $\alpha = 0$. Subtracting βl^2 we arrive at the integral of the form $J' = J_1$.

Case 2. Let $\alpha \neq 0$. In this case it is more convenient to view $H' = H = \frac{1}{\alpha}J - \frac{\beta}{\alpha}l^2 = H'_0 + H'_1$, $H'_2 \equiv 0$.

In other words, a nondegenerate commutative algebra of quadratic integrals is generated by the forms M^2, l^2 and either 1) by forms

$$H = \langle AM, M \rangle + 2\langle BM, l \rangle + \langle Cl, l \rangle \text{ and } J = \langle DM, l \rangle$$

or 2) by forms

$$H = \langle AM, M \rangle + 2\langle BM, l, \rangle, \quad J = 2\langle DM, l \rangle + \langle Cl, l \rangle.$$

Consider these cases separately.

1) The condition $\{H, J\} = 0$ is equivalent to the system of equations

(12)
$$\begin{cases} D_1(A_2 - A_3) = B_2 D_3 - B_3 D_2 & (1\ 2\ 3) \\ D_1(C_2 - C_3) = B_2 D_3 - B_3 D_2 & (1\ 2\ 3) \end{cases}$$

This system implies that $A = C + \lambda E$ and there holds the condition (10) (which coincides with (11)) for $k = 1$. The case $A = C$ corresponds to the forms diagonal in the standard matrix prepresentation so(4), and the condition (10), as can be easily verified, distinguishes among them metrics of the type φ_{abD} which are already well known to the reader.

For $\lambda \neq 0$ we obtain the trivial extension of the family of metrics of the type φ_{abD} up to the six-parameter family of metrics which are already nondiagonal.

The Lax representation for these metrics has been found in the paper [76], the explicit formulae for the solutions in theta-functions of Prym manifolds have been recently derived by Babenko. A detailed algebro-geometric analysis of these systems is contained in the paper by Haine [220].

2) In case 2, the condition $\{H, J\} = 0$ leads to the system of equations

(13)
$$\begin{cases} D_1(A_2 - A_3) = B_2 D_3 - B_3 D_2 & (1\ 2\ 3) \\ -B_1(C_2 - C_3) = B_2 D_3 - B_3 D_2 & (1\ 2\ 3) \end{cases}$$

In the assumption that $B_i \neq 0$, this system has the following solution:

$$B_i = b_i, \quad A_1 = -\frac{B_2 B_3}{B_1} + a, \quad D_1 = \frac{\lambda}{B_1}, \quad C_1 = -\frac{\lambda B_1}{B_2 B_3} + c, \quad (1\ 2\ 3).$$

The corresponding forms are given by

$$2H = \sum_{j=1}^{3}(\gamma b_j^{-2} M_j^2 + 2b_j M_j l_j) + aM^2,$$

$$2J = \sum_{j=1}^{3}(2b_j^{-1} M_j l_j + \gamma^{-1} b_i^2 l_j^2) + cl^2, \quad \gamma = -b_1 b_2 b_3.$$

The general integral of the algebra has the form $H + J$ and depends on six parameters $b_1, b_2, b_3, a, c, \lambda$. Note that the system also has the trivial solution $B = D = 0$, A and C may be arbitrary numbers corresponding to the separation: $H = \langle AM, M \rangle$, $J = \langle Cl, l \rangle$ which also depends on the six parameters.

As has already been pointed out, the second case of integrability was first suggested by Steklov [213] and since then has been repeatedly rediscovered in various forms, (see [214], [209], [15], [215]). In the form presented here the algebra is given in the paper [209], while in [15], [215], it is given as relations for the coefficients of the quadratic forms generating the algebra. The Lax representation with the special parameter for these systems was found in [212].

These integrable cases are related to the well-known integrable cases of the Kirchhoff–Thompson equations of motion of a rigid body in a liquid which are described by the Euler equations on the Lie algebra of a group of motions of the three-dimensional space $E(3)$. As mentioned by Novikov in [97], when the algebra so(4) is contracted to the algebra $e(3)$, the family of metrics φ_{abD} (case 1) reduces to the Clebsch case. A similar observation for the second case of integrability is made in [221]. This case reduces to the familiar Steklov–Lyapunov–Kolosov family. Quite recently, Babenko and Golod [223] have independently discovered the following fact. These systems are not only transformed into one another when the Lie algebra is contracted, but they are simply equivalent, that is, coincide after the change of variables.

In conclusion we discuss integrable cases with an integral of degree higher than two. Bogoyavlensky has revealed [158] the following case of particular integrability, that is, integrability under certain conditions imposed on initial data. Let

$$2H = \sum_{i=1}^{3} \alpha_i (M_i^2 + l_i^2) + 2\beta_i M_i l_i,$$

where

$$\beta_i = \alpha - 3\alpha_i, \qquad \alpha = \alpha_1 + \alpha_2 + \alpha_3.$$

Then on the one-parameter set of orbits given by the condition $M^2 = l^2$ there exists an additional integral

$$J = [(\alpha_2 - \alpha_3)(M_1 - l_1)^2 + (\alpha_3 - \alpha_1) \cdot (M_2 - l_2)^2 + (\alpha_2 - \alpha_1)(M_3 - l_3)^2]^2$$
$$- 4(\alpha_2 - \alpha_3)(\alpha_3 - \alpha_1) \cdot (M_1 - l_1)^2 \cdot (M_2 - l_2)^2.$$

Notice that for the indicated parameters α and β the relations (10) and (11) are fulfilled for $k = m = 3$ although the theorem is formally not applicable to the cases of particular integrability.

The case of general integrability with an integral of fourth degree is announced in the recent paper by Adler and van Moerbeke [215]. The theorem formulated in [215] asserts that this is the only possible (besides quadratic families) case of algebraic (see above) integrability of the Euler equations on so(4).

In a number of astronomical problems, it is necessary to analyze rotation of a rigid body T around the centre of mass in a gravitational field of other bodies. Particularly important is the case where the distance to an attracting object is much larger than the dimensions of the rigid body T. In a special case of an axisymmetric gravitational field, integrability of rigid-body rotation was proved by Brun (1893). A complete solution of the problem is given in [308], where the following result is proved.

THEOREM 4.3.2 (BOGOYAVLENSKY). *Rotation of an arbitrary rigid body T, which has small dimensions, around an immobile centre of mass O in a gravitational field of an arbitrary object V is described by a certain universal completely Liouville-integrable dynamic system. Integrability of the dynamics of a rigid body is realized in Riemannian theta-functions.*

In concrete astronomical problems, moving objects do not have immobile points and their translational and rotary motions are interrelated. The following theorem, which is proved in [309] strengthens Theorem 4.3.2.

THEOREM 4.3.3 (BOGOYAVLENSKY [309]). *The translational-rotary motion of an arbitrary rigid body T in a Newtonian field with an arbitrary (inhomogeneous) quadratic potential is determined by a Liouville-integrable Hamiltonian system. The dynamics of the centre of mass O is integrated in elementary functions; rotation of a rigid body around the centre of mass is integrated in Riemannian theta-functions.*

This theorem is applicable for studying translational-rotary motion of an arbitrary rigid body in the neighbourhood of points of Newtonian potential extremum, where quadratic approximation is sufficiently accurate.

§4 Duplication of Integrable Analogues of the Euler Equations by Means of an Associative Algebra with Poincaré Duality

4.1 Algorithm for Constructing Integrable Lie Algebras

The results of this subsection rest on the following idea. Suppose a Lie algebra is give, on the orbits of which there exists a maximal linear commutative algebra of functions (that is, a complete involutive family).

How should a series of other Lie algebras with the same property be constructed? In other words, what is the way to "duplicate" integrable Lie algebras (and integrable Hamiltonian systems)? It turns out several important algebraic mechanisms exist that enable such new Lie algebras to be effectively constructed. One of such mechanisms was first proposed by Trofimov and later on developed by Brailov and Le Ngok Tyuen.

A commutative, associative, graduated algebra $A = A_0 \oplus \cdots + A_n$ with a unit element $\varepsilon_1 \in A_0$ will be called a *connected algebra with Poincaré duality* if $\dim A_0 = 1$ and if on the algebra A there exists a linear functional α, identically equal to zero on A_i for $i < n$, such that the bilinear symmetric form $\alpha(a \cdot b)$ (where $a, b \in A$) is nondegenerate.

Let G be a Lie algebra. The tensor product $G_A = G \otimes A$ is a Lie algebra whose commutator is given by the relation $[g \otimes a, h \otimes b] = [g, h] \otimes ab$ for any $g, h \in G$.

Let G be a Lie algebra and A a connected algebra with Poincaré duality. Let α be a functional such that the quadratic form $\alpha(a^2)$, $a \in A$ is nondegenerate. Examine the restriction of this form to the space $A_{n/2}$. If n is even, then replacing in case of necessity α by $-\alpha$, we may assume that the number of "pluses" in the quadratic form $\alpha(a^2)$, $a \in A_{n/2}$ is not less than the number of "minuses". Therefore, there exists a basis $\gamma_1, \ldots \gamma_k$ of the space $A_{n/2}$ such that $\alpha(\gamma_i \gamma_{\tau(i)}) = \delta_{ij}$ for all $i, j = 1, \ldots k$, where τ is a permutation such that τ^2 is an identical permutation. Since the scalar product $\alpha(ab)$ (where $a, b \in A$) is nondegenerate, it follows that $\dim A_i = \dim A_{n-i}$ for any $i = 1, \ldots, n$. Choose a special basis B in the algebra A. First choose a basis in A_0. Since $\dim A_0 = 1$ then $\gamma_1^0 = \varepsilon_1$ forms a basis in A_0 which consists of a single element. Next choose arbitrary bases (γ_i^p) in the spaces A_p for $p \langle \frac{n}{2}$. If $n = 2p$ is even then in the space A_p choose the basis (γ_i^p) so that $\alpha(\gamma_i^p \gamma_{\tau(i)}^p) = \delta_{ij}$. In the spaces A_p with graduation $p > \frac{n}{2}$ choose the bases (γ_i^p) in a way that $\alpha(\gamma_i^p \gamma_j^p) = \delta_{ij}$, where $p + q = n$, $q < \frac{n}{2}$ and, therefore, the elements γ_j^q have been defined before. Numerate the elements γ_i^p ($p = 0, \ldots, n$; $i = 1, \ldots, \dim A_p$) by integers from 1 to N with the result that for any p greater that q and any i, j the number of the element γ_i^p be greater than the number of the element γ_j^q. The element with the number s will be denoted by ε_s. Finally we obtain a homogeneous basis $B = (\varepsilon_1, \ldots, \varepsilon_N)$ of the algebra A. Homogeneity of the basis means that $\varepsilon_1 \in A_{d(i)}$ for each $i = 1, \ldots, N$. It is readily seen that $\alpha(\varepsilon_i \varepsilon_{\tau(i)}) = \delta_{ij}$ for any i, j from 1 to N and for an appropriate permutation τ of the set $(1, \ldots N)$. From the definition of the basis B if also follows that $d(i) + d(\tau(i)) = n$. Let $T: A \to A$ be a linear operator such that $T(\varepsilon_i) = \varepsilon_{\tau(i)}$ for each $i = 1, \ldots, N$. From the definition of the basis B it follows that τ^2 is an identical permutation. Hence, T^2 is an identical operator. Let $\langle \, , \, \rangle$ be such a scalar product in the algebra A that $\langle \varepsilon_i, \varepsilon_j \rangle = \delta_{ij}$ for any $i, j = 1, \ldots, N$. With respect to this product we have

$$(1) \qquad \langle Ta, b \rangle = \alpha(ab)$$

for any $a, b \in A$. Let $m = \dim G$ and $x_1, \ldots, x_m$ be linear coordinate functions on G^*. The functions $x_1, \ldots, x_m$ can be interpreted as elements of G. On the contrary, the elements $x_i \otimes \varepsilon_j$ of the Lie algebra G_A will be interpreted as linear coordinate functions on G_A^* and denoted by x_i^j. A polynomial function f on G^* such that $f(0) = 0$ in the coordinates $x_1, \ldots, x_m$, is written as

$$(2) \qquad f = \sum_{k=1}^{\infty} \sum_{i_1, \ldots, i_k} f_{i_1 \ldots i_k} x_{i_1} \cdots x_{i_k}.$$

Henceforth we widely use multiindex notation: $I = (i_1, \ldots, i_k)$ is a multiindex, $x_I = x_{i_1} \ldots x_{i_k}$ the product of linear functions $f_I = f_{i_1 \ldots i_k}$ a constant coefficient at the monimial x_I in the sum (2). Besides, summation sign over repeated indices and multiindices is omitted everywhere. The short notation is $f = f_I x_I$. For any element $a \in A$ define the polynomial function

$$(3) \qquad f^a = \langle T_a, \varepsilon_J \rangle f_I x_I^{\tau(J)}$$

on the dual space G_A^*. We will explain the multiindex notation in formula (3). Here $\varepsilon_J = \varepsilon_{i_j} \ldots \varepsilon_{j_k}$ is the product of the basis elements of the algebra A, and

§4 Duplication of Integrable Analogues

$x_I^{\tau(J)} = x_{i_1}^{\tau(j_1)} \ldots x_{i_k}^{\tau(j_k)}$ is the product of the linear functions on G_A^*. Let, for instance, $f = x_i$ be a linear coordinate function and $a = \varepsilon_j$. Then

$$f^a = x_i^{\varepsilon_j} = \langle T\varepsilon_j, \varepsilon_k \rangle x_i^{\tau(k)} = \langle \varepsilon_{\tau(j)}, \varepsilon_{\tau(k)} \rangle x_i^k = x_i^j.$$

For any function f the function f^{ε_j} will be henceforth also denoted by $f^{(j)}$. Let F be a certain family of functions on G^*. Let us agree to denote the family $\{f^{(j)} : f \in F, \varepsilon_j \in B\}$ by F_B.

THEOREM 4.4.1 (BRAILOV). *Let a set F of independent polynomial functions on the dual space G^* to the Lie algebra G be closed with respect to the Poisson bracket. Then:*

a) The set F_B is closed with respect to the Poisson bracket and for any $f, g \in F$ and $a, b \in A$ there holds the identity

(4) $$\{f^a, g^b\} = \{f, g\}^{ab}$$

b) The set F_B consists of independent functions.

c) The index satisfies the inequality $\operatorname{ind} F_B \leqslant N \cdot \operatorname{ind} F$.

COROLLARY 4.4.1. *Suppose that for the Lie algebra G of index r there exists a set $J_1, \ldots J_r$ of independent invariants. Then $J_1^{(1)}, \ldots J_r^{(N)}$ is a set of independent invariants of the Lie algebra G_A and the index is of the form $\operatorname{ind} G_A = N \cdot r$.*

PROOF: Let g belong to G and l_g be a linear function on G^* such that $l_g(x) = x(g)$ for any $x \in G^*$. Since J_i is an invariant of the algebra G, it follows that for any $g \in G$ we have $\{J_i, l_g\} = 0$. By virtue of Theorem 4.4.1, a) we have $\{J_i^a, l_{g \otimes b}\} = \{J_i, l_g\}^{ab} = 0$. Consequently, $J_i^{(j)}$ is an invariant of the Lie algebra G for any i, j. Independence of the invariants $J_i^{(j)}$ follows from Theorem 4.4.1, b). Independence of the invariants $J_i^{(j)}$ implies validity of the inequality $\operatorname{ind} G_A \geqslant N \operatorname{ind} G$. The inverse inequality follows from Theorem 4.4.1, b).

COROLLARY 4.4.2. *If G is a semisimple Lie algebra and Q is the Killing form then $Q^{(1)}$ is a nondegenerate bilinear symmetric form invariant under all inner automorphisms of the Lie algebra G_A, and therefore under identification of G_A^* and G_A by means of $Q^{(1)}$ the Hamiltonian equation $\dot{x} = \operatorname{ad}_{dH(x)}(x)$ on G_A^* is written on G_A in the Lax form $\dot{X} = [X, dH(X)]$.*

COROLLARY 4.4.3. *If G is a semisimple Lie algebra of rank r then $\operatorname{ind} G_A = N \cdot r$.*

COROLLARY 4.4.4. *The equation $\dot{M} = [M, \Omega]$, where*

$$M, \Omega \in so(n)_A, \quad \Omega = \sum_{i=1}^{N} \Omega_i \otimes \varepsilon_i, \quad M = \sum_{i=1}^{N} (\Omega_i I + I\Omega_i) \otimes \varepsilon_i,$$

is completely integrated by a set of integrals F_B, where F is a complete set of integrals of the Euler equation on the Lie algebra $so(n)$.

PROOF: It suffices to prove that

(5) $$\dim so(n)_A + \operatorname{ind} so(n)_A = \dim F_B + \operatorname{ind} F_B,$$

where $\dim F_B$ is the number of functions in the set F_B. We will prove that both sides of the equality (5) are equal to $q = N(\dim \operatorname{so}(n)) + \operatorname{ind} \operatorname{so}(n))$. The left-hand side of (5) is equal to q by virtue of Corollary 4.4.3. Consider the linear subspace generated by the set of functions F. This linear subspace is also a reductive Lie algebra with respect to the Poisson bracket. Hence, $\operatorname{ind} F_B = N \cdot \operatorname{ind} F$. Since $\dim \operatorname{so}(n) + \operatorname{ind} \operatorname{so}(n) = \dim F + \operatorname{ind} F$, then the right-hand side of (5) is also equal to $N(\dim F + \operatorname{ind} F) = q$.

COROLLARY 4.4.5. *If the index* $\operatorname{ind} G$ *is equal to zero, it follows that* $\operatorname{ind} G_A$ *is also equal to zero.*

Lie algebras of index 0 are sometimes called *Frobenius Lie algebras*. In recent years such Lie algebras have been applied (see [224]) in constructing solutions of the Yang–Baxter equations.

PROOF OF THEOREM 4.4.1: We begin with item a). It is necessary that we can differentiate functions of the form f^a. We have:

$$\frac{\partial f^a}{\partial x_i^{\tau(j)}} = \alpha(a\varepsilon_J) f_I \frac{\partial x_I^{\tau(J)}}{\partial x_i^{\tau(j)}}$$

$$= \sum_{\substack{K \bullet L = I \\ M t P = J}} \alpha(a\varepsilon_J) f_I x_K^{\tau(M)} \frac{\partial x_s^{\tau(t)}}{\partial x_i^{\tau(j)}} x_L^{\tau(P)}$$

$$= \alpha(a\varepsilon_j \varepsilon_M \varepsilon_P) f_{K \bullet L} x_K^{\tau(M)} \frac{\partial x_s}{\partial x_i} x_L^{\tau(P)}$$

(6) $$= \left(\frac{\partial f}{\partial x_i}\right)^{a\varepsilon_j}$$

LEMMA 4.4.1. *For any functions*

$$f = f_I x_I, \qquad g = g_J x_J, \qquad h = h_K x_K$$

and elements $a, b \in A$ *we have*

(7) $$(fgh)^{ab} = f^{aT(\varepsilon_p)} g^{bT(\varepsilon_q)} h^{\varepsilon_p \varepsilon_q}.$$

PROOF: We have:

$$(fgh)^{ab} = \alpha(ab\varepsilon_L \varepsilon_R \varepsilon_S) f_I g_J h_K x_I^{\tau(L)} x_J^{\tau(R)} x_K^{\tau(S)}$$
$$= \langle \varepsilon_p, a\varepsilon_L \rangle \langle \varepsilon_q, b\varepsilon_R \rangle \alpha(\varepsilon_p \varepsilon_q \varepsilon_s) f_I g_J h_K x_I^{\tau(L)} x_J^{\tau(R)} x_K^{\tau(S)}$$
$$= \alpha(a\varepsilon_{\tau(p)} \varepsilon_L) \alpha(b\varepsilon_{\tau(q)} \varepsilon_R) \alpha(\varepsilon_p \varepsilon_q \varepsilon_s) f_I g_J h_K \cdot x_I^{\tau(L)} x_J^{\tau(R)} x_K^{\tau(S)}$$
$$= f^{a\varepsilon_{\tau(p)}} g^{B\varepsilon_{\tau(q)}} h^{\varepsilon_p \varepsilon_q}.$$

Here we have made use of the fact that for any element $c \in A$ we have $c = \langle \varepsilon_p, c \rangle \varepsilon_p$. Indeed,

$$a\varepsilon_L = \langle \varepsilon_p, a\varepsilon_L \rangle \varepsilon_p; \qquad b\varepsilon_R = \langle \varepsilon_q, b\varepsilon_R \rangle \varepsilon_q.$$

Consequently,
$$\alpha(ab\varepsilon_L\varepsilon_R\varepsilon_s) = \alpha(\langle\varepsilon_p, a\varepsilon_L\rangle\langle\varepsilon_q, b\varepsilon_R\rangle\varepsilon_p\varepsilon_q\varepsilon_s)$$
$$= \langle\varepsilon_p, a\varepsilon_L\rangle\langle\varepsilon_q, b\varepsilon_R\rangle\alpha(\varepsilon_p\varepsilon_q\varepsilon_s).$$

Now prove formula (4):
$$\{f^a, g^b\} = \frac{\partial f^a}{\partial x_i^p}\frac{\partial g^b}{\partial x_j^q}\{x_i^p, x_j^q\}$$
$$= \left(\frac{\partial f}{\partial x_i}\right)^{aT(\varepsilon_p)}\left(\frac{\partial g}{\partial x_j}\right)^{bT(\varepsilon_q)}\{x_i, x_j\}^{\varepsilon_p\varepsilon_q}$$
$$= \left(\frac{\partial f}{\partial x_i}\frac{\partial g}{\partial x_j}\{x_i, x_j\}\right)^{ab} = \{f, g\}^{ab}.$$

b) Let the set F consist of r independent functions, $F = (F_1, \ldots, f_r)$. It is necessary to prove independence of $N \cdot r$ functions : $f_1^{(1)}, \ldots, f_r^{(N)}$. Suppose for definiteness and the functions $f_1, \ldots, f_r$ are independent in the first r coordinates. To say it differently,

(8) $$\det\left(\frac{\partial f_j}{\partial x_i}\right) \neq 0$$

at some point $y \in G^*$. For any p, q from 1 to N define an $r \times r$ matrix
$$M_{pq} = \left(\frac{\partial f_j^{(q)}}{\partial x_i^p}\right).$$

Prove that for $p < q$ the matrix M_{pq} consists of zeros. Indeed, let $d(p)$ be less than $d(q)$. Then $d(\tau(p))$ is greater than $d(\tau(q))$ and $\langle\varepsilon_{\tau(q)}, a\varepsilon_{\tau(p)}\rangle$ is equal to zero for any $a \in A$. That is why if $\langle\varepsilon_{\tau(q)}, \varepsilon_J\rangle$ is not equal to zero, then the product $x_I^{\tau(J)}$ of linear functions does not contain x_i^p for any i from 1 to m. Hence, for $d(p) < ed(q)$ either $< T(\varepsilon_q), \varepsilon_J > = 0$ or $\frac{\partial x_I^{\tau(J)}}{\partial x_i^p} = 0$. Therefore, for $d(p) < d(q)$ the matrix M_{pq} is zero. Let $d(p)$ be equal to $d(q)$. Since $\dim A_0 = 1$ then for $p \neq q$ the matrix M_{pq} is also zero. Thus, for $p < q$ the matrix M_{pq} consists of zeros. Consider the case $p = q$. We have

(9) $$(M_{pp})_{ij} = \frac{\partial f_j^{(p)}}{\partial x_i^p} = \left(\frac{\partial x_i}{\partial f_j}\right)^{\varepsilon_p T(\varepsilon_p)} = \left(\frac{\partial f_j}{\partial x_i}\right)^{(N)}$$

Find the expression for the function $g^{(N)}$ for an arbitrary function $g = g_I x_I$. By definition
$$g^{(N)} = g^{\varepsilon_N} = \langle T(\varepsilon_N), \varepsilon_J\rangle \quad g_I x_I^{\tau(J)} = \langle\varepsilon_1, \varepsilon_J\rangle g_I x_I^{\tau(J)}.$$

Since $\langle\varepsilon_1, \varepsilon_J\rangle$ is not equal to zero only when $\varepsilon_J = \varepsilon_1 \ldots \varepsilon_1$, then

(10) $$g^{(N)} = g_{i_1 \ldots i_k} x_{i_1}^N \ldots x_{i_k}^N.$$

Let $M = (M_{pqij})$ be a matrix composed of matrices M_{pq}. The matrix M is a box-triangular, its diagonal containing the boxes

$$M_{pp} = \left(\left(\frac{\partial f_j}{\partial x_i}\right)^{(N)}\right).$$

We have assumed that $y \in G^*$ is a point such that the inequality (8) is fulfilled. By virtue of (9) and (10), at each point $z \in G_A^*$ such that its higher coordinates $x_i^N(z) = x_i(y)$, we have $\det M = \sqcap_p \det M_{pp} \neq 0$. Thus, the functions $f_1^{(1)}, \ldots f_r^{(N)}$ are independent.

c) By definition, ind F_B is the minimal dimension of the Lie algebra $K_x F_B$. Therefore, to prove the inequality ind $F_B \leqslant N$ ind F it suffices to indicate a point $z \in G_A^*$ for which $\dim K_z F_B \leqslant N$ ind F. Let $y \in G^*$, $\dim K_y F =$ ind F and $z \in G_A^*$ be such a point that all its higher coordinates $x_i^N(z)$ are equal to $x_i(y)$. Consider $Nr \times Nr$ matrices $L = (\{f_i^{(p)}, f_j^{(q)}\}(z))$ and $\tilde{L} = (\{f_i^{\epsilon_p}, f_j^{T(\epsilon_q)}\}(z))$. The matrix $\tilde{L}$ contains the boxes $L_{pq} = (\{f_i^{\epsilon_p}, f_j^{T(\epsilon_q)}\}(z))$ which consists of zeros when p is greater than q. Formulae (4), (10) imply that

$$(L_{pp})_{ij} = \{f_i^{\epsilon_p}, f_j^{T(\epsilon_q)}\}(z) = \{f_i, f_j\}^{\epsilon_p T(\epsilon_q)}(z) = \{f_i, f_j\}^{(N)} = \{f_{i,j}\}(y).$$

The matrix $\tilde{L}$ is therefore box-triangular and its diagonal contains equivalent boxes $L_{pp} = (\{f_i, f_j\}(y))$. Hence, rank $L =$ rank $\tilde{L} \geqslant$ rank$(\{f_i, f_j\}(y))$. Thus,

$$\text{ind } F_B \leqslant \dim K_z F_B = Nr - \text{rank } L \leqslant N(r - \text{rank}(\{f_i, f_j\}(y))) = N \text{ ind } F.$$

This completes the proof of Theorem 4.4.1.

4.2 Frobenius Algebras and Extensions of Lie Algebras

4.2.1. Statement of the Problem. Let A be a commutative, associative (finite-dimensional) algebra with unity. Consider a tensor product $G \otimes A$, which is an extension of a Lie algebra G, with the following commutator:

$$[\xi \otimes a, \eta \otimes b] = [\xi, \eta] \otimes ab; \xi, \eta \in G; a, b \in A.$$

It is readily seen that this operation transforms the space $G \otimes A$ into a Lie algebra which we further on denote by G_A.

Fomenko and Trofimov have put the following question: when do complete involutive sets of functions (that is, maximal linear commutative algebras of functions) exist on a dual space G_A^* and what is the way to find them? It turned out that in this problem the leading role is played by Frobenius algebras. But as has been notified by Brailov, the function extension method exploited in this case may be formulated without the assumption that the algebra A possesses Frobenius properties.

4.2.2. The Extension Method of Polynomial Functions. Although the further applications of the extension method refer to functions on dual spaces G^* to the

Lie algebras G, in the present subsection the structure of the Lie algebra on G is of no importance whatsoever. We assume G to be a vector space, A an arbitrary associative, commutative algebra with unity.

Let $\langle\,,\,\rangle$ be an arbitrary nondegenerate bilinear symmetric form on A. Let us agree to denote the dimension of A by N. Two bases (ε_p) and (ε^p) in A are called conjugate if $\langle \varepsilon^p, \varepsilon_q \rangle$ is equal to δ_q^p for any $p, q = 1, \ldots, N$. This immediately implies that

$$a = \sum_p \langle a, \varepsilon_p \rangle \varepsilon^p$$

for each element $a \in A$. Choose an arbitrary coordinate system $x_1, \ldots, x_n$ on G^*. A polynomial function f on G^* will then be written in the form

$$f = \sum_{k=0}^{\infty} \sum_{i_1 \ldots i_k} f_{i_1 \ldots i_k} x_{i_1} \ldots x_{i_k}.$$

The linear functions x_i on G^* can be treated as elements of G. Thus, for any $i = 1, \ldots, n; p = 1, \ldots N$ the product $x_i \otimes \varepsilon^p \in G_A$ is equal to the product $G \otimes A$. The element $x_i \otimes \varepsilon^p$ is a linear function on G_A^*, and treated as a linear function is also denoted by $x_i^p = x_i \otimes \varepsilon^p$. All the functions $x_i^p (i = 1, \ldots, n; p = 1, \ldots, N)$ form the coordinate system on G_A^*, and we can express the arbitrary function G_A^* in terms of x_i^p.

For each polynomial function f on G^* define an A-valued extended function $\tilde{f}$ on G_A^* by the formula

$$\tilde{f} = \sum_{k=0}^{\infty} \sum_{i_1 \ldots i_k} f_{i_1 \ldots i_k} x_{i_1}^{j_1} \ldots x_{i_k}^{j_k} \varepsilon_{j_1} \ldots \varepsilon_{j_k}.$$

We will show that the function $\tilde{f}$ does not depend on the choice of the basis (ε_p). Indeed, let $(\varepsilon_{p'})$ be another basis, $(\varepsilon^{p'})$ a basis conjugate to it, and $(\lambda_p^{p'})$ the transformation matrix from (ε_p) to $(\varepsilon_{p'})$, $\varepsilon_p = \lambda_p^{p'} \varepsilon_{p'}$. (Summation signs over repeated indices are henceforth omitted.) Then

$$\varepsilon^{p'} = \langle \varepsilon^{p'}, \varepsilon_p \rangle \varepsilon^p = \langle \varepsilon^{p'}, \lambda_p^{q'} \varepsilon_{q'} \rangle \varepsilon^p = \lambda_p^{p'} \varepsilon^p,$$

and therefore the sums of the form $x^p \varepsilon_p$ are invariant under the change of th basis because

$$x^{p'} \varepsilon_{p'} = (x \otimes \lambda_p^{p'} \varepsilon_{p'}) = (x \otimes \varepsilon^p) \lambda_p^{p'} \varepsilon_{p'} = x^p \varepsilon_p.$$

Since the right-hand side of the formula defining the function $\tilde{f}$ consists entirely of the product of such kind of sums, the definition of the function $\tilde{f}$ does not depend on concrete choice on the basis of (ε_p).

Now establish the dependence of the function $\tilde{f}$ on the choice of the form $\langle\,,\,\rangle$. Let $\langle\,,\,\rangle'$ be another nondegenerate bilinear symmetric form. Let (ε_p) be a basis in A, (ε^p) and $(\varepsilon^{p'})$ be bases conjugate to (ε_p) with respect to the forms $\langle\,,\,\rangle$ and $\langle\,,\,\rangle'$, respectively. Then passing over from the form $\langle\,,\,\rangle$ to the form $\langle\,,\,\rangle'$ in the definition of the function $\tilde{f}$ reduces to a linear change of its arguments $x_i^p \to x_i^{p'}$, where x_i^p is equal to $x_i \otimes \varepsilon^p$ and $x_i^{p'} = x_i \otimes \varepsilon^{p'}$.

Eventually, we would like to make the last remark on the definition of functions of the form $\tilde{f}$. We will show that of greater importance from the point of view of obtaining interesting examples of functions $\tilde{f}$ are undecomposable algebras A, that is, such algebras which cannot be represented as the direct sum of algebras of smaller dimensions. Indeed, suppose that A is equal to $A_1 \otimes A_2$. We have shown above that the functions $\tilde{f}$ do not depend on the choice of the basis (ε_p), and therefore we may choose the basis (ε_p) in the most convenient way. First choose the basis $(\varepsilon_p)_{p=1,\ldots,s}$ in A_1 and $(\varepsilon_p)_{=s+1,\ldots,N}$ in A_2. The set $(\varepsilon_p) = (\varepsilon_p)_{p=1,\ldots N}$ is obviously the basis in A. For $1 \leqslant p \leqslant s$, $s+1 \leqslant q \leqslant N$ the product $\varepsilon_p \varepsilon_q$ is zero, and therefore in the expression for $\tilde{f}$ there remain only terms in which all the indices are either simultaneously less than $s+1$ or simultaneously greater than s. Hence, $\tilde{f} = \tilde{f}_1 + \tilde{f}_2$, where the function $\tilde{f}_1$ is constructed by the algebra A_1 and the function $\tilde{f}_2$ by the algebra A_2. We may say that in the case of decomposable algebras A the variables entering the function $\tilde{f}$ are separated.

In the formulations of principal theorems on extended functions, instead of A-valued functions $\tilde{f}$ we consider functions f^a of the form $f^a = \langle a, \tilde{f} \rangle$, $a \in A$. In the case $a = \varepsilon^p$ we employ also the notation $f^p = \langle \varepsilon^p, \tilde{f} \rangle$ which will not lead to misunderstanding because the integers p are not elements of A.

THEOREM 4.4.2 (BRAILOV). *For functionally independent functions* $f_1, \ldots, f_r$ *all* Nr *functions* $f_1^1, \ldots, f_r^N$ *are also functionally independent.*

To prove this, we will need three lemmas. First of all we will need the rule for differentiating functions of the form $\tilde{f}$.

LEMMA 4.4.2.
$$\frac{\partial \tilde{f}}{\partial x_i^p} = \varepsilon_p \left(\widetilde{\frac{\partial f}{\partial x_i}} \right).$$

PROOF: We have
$$\frac{\partial \tilde{f}}{\partial x_i^p} = \sum_{k=0}^{\infty} \sum_{i_1,\ldots,i_k} \sum_{s=1}^{k} f_{i_1\ldots i_k} x_{i_1}^{j_1} \ldots x_{i_{s-1}}^{j_{s-1}} \frac{\partial x^{j_s}}{\partial x_i^p}$$
$$\times \tau x_{i_{s+1}}^{j_{s+1}} \ldots x_{i_k}^{j_k} \varepsilon_{j_1} \ldots \varepsilon_{j_k} = \varepsilon_p \left(\widetilde{\frac{\partial f}{\partial x_i}} \right),$$

which implies the lemma.

Let $x \in G^*$ be a point with coordinates $x_1, \ldots, x_n$ and $x^1 \in G_A^*$ a point with coordinates $x_1^1, \ldots, x_n^N$ such that $x_i^1 = x_i$, $x_i^p = 0$ for any $i = 1, \ldots, n$ and $p > 1$.

LEMMA 4.4.3. *Let* (ε_p) *be a basis in* A *such that* ε_1 *is a unit element of the algebra* A. *Then* $\tilde{f}(x^1) = f(x)\varepsilon_1$.

PROOF: In the sum determining the value of the function $\tilde{f}$ at the point x^1 it is only the terms of the form $f_{i_1} \ldots i_k x_{i_1}^1 \ldots x_{i_k}^1 (\varepsilon_1)^k$ that are not equal to zero. Since $(\varepsilon_1)^k = \varepsilon_1$, the lemma follows.

LEMMA 4.4.4.
$$\frac{\partial f^q}{\partial x_i^p}(x^1) = \delta_p^q \left(\widetilde{\frac{\partial f}{\partial x_i}}(x)\right).$$

PROOF: Using Lemmas 4.4.2 and 4.4.3, we find
$$\left(\frac{\partial}{\partial x_i^p} f^q\right)(x^1) = \langle \varepsilon^q, \varepsilon_p \widetilde{\left(\frac{\partial f}{\partial x_i}\right)}(x^1)\rangle = \langle \varepsilon^q, \varepsilon_p \varepsilon_1\rangle \frac{\partial f}{\partial x_i}(x) = \delta_p^q \frac{\partial f}{\partial x_i}(x),$$
as required.

PROOF OF THEOREM 4.4.2: Suppose for definiteness that the functions f_j are independent in the first r coordinates. Then at a certain point $x \in G^*$ the determinant $\det L$ of the matrix
$$L = \left(\frac{\partial f_j}{\partial x_i}(x)\right)$$
is nonzero. Consider the matrix
$$M = \left(\frac{\partial f_j^q}{\partial x_i^p}(x^1)\right)$$
in partial derivatives of the functions f_j^q calculated at the point $x^1 \in G_A^*$ with coordinates $x_i^p = \delta_1^p x_i$, where x_i are coordinates of the point $x \in G^*$ at which $\det L \neq 0$. Lemma 4.4.4 implies that the matrix M is a box-diagonal, and its diagonal contains identical boxes L. Thus, $\det M = (\det L)^N \neq 0$ and all Nr functions $f_1^1, \ldots, f_r^N$ are functionally independent.

4.2.3. Connection of Poisson brackets on dual spaces to Lie algebras G and G_A for Frobenius algebras A. For any Lie algebra G and at least for several algebras A there exists profound connection between Poisson brackets for the dual spaces G^* and G_A^*. This connection was first discovered by Trofimov [135] who proved that when A is the algebra of truncated polynomials, involutivity of functions f and g on G^* entails (for any $a, b \in A$) involutivity of functions f^a and g^b on G_A^*. This result was strengthened by Brailov in two directions. Firstly, in the paper [19] it was proved that instead of algebras of truncated polynomials one can use Poincaré duality algebras and, in particular, the rings of real cohomologies of even dimension of any oriented manifolds. Thus, the class of algebras A, for which there exists connection between the Poisson brackets on G^* and G_A^* was considerably extended. Secondly, in the paper [206], devoted among other things to duplication of systems integrable in "noncommutative' sense, the formula was first obtained. The final result in this direction was established by Le Ngok Tyeuen [278] who proved equivalence between the identity (1) and Frobenius properties of the algebra A. It should however be noted that the original formulations and proofs of Le Ngok Tyeuen exhibit an explicit dependence on the basis (ε_p) chosen in the algebra A in the rather special way. Brailov suggested formulations and proofs invariant with respect to the basis (ε_p). We present them below.

We remind the reader of the definition of the Frobenius algebra. Let $\langle\,,\,\rangle$ be a symmetric bilinear form on the algebra A. The form $\langle\,,\,\rangle$ is called "invariant" if $\langle a, bc\rangle$ is equal to $\langle ab, c\rangle$ for any $a, b, c, \in A$. An algebra A is called a Frobenius algebra if it possesses at least one nondegenerate "invariant" form $\langle\,,\,\rangle$.

THEOREM 4.4.3 (LE NGOK TYEUEN). *Let G be a noncommutative Lie algebra, A an associative commutative algebra with unity and with a nondegenerate bilinear symmetric form $\langle\,,\,\rangle$ which helps to define extended functions f^a on G_A^* for polynomial funcitons f on G^*. Then:*

1) If A is a Frobenius algebra and $\langle\,,\,\rangle$ is an "invariant" form then there holds the equality (1) for any polynomial functions f, g on G^* and for any elements $a, b \in A$.

2) Inversely, if the equality (1) holds for any f, g, a, and b then A is a Frobenius algebra and the nondegenerate "invariant" form on it is the form $\langle a, b \rangle = \langle \varepsilon_1, ab \rangle$, where ε_1 is the unit element of the algebra A.

Notice that the theorem does not assert "invariance" of the original form $\langle\,,\,\rangle$, for which there holds the identity (1). The form $\langle\,,\,\rangle$ may, generally speaking, be not "invariant", as in the example which follows. Let A be an algebra with two generators ε and ρ and with relations

$$\varepsilon^2 = \varepsilon, \qquad \rho^2 = \rho, \qquad \varepsilon\rho = 0.$$

In this algebra there is a unit element $\varepsilon + \rho$. Let the form $\langle\,,\,\rangle$ be given by the equalities:

$$\langle \varepsilon, \varepsilon \rangle = \langle \rho, \rho \rangle = 0, \qquad \langle \varepsilon, \rho \rangle = 1.$$

It can be verified that for the form $\langle\,,\,\rangle$ there holds the equality (1). But the form $\langle\,,\,\rangle$ is not "invariant". Indeed, if it were an "invariant" form, we would have

$$1 = \langle \varepsilon, \rho \rangle + \langle \varepsilon\varepsilon, \rho \rangle = \langle \varepsilon, \varepsilon\rho \rangle = 0.$$

PROOF: Using Lemma 4.4.2, we rewrite the identity (1) as

$$\langle a, \varepsilon_p \widetilde{\left(\frac{\partial f}{\partial x_i}\right)} \rangle \langle b, \varepsilon_q \widetilde{\left(\frac{\partial g}{\partial x_j}\right)} \rangle \langle \varepsilon^p \varepsilon^q, \widetilde{\{x_i, x_j\}} \rangle = \langle a, b \frac{\widetilde{\partial f}}{\partial x_i} \frac{\widetilde{\partial g}}{\partial x_j} \widetilde{\{x_i, x_j\}} \rangle.$$

Since the functions f and g are arbitrary, the relation obtained is equivalent to the identity

(2) $$\langle a, \varepsilon_p \xi \rangle \langle b, \varepsilon_q \eta \rangle \langle \varepsilon^p \varepsilon^q, \varsigma \rangle = \langle ab, \xi\eta\varsigma \rangle$$

where ξ, η, ς are arbitrary elements of the algebra A.

Suppose now that $\langle\,,\,\rangle$ is an "invariant" form. Then

$$\langle a, \varepsilon_p \xi \rangle \langle b, \varepsilon_q \eta \rangle \langle \varepsilon^p \varepsilon^q, \varsigma \rangle = \langle \varepsilon_p, a\xi \rangle \langle \varepsilon_q, b\eta \rangle \langle \varepsilon^p \varepsilon^q, \varsigma \rangle$$
$$= \langle (a\xi)(b\eta), \varsigma \rangle = \langle ab, \xi\eta\varsigma \rangle.$$

This proves assertion 1) of the theorem.

Assume now that there hold the equivalent identities (1) and (2). Then substituting $\eta = \varepsilon_1$ into the identity (2), we obtain

(3) $$\langle a, \varepsilon_p \xi \rangle \langle b, \varepsilon_q \rangle \langle \varepsilon^p \varepsilon^q, \varsigma \rangle = \langle a, \varepsilon_p \xi \rangle \langle \varepsilon^p b, \varsigma \rangle = \langle ab, \xi\varsigma \rangle.$$

If now we substitute $a + \varsigma = \varepsilon_1$, $\xi = \varepsilon_i$, $b = \varepsilon^j$ into (3), we derive

(4) $\qquad \langle \varepsilon_1, \varepsilon_i \varepsilon_p \rangle \langle \varepsilon^p \varepsilon^j, \varepsilon_1 \rangle = \langle \varepsilon^j, \varepsilon_i \rangle = \delta_i^j.$

Define the matrices (λ_{ij}) and (λ^{ij}) by formulae $\lambda_{ij} = \langle \varepsilon_1, \varepsilon_i \varepsilon_j \rangle$ and $\lambda^{ij} = \langle e_1, \varepsilon^i \varepsilon^j \rangle$. From (4) it follows that the product of the matrices (λ_{ij}) and (λ^{ij}) is the identity matrix. Consequently, both matrices (λ_{ij}) and (λ^{ij}) are nondegenerate. Hence, the bilinear form $\langle a, b \rangle' = \langle \varepsilon_1, ab \rangle$ is nondegenerate. The form $\langle , \rangle'$ is "invariant" because $\langle ab, c \rangle' = \langle \varepsilon_1, abc \rangle = \langle a, bc \rangle'$. This implies assertion 2) and completes the proof of the theorem.

4.2.4. The index of the Lie algebra G_A. The definition of the index $\operatorname{ind} G$ for any Lie algebra G is given in the book by Dixmier [204]. The importance of the index in the theory of integrable systems on Lie algebras is explained by the fact that the index of a Lie algebra coincides with codimension of the orbit of the coadjoint action of general position in G^*. We recall the reader that a set of independent commuting (by the Poisson bracket) functions $f_1, \ldots, f_m$ on G^* is called complete involutive if $m = 1/2(\dim G + \operatorname{ind} G)$. Such a definition is motivated by the fact that the Euler equation possessing a complete involutive set of integrals is completely Liouville integrable when restricted to the orbit of general position in G^*. To check completeness of the sets of commuting functions on G_A^* it is necessary to know the index of the Lie algebra G_A.

THEOREM 4.4.4 (BRAILOV). *For any Lie algebra G and for any commutative associative Frobenius algebra A, of dimension N, with unity the index $\operatorname{ind} G_A$ is equal to $N \operatorname{ind} G$.*

In view of the importance of Theorem 4.4.4 we make a short digression devoted to the history of the question. It is readily seen that it is suffices to prove Theorem 4.4.4 only for irreducible algebras A, that is, for such algebras which can be represented neither in the form $A = A_1 \oplus A_2$ nor in the form $A = A_1 \otimes A_2$, where A_1, A_2 are algebras of smaller dimension. That is why, when enumerating the results obtained we will restrict ourselves to mention only irreducible algebras A. The first work in this direction, which is important for us, is the paper by Takiff [129]. This paper completely describes the ring of invariants for a semisimple Lie algebra G and for an algebra $A = k[\varepsilon]/(\varepsilon^2)$. In particular, for such G and A there holds the equality $\operatorname{ind} G_A = 2 \operatorname{ind} G$. In the case $A = k[\varepsilon]/(\varepsilon^2)$ the Lie algebra G_A is a semidirect extension of G via the adjoint representation. Therefore, applying the Rais'es formula [205] for the index of semidirect extension, we derive the equality $\operatorname{ind} G_A = 2 \operatorname{ind} G$ already for any Lie algebra G. Subsequent results obtained by Trofimov and Brailov enabled Theorem 4.4.4 to be proved for Lie algebras G possessing a sufficient amount of functionally independent polynomial invariants $J_1, \ldots, J_r (r = \operatorname{ind} G)$ in the case where $A = k[\varepsilon]/(\varepsilon^n)$ is an algebra of truncated polynomials [135] and where A is an algebra with Poincaré duality [19]. In the presence of invariants $J_1, \ldots, J_r$, Theorem 4.4.4 is proved by Le Ngok Tyeuen [278] also for Frobenius algebras A. Elashvili [279] has formulated Theorem 4.4.4 in the general case of the hypothesis.

PROOF: Let $\xi \in G^*$ be a regular element, that is, such an element that an orbit passing through it has maximal dimension. Then for a set x_i of linear coordinate functions on G^* the rank of the matrix $M = (\{x_i, x_j\}(\xi))$ is equal to $n - r$, where $n = \dim G$, $r = \operatorname{ind} G$. Let x_i^p be coordinates on the dual space G_A^*. Fix a point ξ^1 on G_A^* by setting its coordinates x_i^p equal to $x_i^p = \delta_1^p x_i$, where x_i are coordinates of the point ξ. All the linear functions of the form $x_j^{[q]} = x_j^{\varepsilon_q}$ on G_A^* form a coordinate system, which is generally speaking distinct from (x_i^p). But the ranks of the matrices $L = (\{x_i^p, x_j^q\}(\xi^1))$ and $L' = (\{x_i^p, x_j^{[q]}\}(\xi^1)$ coincide because one matrix is obtained from the other through a nondegenerate linear transformation. Applying Theorem 4.4.3 and Lemma 4.4.3 we come to

$$(\{x_i^p, x_j^{[q]}\}(\xi^1) = \{x_i, x_j\}^{\varepsilon_p \varepsilon_q}(\varepsilon^1) = \langle \varepsilon_p \varepsilon^q, \varepsilon_1 \rangle \{x_i, x_j\}(\xi) = \delta_p^q \{x_i, x_j\}(\xi).$$

The matrix L' is therefore box-diagonal and its diagonal contains identical boxes coinciding with the matrix M. Hence $\operatorname{rank} L = \operatorname{rank} L' = N \operatorname{rank} M = N(n - r)$. Since $\operatorname{rank} L$ is exactly the dimension of such an orbit of the coadjoint action which passes ξ^1, we arrive at the estimate $\operatorname{ind} G_A \leqslant Nr$. For obtaining the inequality $\operatorname{ind} G_A \geqslant Nr$ we suppose that on G^* there exist r functionally independent invariants $J_i (i = 1, \ldots, r)$, that is, polynomial functions constant on the orbits. Then the extended functions $J_i^p (p = 1, \ldots, N)$ will be invariant functions on G_A^*. Indeed, invariance of any function J_i^p is equivalent to the fact that it commutes in the sense of Poisson bracket with all coordinate functions x_j^q. We have

$$\{J_i^p, x_j^q\} = \{J_i, x_j\}^{\varepsilon^p \varepsilon^q} = 0$$

because J_i is an invariant, and therefore J_i commutes with all coordinate functions x_j. Thus J_i^p is an invariant. By Theorem 4.4.2, all invariants $J_i^p (i = 1, \ldots, r; p = 1, \ldots N)$ are functionally independent. Having Nr independent invariants, we are led to the inequality $\operatorname{ind} G_A \geqslant Nr$. The inequality $\operatorname{ind} G_A \leqslant Nr$ has already been obtained above. Thus, if the orbits are polynomially separable then $\operatorname{ind} G_A = N \operatorname{ind} G$, and the theorem follows. Unfortunately, r functionally independent polynomial invariants do not exist for every Lie algebra G of index r. Nonetheless, in this general case all of the presented scheme of reasoning is preserved if instead of polynomial invariants one uses local analytic invariants.

Let ξ be a certain point in G^* with coordinates $\xi_1, \ldots, \xi_n$. Set the coordinates x_i', centered at the point ξ, by the formula $x_i' = x_i - \xi_i$, $i = 1, \ldots, n$. Functions f analytic at the point ξ can be written in the form of series

$$f = \sum_{k=0}^{\infty} f_{i_1 \ldots i_k} x_{i_1}' \ldots x_{i_k}' \stackrel{\text{def}}{=\!=} f_I x_I',$$

where

$$f_I = f_{i_1 \ldots i_k}, \quad x_I' = x_{i_1}' \ldots x_{i_k}'.$$

If the series $f_I x_I'$ converges in a ball $B_\delta(\xi)$ of radius δ centered at the point ξ and the function $f(x) = f_I x_I'$ is constant on the connectedness components of the intersections $O \cap B_\delta(\xi)$ of all the orbits O of the coadjoint action with the ball $B_\delta(\xi)$, then such a function is called a local invariant at the point ξ. Since the coadjoint action is analytic, it follows that for any regular point ξ there exist analytic

§4   *Duplication of Integrable Analogues*   243

functions $f_1, \ldots, f_r$ which are functionally independent and are local invariants at the point ξ. Define the extensions of analytic functions by the formula which has been used to define the extensions of polynomial functions, but change the usual coordinates x_i (that is, zero-centered coordinates) by the coordinates x'_i centered at the point ξ.

It can be easily verified that Theorems 4.4.2 and 4.4.3 are valid for the extensions of analytic functions because all the arguments (including Lemmas 4.4.2, 4.4.3, and 4.4.4) are preserved when we proceed from polynomial functions to analytic ones if instead of the coordinates x_i we use the coordinates x'_i centered at the point in the neighbourhood of which these analytic functions are defined. The extensions of the functions $f_1^1, \ldots, f_r^N$ (local invariants $f_1, \ldots, f_r$) defined in the neighbourhood of the point $\xi^1 \in G_A^*$ with coordinates $x_i^p(\xi^1) = \delta_1^p \xi_i$ are therefore funcionally independent and in involution with all the functions.

Thus, for any $i = 1, \ldots, r$ and any $p = 1, \ldots, N$ the function f_i^p is a local invariant at the point ξ^1.

Regular elements are everywhere dense, for which reason in any arbitrarily small neighbourhood of the point ξ^1 there exists a regular point η. The codimension of the orbit passing through η is not less than Nr because in the neighbourhood of ξ^1 there exist Nr local invariants. From this we obtain $\operatorname{ind} G_A \geqslant Nr$. We have already proved the converse inequality, and therefore in the very general case $\operatorname{ind} G_A$ is also equal to Nr. This completes the proof of the theorem.

4.2.5. Construction of Complete Involutive Sets by Means of Frobenius Algebras. The results obtained above make it possible to prove the following theorem which is immediately related to the problems of Hamiltonian geometry and mechanics.

THEOREM 4.4.5 (BRAILOV). *If A is a commutative, associative, Frobenius algebra with unity and $f_1, \ldots, f_m$ a complete involutive set of independent functions on G^*, then the set of extended functions $f_1^1, \ldots, f_m^N$ on G_A^* is also a complete involutive set.*

PROOF: Since $f_1, \ldots, f_m$ is a complete set then $m = \frac{1}{2}(\dim G + \operatorname{ind} G)$. By Theorem 4.4.2, the set $f_1^1, \ldots, f_m^N$ consists of independent functions. These functions are pairwise involutive by virtue of Theorem 4.4.3. By Theorem 4.4.4, $\operatorname{ind} G_A$ is equal to $N \operatorname{ind} G$. Therefore,

$$Nm = \frac{1}{2}N(\dim G + \operatorname{ind} G) = \frac{1}{2}(\dim G_A + \operatorname{ind} G_A),$$

and $f_1^1, \ldots, f_m^N$ is a complete involutive set of functions. This implies the theorem.

4.3 Maximal Linear Commutative Algebras of Functions on Contractions of Lie Algebras

Let $G = H \oplus V$ be a decomposition of a Lie algebra into the direct sum of linear subspaces possessing the property:

$$[H, H] \subset H, \quad [H, V] \subset V, \quad [V, V] \subset H.$$

In other words, G is a symmetrically graduated Lie algebra. Consider an algebra $\tilde{G} = H \oplus \varepsilon V$, where $\varepsilon^2 = 0$.

The commutator in $\tilde{G}$ is set as follows:

$$[h_1 + \varepsilon v_1, h_2 + \varepsilon v_2] = [h_1, h_2] + \varepsilon([h_1, v_2] + [h_2, v_1]) \in H \oplus \varepsilon V = G.$$

Let us agree to call the algebra $\tilde{G}$ the *contraction of the Lie algebra G*. Consider the decomposition of the dual space $G^* = H^* \oplus V^*$, where $V^* = H^\perp$, $H^* = V^\perp$. The algebras G and $\tilde{G}$ are isomorphic as linear spaces, and therefore G^* and $\tilde{G}^*$ are identified.

Let f be an arbitrary smooth function on G^*. At a fixed point, its differential df can be regarded as an element of one of the algebras G or $\tilde{G}$. If $df = f_h + f_v \in H \oplus V$ then $\widetilde{df} = f_h + \varepsilon f_v \in H \oplus \varepsilon V$ and vice versa. Thus, two Poisson brackets

$$\{f, g\}(x) = \langle x, [df, dg] \rangle$$

and

$$\widetilde{\{f, g\}}(x) = \langle x, [\widetilde{df}, \widetilde{dg}] \rangle$$

are given on the space G^*.

We will present the result obtained by Reyman [117] needed in what follows.

THEOREM 4.4.6 (REYMAN). *Let f and g be invariants of a Lie algebra G. Then for any $\lambda, \mu \in \mathbb{R}$, $a \in V^*$ the functions*

$$f^{\lambda, a}(h + v) = f(v + \lambda h + \lambda^2 a)$$

and

$$g^{\mu, a}(h + v) = g(v + \mu h + \mu^2 a),$$

where $h \in H^*$, $v \in V^*$ are involution with respect to the Poisson bracket on $\tilde{G}^*$, that is,

$$\widetilde{\{f^{\lambda, a}, g^{\mu, a}\}} = 0.$$

If $f_1, \ldots, f_s$ is a complete set of invariants of the algebra G then the family of functions

$$f_i^{\lambda, a}, \ i = 1, \ldots, s; \quad \lambda \in \mathbb{R}$$

is involutive on $\tilde{G}^*$.

A similar assertion for an algebra of the type $\Omega(G) = G \oplus_{\text{ad}} G$ can be deduced with the help of the methods used by Adler, van Moerbeke, and Ratiu [4], [118]. The involutive set

$$f_i^{\lambda, a}, \ i = 1, \ldots, s, \lambda \in \mathbb{R}$$

is not, generally speaking, complete. Calculations show that it will not be complete, for instance, in the case of the Lie algebra

$$\tilde{G} = \text{so}(4) \oplus_\rho \mathbb{R}^4,$$

where ρ is a standard representation of the algebra $\text{so}(4)$ on $\mathbb{R}^4$. Here $G = \text{so}(5)$.

§4. Duplication of Integrable Analogues

The case of the Lie algebra $\Omega(G)$. Let G be a semisimple Lie algebra. Examine the tensor extension of the algebra G of the form

$$\Omega(G) = G \otimes \mathbb{R}[x]/_{(x^2)} = G \oplus \varepsilon G, \quad \varepsilon^2 = 0;$$
$$G^{\mathbb{C}} = G \otimes \mathbb{C} = G \oplus iG, \quad i^2 = 0.$$

It is clear that the Lie algebra $\Omega(G)$ is contraction of the Lie algebra $G^{\mathbb{C}}$. Identify $\Omega^*(G)$ with $\Omega(G)$ and $G^{\mathbb{C}}$ with $(G^{\mathbb{C}})^*$ by means of nondegenerate scalar products

$$\langle x_1 + \varepsilon y_1, x_2 + \varepsilon y_2 \rangle = (x_1, x_2) - (y_1, y_2),$$
$$\langle x_1 + i y_1, x_2 + i y_2 \rangle = (x_1, x_2) - (y_1, y_2),$$

where $(\,,\,)$ is the Killing form on G. Consider the invariants of the coadjoint representation of the algebra $G^{\mathbb{C}}$, namely: $f_1, \ldots, f_s$. As above, for any $a \in G$, for any $\lambda \in \mathbb{R}$ consider functions on $\Omega^*(G)$ of the form

$$f_j^{\lambda,a}(x + \varepsilon y) = f_j(\lambda x + i(y + \lambda^2 a)).$$

By virtue of Theorem 4.4.6, for a fixed vector $a \in G$ these functions mutually commute with respect to the canonical Kirillov bracket on $\Omega^*(G)$. The theorem which establishes completeness of the set of such functions in the case of the Lie algebra $\Omega(G)$.

THEOREM 4.4.7 (BOLSINOV). *The functions*

$$f_j^{\lambda,a}, \; i = 1, \ldots, s, \lambda \in \mathbb{R}$$

form a complete involutive family of functions on $\Omega^*(G)$ for any element $a \in G$ of general position.

PROOF: First point out an explicit relationship between the invariants of the algebra G and the invariants of the algebra $G^{\mathbb{C}}$. Let $h_1, \ldots, h_m$ be a complete independent set of invariants of the algebra G where $(m = \mathrm{ind}\, G)$. By virtue of semisimplicity of G we may assume that $h_1, \ldots, h_m$ are homogeneous polynomials. To obtain a complete set of invariants of the Lie algebra $G^{\mathbb{C}}$, one should consider the same polynomials, but already of the complex variables, and then take their real and imaginary parts. More precisely, if $z = x + iy \in G \oplus iG \approx (G^{\mathbb{C}})^*$ then the functions

$$f_\alpha(z) = \operatorname{Re} h_\alpha(z), \quad 1 \leqslant \alpha \leqslant m, \quad z = (z_1, \ldots, z_n);$$
$$f_{\beta+m}(z) = \operatorname{Im} h_\beta(z), \quad 1 \leqslant \beta \leqslant m$$

are functionally independent and form a complete set of invariants of the Lie algebra $G^{\mathbb{C}}$, in particular,

$$s = 2m = \mathrm{ind}_{\mathbb{R}}(G^{\mathbb{C}}).$$

For the sake of convenience consider somewhat different invariants, namely,

$$f_j(z) = \operatorname{Re}\left(h_j(z) \cdot i^{-\deg h_j}\right),$$
$$f_{m+j}(z) = \operatorname{Re}\left(h_j(z) \cdot i^{-\deg h_j + 1}\right),$$

where $\deg h_j$ is the degree of the homogeneous polynomial h_j. It is clear that the functions f_j and f_{m+j} are indeed the real and imaginary parts of the function h_j, but may be with other signs.

To prove completeness of the family $f_i^{\lambda,a}$, it suffices to show that the differential of the functions $f_i^{\lambda,a}$ generate a subspace of dimension

$$\frac{1}{2}(\dim \Omega(G) + \operatorname{ind} \Omega(G))$$

at a certain point

$$x + \varepsilon y \in G \oplus \varepsilon G \approx \Omega^*(G).$$

The idea of the proof consists in the choice of the point $x + \varepsilon y$ so as to make the explicit form of these differentials be much simpler. Then the problem turns out to reduce to the result of Fomenko and Mischenko concerning completeness of the involutive family of shifts of invariants in the case of a semisimple Lie algebra.

Consider the differentials at the point $x = 0$, $y = y_0$. A concrete choice of the element y_0 is made later on. We do not distinguish between the functions on the spaces $\Omega^*(G)$ and $(G^{\mathbb{C}})^*$ assuming

$$f_j^{\lambda,a}(x + iy) = f_j^{\lambda,a}(x + \varepsilon y).$$

LEMMA 4.4.5. *The differentials of the functions $f_j^{\lambda,a}$ at the point $0 + iy_0 \in G \oplus iG$ are of the following explicit form:*

$$df_j^{\lambda,a} = i dh_j(y_0 + \lambda^2 a) \in iG, \qquad df_{m+j}^{\lambda,a} = \lambda dh_j(y_0 + \lambda^2 a) \in G.$$

The proof reduces to a direct (although cumbersome) calculation which we leave to the reader.

Let us go back to the proof of Theorem 4.4.7. Lemma 4.4.5 implies that the differentials of the functions $f_j^{\lambda,a}$ (for $\lambda \neq 0$, $j = 1, \ldots, 2m$) at the point $0 + iy_0 \in G \oplus iG$ generate a subspace of the form

$$L = \lambda A + iA = A + iA \subset G \oplus iG,$$

where $A \subset G$ is the subspace generated by the differentials of the shifts of invariants of the Lie algebra G at the point y_0. If the algebra G were a complex Lie algebra then by virtue of regularity of $a \in G$ one could choose y_0 with the result that

$$\dim A = \frac{1}{2}(\dim G + \operatorname{ind} G).$$

This statement is a reformulation of the theorem of Fomenko and Mischenko on completeness of the family of shifts of invariants in the case of a semisimple complex Lie algebra. See [91]. It can be easily show that the *shifts of invariants of a real semisimple Lie algebra also form a complete involutive family of functions.*

LEMMA 4.4.6. *Let G be a real semisimple Lie algebra, $f_1, \ldots, f_m$ a complete set of invariants of the algebra G, and $a \in G$ a regular element (of general position). Then the functions*

$$f_j^\lambda(x) = f_i(x + \lambda a), \; i = 1, \ldots, m, \lambda \in \mathbf{R}$$

form a complete involutive set of functions on G^*.

PROOF: Consider a complexification $G^{\mathbf{C}}$ of the algebra G. In doing so, we treat $G^{\mathbf{C}}$ as a complex Lie algebra. Note that we have viewed it above as a real algebra. Obviously,

$$\dim_{\mathbf{R}} G = \dim_{\mathbf{C}} G^{\mathbf{C}}$$

and

$$\mathrm{ind}_{\mathbf{R}} G = \mathrm{ind}_{\mathbf{C}} G^{\mathbf{C}}.$$

Regularity of the element $a \in G$ causes regularity of this element in the complex algebra into which the initial algebra is embedded as a real subspace. Therefore, the shifts of complex invariants by the element a form complete family on $(G^{\mathbf{C}})^*$, that is, $h_i(z + \lambda a)$, where $h_1, \ldots, h_m$ are complex invaiants of the algebra $G^{\mathbf{C}}$. Let $e_1, \ldots, e_k$ be the basis in G. We will regard it also as the basis in $G^{\mathbf{C}}$. In this case the invariants $h_1, \ldots, h_m$ of the algebra $G^{\mathbf{C}}$ can be chosen as polynomials with real coefficients. This is readily proved by singling out the real and imaginary parts of the polynomials. Let $h_1, \ldots, h_m$ have real coefficients. Then restriction of these polynomials to G are real invariant polynomials. It is also obvious that the shifts of the polynomials $h_1, \ldots, h_m$ and $h_1|_G, \ldots, h_m|_G$ by the element $a \in G$ are one and the same polynomials, but in the first case of complex and in the second case of real variables. From the results of [89], [91], [92] it follows that from the family $h_j^\lambda(z) = h_j(z + \lambda a)$ one can choose r functionally independent polynomials $g_1, \ldots, g_r$, where $r = \frac{1}{2}(\dim G + \mathrm{ind}\, G)$. The polynomials $g_1, \ldots, g_r$ restricted to the real subspace G are stated to remain functionally independent. Indeed, independence of $g_1, \ldots, g_r$ on $G^{\mathbf{C}}$ means that the rank of the matrix

$$A_z = \left(\frac{\partial g_k}{\partial z_j}\right)$$

is equal to r almost everywhere on $G^{\mathbf{C}}$. It is required to show that the rank of the matrix

$$A_x = \left(\frac{\partial g_k|_G}{\partial x_j}\right)$$

is also equal to r. The elements of the matrices A_z and A_x are one and the same polynomials (in the first case of $z_1, \ldots, z_k$, in the second case of $x_1, \ldots x_k$). From this it is immediate that the ranks of the matrices actually coincide. This proves the lemma.

Going back to the proof of Theorem 4.4.7 we have: there exists an element y_0 such that $\dim A = r = \frac{1}{2}(\dim G + \mathrm{ind}\, G)$, whence

$$\dim L = \dim(A \oplus iA) = 2 \dim A = \dim G + \mathrm{ind}\, G$$
$$= \frac{1}{2}\left(\dim \Omega(G) + \mathrm{ind}\, \Omega(G)\right).$$

Here we have employed the formula $\mathrm{ind}(G \otimes R[x]/(x^2)) = \mathrm{ind}\, G \cdot \dim(\mathbb{R}[x]/(x^2)) = 2\,\mathrm{ind}\, G$ (see, for instance, [135]). This proves the theorem.

Trofimov has proposed an algorithm for constructing functions in involution on tensor extensions $G \otimes A$ (see the previous subsections). In particular, he has constructed complete involutive sets on $\Omega^*(G)$. The family of functions in involution, constructed by Bolsinov in Theorem 4.4.7, *differs from the family* constructed in the paper [135]. In the case of Theorem 4.4.7, the functions $f_j^{\lambda,a}$ are inhomogeneous, this inhomogeneity being essential in the sense that the homogeneous components of the functions $f_i^{\lambda,a}$ do not already form an involutive family.

The Euler equations on the algebra $\Omega^*(G)$. Let G be a semisimple Lie algebra, as above. Recall the construction of the many-parameter family of operators $\varphi_{abD} : G \to G$ which was proposed in the paper [91] and is a *particular case of the general scheme of sections operators* deduced by the present author. We will describe this constructions for the case of an arbitrary real semisimple Lie algebra. Let H be the Cartan subalgebra of the Lie algebra G and $H^\perp$ an orthogonal complement of H with respect to the Killing form. For any element $a \in H$ the subspace $H^\perp$ is invariant under ad_a, and if a is an element of general position then $\mathrm{ad}_a : H^\perp \to H^\perp$ is an isomorphism. Let

$$x = x_1 + x_2 \in G,$$

where

$$x_1 \in H^\perp, \qquad x_2 \in H.$$

Let a, b belong to H. Then the operator

$$\varphi_{abD}(x) = \mathrm{ad}_a^{-1}\, \mathrm{ad}_b\, x_1 + D x_2,$$

where $D : H \to H$ is an arbitrary self-conjugate operator, is well defined. It is readily seen that the operator φ_{abD} is self-conjugate.

Consider a Hamiltonian

$$F(x + \varepsilon y) = \frac{1}{2}\langle \varphi_{abD}(x), x\rangle - \langle b, y\rangle$$

on

$$\Omega^*(G) \approx \Omega(G).$$

Recall that by the Euler equations on the dual space of an arbitrary Lie algebra L we mean the equations:

$$\dot z = \mathrm{ad}_{dF(z)}(z), \qquad z \in L^*, \qquad dF \in (L^*)^* = L.$$

When $\Omega^*(G)$ and $\Omega(G)$ are identified then the Euler equations have in our case the following explicit form

(1)
$$\begin{aligned}\dot x &= [\varphi_{abD}(x), x] - [b, y] \\ \dot y &= [\varphi_{abD}(x), y]\end{aligned}$$

where $x, y, \varphi_{abD}(x)$, $b \in G$. In the case $G = so(3)$, the system (1) acquires the form of the *classical Lagrange case of motion of a rigid body in the gravity field*. Indeed, the *classical equations* have the form

$$\dot M = [M, \omega] + [mgr, \gamma], \quad \dot\gamma = [\gamma, \omega],$$

where M is the moment of inertia of the rigid body, r coordinates of the centre of gravity, γ the direction of the gravitational force, ω the angular velocity. All the vectors are viewed in the coordinate system rigidly connected with the rigid body. As usual, we identify the algebra $so(3)$ with $\mathbf{R}^3$, the commutator transforming into the vector product of vectors.

As the Cartan subalgebra H in the algebra $so(3)$ consider the linear subspace generated by an element $e_{23} \in so(3)$. Suppose that

$$a = \alpha e_{23} \in H, \quad b = \beta e_{23} \in H, \quad \alpha, \beta \neq 0$$

and the operator $D : H \to H$ is multiplication by the number $D(e_{23}) = de_{23}$. Then $\varphi_{abD} : so(3) \to so(3)$ is a diagonal operator in the standard basis e_{12}, e_{13}, e_{23} with eigenvalues $-\frac{\beta}{\alpha}, -\frac{\beta}{\alpha}, d$. Physically this means that the inertia ellipsoid has the axis of symmetry passing through e_{23}. The condition $mgr = -b = -\beta e_{23}$ implies that the centre of gravity of the rigid body is located on the axis of symmetry. Thus, the classical Lagrange case has turned out included into the family of multidimensional Hamiltonian systems on dual spaces to Lie algebras of the type $\Omega(G)$.

THEOREM 4.4.8 (BOLSINOV). *The system of Euler equations (1) is completely Liouville integrable on orbits of general position of the coadjoint representation* $\mathrm{Ad}^*(\Omega(G))$.

PROOF: The proof consists in verification of the fact that the functions $f_j^{\lambda,a}(x+\varepsilon y)$ are integrals of the system (1). Their involutivity follows from Theorem 4.4.6 and completeness from Theorem 4.4.7.

Let f be invariant in an algebra G^C. Show that for any λ the function

$$f^{\lambda,a}(x + \varepsilon y) = f(\lambda x + i(y + \lambda^2 a))$$

is a first integral of the system (1). Calculate the Poisson bracket of the functions $f^{\lambda,a}$ and F on $\Omega^*(G)$. Let $z = \lambda x + i(y + \lambda^2 a)$. Then

$$\{\widetilde{f^{\lambda,a}}, F\} = \langle x, [\lambda f_x(z), \varphi_{abD}(x)]\rangle \\ - \langle y, [\lambda f_x(z), b]\rangle - \langle y, [f_y(z), \varphi_{abD}(x)]\rangle.$$

Since f is the invariant of an algebra then there holds the identity

$$\mathrm{ad}^*_{df(z)} z = \mathrm{ad}_{df(z)} z \equiv 0.$$

We make use of the fact that the scalar product introduced on G^C is invariant, and therefore $\mathrm{ad}^* = \mathrm{ad}$. Let us analyze this identity in more detail. We have

$$\mathrm{ad}_{df(z)} z = [f_x(z) + if_y(z), \quad \lambda x + i(\lambda^2 a + y)] = 0.$$

Consequently

$$[f_x(z), \lambda x] - [f_y(z), \lambda^2 a + y] = 0;$$
$$[f_x(z), \lambda^2 a + y] + [f_y(z), \lambda x] = 0.$$

Employ the derived relations:

$$\begin{aligned}\langle y, [\lambda f_x, b]\rangle &= \langle [y, \lambda f_x(z)], b\rangle \\ &= \langle [y + \lambda^2 a, \lambda f_x(z)], b\rangle - \langle [\lambda^2 a, \lambda f_x(z)], b\rangle \\ &= \langle -[\lambda^2 x, f_y(z)], b\rangle - \langle \lambda f_x(z), [b, \lambda^2 a]\rangle \\ &= \langle f_y(z), [\lambda^2 x, b]\rangle.\end{aligned}$$

We have used the condition that $[a, b] = 0$. Similarly:

$$\langle x, [\lambda f_x(z), \varphi_{abD}(x)]\rangle = \langle \lambda^2 a + y, [f_y(z), \varphi_{abD}(x)]\rangle.$$

Continue the calculation of the bracket $\{\widetilde{f^{\lambda, a}}, F\}$ substituting the expression obtained:

$$\begin{aligned}\{f^{\lambda,a}, F\} &= \langle \lambda^2 a + y, [f_y(z), \varphi_{abD}(x)]\rangle - \langle f_y(z), [\lambda^2 x, b]\rangle - \langle y, [f_y(z), \varphi_{abD}(x)]\rangle \\ &= \langle \lambda^2 a, [f_y(z), \varphi_{abD}(x)]\rangle - \langle f_y(z), [\lambda^2 x, b]\rangle \\ &= \langle [\varphi_{abD}(x), \lambda^2 a], f_y(z)\rangle - \langle [\lambda^2 x, b], f_y(z)\rangle \\ &= \langle [\operatorname{ad}_a^{-1} \operatorname{ad}_b(x_1) + D(x_2), \lambda^2 a] - [\lambda^2(x_1 + x_2), b] f_y(z)\rangle \\ &= \langle -\lambda^2 \operatorname{ad}_a \operatorname{ad}_a^{-1} \operatorname{ad}_b(x_1) + \lambda^2 [D(x_2), a] - \lambda^2 [x_1, b] - \lambda^2 [x_2, b], f_y(z)\rangle \\ &= \langle -\lambda^2 [b, x_1] - \lambda^2 [x_1, b], f_y(z)\rangle \\ &= 0.\end{aligned}$$

We have made use of the fact that a, b, x_2, and $D(x_2)$ all belong to H and, therefore, commute with each other. This proves the theorem.

§5 The Orbit Method in Hamiltonian Mechanics and Spin Dynamics of Superfluid Helium-3

Helium is one of the most widespread elements in the universe. At atmospheric pressure and room temperature this is a colourless transparent gas much lighter than air. It has several isotopes; the most familiar among them is helium-4 with atomic weight 4 and helium-3 with atomic weight 3. Huge amounts of helium-4 are contained in stars, on the Earth it is encountered in natural gases and several minerals; its total store is not large, it amounts to about $7.10^{10} m^3$. Helium-3 is obtained in laboratory as a product of nuclear reactions.

In 1938 Kapitsa [257] discovered that at a temperature of $217°$ K (recall that 0 degrees centigrade corresponds to about $273°$ K) helium-4 transforms into a superfluid state characterized by zero viscosity and the ability to flow without resistance through thin slits and capillaries. The sequential analysis of helium-4 for superfluidity carried out by Landau [258], [259] was the basis of the modern quantum liquid theory.

The low-temperature properties of helium-3 have been investigated since early 1950s. At atmospheric pressure helium-3 remains liquid at an arbitrarily low temperature; in order that it become solid one should apply pressure of about 35 atmospheres. Although helium-3 thus has a tendency to remain liquid, it becomes superfluid with great difficulty. In 1959 Pitayevsky predicted [260] that at sufficiently low temperatures helium-3, however, transforms into a superfluid state. In 1972, after great experimental difficulties were overcome, Osheroff, Richardson, and Lee experimentally discovered superfluidity of helium-3 at a temperature of about 10^{-3} K [261]. Shortly after, Richardson and Lee [262] revealed that depending on temperature and pressure helium-3 may be in various superfluid states or phases which had been predicted before in the theoretical papers by Anderson and Morel [263] and Balian and Werthamer [264].

Superfluid phases of helium-3 possess many unusual properties, in particular, of great interest are the associated magnetic phenomena caused by the fact that the atomic nucleus of helium-3 consists of two protons and one neutron (in helium-4 of two protons and two neutrons), due to which it has a nonzero magnetic moment. The presence of such elementary "magnetics" leads to the fact that a macroscopic sample of helium-3 can be magnetized. If one ignores the difference in the structure of different parts of a sample (this approximation is often quite lawful), then magnetization can be described by one vector common for the whole sample. The study of magnetization dynamics is essential for understanding superfluid state, in particular, for identification of superfluid phases. For this, one usually employs the so-called hydrodynamic approximation which assumes the characteristic time of the process to be much larger than a certain time τ determined by the macroscopic structure of the body. For helium-3, for not very low temperatures τ of the order of $10^{-7} - 10^{-8}$ sec. In a real laboratory situation, the duration of the process is neither too large: it seldom exceeds 10^{-2} sec. Thus, the phenomena under consideration proceed at temperatures somewhere near 10^{-3} K, pressures not exceeding 35 atmospheres, and the characteristic times ranging between 10^{-6} and 10^{-2} sec.

The theoretical description of magnetization dynamics or, using the conventional term, the spin dynamics of superfluid helium-3 is most clearly formulated in the framework of the Hamiltonian formalism. We follow here the fundamental work by Leggett [265]. First of all, one should have a quantity describing the presence of superfluid state. Within the general approach by Landau [258], such a quantity is the so-called order parameter A, which for superfluid helium-3 is a complex 3×3 matrix. The values of the order parameter A corresponding to a certain fixed superfluid phase Φ belong to a certain set X_Φ in the space of all complex 3×3 matrices. There holds the rule which is a consequence of the so-called spin-orbital invariance: if a matrix A belongs to X_Φ, then all matrices of the form

$$A' = e^{i\varphi} R_1 A R_2,$$

where R_1, R_2 are matrices of three-dimensional rotations, also belong to X_Φ. Thus, the superfluid phase of helium-3 is given by an orbit or, in a more general case, by a stratum of orbits of the group

(1) $$G = so(3) \times so(3) \times u(1)$$

At the present time the A-phase predicted by Anderson and Morel [263] and the B-phase predicted by Balian and Werthamer [264] are thoroughly investigated; both of them correspond not to strata, but to individual orbits. The order parameter for the A-phase is of the form

$$A_{ij} = \Delta d_i(\Delta'_j + \sqrt{-1}\Delta''_j), \quad \Delta = \text{const}; \; i,j = 1,2,3 \tag{2}$$

where d_i, Δ'_j, Δ''_j are unit vectors, the vectors Δ'_j and Δ''_j being perpendicular to each other. The order parameter for the B-phase is the form

$$A_{ij} = \frac{\Delta}{\sqrt{3}} e^{i\varphi} R_{ij}, \quad \Delta = \text{const}; \; i,j = 1,2,3 \tag{3}$$

where R_{ij} is the matrix of three-dimensional rotation. The question of the existence of phases corresponding to orbit strata is discussed in the literature from time to time.

One should bear in mind that the order parameter A depends in the general case both on time and space even if the entire volume of the sample of helium-3 is in one and the same superfluid phase because the matrix A may acquire values at the distinct points of the corresponding orbit. However, one can often assume to a good accuracy that the structure of superfluid state is the same in all parts of the sample, and therefore A depends on time only, $A = A(t)$. This approximation substantially simplifies the analysis in the framework of Hamiltonian mechanics (the general approach to the study of the physics of condensed media in the framework of Hamiltonian formalism was proposed by Dzyaloshinskii and Volovikov [266]).

Bearing in mind the above agreements, the spin dynamics of superfluid helium-3 can be formulated within the usual Hamiltonian mechanics, and it appears to be part of the general theory of dynamical systems on orbits of Lie groups. In what follows we ignore the physically important dissipation effects (there may, in fact, be taken into account) and examine a conservative Hamiltonian system, referred to as Leggett equations, which is given by:

1. dynamic variables, namely, the matrix of the order parameter A_{ij} and the vector S_i corresponding to magnetizations;

2. Poisson brackets for dynamic variables

$$\begin{aligned} \{S_i, S_j\} &= \varepsilon_{ijk} S_k, \\ \{S_i, A_{jm}\} &= \varepsilon_{ijk} A_{km}, \\ \{A_{ij}, A_{km}\} &= 0, \end{aligned} \tag{4}$$

the repeated indices imply summation;

3. Leggett Hamiltonian

$$H = \frac{1}{2}\vec{S}^2 - \vec{H}\cdot\vec{S} + A_{ii}A^*_{jj} + A_{ij}A^*_{ji} - \frac{2}{3}A_{ij}A^*_{ij}. \tag{5}$$

It is assumed that appropriate variables are chosen in which numerous dimensional coefficients have disappeared.

The equations of dynamic variables can be obtained using the general rule for calculating time derivatives:

$$\frac{d}{dt}X = \{X, H\}.$$

It is essential that the orbits of the spin-orbital symmetry group are invariant with respect to the equations obtained. The form of the Poisson brackets (4) implies, in particular, that there always exist the integrals of motion

(6) $$S_i A_{ij}, \quad A_{ij}A_{ij}.$$

The Leggett equations, a complicated Hamiltonian system, have been investigated in a series of papers within the recent decade.

We will only dwell here on several results characterizing these equations as a conservative Hamiltonian system. Note that the term $-\vec{H} \cdot \vec{S}$ in the Leggett Hamiltonian (5) corresponds to the external magnetic field applied to a sample of helium-3. In experiment, the field may be either completely switched off or be very large: 500 Gauss and larger (for comparison, the magnetic field of the Earth is about 0.5 Gauss). If $h = |\vec{H}|$ is large, then the Leggett equation involves a large parameter h, which offers the opportunity for constructing asymptotics in h (in this connection see the review by Fomin [267]).

The case $\vec{H} = 0$ corresponds to the fact that the system is magnetized by the external field and then the field is quickly switched off leaving the system to evolve by itself. Superfluid phases may thus be identified from the behaviour of magnetization and characteristic frequencies of motion. For the B-phase the Leggett equations were completely integrated in quadratures by Maki and Ebisawa [268] under the condition $\vec{H} = 0$. The condition $\vec{H} = 0$ is apparently essential for complete integrability. But this is not the case with the A-phase; several particular cases of complete integrability of the Leggett equations were found by Golo [269] who made use of the coincidence between the Leggett equations for the A-phase and the equations of motion of a rigid body in an ideal liquid. It is required that the external magnetic field be collinear to the vector

$$\vec{l} = \vec{\Delta}' \times \vec{D}''$$

where $\vec{\Delta}'$, $\vec{\Delta}''$ are the vectors from formula (2) which sets up the order parameter of the A-phase. The same problem was solved by Poluektov who applied to another method using Euler angles. See [270]. Vollhardt [271] studies a particular case of the Leggett equations for the A-phase with $\vec{H} = 0$.

One should bear in mind that several special solutions of the Leggett equations appear as very essential for theoretical analysis of experiments. For $\vec{H} = 0$ such a solution (the so-called wall pinned (WP) mode) was deduced by Brinkman [272]. Following [272], we consider the parametrization of three-dimensional rotation with the help of angle and axis,

$$\theta, \vec{n}, \vec{n}^2 = 1,$$

$$R_{ij} = \cos\theta\,\delta_{ij} + (1 - \cos\theta)n_i n_j - \sin\theta\,\varepsilon_{ijk}n_k$$

In terms of $\theta, \vec{n}$ the exact solution found by Brinkman has the form

(7)
$$\vec{S} = \text{const}, \quad \theta = \text{const},$$
$$\vec{n} = \vec{u}\cos(wt + \psi_0) + \vec{v}\sin(wt + \psi_0),$$
$$\vec{u}\cdot\vec{S} = \vec{v}\cdot\vec{S} = 0, \quad w = \text{const}.$$

For the case $\vec{H} \neq 0$, Brinkman and Smith [273] obtained two exact stable periodic solutions

(8a)
$$\theta = \arccos(-1/4),$$
$$(\vec{S} - \vec{H})\cdot\vec{n} = 0, \quad \vec{H}\cdot\vec{n} = \text{const}, \quad |\vec{H}| = |\vec{S}|,$$
$$(\vec{S} - \vec{H})\cdot\vec{H} = -2\sin^2\frac{\theta}{2}[\vec{H}^2 - (\vec{H}\cdot\vec{n})^2].$$

(8b)
$$\arccos(-1/4) \leq \theta \leq 2\pi - \arccos(-1/4),$$
$$(\vec{S} - \vec{H})\cdot\vec{n} = 0, \quad \vec{H}\cdot\vec{n} = 0,$$
$$(\vec{S} - \vec{H})^2 = [\vec{H}\cdot(\vec{S} - \vec{H})]^2/(H\sin\frac{\theta}{2})^2,$$
$$(\vec{S} - \vec{H})\cdot\vec{H} = -\vec{H}^2\sin^2\frac{\theta}{2}\left[1 \pm \left(1 - \frac{64}{15}\vec{H}^{-2}\left(\cos\theta + \frac{1}{4}\right)\right)^{1/2}\right],$$

in both cases the vectors $\vec{S}$ and $\vec{n}$ rotating around the axis collinear to the vector $\vec{H}$ at a constant angular velocity.

Concluding we will dwell on a curious phenomenon in the B-phase spin dynamics. The Leggett equations in the variables $\vec{S}, \vec{n}, \theta$ are a system of seven equations (the relation $\vec{n}^2 = 1$ is satisfied automatically), $\vec{S}$ and $\vec{n}$ transforming as vectors in three-dimensional rotations. Making use of this fact, Golo was the first [269] to apply Pohlmeyer's method [274] of reduction of equivariant dynamical systems to the Leggett equations. To this end he introduced the variables

$$S_\| = \vec{S}\cdot\vec{n}, \quad S_\perp = \sqrt{S^2 - S_\|^2}, \quad \theta.$$

From the general Poisson brackets (4) it follows that they form a closed subalgebra of Poisson brackets

$$\{S_\|, \theta\} = -1$$
$$\{S_\|, S_\perp\} = \frac{1}{2}S_\perp \operatorname{ctg}\frac{\theta}{2}$$
$$\{S_\perp, \theta\} = 0$$

In case the external magnetic field $\vec{H} = 0$ is absent, it follows that the total system of the seven Legett equations is reduced to three equations

$$\frac{d}{dt}S_{\shortparallel} = \frac{1}{2}S_{\perp}^2 \operatorname{ctg}\frac{\theta}{2} + \frac{16}{15}\sin\theta(\cos\theta + \frac{1}{4}),$$
$$\frac{d}{dt}S_{\perp} = -\frac{1}{2}S_{\shortparallel}S_{\perp}\operatorname{ctg}\frac{\theta}{2}$$
$$\frac{d}{dt}\theta = S_{\shortparallel}.$$

In reality, taking into account a character of the relation $\vec{n}^2 = 1$, we deal with the reduction of only six equations to three. For $\vec{H} \neq 0$ Novikov applied Pohlmeyer's method and obtained reduction of six equations to five [275]. For details, see also the paper of Novikov, Dubrovin, and Krichever "Integrable systems" (Moscow, VINITI, *Dynamical Systems*, 1985, vol. 4).

CHAPTER 5

NONINTEGRABILITY OF CERTAIN CLASSICAL HAMILTONIAN SYSTEMS

§1 The Proof of Nonintegrability by the Poincaré Method

1.1 *Perturbation Theory and the Study of Systems Close to Integrable*

The subject matter of the book is to reveal mechanisms controlling integrability or nonintegrability of Hamiltonian systems. As we will see, the systems of general form are, generally speaking, nonintegrable in the sense of Liouville. This means that the "majority" of Hamiltonians generate systems for which the number of "good" first integrals is insufficient. Therefore, the integral trajectories of such systems do not lie on Liouville tori but are everywhere dense on substantially more complicated subsets. This effect is especially vivid in the study of the equations of motion of a rigid body. Roughly speaking, *if a rigid body is dynamically nonsymmetric* (does not possess any axes, symmetry planes, etc.), *then its equations of motion are not integrated in any reasonable sense*. In particular, for these equations there exist no analytic integrals (except for the trivial integrals indicated above). As we have seen, these equations are interpreted as a Hamiltonian vector field on four-dimensional surfaces M_{23} embedded into the six-dimensional Euclidean space $\mathbb{R}^6(K,e)$. Of course, on this surface there always exists one energy integral H. Consequently, the trajectories of a system move actually along three-dimensional surfaces which are common level surfaces of three integrals $f_1 = H, f_2, f_3$.

For this Hamiltonian system to be completely integrable on a four-dimensional symplectic manifold M_{23}, one should find one more integral, besides the energy integral H, which would be in involution with it. Thus, the integration problem reduces to finding *another, fourth integral*. This is just the main difficulty encountered by integrating equations of motion of a rigid body.

Thus, in the space of all Hamiltonians there exist open domains that sometimes occupy almost the whole space and consist of Hamiltonians f of general position for which the corresponding Hamiltonian systems sgrad f are not integrable in the sense of Liouville (nor in any other reasonable sense). The described picture is not a rigorously proved theorem because in the formulation presented above many objects need a correct specification, and this is not always possible.

Nevertheless, the available results "of negative character," i.e., those declaring nonintegrability of many concrete types of systems, make it possible to think of the formulated principle as a certain experimental observation. In other words, integrable cases fill up the set "of measure zero" in the space of all systems.

From this it is already clear that the search for integrable systems is an exceedingly complicated problem because in the boundless variety of all possible Hamiltonians one should somehow "guess" or algorithmically reveal those rare cases where certain additional symmetries of the system provoke the appearance of a sufficient number of integrals. That is the reason why we have narrated about the system integration problems in the first part of the book.

We will now demonstrate that a Hamiltonian taken "by rule of thumb" most often generates a nonintegrable system, in any case if the description of motion of a three-dimensional heavy rigid body is meant.

Consider a symplectic manifold M^{2n} with the form ω and let H_0 be the Hamiltonian of a certain Liouville integrable Hamiltonian system $v_0 = \text{sgrad } H_0$. Suppose that T^n is one of the compact connected level surfaces of a set of independent integrals $f_1, \ldots, f_n$ which are in involution in the neighbourhood of T^n. We know that T^n is a torus. In an open neighbourhood U of the torus T^n, consider regular curvilinear action-angle coordinates $s_1, \ldots, s_n, \varphi_1, \ldots, \varphi_n$, where φ_i are angular variables on the torus and s_i coordinates along the normal to the torus. For the sake of brevity, introduce the following vector notation: $s = (s_1, \ldots, s_n)$, $\varphi = (\varphi_1, \cdots \varphi_n)$. Represent the neighbourhood U as the direct product $U = D^n \times T^n$, where $D^n = D$ is an open domain in $\mathbb{R}^n(s)$, for instance, homeomorphic to a sufficiently small open disk. Thus, we have separated regular coordinates in the neighbourhood U into two groups: $U(s, \varphi) = D^n(s) \times T^n(\varphi)$.

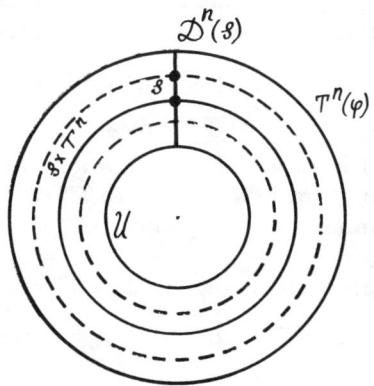

Figure 69

The integral trajectories of the system $v_0 = \text{sgrad } H_0$ lie on the torus T_n (and on the tori close to this one, which have the form $s \times T^n$, where $s \in D^n$) forming a rectilinear winding (Fig. 69). As is known from the Liouville theorem, the Hamiltonian H_0 in the coordinates s, φ depends only on the variables s, that is, H_0 is equal to $H_0(s)$ in the neighbourhood U.

Now consider *perturbation of the initial Hamiltonian system* by means of perturbation of its Hamiltonian. Consider a family of Hamiltonian systems $v_\varepsilon = $ sgrad $H(s, \varphi, \varepsilon)$, where $H(s, \varphi, \varepsilon)$ is a *real-analytic function* defined on the direct product $U \times (-\varepsilon_0, \varepsilon_0)$ and such that for $\varepsilon = 0$ we obtain the initial Hamiltonian, that is, $H(s, \varphi, 0) = H_0(s)$.

The Hamiltonian equations $v_0 = $ sgrad H_0 in the coordinates s, φ are written as

$$\dot{s} = -\frac{\partial H_0}{\partial \varphi} = 0, \qquad \dot{\varphi} = \frac{\partial H_0}{\partial s} = \omega(\sigma),$$

where

$$\omega(s) = (\omega_1(s), \ldots, \omega_n(s)).$$

These equations are seen to be *explicitly integrated*. If the point $s = s^0$ is fixed, it determines a certain torus in the neighbourhood U on which the vector $\omega(s^0)$ is constant (does not depend on φ), and therefore the equations of motion are integrated as follows: $s(t) = s^0$, $\varphi(t) = \varphi^0 + \omega(s^0)t$.

The equations of the perturbed Hamiltonian system $v = $ sgrad H are given by

$$\dot{s} = -\frac{\partial H}{\partial \varphi}, \qquad \dot{\varphi} = \frac{\partial H}{\partial s}, \quad \text{where } H(s, \varphi) = H_0(s) + \varepsilon H_1(s, \varphi) + \ldots.$$

Assuming the perturbation parameter ε to be small, we arrive at the *problem of integrating a perturbed system in the neighbourhood of a completely integrable system*.

The *Poincaré method* allows us in some cases to prove *nonintegrability* of a perturbed system in the neighbourhood of an integrable system.

The leading role in the study of a system for nonintegrability is played by the expansion term $H_1(s, \varphi)$. It is sometimes referred to as the *disturbing function*. Expand the disturbing function into a multiple Fourier series

$$H_1(s, \varphi) = \sum_{m \in \mathbb{Z}^n} H_m(s) e^{i \langle m, \varphi \rangle}.$$

Here $m = (m_1, \ldots, m_n)$ is an integer-valued vector running over all the sites of an integer-valued lattice $\mathbb{Z}^n$ or rank n, that is, $\mathbb{Z}^n = \mathbb{Z} \times \ldots \times \mathbb{Z}$ (n times). By $\langle m, \varphi \rangle$ we mean the usual scalar product of the vectors m and φ, that is, $\langle m, \varphi \rangle = \sum_{k=1}^n m_k \varphi_k$.

That is why the function H_1 is expanded as follows:

$$H_1(s, \varphi) = \sum_{(m_1, \ldots, m_n)} H_{m_1 \ldots m_n}(s) e^{i(m_1 \varphi_1 + \cdots + m_n \varphi_n)}.$$

Consider a domain ball D^n transversal to a Liouville torus (see above).

DEFINITION 5.1.1: . By the *Poincaré set* P in the domain D^n we mean the set of all points $s \in D^n$ for which there exist $n-1$ linearly independent integer-valued vectors $a_1, \ldots, a_{n-1} \in \mathbb{Z}^n$ such that there hold the following conditions:

1) All scalar products $\langle a_k, \omega(s) \rangle$ are equal to zero, $1 \leqslant k \leqslant n-1$;
2) $H_{a_k}(s) \neq 0$.

Here $\omega(s) = (\omega_1(s), \ldots, \omega_n(s))$ is a vector setting components of the vector field $v_0 = \text{sgrad}\, H_0$ on the torus $T^n(s)$. It is natural to call this vector the *frequency vector*. The numbers $\omega_1, \ldots, \omega_n$ are *frequencies of a uniform motion of a trajectory along the torus T^n*. A torus with the set of frequencies $\omega_1, \ldots, \omega_n$ is called *nonresonant* if the equality $k_1 \omega_1 + \cdots + k_n \omega_n = 0$ with integers $k_1, \ldots, k_n$ implies that all k_i are equal to zero. It is obvious that *integral trajectories are everywhere dense on nonresonant tori*.

In the resonant case, integral trajectories are everywhere dense on tori of smaller dimension. We recall that precisely such a situation characterizes those Hamiltonian systems which admit *noncommutative integration* (see above).

Denote by $A(V)$ the class of functions analytic in the domain $V \subset \mathbf{R}^n$. A set $N \subset V$ is called a *uniqueness set for the class of functions $A(V)$* if any analytic function which is equal to zero on the set N becomes identical zero everywhere on the domain V. In particular, two analytic functions coinciding on the uniqueness set N automatically coincide on the entire domain V. For instance, the set N of points of a one-dimensional interval Δ^1 on the real axis $\mathbf{R}^1$ is the uniqueness set for the class $A(\Delta^1)$ if and only if it has a limit point within the interval Δ^1.

Consider an unperturbed Hamiltonian system $v_0 = \text{sgrad}\, H_0$. It is called *nondegenerate* in the neighbourhood U of an invariant torus T^n if $\det\left(\frac{\partial^2 H_0}{\partial s_i \partial s_j}\right) \neq 0$ in a domain D^n, where $U \approx D^n \times T^n$ (see above). Recall that the Hamiltonian H_0 may be assumed to depend only on the variables s, that is, $H_0 = H_0(s_1, \ldots, s_n)$. Let $s^0 = (s_1^0, \ldots, s_n^0)$ be a fixed point from the domain $D^n \subset U$.

THEOREM 5.1.1 (SEE [61]). *Suppose that an unperturbed Hamiltonian system $v_0 = \text{sgrad}\, H_0$ is nondegenerate in a neighbourhood U of an invariant torus T^n. Let a point $s^0 \in D^n$ be a noncritical point of the Hamiltonian $H_0(s)$ and let in any neighbourhood V of this point a Poincaré set N be the uniqueness set for the class $A(V)$. Then the perturbed Hamiltonian equations $v = \text{sgrad}\, H$, that is,*

$$\dot{s} = -\frac{\partial H}{\partial \varphi},$$
$$\dot{\varphi} = \frac{\partial H}{\partial s},$$
$$H(s) = H_0(s) + \varepsilon H_1(s, \varphi) + \ldots$$

do not have such an integral f which is independent of the Hamiltonian H and can be represented as a formal power series $f(s, \varphi) = \sum_{k \geqslant 0} f_k(s, \varphi) \varepsilon^k$ with coefficients analytic in the domain $U = D^n \times T^n$.

The following versions of Theorem 5.1.1 also prove to be instructive for applications.

THEOREM 5.1.2 (SEE [61]). *Let the Hamiltonian $H_0(s)$ of an unperturbed system be nondegenerate in the domain D^n and let the Poincaré set N be everywhere dense in the domain D^n. Then the Hamiltonian equations of a perturbed system $v = \text{sgrad}\, H$ do not have a formal integral $f(s, \varphi) = \sum_{k \geqslant 0} f_k \varepsilon^k$ (independent of the*

Hamiltonian $H(s,\varphi))$ with smooth coefficients $f_k(s,\varphi)$ set in the neighbourhood $U = D^n \times T^n$ of an invariant torus T^n.

Claims similar in the form are also valid for nonautonomous Hamiltonian systems. For details see, for instance, [61].

1.2 *Nonintegrability of the Equations of Motion of a Dynamically Nonsymmetric Rigid Body with a Fixed Point*

The results described above can be used in proving assertion formulated in the title of the present subsection. First of all, we will give a more precise formulation of the problem.

Poincaré was the first to put forward the question of whether in the problem of a heavy rigid body motion around a fixed point there exists one more additional integral distinct from the three known integrals (and independent of them) [104]. For the Hamiltonians H close to the Hamiltonian H_0, which corresponds to the completely integrable Euler case, a negative answer was given by Kozlov in [161]. Let us dwell on this question in more detail. The general equations of motion of a heavy rigid body are written above in Ch. 1, 1.3). As has been noticed, they always have three integrals: the total energy

$$f_1 = H = \frac{1}{2}\langle K, h^{-1}(K)\rangle + m\langle r, e\rangle,$$
$$f_2 = \langle K, e\rangle, \quad \text{and} \quad f_3 = \langle e, e\rangle.$$

The second integral f_2 corresponds to the fact that the projection of angular momentum to a vertical is always constant. This integral corresponds to the group of rotations of a rigid body around the vertical axis. As we have noted, by fixing the values of two integrals, f_2 and f_3, one can lower the order of a system by two and it becomes a vector field on four-dimensional level surfaces $M_{23} \subset \mathbf{R}^6$. Moreover, on these surfaces the obtained system turns to be Hamiltonian. Its Hamiltonian is the total energy of the body with a fixed value of the projection of the angular momentum K to the vertical $\langle K, e\rangle = $ const. The Hamiltonian H can be represented in the form $H = H_0 + \varepsilon H_1$, where H_0 is the kinetic energy, the Hamiltonian of a completely integrable Euler problem on inertial motion of a rigid body (provided that the body is fixed in the centre of mass) and εH_1 the potential energy of the body in a homogeneous gravity field.

Here ε is the product of the body weight by the distance from the centre of mass to the point of support (fixation) of the rigid body. We assume the parameter ε to be small, that is, we examine the motion of a rigid body caused by small perturbation of the integrable Euler case. From the mechanical point of view this is equivalent to the study of fast motions of a rigid body in a moderate field of force.

THEOREM 5.1.3 (SEE [61], [161]). *If a heavy rigid body of the general form is dynamically nonsymmetric then its equations of motion $v = \operatorname{sgrad} H$ (where the Hamiltonian $H = H_0 + \varepsilon H_1$ is a small perturbation of the Hamiltonian H_0 of the integrable Euler case) do not have a formal additional fourth integral $\sum_{k \geq 0} f_k \varepsilon^k$ which has analytic coefficients on the four-dimensional level surface M_{23} and is independent of the function $H = H_0 + \varepsilon H_1$.*

It is not yet clear what is going on "far" from the integrable Euler case, that is, for a rigid body strongly different from the body which satisfies the Euler conditions (see Ch. 1). The point is that the available "real methods" of the analysis of Hamiltonian systems are so far effective only in a small neighbourhood of rare integrable cases. The "complex methods" (which we omit for want of space) do not yet permit a complete description of real cases of nonintegrability.

1.3 Separatrix Splitting

Let V^n be a smooth manifold which we identify with the configuration space (position space) of a certain Hamiltonian system and let $M^{2n} = T^*V$ be a cotangent bundle to V^n naturally identified with the phase space of the system, which (the phase space) is a symplectic manifold. The points from T^*V have the form (x, ξ), where $x \in V^n$ and $\xi \in T_x^*V^n$, that is, ξ is a covector, a linear function on T_xV^n. Suppose that on the phase space M^{2n} a Hamiltonian $H = H(x, \xi, t)$ is given which is, generally speaking, time-(t-)dependent. Then we may consider an extended space $K^{2n+2} = T^*V \times \mathbb{R}^2(E, t)$, with coordinates (x, ξ, E, t), on which (the space) the equations of motion will again be Hamiltonian:

$$\dot{x} = \frac{\partial Q}{\partial \xi}, \qquad \dot{\xi} = -\frac{\partial Q}{\partial x},$$

$$\dot{E} = \frac{\partial Q}{\partial t}, \qquad \dot{t} = -\frac{\partial Q}{\partial E},$$

where

$$Q(x, \xi, E, t) = H(x, \xi, t) - E, \quad x \in V, \ \xi \in T_x^*V.$$

Let the Hamiltonian H be periodic in t with the period 2π and depend on a certain parameter ε, that is, H is equal to $H(x, \xi, t, \varepsilon)$. Suppose that for $\varepsilon = 0$ the Hamiltonian $H(x, \xi, t, 0) = H_0(x, \xi)$ does not depend on time and satisfies the following four conditions:

1) On a $2n$-dimensional manifold $T^*V^n = M^{2n}(x, \xi)$ the Hamiltonian $H_0(x, \xi)$ has at least two critical points (x_-, ξ_-) and (x_+, ξ_+) at which the eigenvalues of the linearized Hamiltonian system

$$\dot{\xi} = -\frac{\partial H_0}{\partial x}, \qquad \dot{x} = \frac{\partial H_0}{\partial \xi}$$

are real and nonzero. In particular, the 2π-periodic solutions

$$(x_-, \xi_-) = (x_-(t), \xi_-(t))$$

and

$$(x_+, \xi_+) = (x_+(t), \xi_+(t))$$

are of hyperbolic type. It should be emphasized that in the problems of nonintegrability of Hamiltonian systems an important role is played by the critical points of the Hamiltonian, that is, points at which sgrad $H_0 = 0$ (and therefore sgrad $H_0 = 0$.

2) By Λ_+ (respectively, Λ_-) we denote a stable (respectively, unstable) separated manifold of the critical point (x_+, ξ_+) (respectively, (x_-, ξ_-)), that is, a

manifold consisting of separatrices tending to the point (x_+, ξ_+) as $t \to +\infty$ (respectively, consisting of separatrices tending to the point (x_-, ξ_-) as $t \to -\infty$). If the critical points of a Hamiltonian are nondegenerate then near the point (x_+, ξ_+) (respectively, (x_-, ξ_-)) the separatrix manifolds may be assumed to be disks. In this case they are called separatrix disks, respectively, stable or unstable.

Along stable separatrix surfaces, integral trajectories tend to the critical point (with increasing time), while along unstable separatrix surfaces they move away from the critical point (with increasing time). Condition 2 consists in the fact that $\Lambda_+ = \Lambda_-$, that is, the stable separatrix surface Λ_+ of the point (x_+, ξ_+) coincides with the unstable separatrix surface Λ_- of the point (x_-, ξ_-). From this it follows, in particular, that $H_0(x_+, \xi_+) = H_0(x_-, \xi_-)$.

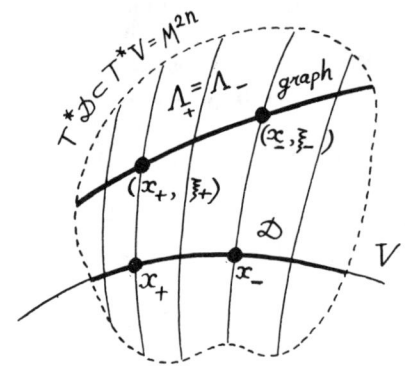

Figure 70

3) Suppose that in a configuration space V^n there exists a domain $D = D^n$ containing both points x_+ and x_- and possessing the following property. Consider a domain $T^*D \subset T^*V = M^{2n}$. It contains both critical points (x_+, ξ_+) and (x_-, ξ_-) of the Hamiltonian H (Fig. 70). Condition 3 requires that the part of the separatrix surface $\Lambda_+ = \Lambda_-$ contained in the domain T^*D be represented in the form of an n-dimensional graph of the gradient of a certain analytic function defined on the domain D. This means that the separatrix surface is represented in the form $\xi = \frac{\partial S_0}{\partial x}$, where S_0 is a real analytic function of x. Such n-dimensional surfaces are called *Lagrangian*. We will not describe their properties here because they are not needed for the formulation of the *separatrix splitting effect*.

Under the above assumptions there appears a certain vector field naturally defined on the domain D which is described on D by the following system of differential equations: $\dot{x} = \frac{\partial}{\partial \xi} H_0(\xi(x), x)$, where $\xi = \xi(x) = \frac{\partial S_0(x)}{\partial x}$ sets the described graph. In small neighbourhoods of the points x_+ and x_- the integral trajectories of this system tend to x_+ as $t \to +\infty$ and to x_- as $t \to \infty$ (Fig. 71).

4) Finally, require that the differential equation $\dot{x} = \frac{\partial}{\partial \xi} H_0(\xi(x), x)$ have in the domain D^n a separatrix $x_0(t)$ joining the point x_- and x_+, that is, going (as t increases) from the point x_- to the point x_+ (Fig. 71).

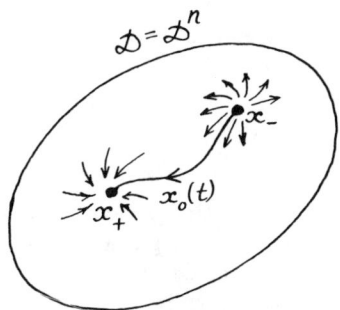

Figure 71

Thus, the initial $2n$-dimensional system determines (under the above assumptions) a certain n-dimensional system on the domain D^n. The properties of the latter have a substantial effect on integrability or nonintegrability of the initial system. A system generated by a Hamiltonian $H_0(x, \xi)$ may be regarded as an unperturbed system. In applications, it is most often chosen to be completely integrable. A system sgrad $H(x, \xi, t, \varepsilon)$ may therefore be interpreted as perturbation of a system with the Hamiltonian $H_0(x, \xi)$. Thus, for studying integrability or nonintegrability of a perturbed system we may apply the Poincaré method (see 1.1). Now we can describe the important *effect of separatrix splitting, which was first discovered by Poincaré and applied by him to motivate nonintegrability of some systems.*

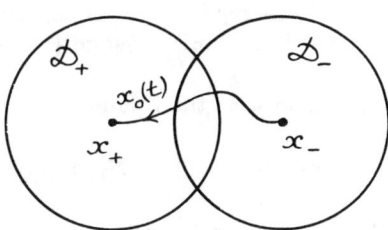

Figure 72

Consider the small perturbation $H(x, \xi, t, \varepsilon)$ of a Hamiltonian $H_0(x, \xi)$. For small values of the perturbation parameter ε the separatrix surfaces Λ_+ and Λ_- do not vanish but transform into certain perturbed surfaces Λ_+^ε and Λ_-^ε which, of course, do not necessarily coincide. To describe these events more exactly, consider open neighbourhoods D_+^n and D_-^n of some points X_+ and X_-, respectively, in a configuration space V^n (Fig. 72). Then in the domains $D_+ \times \mathsf{R}(t)$ and $D_- \times \mathsf{R}(t)$

the equation of separatrix surfaces is written as follows:

$$\xi = \frac{\partial S_+}{\partial x}$$

and

$$\xi = \frac{\partial S_-}{\partial x},$$

respectively, where $S_\pm(x,t,\varepsilon)$ is a function periodic in t (with a period 2π), analytic for all $x \in D^\pm$ and for small values of a parameter ε. For $\varepsilon = 0$ the separatrix surfaces $\Lambda_+^0 = \Lambda_+$ and $\Lambda_-^0 = \Lambda_-$ coincide. In the general case, for small values of the parameter $\varepsilon \neq 0$ the separatrix surfaces Λ_+^ε and Λ_-^ε treated as subsets in the direct product $T^*(D_+ \cap D_-) \times \mathbf{R}$ do not already coincide (Poincaré).

DEFINITION 5.1.2: *The phenomenon described above is called the splitting of separatrix surfaces (or the separatrix splitting).*

THEOREM 5.1.4 (POINCARÉ [162]). *If*

$$H(x, \xi, t\varepsilon) = H_0(x, \xi) + \varepsilon H_1(x, \xi, t) + \ldots$$

and

$$H_1(x_+, \xi_+, t) = H_1(x_-, \xi_-, t)$$

and if, besides

$$\int_{-\infty}^{+\infty} \{H_0, H_1\}(x_0(t), \xi(x_0(t))) \, dt \neq 0,$$

then for small values of the parameter $\varepsilon \neq 0$ the perturbed separatrix surfaces Λ_+^ε and Λ_-^ε do not coincide.

As was noted by Poincaré, the separatrix splitting turns out to be an obstacle for integrability of a perturbed Hamiltonian system (viewed in the neighbourhood of an integrable unperturbed system).

Consider a perturbed system with the Hamiltonian

$$H(x, \xi, t, \varepsilon) = H_0(x, \xi) + \varepsilon H_1(x, \xi, t) + \ldots$$

satisfying conditions 1–4. In particular, the unperturbed system is assumed to have two hyperbolic equilibrium positions (x_+, ξ_+) and (x_-, ξ_-) joined by a separatrix which is sometimes called a doubly asymptotic solution: $t \to (x_0(t), \xi_0(t))$, $t \in \mathbf{R}$. The theorem that follows was proved by Bolotin and generalizes some familiar techniques for the proof of nonintegrability.

THEOREM 5.1.5 (SEE [61]). *Suppose the following conditions are fulfilled:*

a) $\int_{-\infty}^{+\infty} \{H_0, \{H_0, H_1\}\}(x_0(t), \xi_0(t), t) \, dt \neq 0$.

b) *For small values of the parameter ε a perturbed system also has a doubly asymptotic solution (separatrix) $t \to (x_\varepsilon(t), \xi_\varepsilon(t))$ close to the solution $t \to (x_0(t), \xi_0(t))$.*

§1. The Proof of Nonintegrability

Then for small values $\varepsilon \neq 0$ in any neighbourhood of closure of the trajectory $(x_\varepsilon(t), \xi_\varepsilon(t))$ the Hamiltonian equations $v = \operatorname{sgrad} H$ do not have a complete set of independent integrals which are in involution.

It is just the separatrix splitting that allows us to prove nonintegrability of the equations of motion of a dynamically nonsymmetric rigid body with a fixed point in the neighbourhood of the integrable Euler case (see subsection 1.2 of the present section). In the problem of fast motion of a nonsymmetric rigid body the Hamiltonian is the form $H = H_0 + \varepsilon H_1$, where

$$H_0 = \frac{1}{2}\langle AK, K\rangle, \qquad H_1 = x_0 e_1 + y_0 e_2 + z_0 e_3.$$

Here $r = (x_0, y_0, z_0)$ is the radius vector of the centre of mass of a rigid body,

$$e = (e_1, e_2, e_2)$$

and

$$A = \begin{pmatrix} a_1 & 0 & 0 \\ 0 & a_2 & 0 \\ 0 & 0 & a_3 \end{pmatrix};$$

the numbers a_1, a_2, a_3 are inverse of the principal moments of inertia of the rigid body. For $\varepsilon = 0$ we obtain a completely integrable Euler case because $H = H_0$. In this unperturbed problem, on all noncritical three-dimensional level surfaces M_{123} given by the three integrals:

$$f_1 = H_0 = c_1 = \text{const} > 0, \qquad f_2 = c_2 = \text{const}, \qquad f_3 = c_3 = 1 = \text{const},$$

there exists two unstable periodic solutions. Namely, if $a_1 < a_2 < a_3$ then

$$K_1 = K_3 = 0, \qquad K_2 = K_2^0 = \pm\sqrt{\frac{2c_1}{a_2}}, \qquad e_2 = e_2^0 = \pm\frac{c_2}{K_2^0},$$

$$e_1 = \alpha \cos(a_2 K_2^0) t, \qquad e_3 = \alpha \sin(a_2 K_2^0) t, \qquad \alpha^2 = 1 - \left(\frac{c_2}{K_2^0}\right)^2.$$

The inequality $\langle K, e\rangle^2 \leqslant \langle K, K\rangle\langle e, e\rangle$ and independence of the integrals f_1, f_2, f_3 on the common level surface M_{123} imply that $\alpha^2 > 0$. Stable and unstable separatrix surfaces of the two indicated periodic (in t) solutions can be set as intersections of a three-dimensional manifold M_{123} by hyperplanes of the form

$$K_1\sqrt{a_2 - a_1} \pm K_2\sqrt{a_3 - a_2} = 0.$$

The splitting of these separatrix surfaces is studied in [164], [165]. The behaviour of the solutions of a perturbed problem was studied using computer in the interesting paper [166]. In the schemes obtained in the calculations and showing the behaviour of integral trajectories it is clearly seen that the invariant curves of the unperturbed problem become chaotic in the neighbourhood of separatrices. This phenomenon is just the reason for nonintegrability of the equations of motion of a heavy nonsymmetric rigid body with a fixed point (near the integrable Euler case).

1.4 Nonintegrability in the General Case of the Kirchoff Equations of Motion of a Rigid Body in an Ideal Liquid

In the previous subsection we have mentioned several general methods for proving nonintegrability of Hamiltonian systems sgrad H, which are small perturbations of integrable system. In the autonomous case, the splitting condition for separatrix surfaces located at a certain fixed energy level can be written in the following form:

$$\int_{-\infty}^{+\infty} \{f_0, H_1\}\, dt \neq 0,$$

where f_0 is the integral of an unperturbed system and $H = H_0 + \varepsilon H_1$. Consider the Kirchhoff equations

$$\begin{cases} \dot{K} = [K, \omega] + [e, u], \quad \dot{e} = [e, \omega], \omega = \frac{\partial H}{\partial K}, \quad u = \frac{\partial H}{\partial e}, \\ H = \tfrac{1}{2}\langle AK, K\rangle + \langle BK, e\rangle + \tfrac{1}{2}\langle Ce, e\rangle. \end{cases}$$

As is known (see Ch. 1), these equations describe the motion of a rigid body in an ideal unbounded liquid. The matrix

$$A = \begin{pmatrix} a_1 & 0 & 0 \\ 0 & a_2 & 0 \\ 0 & 0 & a_3 \end{pmatrix}$$

is diagonal and the matrices B and C may be assumed symmetric.

THEOREM 5.1.6 (SEE [167], [61]). *Let the numbers a_1, a_2, a_3 be pairwise distinct. If the Kirchoff equations have an additive (fourth) integral independent of the three classical integrals $f_1 = H$, $f_2 = \langle K, e\rangle$, $f_3 = \langle e, e\rangle$ and analytic integrals on the six-dimensional space $\mathbb{R}^6(K, e)$, then the matrix B automatically turns out to be diagonal*

$$B = \begin{pmatrix} b_1 & 0 & 0 \\ 0 & b_2 & 0 \\ 0 & 0 & b_3 \end{pmatrix}$$

and there holds the following relation:

$$\frac{b_2 - b_3}{a_1} + \frac{b_3 - b_1}{a_2} + \frac{b_1 - b_2}{a_3} = 0.$$

If $B = 0$, then an independent (fourth) integral exists only if the matrix C is diagonal

$$C = \begin{pmatrix} c_1 & 0 & 0 \\ 0 & c_2 & 0 \\ 0 & 0 & c_3 \end{pmatrix}$$

and the following condition is satisfied:

$$\frac{c_2 - c_3}{a_1} + \frac{c_3 - c_1}{a_2} + \frac{c_1 - c_2}{a_3} = 0.$$

Thus, the Kirchhoff equations are nonintegrable in the general case.

It will not be out of place to note that the integrable Clebsch case is determined just by the above condition:

$$\frac{b_2 - b_3}{a_1} + \frac{b_3 - b_1}{a_2} + \frac{b_1 - b_2}{a_3} = 0.$$

The proof of Theorem 5.1.6 also rests upon the separatrix splitting phenomenon. To this end, one should represent the Kirchhoff equations as perturbations of integrable equations. Introduce a small parameter ε replacing in the Kirchhoff equations e by εe. Then on the fixed four-dimensional level surface of two integrals

$$M_{23} = (f_2 = \langle K, e \rangle = c_2; \qquad f_3 = \langle e, e \rangle = c_3)$$

the Kirchhoff equations turn out to be Hamiltonian with the Hamiltonian $H = H_0 + \varepsilon H_1 + \varepsilon^2 H_2$, where H_0, H_1, H_2 are restrictions of the functions $\frac{1}{2}\langle AK, K \rangle$, $\langle BK, e \rangle$, $\langle Ce, e \rangle$ (respectively) to the level surface M_{23}. Smallness of the perturbed parameter ε implies that the constant energy $f_1 = H = c_1$ greatly exceeds the constants c_2 and c_3. For $\varepsilon = 0$ we again arrive at the integrable Euler case of inertial motion of a free rigid body. That is why we may apply the technique used in the preceding subsection.

Note that Theorem 5.1.6 proves nonintegrability of the general Kirchhoff equations not only in a small neighbourhood of the integrable Euler case, but also "far" from it, that is, for the set of matrices A, B, C which occupy an open, everywhere dense region in the entire 15-dimensional parameter space. This is what differs Theorem 5.1.6 from Theorem 5.1.3, in which nonintegrability of the equations of motion of a dynamically nonsymmetric rigid body with a fixed point is proved for the present only in the neighbourhood of the integrable Euler case. This is due to the fact that in the latter case the vector e is a unit vector, and therefore $\langle e, e \rangle = 1$. This hampers multiplication of e by the small parameter ε, that is, does not allow us to apply the technique used in the proof of Theorem 5.1.6.

§2 Topological Obstacles for Complete Integrability

2.1 Nonintegrability of the Equations of Motion of Natural Mechanical Systems with Two Degrees of Freedom on High-Genus Surfaces

Analyze a natural mechanical system with two degrees of freedom. This means that its configuration space is two-dimensional. We assume this space to be a two-dimensional compact orientable real-analytic manifold M^2. From elementary topology such a manifold is known to be diffeomorphic to the sphere S^2 to which g handles are glued (Fig. 73). The number g is usually called the *genus of the surface*. It is also known that this is the only topological invariant of closed orientable connected surfaces, that is to say, two surfaces of the indicated type are diffeomorphic if and only if their genera coincide.

Consider the cotangent bundle T^*M^2 to the manifold M^2. It is well known that the cotangent bundle T^*M^n of an arbitrary smooth manifold M^n can be

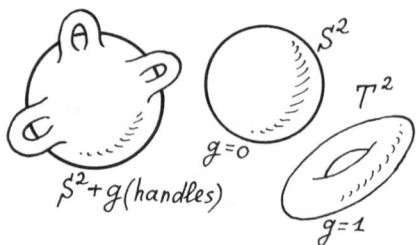

Figure 73

transformed in a natural way into a symplectic $2n$-dimensional manifold. To this end, it suffices to exhibit a symplectic structure on T^*M^n, that is, a closed nondegenerate exterior 2-form ω. Examine a natural projection $p : T^*M \to M$, where $p(x,\xi) = x$, $x \in M\varepsilon \in T^*_x M$. The manifold T^*M is a fibred space whose basis is M and fibres $p^{-1}(x)$ are the cotangent spaces $T^*_x M$.

Define on T^*M a certain natural 1-form α. Let $a \in T_y(T^*M)$ be a tangent vector to the manifold T^*M at the point $y = (x,\xi)$. Since p maps T^*M into M, its differential dp maps $T_y(T^*M)$ into $T_{p(y)}M$, where $p(y) = x$ and $y = (x,\xi)$. To set the form α, we put, by definition, $\alpha(a) = \langle \xi, \mathrm{d}p(a)\rangle$, where $\langle,\rangle$ stands for the value of the covector ξ on the vector $\mathrm{d}p(a)$. This quantity is well defined because $\mathrm{d}p(a) \in T_x M$ and $\xi \in T^*_x M$. Recall that the covector ξ is a linear functional (linear form) on the tangent space $T_x M$. In other notation the value of the covector on the vector can be written as $\xi(\mathrm{d}p(a))$. Eventually, as the required exterior 2-form ω we take the following form: $\omega = d\alpha$, where d is the operation of outer differentiation. It is clear that such a form is closed because $d^2 \equiv 0$.

LEMMA 5.2.1. *The closed differential 2-form ω defined above is nondegenerate and therefore sets the symplectic structure on a $2n$-dimensional manifold T^*M.*

PROOF: Introduce on M^n local regular coordinates $x_1, \ldots, x_n$. Then the covector $\xi \in T^*_x M$, which is the 1-form on $T_x M$, is given by conjugate coordinates $\xi_1, \ldots, \xi_n$. From the definition of the form ω it follows that $\omega = d\alpha = d\xi_1 \wedge dx_1 + \cdots + d\xi_n \wedge dx_n$. This means that the form ω is nondegenerate for its matrix has the canonical symplectic form $\begin{pmatrix} 0 & E \\ -E & 0 \end{pmatrix}$. This proves the lemma.

Thus, the phase space of a mechanical system has a natural symplectic structure, which fact will be used for our further purposes. In our concrete example of a system with two degrees of freedom, the cotangent bundle T^*M^2 has a structure of a four-dimensional real-analytic symplectic manifold. The motion of the system is described by the Hamiltonian equations sgrad F, where the Hamiltonian F will be thought of as a real analytic function on T^*M. The Hamiltonian will be taken in the following form: $F(x,\xi) = K(x,\xi) + U(x)$, where for all $x \in M$ the function $K(x,\xi)$ is a quadratic form in the variables $\xi \in T^*_x M$ and the function $U(x)$ depends only on $x \in M$. The functions $K(x,\xi)$ and $U(x)$ will be treated as real-analytic on the manifolds T^*M and M, respectively. The quadratic form $K(x,\xi)$ is usually identified with the kinetic energy of the system and the function

$U(x)$ with the potential energy (and is called a potential).

Then the Hamiltonian equations of the system (with respect to the symplectic structure introduced above) will be canonically written as

$$\dot{\xi} = -\frac{\partial F}{\partial x}, \qquad \dot{x} = \frac{\partial F}{\partial \xi}.$$

It solutions ar represented by integral trajectories in the manifold T^*M. The Hamiltonian $F(x, \xi)$ is an integral of the system. The question arises: in which cases is such a system integrable?

If the configuration space is not very complicated in its construction, such systems often admit complete integration. Such examples are, in particular, the inertial motion of a material point along a two-dimensional sphere or a two-dimensional torus given in standard metrics. We see that the surface M^2 has here a low genus, namely zero (in the case of a sphere) and unity (in the case of a torus). If the Riemannian metric on a sphere S^2 and on a torus T^2 is not standard, the corresponding Hamiltonian system may not admit complete integration (see Ch. 2). Nevertheless, one can describe all Riemannian metrics on a sphere and on a torus for which complete integration is possible (in principle). Details are given in the sequel. Now we would like to emphasize that the case of a sphere and a torus admits, "in principle," complete integration (for some metrics). Roughly speaking, simple metrics determine nonintegrable systems. In our case the obstacle for integrability lies in the properties of metrics, i.e., is of a metric but not of a topological character. It turns out (and we will presently deal with it) that if a configuration space is topologically more complicated, that is, if the number of handles is greater than unity, then there appear purely topological obstacles that forbid analytic Hamiltonian systems to have a sufficient number of analytic commuting integrals on cotangent bundle. In other words the specified Hamiltonian systems do not admit complete integration if the configuration space is of high genus (higher than unity).

THEOREM 5.2.1 (KOZLOV [61]). *Let a real-analytic configuration space M^2 have a genus higher than unity, i.e. is homeomorphic to a sphere with g handles, where $g > 1$. Then the Hamiltonian canonical equations*

$$\dot{\xi} = -\frac{\partial F}{\partial x}, \quad \dot{x} = \frac{\partial F}{\partial \xi},$$

where $F(x, \xi)$ is a real-analytic Hamiltonian on T^*M^2 do not have a first integral which is real-analytic on T^*M^2 and functionally independent of the energy integral (that is, of the Hamiltonian) $F(x, \xi)$.

In the case of a two-dimensional configuration space, the problem of complete Liouville integrability reduces to finding one more (second) additional integral independent of the Hamiltonian. We think, as usual, that independence holds everywhere on the manifold. In the case of analytic functions, it suffices to require that they be independent at one point. Then they will be independent almost everywhere on T^*M.

If we weaken the assumptions of Theorem 5.2.1 and restrict our consideration to smooth manifolds M^2 and smooth Hamiltonians, then the assertion of Theorem

5.2.1 will, generally speaking, invalid. More precisely, one can easily construct a smooth Hamiltonian $F(x, \xi) = K + U$ on T^*M^2 such that the canonical Hamiltonian equations on T^*M^2 have an additional (second) smooth integral independent of the function $F(x, \xi)$ on a certain open subset (*which is not*, however, everywhere dense in T^*M^2), see [61]. At the same time it is not clear whether one can construct such a natural "mechanical" smooth Hamiltonian for which there exists a second additional smooth integral independent of the Hamiltonian *almost everywhere*, that is, on an open *everywhere dense* subset in T^*M^2, where the genus $(M^2) > 1$. Such an integral (if it does exist) cannot be Bott (see below).

Theorem 5.2.1 is a consequence of a more general assertion of Kozlov concerning nonintegrability of equations of motion for fixed sufficiently large values of total energy (see Theorem 5.2.2 below).

Consider a four-dimensional manifold T^*M^2 and a Hamiltonian system sgrad F, where $F = K + U$. Level surfaces $Q_h = ((x, \xi) \in T^*M : F(x, \xi) = h = \text{const})$ are set by the equation $K + U = h$ and are three-dimensional analytic submanifolds in T^*M^2 if the energy constant h is sufficiently large. Namely, it suffices to assume that $h > \max U$, where the maximum of the potential energy U is taken throughout the manifold M^2. Under this assumption we have: $K = h - U > 0$ on the level surface $F = h$. Since K is a positive definite quadratic form on each tangent space, then at each point $x \in M^2$ the equation $K(x, \xi) = h + U(x) > 0$ defines a circle in the tangent plane $T_x M^2$ (Fig. 74). It is clear that we have essentially used the assumption that the energy constant h is sufficiently large. For small h the solution of the equation $K = h - U = \text{const}$ can for instance, degenerate, and instead of a circle in the tangent plane we will have a point. Thus, for sufficiently large h the level surface F of the Hamiltonian turns out to be a three-dimensional analytic manifold Q fibring over the manifold M^2 with circle as a fibre. Since the vector field sgrad F is tangent to the surface $Q_h = (F = h)$, we obtain on Q_h^3 an analytic system of differential equations.

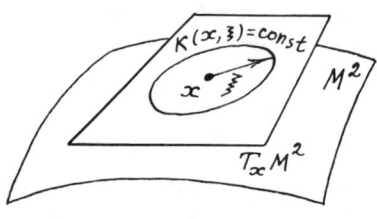

Figure 74

THEOREM 5.2.2 (SEE [61]). *If the genus of an analytic surface M^2 (the configuration space of the initial system) is higher than unity, then for all $h > \max U$ the analytic flow* sgrad $F|_Q$ *on the three-dimensional constant-energy manifold Q_h^3 does not have an additional inconstant real-analytic integral.*

In the smooth case, under the assumptions of Theorem 5.2.1 and 5.2.2 one can state the absence of new (additional) smooth integrals satisfying certain limitations.

THEOREM 5.2.3 (SEE [61]). *If the genus of a smooth surface M^2 is higher than unity, then for all sufficiently large h, that is for $h > \max U$, the flow sgrad $F|_{Q_h}$ does not have such a smooth first integral f which is defined on the smooth level surface Q_h^3 and satisfies the following conditions:*

1) The integral f interpreted as a smooth function on the energy level surface Q_h^3 has only a finite number of critical values (that is, such values to which there correspond critical points of the function f on Q_h^3).

2) The points $x \in M^2$, for which the sets $(f = c = \text{const})$ are finite or coincide with the whole fibre, i.e. with the circle S_x^1, are everywhere dense in M_2. Here $S_x^1 \subset T_x M^2$.

If the surface M^2 and the Hamiltonian F are analytic, then both conditions 1 and 2 of Theorem 5.2.3 are automatically fulfilled (the property 1 requires special proof), and therefore in an analytic case Theorem 5.2.3 immediately implies Theorem 5.2.1. More generally, if a compact orientable surface M^2 is nonhomeomorphic to a sphere and to a torus then the above-mentioned equations of system motion do not have a new integral which is a smooth function on T^*M^2 analytic for fixed $x \in M$ on cotangent two-dimensional planes T_x^*M and having only a finite number of distinct critical values. The number of critical points is not necessarily finite. Functions polynomial in momenta are an example of integrals analytic in the momenta ξ.

It is readily seen that a smooth function on a compact smooth closed manifold has a finite number of critical values if all of its critical points are isolated (then their number is finite) or if they fill nondegenerate critical submanifolds (in this case their number is also finite). The latter type of functions we have called Bott functions (Ch. 2).

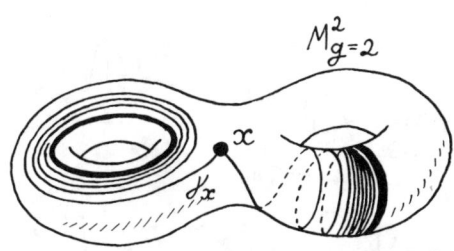

Figure 75

We will not prove this theorem because below we present the proof of a more general assertion whose particular case is Theorem 5.2.3. We only note that one should investigate the properties of the fundamental group of the manifold. It turns out (see [168]) that there always exist geodesic rays γ_x *emanating* from a point $x \in M^2$ and asymptotically approaching a certain closed geodesic from a given homotopy class (Fig. 75). Next, an important role in the proof is played by the natural mapping $f_i : \Delta_i \times T^2 \to D_i = \cup_{\xi \in \Delta_i} T_\xi^2$, where Δ_i are open intervals into which the critical momenta partition the circle S_x^1. See Fig 76.

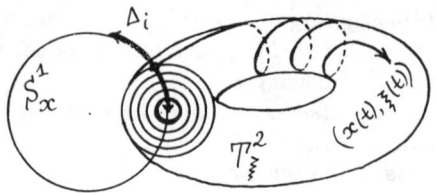

Figure 76

Theorem 5.2.1 can be reformulated as follows. Let M^2 be a real-analytic two-dimensional compact closed connected manifold endowed with an arbitrary real-analytic Riemannian metric. If the genus of a manifold is higher than unity then the geodesic flow of this metric (as the Hamiltonian flow on a four-dimensional manifold T^*M^2) is not completely Liouville integrable, that is, does not have an additional (second) integral which is independent of the energy integral and is in involution with it.

2.2 Nonintegrability of Geodesic Flows on High-Genus Riemann Surfaces with Convex Boundary

Let M^2 be a connected compact two-dimensional analytic Riemannian manifold with the boundary ∂M homeomorphic to a circle (that is, the boundary is connected). The way of obtaining such a manifold is well known from elementary topology. One should remove from a sphere with g handles a certain number of sufficiently small two-dimensional open nonintersecting disks (Fig. 77). In other words, such a manifold is homeomorphic to a two-dimensional flat domain (whose boundary is connected) with g handles.

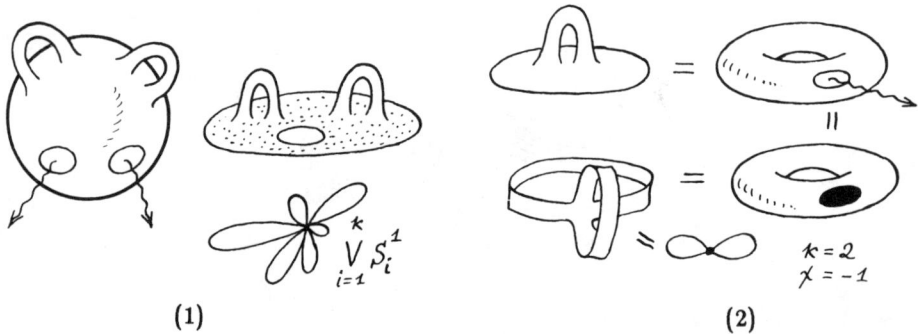

Figure 77

Consider the Euler characteristic $\chi(M^2)$ of such a surface. By definition, it is equal to $1 - \operatorname{rank} H_1(M, \mathbb{Z})$. We assume the surface M to be connected from the homotopic point of view, a surface with a non-empty boundary ∂M is homotopically equivalent to the so-called "bouquet" of several circles, i.e. is deduced from a finite set of circles by way of their gluing to one another at one point (Fig. 77).

The number k of such circles is precisely equal to the rank of the one-dimensional integer-valued homology group $H_1(M, \mathbb{Z})$. Moreover, this group is free (in the case of a non-empty boundary) and is isomorphic to $\mathbb{Z} \oplus \cdots \oplus \mathbb{Z}$ (k times) $\approx \mathbb{Z}^k$. For instance, removing one disk from a two-dimensional torus, we obtain with one hole homotopically equivalent to a bouquet of two circles, that is, in our example $k = 2$ and $\chi(M) = 1 - 2 = -1$ (Fig. 77).

Note that removing disks one by one from the surface M with a non-empty boundary, we, generally speaking, change of course the Euler characteristic of the surface.

Consider on M the geodesics defined by a prescribed Riemannian metric. The boundary ∂M (generally speaking, disconnected) is called *locally geodesically convex* if any two sufficiently close points x and y that lie on the boundary are joined by a single minimal geodesic wholly lying within the manifold M (Fig. 78).

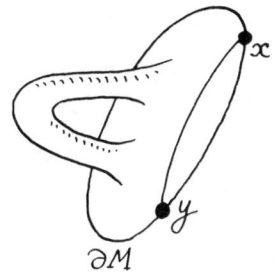

Figure 78

Consider a cotangent bundle T^*M and let H be a quadratic form setting the Riemannian metric on M. At each point $x \in M$ it sets a positive definite form

$$H(x, \xi) = \sum_{ij} g_{ij}(x) \xi_i \xi_j.$$

The geodesic flow has the function $H(x, \xi)$ as a first integral. Examine three-dimensional level surfaces Q_h^3 of the integral H, that is, $Q_h^3 = ((x, \xi) \in T^*M, H(x, \xi) = h = \text{const})$. As we already know (see §2.1), these surfaces fibre over the surface M with a circle as a fibre provided that h is greater than zero. Fix, for simplicity, the value of h to be equal to zero and examine a three-dimensional manifold Q_1^3.

THEOREM 5.2.4 (BOLOTIN). *Let M^2 be a connected Riemannian compact two-dimensional real-analytic manifold with a locally geodesically convex boundary and such that $\chi(M) < 0$. Then the geodesic flow of the Riemannian metric on the three-dimensional manifold of constant energy $h = 1$ does not have an analytic first integral (which is independent of the energy integral and is in involution with it on T^*M).*

If the boundary of the manifold is empty, we arrive at Theorem 5.2.1 from §2.1. Thus, Theorem 5.2.4 generalizes Theorem 5.2.1.

The proof of Theorem 5.2.4 rests on the existence of an infinite number of hyperbolic closed geodesics. Instead of the homology group in [61] for the proof of Theorem 5.2.3 (see §2.1. above) in the case of an empty boundary ∂M one should take here another topological invariant of the surface. Let $\Gamma(M)$ be a set of free homotopy classes in M and $\prod(M) \subset \prod(M)$ a set of free homotopy classes containing curves multiple of simple closed curves. If the characteristic of the surface is negative then the set $\prod(M)$ is infinite. Indeed, in this case rank $H_1(M) > 1$, and therefore on the surface there exist at least two independent single cycles, and their various linear combinations with mutually simple coefficients generate an infinite set $\prod(M)$. Note that the fundamental group $\pi_1(M)$ of the indicated surface is a free group with $1 - \chi(M)$ generators, that is, the number k of generators in $\pi_1(M)$ is equal to the rank of $H_1(M, \mathbb{Z})$. The one-dimensional homology group $H_1(M, \mathbb{Z})$ is isomorphic here to a free Abelian group $\mathbb{Z}^k$.

Thus, Theorem 5.2.4 will be valid if we prove the following assertion.

LEMMA 5.2.2. *If there exists an inconstant analytic first integral of the geodesic flow on Q_1^3, then the set $\prod(M)$ is finite.*

PROOF: Let f be an analytic first integral of the geodesic flow on Q_1^3. Then on the compact Q_1^3 the integral f has a finite set $\sum$ of critical values. We put

$$W = Q_1^3 \backslash f^{-1} \sum .$$

Let $\Lambda \subset \Gamma(M)$ be a set of homotopy classes containing images of closed trajectories of the geodesic flow in W under the projection $\pi : Q_1^3 \to M$. The set $\prod(M) \backslash \Lambda$ is finite.

Indeed, any element of the set $\prod(M) \backslash \Lambda$ is a homotopy class of curves which are not images under the mapping π of the non-self-intersecting closed trajectory of the geodesic flow on Q_1^3, on which the integral f takes up a regular value. There is a finite number of such homotopy classes. The proof presented by Kozlov in [61] for the case of an empty ∂M rests on Gaydukov's existence theorem [168] for asymptotic geodesics to closed geodesics from the indicated homotopy classes. This proof is also applicable in the general case, if instead of Gaydukov's theorem we use its extension to Riemannian manifolds with convex boundaries. We leave the details to the reader.

It remains to show that the set $\prod(M) \cap \Lambda$ is finite. For any point $x \in W$, let $\Gamma_x \subset W$ be such a connected component of the set $(f = f(x))$ that contains the point x. The difficulty is at as distinct from the case of an empty ∂M, the Liouville theorem is not applicable here: not all the surfaces Γ_x (where $x \in W$) are tori.

LEMMA 5.2.3. *Let $V \subset W$ be a connected component of W containing a periodic trajectory of geodesic flow. Then Γ_x is a torus for all $x \in V$.*

PROOF: By the Liouville theorem, Γ_x is a torus if $\Gamma_x \cap Q_1^3 = \emptyset$. By virtue of convexity of ∂M, there exists such a neighbourhood U of the boundary ∂Q_1^3 in Q_1^3 that each trajectory of the geodesic flow on Q_1^3 coming into U leaves Q_1^3 transversally to ∂Q_1^3. For this reason Γ_x is a torus if and only if $\Gamma_x \cap U = \emptyset$. From this and from the implicit function theorem it follows that the set of points $x \in V$, such that Γ_x is a torus, is open and closed in V. Show that this set is non-empty.

Let $g^t : Q_1^3 \to Q_1^3$ is a geodesic flow, $f^s : Q_1^3 \to Q_1^3$ a phase flow of the integral f and let $x_0 \in V$ be such a point that $t \to g^t x_0; t \mod \tau$ is a periodic trajectory. Then Γ_{x_0} is a torus. By the Liouville theorem it suffices to show that the mapping $(t, s) \to g^t f^s x_0$ is specified for all $(t, s) \in \mathbb{R}^2$.

If $t \to g^t(f^s x_0)$ is specified for all $t \mod \tau$, then $t \to g^t(f^s x_0)$ is a periodic trajectory, so that $g^t f^s x_0 \notin U$. Consequently, the set of values $s \in \mathbb{R}$, such that $g^t f^s x_0$ is specified for all $t \mod \tau$, is open and closed in $\mathbb{R}$ as required.

From the lemma and from finiteness of the set of critical values of f it follows that the number of connected components of W containing periodic trajectories of geodesic flow is finite. Each of the components is homotopically equivalent to a torus. Let $V_1, \ldots, V_n$ be the indicated connected components of W. Then $\pi_1(V_i)$ is a commutative group, so that under the projection $\pi : Q_1^3 \to M$ the image $\pi_* \pi_1(V_i)$ is a cyclic subgroup of the free group $\pi_1(M)$. Therefore, $\prod(M) \cap \pi_* \Gamma(V_i)$ is a one-element set and $\bigcup_i \prod(M) \cap \pi_* \Gamma(V_i) = \prod(M) \cap \Lambda$ is a finite set. This completes the proof.

2.3 Nonintegrability of the Problem of n Gravitating Centres for $n > 2$

Let $z_1, \ldots, z_n$ be distinct points of a complex plane C. The Hamiltonian function of the problem of n centres is of the form

$$H = \frac{1}{2}|p|^2 + V(z); \qquad (z, p) \in T^* U = U \times \mathbb{C},$$

where $U = \mathbb{C} \setminus (z_1, \ldots, z_n)$ is the configuration space and V the gravitational potential of the point $z \in U$ by the points $z_1, \ldots, z_n$, that is, $V(z) = -\sum_{i=1}^n \mu_i/|z - z_i|$; $\mu_i > 0$. Since $V < 0$ then for $h > 0$ the level $(H = h)$ of the energy integral H is an analytic hyperplane in the phase spacing Q_1^3.

THEOREM 5.2.5 (BOLOTIN [169]). *Let $n > 2$. Then for $h \geqslant 0$ the problem of n centres does not have analytic first integrals on the energy integral level $(H = h)$.*

Thus, the problem of n gravitating centres is Liouville integrable in the domain $(H \geqslant 0)$ only in the Kepler and Euler cases.

The question of integrability of the problem of n gravitating centres in the domain $(H < 0)$ remains open. The classical Poincaré methods of perturbation theory are apparently inapplicable for the proof of nonintegrability of this problem. The proof of Theorem 5.2.3 rests on the existence of an infinite number of unstable periodic solutions of non-negative energy level.

Reduce the problem to the study of geodesic flow. For our further purposes we make use of the following properties of the potential V.

1) $V(z) < 0$ for $z \in U$.
2) $V(z) + \frac{1}{2}\langle z, V_z \rangle \langle 0$ for $|z| \to \infty$.
3) The function $\varsigma \to V(z_i + \varsigma^2)|\varsigma|^2$ is continued up to the analytic function in the neighbourhood of $\varsigma = 0$ which is distinct from zero at the point 0.

Let $h \geqslant 0$. The Levi–Civita regularization reduces the phase flow of the problem of n attracting centres at the energy level $(H = h)$ to the geodesic flow on the Riemann surface.

LEMMA 5.2.4. *There exists a Riemann surface M, a holomorphic mapping $\pi : M \to \mathbb{C}$, and a complete Riemannian metric g_h on M, such that:*
  1) *$\pi^{-1}(z_i)$ consists of one point, $i = 1, \ldots, n$;*
  2) *$\pi : \tilde{U} = \pi^{-1}(U) \to U$ is a two-sheeted covering;*
  3) *$\pi_*^{-1}(H = h) = T_1^* \tilde{U} := (g_h = 1) \cap T^* \tilde{U}$.*
  Here $\pi_* : T^* \tilde{U} \to T^* U$ is a mapping induced by π.

PROOF: Let M be a Riemann surface of the function

$$\sqrt{(z - z_1) \cdots (z - z_n)}$$

and $\pi : M \to \mathbb{C}$ a projection. Then $z = \pi$ is a coordinate on $\tilde{U} = \pi^{-1}(U) \subset M$ and $\sqrt{z - z_i}$ a coordinate in the neighbourhood of the point $\pi^{-1}(z_i) \in M$. The first two assertions of the lemma are obvious. Prove now the third assertion. Specify a Jacobi metric

$$g(z, p) = |p|^2 / 2(h - V(z))$$

on $T^* U$ by the condition $(H = h) = (g = 1)$ and lift it by the mapping $\pi : \tilde{U} \to U$ up to the Riemannian metric $\pi^* g$ on $\tilde{U}$. Show the Riemannian metric obtained continues up to the analytic Riemannian metric in the neighbourhood of the point $\pi^{-1}(z_i)$. Let p_z, p_ς be momenta conjugate to the coordinates z and $\varsigma = \sqrt{z - z_i}$. Then $p_z = p_\varsigma / 2\varsigma$. Therefore, in the coordinates ς, p_ς the metric $\pi^* g$ is the form

$$\pi^* g = \frac{|p_z|^2}{2(h - V(z))} = \frac{|p_\varsigma|^2}{8(h|\varsigma|^2 - V(z_i + \varsigma^2)|\varsigma|^2)}.$$

By virtue of the property 3 of the potential V, this formula defines the analytic Riemannian metric in the neighbourhood of the point $\pi^{-1}(z_i)$. Completeness of this metric follows from the properties 1 and 2 of the potential.

From the lemma just proved there follows the Maupertuis principle. The mapping $\pi_* : T_1^* \tilde{U} \to T^* U$ carries the trajectories of the geodesic flow of the metric g_h on $T_1^* \tilde{U}$ into the trajectories of the problem of n gravitating centres on the energy level $(H = h)$. Find the connection between the integrals of the problem of n gravitating centres and of the geodesic flow on $T_1^* M = Q_1^3$.

Let f be an analytic first integral of the n-body problem on the energy level $(H = h)$. Then there exists an analytic first integral $\tilde{f}$ of the geodesic flow on $T^* M$ such that $\tilde{f} = f \circ \pi_*$ on $T^* \tilde{U}$.

PROOF: By the Maupertuis principle, the function $f \circ \pi_*$ on $T_1^* \tilde{U}$ is an analytic first integral of restriction of the geodesic flow

$$g^t : T_1^* M \to T_1^* M$$

to $T_1^* \tilde{U}$. Here $Q_1^3 = T_1^* M$. Define the function $\tilde{f}$ on $T_1^* M = Q_1^3$ setting $\tilde{f} = f \circ \pi_*$ on $T_1^* \tilde{U}$ and

$$\tilde{f} = f \circ \pi_* \circ g^t$$

on $g^{-t} T_1^* \tilde{U}$, where $t > 0$ is so small that

$$g^{-t} T_1^* \tilde{U} \cup T_1^* \tilde{U} = T_1^* M.$$

Then $\tilde{f}$ is a well defined analytic first integral of the geodesic flow on T_1^*M.

To complete the proof, it remains to show that the geodesic flow on T_1^*M does not have analytic first integrals.

The proof is based on topological considerations. Applying the Riemann–Hurwitz formula to the mapping $\pi : M \to \mathbb{C}$ we deduce that the Euler characteristic

$$\chi(M) = 2 - n < 0$$

for $n > 2$.

By Theorem 5.2.3 (see [61]), the geodesic flow on a compact Riemann surface with a negative Euler characteristic does not have analytic first integrals. The surface M is noncompact, but M contains a compact submanifold with a geodesically convex boundary and homotopy equivalent to M.

Indeed, let $D \subset \mathbb{C}$ be a disk of a sufficiently large radius centered at the point $0 \in \mathbb{C}$. Then the set $\pi^{-1}(D)$ is compact and homotopy equivalent to M. Show that a geodesic tangent to $\partial \pi^{-1}(D)$ at a certain point has no other common points with $\pi^{-1}(D)$. By the Maupertuis principle, the image of a geodesic under the mapping $\pi : M \to \mathbb{C}$ is the trajectory of the motion (with the energy h) tangent to ∂D. Along such a motion we have

$$\frac{1}{2}(|z|^2)'' = 2(h - V(z)) - \langle z, V_z \rangle < 0; \qquad |z| \to \infty \quad \text{as required.}$$

Thus, the theorem reduces to the extension of the results of the paper [61] to surfaces with convex boundaries. But this theorem has already been proved above, in §2.2. This completes the proof of Theorem 5.2.5.

2.4 Nonintegrability of Several Gyroscopic Systems

We present the results obtained by developing the ideas expressed in §§2.1 and 2.2.

A mechanical system with the configuration space M and the kinetic energy T will be called *gyroscopic* if besides the potential forces given by the potential energy V it is affected by the *gyroscopic forces* given by the differential 2-form Γ on M. In local coordinates Γ, is a *skew-symmetric matrix of gyroscopic forces*. The differential form Γ is usually assumed to be closed: $d\Gamma = 0$. Then the gyroscopic system is a Hamiltonian system with the Hamiltonian function $H = T + V$ on the phase space T^*M and with the Hamiltonian structure $\Omega + \pi^*\Gamma$, where Ω is a standard Hamiltonian (symplectic) structure on T^*M and $\pi : T^*M \to M$ a projection.

To apply the variational calculus methods, one should know when a gyroscopic system is *Lagrangian*, i.e is described by the Lagrange equations with the Lagrange function L on TM. For this, it is necessary and sufficient that the form of the gyroscopic forces be exact, that is, $\Gamma = d\Lambda$ where Λ is the differential 1-form on M.

Then the symplectic structure on T^*M is of the form $\Omega + \pi^*\Gamma = d(p \cdot dq + \pi^*\Lambda)$, where $p \cdot dq$ is the canonical 1-form on T^*M. The transformation $f : T^*M \to T^*M; p \in T_q^*M \to p + \Lambda(q) \in T_q^*M$ carries $\Omega + \pi^*\Gamma$ into the standard Hamiltonian structure on T^*M, that is, $f^*\Omega = \Omega + \pi^*\Gamma$. The corresponding Hamiltonian function $H_\Lambda = f^*H$ on T^*M has the form

$$H_\Lambda(p, q) = T(p + \Lambda(q), q) + V(q); \qquad p \in T_q^*M,$$

and the Lagrange function L_Λ on TM obtained by the Lagrange transformation is

$$L_\Lambda(\dot{q},q) = T(\dot{q},q) - V(q) - \Lambda(q) \cdot \dot{q}; \qquad \dot{q} \in T_q M.$$

Let the gyroscopic system under consideration have two degrees of freedom. Then the condition $d\Gamma = 0$ is automatically satisfied, and by the de Rham theorem the gyroscopic system is Lagrangian if M is noncompact or if M is compact and $\iint_M \Gamma = 0$. If M is compact and $\iint_M \Gamma \neq 0$, then the gyroscopic system is not Lagrangian. In what follows, the latter three cases are considered separately.

EXAMPLE 1: For a gyroscopic system which arises on the Poisson sphere M, we have in the rigid-body dynamics, after the removal of the area integral, $\Gamma = \sigma dS$, where σ is the *value of the area constant* and dS the *element of the area on the sphere* M. Therefore, $\iint_M \Gamma = 4\pi\sigma$, so that for $\sigma \neq 0$ the gyroscopic system is not Lagrangian.

The results formulated below were obtained by Bolotin. We may assume without loss of generality that the configuration space M is an analytic manifold. Bearing in mind applications to celestial mechanics, we assume the potential energy V to have singularities at the points of a finite set $\sum \subset M$. Then the phase space of the gyroscopic system is $T^*(M \backslash \sum)$.

Let $H = T + V$ be the energy integral on $T^*(M \backslash \sum)$. Bolotin has shown that for Lagrangian gyroscopic systems with two degrees of freedom the presence of $n > 2\chi(M)$ Newton-type singularities of the potential energy hampers the existence of analytic first integrals on $T^*(M \backslash \sum)$ independent of the energy integral H.

In the case $n = 2\chi(M)$, the presence of nontrivial gyroscopic forces (for instance, $\iint_M \Gamma \neq 0$) hampers the existence of first integrals of class C^1 on $T^*(M \backslash \sum)$ polynomial with respect to momentum and independent of H. Here $\chi(M)$ is the Euler characteristic of M. Exact formulations are presented hereafter.

EXAMPLE 2: The finite many-body problem. Let n points $z_1,\ldots,z_n$ be fixed in a plane M rotating around a point $0 \in M$ at a constant angular velocity ω and let a point $z \in M$ move under the action of gravitational attraction by the points $z_1,\ldots,z_n$. The potential energy

$$V(z) = -\left(\sum_{i=1}^n \mu_i/|z - z_i| + \omega^2|z|^2/2\right); \mu_i > 0$$

has the singularities on the set $\sum = \{z_1,\ldots,z_n\} \subset M$. The matrix of the gyroscopic forces is given by $2\begin{pmatrix} 0 & -\omega \\ \omega & 0 \end{pmatrix}$, and the form of the gyroscopic forces is given by $\Gamma = 2\omega dS$, where dS is an area element on M.

This gyroscopic system is integrable for $n = 1$ and for all ω (the Kepler problem), as well as for $n = 2$ and $\omega = 0$ (the Euler problem). For $n = 2$ and $\omega \neq 0$ (the finite three-body problem) Poincaré proved non-existence of first integrals analytic with respect to parameters μ_i and independent of the Jacobi integral H, and Siegel proved non-existence of algebraic first integrals independent of H. When $n > 2$, it turns out that for all ω there exist no analytic first integrals of the many-body problem which are independent of the Jacobi integral. Since $\chi(M) = 1$, these properties of the finite many-body problem are in agreement with the general assertion presented above.

Now proceed to exact formulations. Let the form Γ of gyroscopic forces be a differential form of class C^1 on M. Wherever we deal with first integrals of class C^1, polynomial in momentum, the kinetic energy T is assumed to belong to $C^1(T^*M)$ and the potential energy V to belong to $C^1(M\backslash\sum)$, and wherever we deal with analytic first integrals, T and V are assumed to be analytic functions, that is, $T \in C^\omega(T^*M)$, $V \in C^\omega(M\backslash\sum)$. For analytic gyroscopic systems, the indicated classes of first integrals are related as follows.

LEMMA 5.2.5. *Let*
$$T \in C^\omega(T^*M), \quad V \in C^\omega(M\backslash\sum), \quad \Gamma \in C^\omega.$$
If $F \in C^1(T^*(M\backslash\sum))$ is a first integral of a gyroscopic system polynomial in momenta, then $F \in C^\omega(T^*(M\backslash\sum))$.

This follows from the regularity theorem for analytic elliptic differential operators: if F is a polynomial in momentum, then the condition that the Poisson bracket $\{H, F\}$ is identical zero determines the elliptic differential equation for the coefficients of the polynomial F.

By definition, the potential V has *Newton-type singularities* on the set $\sum = (q_1, \ldots, q_n) \subset M$ if in the neighbourhood of the point q_i, in the coordinates z, conformal with respect to the Riemannian metric T, there holds the relation:
$$V(z) = \frac{\Phi(z)}{|z|},$$
where the function Φ does not have a singularity at the point $z = 0$ (that is, $\Phi \in C^1$ if $V \in C^1$ if $V \in C^1(M\backslash\sum)$) and $\Phi \in C^\omega$ if $V \in C^\omega(M\backslash\sum)$ and $\Phi(0) < 0$.

Let the manifold M be compact. Recall that if $\iint_M \Gamma = 0$ then $\Gamma = d\Lambda$, so that the function V_Λ on $M\backslash\sum$ is defined as follows:
$$V_\Lambda(q) = H_\Lambda(0, q) = T(\Lambda, (q), q) + V(q).$$

PROPOSITION 5.2.1. *Let the manifold M be compact and*
$$\iint_M \Gamma = 0, \quad h > \sup_M V_\Lambda.$$
Then on the hypersurface $(H = h) \subset T^*(M\backslash\sum)$ there exist no analytic functions which are first integrals of the gyroscopic system.

According to the above proposition, we imply here the condition
$$T \in C^\omega(T^*M), \quad V \in C^\infty(M\backslash\sum).$$

PROPOSITION 5.2.2. *Let the manifold M be compact and let $n \geqslant 2\chi(M)$ and $\iint_M \Gamma \neq 0$. Then on $T^*(M\backslash\sum)$ there exist no functions of class C^1, polynomial in momentum, which are first integrals of the gyroscopic system and are independent of the energy integral.*

The conditions $n > 2\chi(M)$ and $n \geqslant 2\chi(M)$ in Propositions 5.2.1 and 5.2.2 cannot be weakened. When $n \leqslant 2\chi(M)$ and $\iint_M \Gamma = 0$ and when $n < 2\chi(M)$ and

$\iint_M \Gamma \neq 0$, in the gyroscopic system there may exist first integrals, polynomial in momenta and independent of the energy H, while when $n = 2\chi(M)$ and $\iint_M \Gamma \neq 0$, there may exist analytic first integrals independent of the energy H.

EXAMPLE 3: Let $M = \mathbb{R}^2/\mathbb{Z}^2$ be a standard two-dimensional torus, the form T be given by the Euclidean metric on a plane, the potential V be identical zero, and the matrix of gyroscopic forces of the form $\begin{pmatrix} 0 & -\omega \\ \omega & 0 \end{pmatrix}$. Then $n = 2\chi(M) = 0$ and $\iint_M \Gamma = \omega \iint_M dS = \omega$. For $\omega = 0$, the constructed gyroscopic system has on T^*M the first integral $F(p,z) = p$ linear in momenta. For $\omega \neq 0$, by virtue of Proposition 5.2.2 there exist no first integrals polynomial in momenta and independent of the energy integral $H = T = \frac{1}{2}|p|^2$. However, for any analytic doubly periodic function f on a plane, the function $F(p,z) = f(p - \Gamma z)$ on T^*M is an analytic first integral of the constructed gyroscopic system.

Note that for $n = 0$, $\chi(M) < 0$, and $\Gamma = 0$, Proposition 5.2.1 coincides with Kozlov's theorem [61] and Proposition 5.2.2 with Kolokol'tsov's theorem [63].

To extend the previous results to the case of a noncompact configuration space M, additional conditions at infinity are needed which are different for the proof of non-existence of integrals analytic and polynomial in momenta. These conditions will be presented in the form convenient for application to the finite many-body problem.

There exists a compact two-dimensional manifold $\bar{M} \supset M$ such that $\bar{M}\setminus M$ is a finite or a countable set of infinitely remote points. Let $U \subset \bar{M}$ be a neighbourhood of $\bar{M}\setminus M$ diffeomorphic to the union of a finite or a countable number of disks. Suppose that the metric determined by a kinetic energy T on the manifold M satisfies the following condition: each closed curve in $U \cap M$, embracing one of the infinitely remote points, cannot be deformed to infinity in $U \cap M$ in the class of bounded-length curves. Represent the form of the gyroscopic forces as $\Gamma = \omega \cdot dS$, where ω is a function on M, and dS is an area element on M. The surface M may always be assumed orientable because in the contrary case it suffices to pass over to an appropriate two-sheeted covering of the nonorientable surface. Suppose that the function ω preserves sign. In case the manifold M is noncompact, the gyroscopic systems always Lagrangian, namely, $\Gamma = d\Lambda$. Suppose, the function V_Λ is bounded form above on the manifold M.

PROPOSITION 5.2.3. *Let a manifold M be noncompact and let the above assumptions be valid. Then for $n > 2\chi(M)$ and $h > \sup_M V_\Lambda$ the gyroscopic system does not have analytic first integrals on the hypersurface $(H = h) \subset T^*(M\setminus \Sigma)$.*

EXAMPLE 4: The finite many-body problem. Here $T = \frac{1}{2}|p|^2$ is the Euclidean metric on a plane and
$$\Gamma = 2 \begin{pmatrix} 0 & -\omega \\ \omega & 0 \end{pmatrix} = 2\omega dS,$$
where dS is an area element on the plane and $\Lambda = \Gamma z \cdot dz$. Therefore,
$$V_\lambda(z) = \frac{1}{2}|\Gamma z|^2 + V(z) = -\sum_{i=1}^{n} \frac{\mu_i}{|z - z_i|}.$$

By virtue of proposition 5.2.3, the finite many-body problem does not have analytic first integrals on the energy level

$$(H = h) \subset T^*(M\backslash\textstyle\sum), \qquad h > 0,$$

for $n > 2$. In reality, this is also true for $h = 0$. When $n = 2$ and $\omega = 0$, this problem is integrable.

Suppose that the metric determined by the kinetic energy T is *Euclidean at infinity*. This means that in M a certain neighbourhood of each of the infinitely remote points of $\bar{M}\backslash M$ is isometric to the neighbourhood of infinity on a Euclidean plane.

PROPOSITION 5.2.4. *Let a manifold M be noncompact and the kinetic energy T be Euclidean at infinity. Let $n > 2\chi(M)$ or $n = 2\chi(M)$, but let the integral $\iint_M \Gamma$ be absolutely convergent and not equal to zero. Then the gyroscopic system does not have on $T^*(M\backslash\sum)$ any first integrals of class C^1, polynomial in momenta and independent of the energy integral.*

By definition, the integral $\iint_M \Gamma$ is absolutely convergent if $\iint_M |\omega| dS < \infty$, where dS is an area element on M and $\Gamma = \omega \cdot dS$. This theorem is not applicable to the finite three-body problem, because for it $n = 2\chi(M) = 2$, and the integral $\iint_M \Gamma = 2\omega \iint_M dS$ diverges.

Let $n = 2\chi(M) > 0$. On the manifold M, a function f, holomorphic in the conformal structure on M given by the Riemannian metric T is existent and unique up to multiplication by a constant. This function has simple zeros at the points of the set $\sum$, does not have other zeros, and has poles of order n at infinity. Let $\widetilde{M}$ be a Riemann surface of the function $\sqrt{f}$ and $\pi : \widetilde{M} \to M$ a projection. Then $\sqrt{f}$ is a single-valued function on $\widetilde{M}$. Define $\iint_M \Gamma/\sqrt{f}$ by the formula

$$\iint_M \Gamma/\sqrt{f} = \frac{1}{2} \iint_{\widetilde{M}} \pi^*\Gamma/\sqrt{f}$$

PROPOSITION 5.2.5. *Let a manifold M be noncompact, the kinetic energy T be Euclidean at infinity, and $n = 2\chi(M) > 0$. Let f be a holomorphic function on M which has simple zeros at the points of the set $\sum$. If the integral $\iint_M \Gamma/\sqrt{f}$ is absolutely convergent and not equal to zero, then the gyroscopic system does not have on $T^*(M\backslash\sum)$ any first integrals of class C^1, polynomial in momenta and independent of the energy integral.*

§3 Topological Obstacles for Analytic Integrability of Geodesic Flows on Non-Simply-Connected Manifolds

We have already seen that in the two-dimensional case non-simply-connectedness of a manifold may affect integrability of a system: if the fundamental group is "sufficiently large," there are no analytic integrals. In this section we generalize Theorems 5.2.2 and 5.2.3 on analytic integrability of geodesic flows of Riemannian metrics on two-dimensional high-genus surfaces. This generalization has been recently obtained by Taymanov and is valid for Riemannian analytic manifolds of arbitrary dimension, whose fundamental groups are "sufficiently large."

Let M^n be a closed manifold. Consider the geodesic flow on M^n which on the cotangent bundle T^*M^n (endowed with a natural symplectic structure) is Hamiltonian with the Hamiltonian $H(x,p) = \frac{1}{2}\sum g_{ij}p_i p_j$, where $x \in M$, p_i are coordinates in the fibre of the cotangent bundle $T^*_x M$ and $g_{ij}(x)$ is the Riemannian metric on M^n.

By the Liouville theorem, for a sufficiently good description of a geodesic flow one should find a set of n functionally independent involutive first integrals of the flow. One integral is already available—this is the Hamiltonian H. A geodesic flow can therefore be always reduced to the constant energy level $H = 1$, which we will denote by L. The momentum mapping $F : L \to \mathbf{R}^{n-1}$, $F(q) = (I_1(q), \ldots, I_{n-1}(q))$, where $I_1, \ldots, I_{n-1}$, $I_n = H$ is a complete set of functionally independent first integrals in involution, is given on a manifold L. We will specify the concept of a *geometrically simple set of first integrals*.

DEFINITION 5.3.1: A complete set of involutive first integrals of a geodesic flow $I_1, \ldots, I_{n-1}$, $I_n = H$, is called *geometrically simple* if:

1) The surface L contains a closed set Γ such that $L\backslash\Gamma$ is open and everywhere dense, has a finite number of linear connectedness components $L\backslash\Gamma = \cup^t_{\alpha=1} U_\alpha$, and on $L\backslash\Gamma$ the momentum mapping F has a maximal rank.

2) The mapping $F : U_\alpha \to F(U_\alpha)$ is fibring into Hamiltonian n-dimensional tori over the domains $F(U_\alpha)$ homeomorphic to $(n-1)$-dimensional disks: $F(U_\alpha) \approx D^{n-1}$.

3) For any point $q \in L$ there exists a neighbourhood $W(q)$ such that $W(q) \cap (L\backslash\Gamma)$ has a finite number of linear connectedness components.

Geometrically simple complete involutive sets of integrals appear to exist only on manifolds with a "sufficiently small" fundamental group (or one-dimensional homology group). More precisely, we have the following assertions.

THEOREM 5.3.1 (TAYMANOV). *If the fundamental group $\pi_1(M^n)$ of a closed Riemannian manifold M^n is not almost commutative, that is, does not contain a commutative subgroup of finite index, then the geodesic flow on M^n does not admit a geometrically simple set of first integrals.*

THEOREM 5.3.2 (TAYMANOV). *If $\dim H_1(M^n, \mathbf{Q}) > \dim M^n$, then the geodesic flow on M^n does not admit a geometrically simple complete set of first integrals.*

If the configuration space is two-dimensional, then Theorem 5.2.2 obviously follows. Note that Theorems 5.3.1 and 5.3.2 are applicable to geodesic flows on cotangent bundles to Riemannian manifolds. It would be of interest to deduce analogous statements for an arbitrary symplectic manifold, that is, to find out what topological invariants hamper the existence of a complete set of involutive integrals. We will develop the results of Ch. 2 in this direction.

Consider the analytic case. Let M^n be an analytic manifold with an analytic metric. In this case, using the results obtained by Gabrielov [173], [174], which are a continuation of a paper by Loyasevich [175], one can prove the fact that a complete set of involutive analytic integrals is geometrically simple. This follows from semi-analyticity of the set of critical points of the momentum mapping $F : L \to \mathbf{R}^{n-1}$ and the subsequent constructiveness (in the sense of [173]) of the set of critical values, the complement of which in $\mathbf{R}^{n-1}$ has a finite number of linear connectedness components. Endow the set of critical values C_1 with an additional constructive set C_2 consisting of "partitions" with the result that the set $\mathbf{R}^{n-1}\backslash(C_1 \cup C_2)$ split into a finite number of disks. Then $\Gamma = F^{-1}(C_1 \cup C_2)$ and the condition 3 follows from the closedness of the family of constructive sets with respect to obtaining a complete pre-image under proper analytic mapping.

Thus, these remarks and Theorems 5.3.1 and 5.3.2 imply the following assertions.

THEOREM 5.3.3. *If the fundamental group $\pi_1(M^n)$ of an analytic closed Riemannian manifold M^n is not almost commutative, then the geodesic flow does not possess a complete involutive set of analytic functionally independent first integrals, i.e. is not Liouville integrable in the class of analytic first integrals.*

THEOREM 5.3.4. *If $\dim H_1(M^n, \mathbb{Q}) > \dim M^n$, then the geodesic flow on a closed analytic manifold M^n does not possess a complete involutive set of analytic functionally independent first integrals.*

REMARK 1: From the topological point of view, the almost commutativeness of the fundamental group $\pi_1(M^n)$ is equivalent to the existence of a finite-sheeted covering $p : \tilde{M} \to M$, where the group $\pi_1(\tilde{M})$ is commutative.

REMARK 2: It is known that if a closed manifold M admits the metric of strictly negative curvature, then the group $\pi_1(M)$ is noncommutative and the maximal commutative subgroup in $\pi_1(M)$ is infinite and cyclic. See [176]. It is clear that such fundamental groups $\pi_1(M)$ are not almost commutative.

Now we proceed to the proof of these theorems.

The general construction.

Take a finite set of indices $\Lambda = (\lambda_1, \ldots, \lambda_s)$ to numerate the vertices of a certain graph. Plot this graph. To this end, we associate in some way or other with each element of the group $g \in G$ an arrow starting from $\mu_1(g)$ and terminating at $\mu_2(g)$ where $\mu_i \in \Lambda$. Obtain an oriented graph $\Lambda'(\mu)$, where each edge of a graph has a label from the group G and to each element $g \in G\backslash\{1\}$ there corresponds an edge for which this element is a label.

From the graph $\Lambda'(\mu)$ obtain the graph $\Lambda(\mu)$ by way of supplement. Namely, if the element g labels the arrow from $\mu_1(g)$ to $\mu_2(g)$, then supplement the graph $\Lambda'(\mu)$ with the arrow from $\mu_2(g)$ to $\mu_1(g)$ with a label g^{-1}.

Examine the paths on the graph $\Lambda(\mu)$ by joining the edges of the graph into chains $\xi = \gamma_k \gamma_{k-1}, \ldots, \gamma_1$, where the terminal point of γ_l coincides with the initial point of γ_{l+1}. Let $\lambda_j \in \Lambda$ and examine the set of loops with the beginning and end at λ_j. Denote this set by M_j.

Let $\xi = \gamma_k \ldots \gamma_1$ be a loop and $m(\gamma_i)$ a label of the edge γ_j. Define $\hat{m}_j : M_j \to G$ as follows: $\hat{m}_j(\xi) = m(\gamma_k), \ldots, m(\gamma_1)$. Since all $\hat{m}_j$ are constructed in a similar manner, the subscript j is henceforth omitted.

CLAIM 5.3.1. *The set $\hat{m}(M_j)$ is a subgroup in G.*

The notation: $\hat{m}(M_j) = H_j$.

CLAIM 5.3.2. *The group G is a union of a finite number of left cosets by subgroups H_i, $i = 1, \ldots, s$.*

PROOF: Let λ_α and λ_β belong to Λ and be joined by the arrow from λ_α to λ_β. Take an arrow for each such pair and denote the associated label by $g_{\alpha\beta}$. Then we have:

$$G = \left(\cup^s_{\substack{\alpha=1 \\ \alpha \neq \beta}} \cup^s_{\beta=1} g_{\alpha\beta} H_\alpha \right) \cup \left(\cup^s_{\alpha=1} H_\alpha \right).$$

Prove that relation. Let another arrow from λ_α to λ_β, with a label $\eta \in G$, be given. Then $g_{\alpha\beta}^{-1} \eta \in H_\alpha$, and therefore, $\eta \in g_{\alpha\beta} H_\alpha$. On the other hand, to any element $g \in G \backslash \{1\}$ there corresponds an edge for which this element is a label, as required.

CLAIM 5.3.3. *If the group G is a group-theoretical union of a finite number of left cosets by subgroups $H_1, \ldots, H_s$, then at least one of these subgroups, a certain H_j, has a finite index in the group G, $[G : H] < \infty$.*

PROOF: We will construct the proof by the number of subgroups. For $s = 1$ the assertion is obvious. Let the assertion be proved for $s = N - 1$. This means that

$$G = \cup_{ij} g_{ij} H_i \quad \text{and} \quad [G : H_N] = \infty \quad (\ast)$$

Then a certain class hH_N does not enter in the notation $(\ast)$

$$hH_k \subset \cup_{i \neq N, j} g_{ij} H_i, \qquad H_k \subset \cup_{i \neq N, j} h^{-1} g_{ij} H_i,$$
$$G = \left(\cup_{i \neq N, j} g_{ij} H_i \right) \cup \left(\cup_{i \neq N, j, l} g_{N_j} h^{-1} g_{il} H_i \right).$$

Thus, we have reduced the problem to the case $s = N - 1$ for which the assertion is true. This completes the proof.

CLAIM 5.3.4. *If the element g has an infinite order, $g \in G$, then for any realization of $\Lambda(\mu)$ there exist k, l such that $g^k \in H_l$, where $k \neq 0$.*

PROOF: In $\Lambda(\mu)$ there exists an infinite number of arrows with labels g^m, $m \in N$, and the number of vertices is finite. Consequently, there exist two arrows, g^{m_1} and g^{m_2}, that join the same vertices λ_{l_1} and l_{l_2}. Then $g^{m_2 - m_1} \in H_{l_1}$, as required.

PROOF OF THEOREM 5.3.1: Let $x \in M^n$ and let V and U be such neighbourhoods of the point x that $x \in V \subset \bar{V} \subset U$, where $\bar{V}$ is the closure of V and over U (and therefore over V and $\bar{V}$) we have a universal covering, and the cotangent bundle is trivial. Let $p : L \to M^n$ be a standard projection, $p^{-1}(\bar{V}) \cap (L \backslash \Gamma) = \cup_\beta Q_\beta$. Introduce the equivalence relation between the near connectedness components : $Q_\alpha \sim Q_\beta$ if $Q_\alpha \subset P$ and $Q_\beta \subset P$, where $P \subset p^{-1}(U) \cap (L \backslash \Gamma)$ and is linearly connected. From each equivalence class choose one element and obtain (from the condition 3) of geometric simplicity) a finite set $(\Phi_1, \ldots, \Phi_s)$. Since the manifold M^n is complete, any element $g \in \pi_1(M)$ can be realized by a geodesic loop with the vertex at a point x. If a loop lies in Γ, then its initial data can be arbitrarily little "stirred" with the result that the new trajectory lying sufficiently close to the geodesic loop will go back to $p^{-1}(\bar{V})$, and if we join its terminal and initial points on M^n with the point x by line segments in $\bar{V}$, we will obtain the loop realizing the element $g \in \pi_1(M, x)$. Of importance here is the triviality of the universal covering over U.

Fix in Φ_i a point $\varphi_i \in \Phi_i$. Above we have associated with each element $g \in G$ a segment of the trajectory of a geodesic flow with the initial point at $Q_{\alpha(g)}$ and the end point at $Q_{\beta(g)}$. Since $Q_{\alpha(g)} \sim \Phi_k$ and $Q_{\beta(g)} \sim \Phi_l$, then we join the end and the beginning with φ_l and φ_k by segments in $p^{-1}(U) \cap (L\backslash\Gamma)$.

We now take $\Phi_1, \ldots, \Phi_s$ as vertices of the graph, and with each $g \in G\backslash\{1\}$ associate an edge beginning at φ_k and terminating at φ_l. It is obvious that a standard set of paths in $L\backslash\Gamma$ allows the loops on the graph to be associated with loops in $L\backslash\Gamma$.

From Claim 5.3.3 it follows that a certain H_j has in $G = \pi_1(M)$ a finite index. From an explicit geometric construction it is clear that

$$H_j \subset p_* \varphi_{\gamma*}(\pi_1(U_\gamma)),$$

where

$$L\backslash\Gamma = \bigcup_{\alpha=1}^t U_\alpha, \quad U_\alpha \approx T^n \times D^{n-1}, \quad \varphi_\gamma : U_\gamma \to L\backslash\Gamma$$

is embedding, $\Phi_j \subset U_\gamma$, H_j is noncommutative, and $p_* \varphi_{\gamma*}(\pi_1(U_\gamma))$ is a homomorphic image of the Abelian group. This is a contradiction which proves the assertion.

PROOF OF THEOREM 5.3.2: Consider a Hurewicz homomorphism $\lambda : \pi_1(M^n) \to H_1(M^n)$ and an embedding $\chi : H_1(M^n) \to H_1(M^n, \mathbb{Q})$. Let g belong to $\pi_1(M)$ and have an infinite order. By virtue of Claim 5.3.4 we have :

$$g^k \in H_l$$

and

$$\chi \circ \lambda(g) \subset p_* \varphi_{\gamma*}(H_1(T^n, \mathbb{Q})),$$

that is,

$$H_1(M^n, \mathbb{Q}) = \cup_{\alpha=1}^t (p\varphi_\alpha)_* H_1(T^n, \mathbb{Q}),$$

which is impossible if $\dim H_1(M^n, \mathbb{Q}) > \dim H_1(T^n, \mathbb{Q}) = \dim M^n$. This proves the theorem.

CLAIM 5.3.5. *Let M^n be a closed analytic manifold with an analytic Riemannian metric and $I_1, \ldots, I_{n-1}, I_n = H$, an involutive set of analytic first integrals of the geodesic flow. Then the set $I_1, \ldots, I_{n-1}, I_n = H$, is geometrically simple.*

PROOF: Following the paper [173], introduce the basic definitions.

1) Let M be a real-analytic manifold. A subset $A \subset M$ is *semianalytic* if in the neighbourhood of each point $x_0 \in M$ it can be represented in the form of a finite union of sets $(f_i(x) = 0, i = 1, \ldots, I; g_j(x) > 0, j = 1, \ldots J)$, where f_i and g_j are real-analytic functions in the neighbourhood of the point $x_0 \in M$.

2) A subset $A \in M$ is called *constructive* if it is the image of a certain relatively compact semianalytic set $B \subset N$ under a certain analytic mapping of manifolds $N \to M$.

In the paper [174] constructive sets are referred to as P- sets.

We will present several important facts concerning constructive and semianalytic sets.

3) Operations of finite combination, intersection and product, obtaining image and pre-image under a proper analytic mapping do not overstep the limits of the class of constructive sets [174].

4) The class of constructive sets is closed with respect to closures and additions, [174].

5) Each constructive set is a union of a finite number of constructive sets each of which is given by a system of analytic equations $(f_i(x) = 0, i = 1, \ldots, I)$ in the neighbourhood of any of its points, [174].

6) If M is a constructive set in $\mathbb{R}^N$, then the number of connectedness components of M is finite, [174].

7) Any constructive set is a finite union of connected constructive sets. Any connected constructive set is linearly connected, [174].

The properties 3-7 are proved in [174], but in the form suggested above they are included in [173].

8) Let $\{B_\nu\}$ be a locally finite family of semianalytic subsets of a finite-dimensional affine space $\tilde{L}$. Then there exists a locally finite symplical complex K, with $|K| = \tilde{L}$ and with a homeomorphism $\tau : \tilde{L} \to \tilde{L}$ (onto $\tilde{L}$), such that:

a) for any cell $\sigma \in K$ we have $\tau : (\sigma)$, an analytic submanifold in $\tilde{L}$, and $\tau_\sigma : \sigma \to \tau(\sigma)$, an analytic isomorphism;

b) for any cell $\sigma \in K$ and any B_ν we have either $\tau(\sigma) \subset B_\nu$ or $\tau(\sigma) \subset \tilde{L} \backslash B_\nu$.

We will now prove the assertion.

Let $F : L \to \mathbb{R}^{n-1}$ be the momentum mapping, $V \subset L$ the set of critical points of the mapping F (it is semianalytic), and $C_1 = F(V)$ the set of critical values of the mapping F, that is, $C_1 \subset \mathbb{R}^{n-1}$, and C_1 is constructive. Then $F(L) \cap (\mathbb{R}^{n-1} \backslash C_1) = \cup_{j=1}^{s} Y_j$, where Y_j are open domains and ∂Y_j are their boundaries. Note that $\partial Y_j \subset C_1$. Next, $C_1 = \cup_{\alpha=1}^{J} C_1^\alpha$, where C_1^α are locally analytic (see the property 5). Pay attention to another important statement from [174].

In a cube I^m of a space $\mathbb{R}^m$, let a basis $(x' = (x_1, \ldots, x_k), x'' = (x_{k+1}, \ldots, x_m))$ be chosen and M be a k-dimensional constructive subset (that is, under projection onto a certain k-dimensional subspace its image is dense, whereas under projection onto any $(k+1)$-dimensional subspace it is nowhere dense) in I^m. Then there exist open constructive sets Ω_p (where $p = 0, \ldots, T$) in the space (x'), such that:

1) $\cup_p \Omega_p$ is everywhere dense in I^k;

2) $M \cap (\Omega_0 \times (x'')) = \varnothing$;

3) $M \cap (\Omega_p \times (x''))(p = I, \ldots, T)$ coincides with the set $(x' \in \Omega_p, x_{k+j} = \bar{f}_{pj}(x'), j = 1, \ldots, m-k)$, where $\bar{f}_{pj}$ are p-valued analytic functions in Ω_p.

From this it follows that the partition $C_1 = \cup_{\alpha=1}^{J} C_1^\alpha$ can be so chosen that all C_1^α be homeomorphic to the domains in $\mathbb{R}^q$, $q = \dim C_1^\alpha$ and $C_1 = \cup_{\beta=0}^{n-1} \cup_{\alpha=1}^{J_\beta} C_{1,\beta}^\alpha$ where $\dim C_{1,\beta}^\alpha = \beta$ and $\partial C_{1,\beta}^\alpha \subset \cup_{\gamma=0}^{\beta-1} \cup_{\alpha=1}^{J_\gamma} C_{1,\gamma}^\alpha$. This implies triangularity of C_1. One can easily notice that triangulation extends to all Y_j and the "partitions" can be made constructive. Thus, $F(L)$ is triangulable and an $(n-2)$-dimensional skeleton is a constructive set. Here the $(n-2)$-dimensional skeleton

$$\langle F(L) \rangle^{(n-2)} = C_1 \cup C_2, \quad F^{-1}(C_1 \cup C_2) = \Gamma.$$

The properties of geometric simplicity now follow from the properties of the constructive sets 4 and 6.

Some papers by Loyasevich on triangulation of semianalytic sets are also presented in the work by Mather, "Stratifications and mappings," Preprint, Harvard, 1971, pp. 1–68. The definition of the constructive set according to Chevalley given in that work does not coincide with the definition according to Palmadov [173].

§4 Integrability and Nonintegrability of Geodesic Flows on Two-Dimensional Surfaces, Spheres, and Tori

4.1 The Holomorphic 1-Form of the Integral of a Geodesic Flow Polynomial in Momenta and the Theorem on Nonintegrability of Geodesic Flows on Compact Surfaces of Genus $g > 1$ in the Class of Functions Analytic in Momenta

In this section we consider the cases of complete Liouville integrability of geodesic flows on two-dimensional Riemann surfaces. We describe the cases of complete integrability as well as the cases of nonintegrability, that is, when the absence of analytic integrals can be proved.

We restrict our consideration to a special case of discovery of integrals polynomial in momenta. It turns out possible to describe those Riemannian metrics whose geodesic flows admit such integrals.

We recall the reader several definitions. Examine a symplectic manifold T^*M^n. Let local coordinates $q_1, \ldots q_n$ be given on M^n. Then the local coordinates on T^*M can be written in the form $q_1, \ldots, q_n, p_1, \ldots p_n$.

By the *geodesic flow of the Riemannian manifold* M with the metric $ds^2 = \sum g_{ij} dq_i dq_j$ we mean a Lagrangian system in the tangent bundle TM with the Langrange function $L = \sum g_{ij} \dot{q}_i \dot{q}_j$. Identifying TM with T^*M by means of the Riemannian metric, we may assume that the geodesic flow acts in T^*M. The corresponding system appears to be Hamiltonian, the Hamiltonian H being the Legendre transformation of the Lagrange function L, that is, $H = \sum a_{ij} p_i p_j$, where the matrix $A = (a_{ij})$ setting the Hamiltonian H is related to the matrix $G = (g_{ij})$ setting the function L as $A = \frac{1}{4} G^{-1}$.

To the integrals, polynomial in velocities in TM, there correspond the integrals polynomial in momenta (of a respective system) in T^*M.

The results expounded below have been obtained by Kolokol'tsov.

It should be noted that Theorems 5.3.1 and 5.4.3 are, in fact, a reformulation using modern terminology, of the results obtained by Birkhoff [307]. Birkhoff considered not only absolute but also conditional integrals (on a level surface).

THEOREM 5.4.1. *Suppose that in a domain* $U \subset \mathbf{R}^2$ *a Riemannian metric* $ds^2 = \lambda(x,y)(dx^2 + dy^2)$ *of smoothness class* C^1 *is given, such that the local geodesic flow determined by this metric, that is, a Hamiltonian system in* T^*U *with the*

Hamiltonian $H = \frac{1}{2}\lambda^{-1}(p_x^2 + p_y^2)$ has an integral

$$F_n + \sum_{m=0}^{n} b_m(x,y)p_x^{n-m}p_y^m, \quad n \geqslant 1$$

which is homogeneous in momenta. Then the function

$$R(z) = (b_0 - b_2 + \ldots) + i(b_1 - b_3 + \ldots) = \sum_{m=0}^{n} i^m b_m$$

is a holomorphic function of the complex variable $z = x + iy$. Furthermore, if in some other isothermic coordinates (u,v) related to (x,y) by the holomorphic transformation $w(z)$, where $w = u + iv$, the integral F_n is expressed by the formula

$$F_n = \sum_{m=0}^{n} c_m(u,v)p_u^{n-m}p_v^m,$$

then the holomorphic functions

$$R(z) \text{ and } S(w) = (c_0 - c_2 + \ldots) + i(c_1 - c_3 + \ldots) = \sum_{m=0}^{n} i^m c_m$$

are related by the formula $S(w) = (w'(z))^n R(z)$.

The latter formula obviously implies that in the neighbourhood of any point on U, which does not belong to the set of zeros of the function $R(z)$, the holomorphic 1-form

$$w = \frac{dz}{\sqrt[n]{R(z)}}$$

is invariantly defined, which is natural to think as a *holomorphic form associated with a polynomial integral*.

The proof of Theorem 5.4.1 is carried out by direct calculations using the formulae for the Poisson bracket of functions in the canonical symplectic structure of the space T^*U. This theorem suggests an interesting corollary.

THEOREM 5.4.2 (SEE [63]). *On a compact two-dimensional orientable manifold M of genus $g > 1$, a geodesic flow of smoothness class C^1 cannot possess an additional first integral analytic in momenta (continuously differentiable in T^*M) independent of the energy integral.*

Since an integral analytic in momenta can be expanded in a converging series of homogeneous polynomials, which are obviously invariantly defined in T^*M and are also integrals of a given geodesic flow, then the theorem formulated above is, in fact, equivalent to the one that follows.

THEOREM 5.4.2 (1). *Let H be the Hamiltonian of a geodesic flow on a compact orientable two-dimensional manifold M of genus $g > 1$. Then any homogeneous integral, polynomial in momenta, of this flow has the form $F = cH^m$, where $c \in \mathbf{R}$, $m \in \mathbf{Z}$.*

PROOF: Let F_n be a homogeneous polynomial integral of degree n. Then the function $R(z)$, defined by virtue of Theorem 5.4.1 by the integral F_n in arbitrary isothermic coordinates x, y is identical zero. In the contrary case one can construct on M, a globally defined meromorphic tensor field $\frac{dz^n}{R(z)}$, where $dz^n = dz \otimes \cdots \otimes dz$ (this is implied by Theorem 5.4.1) nowhere vanishes, which leads to a contradiction, for on a Riemann surface of genus g the degree of divisor of any tensor field set up in local coordinates by the formula $f(z)dz^n$ with the meromorphic function $f(z)$ is equal to $n(2g - 2)$. Indeed, firstly, the degree of divisor of such a tensor field does not depend on the choice of representative because all such tensor fields are obtained from one another through multiplication by an appropriate meromorphic function and, secondly, multiplying n Abelian differentials in a tensor manner we obtain a representative whose divisor's degree is equal to $n(2g - 2)$ because the degree of the divisor of any Abelian differential on a surface of genus g is known to be equal to $2g - 2$.

Next, it is readily seen that the identical vanishing of the function $R(z)$ is equivalent to the fact that the integral F_n is decomposed into the product of the Hamiltonian H by a homogeneous integral F_{2n-2} of degree $n - 2$. The proof of the theorem is now concluded with induction by n.

A function F on a cotangent bundle T^*M is a first integral of a Hamiltonian system with the Hamiltonian H if and only if the Poisson bracket $\{H, F\}$ is identical zero. If the manifold M is two-dimensional, then the integrals quadratic in momenta are of the form $F = \sum b_{ij}(q_1, q_2)p_i p_j$, where $B = (b_{ij})$ is a symmetric matrix of order two. Make use of the coordinate representative of the Poisson bracket

$$\{H, F\} = \sum_{i=1}^{2} \frac{\partial F}{\partial p_i}\frac{\partial H}{\partial q_i} - \frac{\partial H}{\partial p_i}\frac{\partial F}{\partial p_i}.$$

Equating the coefficients of this polynomial to zero, *we obtain a system of equations for the coefficients of the matrices A and B.* Write these equations in isothermic (conformal) coordinates x, y, in which the Riemannian metric takes the form

$$ds^2 = \lambda(x, y)(dx^2 + dy^2).$$

As is known, such coordinates always exist (locally) for any metric of smoothness class C^2 on a two-dimensional manifold. In these coordinates the function H takes the form $H = \frac{1}{4\lambda}(p_1^2 + p_2^2)$. Given the new function $v = -\ln(4\lambda)$, the system takes the following form

(1)
$$\begin{aligned} b_{11,x} &= b_{11}v_x + b_{12}v_y, & b_{22,y} &= b_{21}v_x + b_{22}v_y, \\ 2b_{12,x} + b_{11,y} &= b_{12}v_x + b_{22}v_y, & & \\ 2b_{12,y} + b_{22,x} &= b_{11}v_x + b_{12}v_y. & & \end{aligned}$$

Comparing the first equation with the fourth and the second with the third, we come to:
$$2b_{12,y} = b_{11,x} - b_{22,x}; \qquad 2b_{12,x} = b_{22,y} - b_{11,y}$$

Thus, $R(z) = (b_{11} - b_{22}) + 2ib_{12}$ is a holomorphic function of the variable $z = x + iy$. If $R_1 = b_{11} - b_{22}$ and $2R_2 = 2b_{12}$ are the real and imaginary parts of the function $R(z)$, then the matrix of the integral is of the form

$$B = \begin{pmatrix} b + R_1 & R_2 \\ R_2 & b \end{pmatrix},$$

where $b = b_{22}$. In this notation, two equations of the system (1) take the form

(2) $$\begin{aligned} b_x &= bv_x + \alpha(x,y) = f(x,y,b), \\ b_y &= bv_y + \beta(x,y) = h(x,y,b), \end{aligned}$$

where
$$\alpha = R_1 v_x + R_2 v_y - R_{1,x}, \qquad \beta = R_2 v_x.$$

Let an arbitrary holomorphic function $R(z)$ be given in a certain coordinate neighbourhood of (x,y). Differentiating the first and second equations of the system (2) with respect to y and x corresponding and equating the right-hand sides of the equations derived, we obtain the following necessary condition for solvability of the system (2) with respect to the function b: $f_y + hf_b = h_x + fh_b$, which is equivalent to

(3) $$\alpha_y + \beta v_x = \beta_x + \alpha v_y$$

The functions α, β, v being replaced by their expressions in terms of the function λ, we deduce the following fundamental equation for the function λ:

(4) $$R_2 \left(\frac{\partial^2 \lambda}{\partial y^2} - \frac{\partial^2 \lambda}{\partial x^2} \right) + R_1 \frac{\partial^2 \lambda}{\partial x \partial y} - 3 \frac{\partial R_2}{\partial x} \frac{\partial \lambda}{\partial x} + 3 \frac{\partial R_2}{\partial y} \frac{\partial \lambda}{\partial y} + 2 \frac{\partial^2 R_2}{\partial y^2} \equiv 0.$$

That is a linear second-order equation hyperbolic at all points x, y except at the zeros of the function R because its discriminant is equal to $R_1^2 + 4R_2^2 = |R|^2$. Let now the condition (4) be fulfilled. Find the function b from the system (2) and thus prove that the condition (4) is also sufficient for solvability of this system with respect to the function b. Expressing b, we derive

$$b = \left(\int_{x_0}^{x} \alpha e^{-v} dx + \int_{y_0}^{y} (\beta e^{-v})(x_0, y) dy + K \right) e^v,$$

where K is a constant. Expressing all the functions in terms of λ, we are ultimately led to :

(5) $$b = \lambda^{-1}[(R_1 \lambda)(x_0, y) - (R_1 \lambda)(x, y) \\ - \int_{y_0}^{y} (R_2 \lambda_x)(x_0, y) dy + \int_{x_0}^{x} (R_2 \lambda_y)(x, y) dx + K]$$

PROPOSITION 5.4.1. *Let $\lambda : D \to \mathbf{R}$ be a positive function of class C^2 given in a domain $D \subset \mathbf{R}^2$. The Riemannian metric $\lambda(x,y)(dx^2 + dy^2)$ determines in T^*D a geodesic flow which has an additional integral F quadratic in momenta if and only if the function λ satisfies (4), where $R = R_1 + 2iR_2$ is a function which is holomorphic in D and is not identical zero, the coefficients of the matrix (b_{ij}) of the integral F being expressed in this case in terms of λ and R by the formula: $b_{12}(x,y) = R_2$, $b_{11}(x,y) = b_{22}(x,y) + R_1$, where $b_{22} = b$ is expressed by formula (5).*

Now let us see how the functions R and λ are transformed under an arbitrary holomorphic change of coordinates. Let $w(z) = u(x,y) + iv(x,y)$ be a holomorphic change of coordinates, $\lambda_1(w) = \lambda_1(u,v)$ the function setting the metric $\lambda_1(u,v)(du^2 + dv^2)$ is isothermic coordinates u, v and $S(w)$ a holomorphic function which is associated with the integral F and corresponds to the coordinates u, v. Then there hold the formulae:

(6) $$\lambda_1(w) \cdot |w'(z)|^2 = \lambda(z)$$
(7) $$S(w)(w'(z))^{-2} = R(z)$$

THEOREM 5.4.3 (SEE [307]). *The geodesic flow in T^*M, where M is a two-dimensional Riemannian manifold, has an additional integral quadratic in momenta if and only if in any isothermic coordinates x, y the function λ setting the metric satisfies equation (4), where $R = R_1 + 2iR_2$ is a holomorphic function of the variable $z = x + iy$, which is not identical zero and under transition to other isothermic coordinates is transformed in accordance with formula (7).*

Consider the solution of the basic equation (4) for an arbitrary holomorphic function $R(z) \not\equiv 0$. Go over to new coordinates, such that $S(w) \equiv 1$. According to formula (7) we then have $w'(z) = \frac{1}{\sqrt{R(z)}}$, which given an explicit expression for such a change in the neighbourhood of a point z_0 such that $R(z_0) \neq 0$. In the coordinates $w = u + iv$ equation (4) takes the form $\frac{\partial^2 \lambda_1}{\partial u \partial v} = 0$, and consequently $\lambda_1(u,v) = f(u) + h(v)$, where f and h are arbitrary functions of class C^2. From formula (6) we eventually find :

$$\lambda(x,y) = \frac{f(u(x,y)) + (h(v(x,y))}{|R(z)|}.$$

Metrics of the form $(f(x) + h(y))(dx^2 + dy^2)$ are called *Liouville* metrics. The latter argument contains, in particular, the classical result that (Birkhoff) in certain appropriate coordinates any Riemannian metric whose geodesic flow has an additional integral quadratic in momenta is of Liouville type. Consequently, the equations for geodesics are integrated in this case by way of separation of variables.

4.2 The Case of a Sphere and a Torus

Setting up a Riemannian metric of class C^2 on a compact two-dimensional orientable manifold transforms this manifold into a Riemann surface. The charts on it are the local charts of the coordinates isothermic for this metric. The Riemannian metric on the sphere S^2 generates a Riemann surface which is homeomorphic to S^2, and therefore is conformally equivalent to the sphere S^2 with the standard complex analytic structure. Hence, for any Riemannian metric on a sphere there exist isothermic coordinates set on the entire sphere without a point. It is natural to refer to such coordinates as global isothermic coordinates. It is also clear that all systems of globally isothermic coordinates are interrelated through linear fractional transformations.

In the assumption that the geodesic flow has an additional integral quadratic in momenta we find out the form of the function $R(z)$ setting up the basic equation (4) in global coordinates. It can be expounded (we leave the proof to the reader) that in the global coordinates $z = x + iy$ there hold the following asymptotic formulae: $\lambda(z) = \frac{a+0(1)}{|z|^4}$ and $R(z) = (c + 0(1))z^4$ as $|z| \to \infty$, where a and c are real and complex constants. Here $R(z)$ is a polynomial whose degree does not exceed four. Next, there always exist global isothermic coordinates in which the polynomial $S(w)$ is of degree not higher than three, where $R(z) = S(w)z^4$. Here $R(z)$ cannot be of degree zero (that is, is not constant), neither can it be linear. To the polynomial $R(z) = z^2$ there correspond metrics with the function $\lambda = f(x^2+y^2)$ whose geodesic flow has an integral linear in momenta (and no other metrics).

In practice, admissible are only two non-equivalent polynomials $R(z)$ which determine the quadratic integral in global isothermic coordinates (these polynomials are not transformed into each other via linear fractional change of coordinates) : 1) $R(z) = z^2$, which corresponds to geodesic flows with a linear integral; 2) $R(z) = 4z^3 - g_2 z - g_3$, where $g_2^3 - 27 g_3^2 \neq 0$, which is the condition for the absence of multiple roots in the cubic polynomial $R(z)$, for otherwise it is reduced to the quadratic one.

The crucial point in the argument of this subsection, which provides the description of the aforementioned Riemannian metrics on the sphere, is the following theorem.

THEOREM 5.4.4 (SEE [63]). *On a two-dimensional sphere of class C^2 whose geodesic flow has an additional integral quadratic in momenta and independent of the energy integral, in some isothermic coordinates $z = x+iy$ set on a sphere without a point (with a point punctured), the metric is necessarily of the form $\lambda(x,y)(dx^2 + dy^2)$, where the function λ has one of the forms exhibited below:*

1) $\lambda = f(x^2 + y^2)$, where f is a positive function of class C^2 such that $f(t) = \frac{a+0(1)}{t^2}$ when $t \to \infty$; such geodesic flows have an integral linear in momenta;

2) $\lambda = \frac{f(u(x,y))+h(v(x,y))}{|4z^3 - g_2 z - g_3|}$, where g_2, g_3 are real constants such that $g_2^3 - 27 g_3^2 > 0$, the quantities u,v being the real and imaginary parts of the transformation $w(z) = P^{-1}(z)$. Here $P(w)$ is the Weierstrass P-function with invariants g_2, g_3 and a pair of periods of the form ω_1 and $i\omega_2$ with real ω_1 and ω_2, while f and h are functions of class C^2 such that a) $f(u) = (u - \frac{k\omega_1}{2})^2(a+0(1))$ when $u \to \frac{k\omega_1}{2}$ and similarly $h(v) = (v - \frac{k\omega_2}{2})^2(a+0(1))$ when $v \to \frac{k\omega_2}{2}$ for any fixed integer k, $a > 0$;

b) their values on the line segments $[\frac{\omega_1}{2}, \omega_1]$, $[\frac{\omega_2}{2}, \omega_2]$ are determined via their values on the line segments

$$[0, \frac{\omega_1}{2}], [0, \frac{\omega_2}{2}]$$

by the formulae

$$f(\frac{\omega_1}{2}+\tau) = f(\frac{\omega_1}{2}-\tau), \qquad \tau \in [0, \frac{\omega_1}{2}],$$
$$h(\frac{\omega_2}{2}+\tau) = h(\frac{\omega_2}{2}-\tau), \qquad \tau \in [0, \frac{\omega_2}{2}];$$

c) the functions f and h are periodic with period ω_1 and ω_2, respectively; it is clear that for such functions f and h the value $f(u(z)) + h(v(z))$ does not depend on the choice of the value of the multivalued function $P^{-1}(z)$.

Inversely, a positive function $\lambda : \mathbf{R}^2 \to \mathbf{R}$, which has one of the two aforementioned forms, sets Riemannian metric on the sphere S^2 whose geodesic flow has an additional integral quadratic in momenta and independent of the energy integral. The formulae for the matrix coefficients of this integral are given in Proposition 5.4.1.

This classification theorem is added with the following two assertions proved recently by Kolokol'tsov.

The *necessary and sufficient condition for C^∞-smoothness of the Riemannian metrics from Theorems* 5.4.4 is fulfillment of the following equations for the derivatives of the functions f and h in the sites of half-period lattice: for any integers k_1, k_2 and any non-negative integer n:

$$f^{(2n+1)}(k_1\frac{\omega_1}{2}) = h^{(2n+1)}(k_2\frac{\omega_2}{2}) = 0,$$
$$f^{(4n+2)}(k_1\frac{\omega_1}{2}) = h^{(4n+2)}(k_2\frac{\omega_2}{2}),$$
$$f^{(4n)}(k_1\frac{\omega_1}{2}) = -h^{(4n)}(k_2\frac{\omega_2}{2}).$$

Furthermore, the condition for real analyticity of these metrics is analyticity of the functions f and h and fulfillment of the equation $f(ix) = -h(x)$.

The metrics from Theorem 5.4.4 corresponding to different pairs of the functions f, h are not equivalent under the action of the group of diffeomorphisms of the sphere S^2. Besides, the group of motions of each metric of the class under consideration consists of four transformations set up in the coordinates u, v by the formula $(u,v) \to (\pm u, \pm v)$ provided that the functions f and h do not coincide, and in case these functions coincide this group has another generating element set up (the same as the transformation) by the formula $(u,v) \to (v,u)$.

REMARK: The classical examples of geodesic flows with quadratic integrals are geodesic flows on standard ellipsoids (generally, on quadrics in $\mathbf{R}^3$) and the geodesic flow of the factormetric on the Poisson sphere which occurs in classical problems of analytical dynamics.

We now investigate a geodesic flow on a torus with an additional integral quadratic in momenta. The corresponding Riemann surface is of genus one and is conformally equivalent to the factor-space $\mathbf{C}/\Gamma$ of the group $\mathbf{C}$ with respect to the translation subgroup Γ generated by two elements $g_1 : z \to z+\omega_1$, $g_2 : z \to z+\omega_2$, where $\text{Im}\,\frac{\omega_1}{\omega_2} \neq 0$. The Riemannian metric set on $\mathbf{C}/\Gamma$ can in an obvious fashion be lifted onto the covering surface, that is, onto $\mathbf{C}$. As a result, we obtain the metric on $\mathbb{C}$ (invariant under the translations g_1 and g_2) whose geodesic flow obviously has an additional quadratic integral which is also invariant. Since the momentum and velocity coordinates are transformed identically under translations, it follows that the function λ setting the metric on $\mathbb{C}$ and the holomorphic function R associated with the integral are invariant under the translations g_1 and g_2. Consequently, rotation may have the result that $R = 1$.

These coordinates are Liouville, and the function in them has the form $\lambda = f(x) + h(y)$. It has been shown above such a function of class C^2 is invariant under g_1 and g_2 only if among its parallelograms of periods there exists a rectangle and the functions f and h are periodic. Inversely, any positive function $\lambda = f(x) + h(y)$ with periodic f and h sets up the metric on a certain torus whose geodesic flow has a first integral

$$\frac{h(y)}{f(x)+h(y)} \cdot p_1^2 - \frac{f(x)}{f(x)+h(y)} \cdot p_2^2.$$

Thus, *we have defined the class of Riemannian metrics on the sphere and torus which admit a polynomial additional integral quadratic in momenta.*

At the same time we would like to make the following instructive remark. The necessary and sufficient conditions described above are not yet enough effective in the following context. We have not yet received the answer to the question of how one can find out whether or not a given metric on a sphere or a torus admits an additional integral. According to Theorem 5.4.4, this will be the case if and only if *in certain isothermic coordinates* (on a sphere with a punctured point) a metric under investigation is reduced to one of the two forms indicated in the theorem. But to answer this question one should find those suitable isothermic coordinates or, conversely, prove that in no isothermic coordinates (on a sphere with a punctured point) does this metric admit the canonical representation of one of the two types specified in the theorem.

In other words, an effective criterion (readily accessible for verification) is needed which permits establishing whether a given metric admits a required form in certain isothermic coordinates.

This property of the metric may probably be "caught" by some invariants and associated with the possibility of isometric metric embedding into the three-dimensional Euclidean space (locally).

4.3 *The Properties of Integrable Geodesic Flows on the Sphere*

A Riemannian manifold M is termed an SC_l-*manifold* if all the geodesics on M are simple (that is, non-self-intersecting) closed curves of length l. Of course, we deal here with single-path geodesics only.

Among two-dimensional orientable closed manifolds it is only the sphere that admits geodesic flows all of whose geodesics are closed. Classical examples of manifolds with such metrics are *Zoll rotation surfaces* and their modifications suggested by Blascheke and Thompson. These modified surfaces are composed of pieces of rotation surfaces.

The first to discuss the question of constructing nontrivial surfaces with closed geodesics was evidently Darboux [189]. We will say that a *Riemannian manifold satisfies* the SC-*property* if there exists a number $l > 0$ such that any geodesic on M is a simple closed curve of length l (or its multiplicities).

Darboux established the condition to be satisfied by the equation for a plane curve in order that the surface obtained by rotating this curve possess the SC-property. He also suggested a beautiful geometric criterion characterizing such curves. Let $\prod$ be a convex surface formed by rotating a plane curve K around the axis L, where K and L belong to a certain plane and the maximal distance from K and L is equal to R. Darboux demonstrated that $\prod$ satisfies the SC condition *if and only if* the length of the arc of the curve K cut off by a straight line Λ parallel to l is equal to the length of the arc of a circle of radius R cut off by a straight line α such that the distance from α to the parallel diameter is equal to the distance between Λ and L.

However, Darboux has not proved the existence of the metric globally defined on S^2 and satisfying the SC-property.

The next step was made in the end of the last century by Tannery who constructed a "pear" all of whose geodesics are closed and have the smallest period l, except for the equator whose smallest period is equal to $l/2$. This pear is an algebraic surface but not a smooth manifold because at one point it has a singularity.

The question of the construction of a nontrivial smooth SC-metric on S^2 was solved in 1902 by Zoll [190] who explicitly constructed a real-analytic SC rotation surface homeomorphic to S^2. A simple generalization of the Zoll construction leads to constructing a family of smooth SC-metrics on S^2 dependent on the functional parameter and invariant under rotations around a certain axis [191]. In his book [192] Blaschke proposed an exquisite modification of the Zoll construction which allows us to construct SC-manifolds which are homeomorphic to the sphere but are not rotation surfaces. Blaschke surfaces are glued of pieces of different SC rotation surfaces by means of films isometric to parts of the standard sphere.

In 1913 in his paper [193] Funk made an attempt to construct a one-parameter family of Riemannian SC-metrics of the form $\varphi(t)g$ on S^2 (where g is a standard metric) represented as the sum of a series with the initial data $\varphi(0) = 1$ and $\frac{d\varphi(t)}{dt}|_{t=0} = h$ for an arbitrary odd function h on S^2, that is, such a function that $h\sigma = -h$ for an antipodal mapping σ of the sphere S^2. However, Funk has not succeeded in proving convergence of the series constructed.

The existence of this type of deformations for each odd initial derivative has been recently proved by Guillemin in his paper [194] where he uses the powerful technique of modern global analysis. From Guillemin's results it follows, in particular, that in a sufficiently small neighbourhood of the standard metric on S^2 there exists a rather rich family of smooth metrics satisfying the SC-condition but not admitting any isometrics except for identical mapping. We would like to note that this is, in fact, the existence theorem because no explicit formulae for these metrics are presented in [194]. The matter is that the proof carried out by Guillemin is based on the implicit function theorem (for functional Frechet type manifolds).

The present state of the problem of describing manifolds with closed geodesics is most exhaustively represented in the book by Besse [191].

Without going into details of this problem for manifolds of dimension larger that two, where there exist many exquisite results but, as was observed by Besse, many fundamental questions remain still open, we only point out the single general condition necessary for a Riemannian manifold to satisfy the SC-condition. Namely, as has been shown by Weinstein [195], the *total volume of a Riemannian*

α-dimensional SC-manifold with a period 2π is equal to an integer multiple of the volume of a standard d-dimensional sphere. An integral equal to the ratio of these volumes is called the *Weinstein number* of an SC-manifold.

It should be emphasized that even in two dimensions the problem of classification and description of SC-manifolds is still far from being completely solved although much has already been done. From the general Bott–Samelson theorem on the topology of SC^m-manifolds (that is, such manifolds all of whose geodesics starting from a certain point m are closed) it follows that among the two-dimensional manifolds it is only on the sphere S^2 and on the projective space $\mathbf{R}P^2$ that there exist Riemannian metrics satisfying the SC-property. Besides, in 1961 Green proved [196] Blaschke's hypothesis using the integral geometry methods for fibre bundle of unit tangent vectors. According to this hypothesis, any *Riemannian SC-metric on $\mathbf{R}P^2$ is isometric to the standard one.* Therefore, among the two-dimensional manifolds the nontrivial SC-structures only exist on the sphere S^2. As far as these structures are concerned, there exist two known general results of Weinstein which are proved, for instance, in the aforementioned book by Besse. The first of these results consists in the fact that the *geodesic flows generated by two such metrics on the sphere S^2 that possess the SC-property are associated through symplectic diffeomorphism*. The second result implies that the *Weinstein number of any SC-manifold with a period 2π homeomorphic to the sphere S^2, is equal to unity*. In other words, the volume of such a manifold is equal to $\frac{4}{3}\pi$, that is, to the volume of a standard sphere.

As shown above (see also [63]), there exists a rich enough family of metrics on the two-dimensional sphere (depending on the functional parameter) *whose geodesic flow has an additional integral quadratic in momenta*. As has been recently demonstrated by Kolkol'tsov in [197], using explicitly formulae *one can specify in this family the class of SC-metrics, that is such metrics that all the geodesics are closed*.

The qualitative behaviour of geodesics can be described in the following manner. On a sphere there exist two pairs of rounding points such that all the geodesics going out of one such point gather at the point opposite to this one. In a bundle of unit-length tangent vectors to the sphere (this three-dimensional fibre bundle PS^2 is a part of the tangent bundle TS^2) the tangent vectors to these geodesics form a two-dimensional film which splits in a natural way the fibre bundle PS^2 (homeomorphic to the three-dimensional projective space) into two full tori. This picture is similar to the structure of geodesic flow on ellipsoid. We will describe this result in more detail.

If the geodesic flow on a sphere admits an additional integral quadratic in momenta and has no linear integral, then in any simply-connected open set not containing four special points, thought of as rounding points, there exist Liouville coordinates, that is, such coordinates in which the metric is written in the form: $ds^2 = (f(u)+h(v))(du^2+dv^2)$. Furthermore, "ripping up" the sphere along geodesic line segments which join "adjacent" rounding points (that is, those which can be joined by a single minimal geodesic) and making an involution, one can represent the sphere as an "infinite cross" (Fig. 79). Here the rectangle $OABC$ corresponds to half the sphere, while the second half is covered by the rectangle $BA'O'C$ which in gluing is identified with the rectangle $AO''C''B$. The points A, B, C, O are identical points denoted by the same letters primed are precisely the aforementioned rounding in the neighbourhood of which there exists no Liouville coordinate system. On the other hand, Liouville coordinates can be introduced on the entire open infinite cross by way of subjecting the generating function to additional periodicity conditions. Namely, denote by a and b the lengths of the line segments $|OC| = |O'C| = a, |AO| = |AO''| = b$. We should emphasize that those are not the lengths of the minimal geodesics to which there correspond these line segments on the sphere by the lengths of segments in the Euclidean coordinate system (u, v) on a plane. Then the functions f and h generating the metric in the Liouville coordinates are defined for all U, v and satisfy the following conditions:

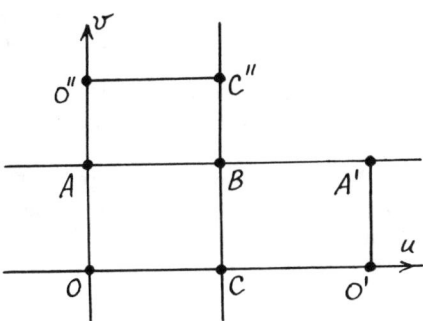

Figure 79

a) the functions f and h are periodic with a period $2a$ and $2b$, respectively;

b) the values of the functions on the line segments $[a, 2a]$, $[b, 2b]$ are determined through their values on the line segments $[0, a]$, $[0, b]$ by the formulae: $f(a + t) = f(a - t)$, $t \in [0, a]; h(b + t) = h(b - t), t \in [0, b]$;

c) there hold the conditions of asymptotic behaviour in the lattice sites: $f(u) = (u - ka)^2(g + O(1))$ when $u \to ka$ and $h(v) = (v - kb)^2(g + O(1))$ when $v \to kb$, $g > 0$, $k \in \mathbb{Z}^+$;

d) besides, the functions f and h are everywhere positive except at the points ka, kb, respectively, and are at least doubly continuously differentiated.

Note that the new vanishing of the expression $f(u) + h(v)$ at the rounding points precisely expresses the fact that at these points the Liouville coordinate system is degenerate.

Inversely, if the functions f and h satisfy the conditions specified above, then after the indicated gluings of the infinite cross we obtain a smooth Riemannian metric on the sphere. The transition to the global coordinate system is realized by the Weierstrass P-function. Conditions a–d are henceforth regarded as fulfilled.

Thus, we have come to the description of the behaviour of geodesics in a Lagrangian system with the Lagrangian $L = \frac{1}{2}(f(u) + h(v))(\dot{u}^2 + \dot{v}^2)$, which corresponds, after the Legendre transformation, to a Hamiltonian system with the Hamiltonian $H = \frac{1}{2} \cdot \frac{p_u^2 + p_v^2}{f(u) + h(v)}$. The straight lines $u = \text{const}$ and $v = \text{const}$ in the infinite cross and the corresponding curves on the sphere will be called, respectively, horizontal and vertical parallels. Investigate now the qualitative behaviour of geodesics. Use the separation of variables for writing equations for geodesics in variables u, v. The motion with a unit velocity is realized when $H = \frac{1}{2}$. The Hamilton–Jacobi equation has the form:

$$\frac{S_u^2 + S_v^2}{h(u) + h(v)} = \text{const} = c_2 = 2H.$$

Denoting now $c_1 = S_u^2 - f(u)c_2 = h(v)c_2 - S_v^2$, we derive the following two equations which set a geodesic in the coordinates u, v:

$$\frac{d}{dt}\left(\int \frac{du}{\pm\sqrt{f(u)c_2 + c_1}} - \int \frac{dv}{\pm\sqrt{h(v)c_2 - c_1}}\right) = 0,$$

$$\frac{d}{dt}\left(\int \frac{f(u)du}{\pm\sqrt{f(v)c_2 + c_1}} + \int \frac{h(v)dv}{\pm\sqrt{h(v)c_2 - c_1}}\right) = 1,$$

The "plus" and "minus" signs are determined here by the signs at p_u and p_v. Hereafter we always assume c_2 to be equal to unity which corresponds to the unit velocity $c_1 = c$. Now from the formulae $p_u^2 = f(u) + c$, $p_v^2 = h(v) - c$ we see that for $c > 0$ the motion proceeds in the strip $h(v) \geq c$ and for $c < 0$ in the strip $f(u) \geq -c$. The geodesics passing through the rounding points have the value $c = 0$. Investigate now the geodesics corresponding to the values $c \neq 0$. It is readily seen that the equations for such a geodesic passing through the point (u_0, v_0) can be written as follows:

$$\int_{u_0}^{u} \frac{d\xi}{\pm\sqrt{f(\xi) + c}} - \int_{v_0}^{v} \frac{d\xi}{\pm\sqrt{h(\xi) - c}} = 0;$$

$$\int_{u_0}^{u} \frac{f(\xi)d(\xi)}{\pm\sqrt{f(\xi) + c}} + \int_{v_0}^{v} \frac{h(\xi)d\xi}{\pm\sqrt{h(\xi) - c}} = t.$$

Writing the closedness condition for the geodesics corresponding to the values $c < 0$ (and replacing c by $-c$), we arrive at the following basic system of equations, where $f_0 = f(u_0) = \max f$:

(*)
$$\int_{v_1(c)}^{v_2(c)} \frac{dx}{\sqrt{h(x) - c}} = \int_0^a \frac{dx}{\sqrt{f(x) + c}}, \quad c \in (0, h_0)$$

$$\int_{u_1(c)}^{u_2(c)} \frac{dx}{\sqrt{f(x) - c}} = \int_0^b \frac{dx}{\sqrt{h(x) + c}}, \quad c \in (0, f_0)$$

$$u_i(c) : f(u_i) = c; \quad v_i(c) : h(v_i) = c.$$

THEOREM 5.4.5. *Let such a Riemannian metric be given on a sphere that the geodesic flow has an additional integral quadratic in momenta which is not the square of the linear integral. Then the following conditions are equivalent:*
  1) *All the geodesics are closed.*
  2) *The corresponding manifold is the SC-manifold all of whose geodesics are simple closed curves of equal length.*

3) The functions f and h, which set up the metric, satisfy the conditions: a) let $f_0 = f(u_0) = \max f(u)$, $h_0 = h(v_0) = \max h(v)$, then the function f on the intervals $(0, u_0)$ and (u_0, a) and the function h on the intervals $(0, v_0)$ and (v_0, b) have continuously differentiable inverse functions; b) the functions f and h satisfy the system $(*)$.

Note that the need of the SC-property for the metrics from Theorem 5.4.4 with all the geodesics closed is not the case with geodesic flow with a linear integral, for it is not difficult to construct such a rotation surface all of whose geodesics are closed but several pairs of geodesics are of distinct lengths.

With an additional condition $a = b$, $f(x) = h(x)$, one can find all the solutions of the system $(*)$. They are expressed by explicit formulae in terms of the Legendre functions. These solutions suffice for constructing such a family of SC-metrics on a sphere which depends on the functional parameter.

CHAPTER 6

A New Topological Invariant of Hamiltonian Systems of Liouville-Integrable Differential Equations. An Invariant Portrait of Integrable Equations and Hamiltonians

§1 Construction of the Topological Invariant

The author has discovered a topological invariant which characterizes Hamiltonian systems of Liouville-integrable differential equations. This chapter is devoted to topological characteristics of Liouville-integrable Hamiltonians. In particular, important ideas of symplectic topology are developed which were first suggested in papers by Novikov, Arnold, Gel'fand, Faddeev, Smale, Moser, and Kozlov. For simplicity we will consider only the case of equations on four-dimensional symplectic manifolds, although some of the results can be extended to a multidimensional case. The construction of the new invariant is based on the "Morse-type" theory for integrable systems of differential equations which was developed by the author in [282]–[284] (and in Chapter 2 of the present book). More precisely, we mean surgery on level surfaces for Bott integrals defined on isoenergy surfaces of integrable systems. Developing these results, we suggest a *new topological invariant of integrable systems* of differential equations: graph Γ, two-dimensional surface P^2 and embedding $k : \Gamma \to P^2$; in the non-resonance case all these geometrical objects do not depend on the choice of additional integral and describe, consequently, the integrable case (Hamiltonian) itself. It turns out that ths invariant can be effectively calculated. As an example, we calculate it for some classical cases of integrability of equations of motion of a heavy rigid body (the cases of Kovalevskaya, Goryachev–Chaplygin, a gyrostat).

For all its undoubted advantages, the classical bifurcational diagram Σ, which usually describes the topology of integrable cases in mechanics, has an essential shortcoming—it depends on the choice of a second integral, i.e., it is not a topological invariant of a system of differential equations. The newly discovered object $\{\Gamma, P, k\}$ is, first, a topological invariant and, second, using this object one can reconstruct the bifurcational diagram (if one fixes a concrete form of a second integral). Thus, the invariant $\{\Gamma, P, k\}$ is a more primary concept than the bifurcational diagram Σ.

Let M^4 be a smooth symplectic manifold and $v = \operatorname{sgrad} H$ be a Hamiltonian system with a smooth Hamiltonian H, which is completely Liouville-integrable on a certain nonsingular compact isoenergy surface Q^3, i.e., a surface set by the equation $H = \operatorname{const}$. Let $f: Q \to \mathbf{R}$ be a second independent integral on Q^3 which commutes with H. Here sgrad is an operation of calculating a skew-symetric gradient. Connected nonsingular level surfaces of the function f on Q are known to be two-dimensional Liouville tori. Suppose that f is a Bott integral on Q, i.e., that its critical points are organized into non-degenerate critical submanifolds in Q. The Bott nature of some known integrals was proved by Oshemkov in [292], [301].

DEFINITION 6.1.1: Let us call the Hamiltonian H *nonresonance* on a given isoenergy surface Q^3 if in Q everywhere dense are Liouville tori on which integral trajectories of a system v form a dense irrational winding.

The experience of concrete known physical systems shows that on four-dimensional manifolds Hamiltonians are mainly non resonance ones on almost all surfaces Q. In a multidimensional case, i.e., on M^{2n} where $n > 2$, examples are known where a Hamiltonian H is "essentially" resonance on almost all surfaces Q. This happens, in particular, when a system is integrable in a noncommutative sense. For an analysis of the main cases of noncommutative integrability see above, and also the review by Trofimov and Fomenko [188]. Recall that in these cases the system is integrated by means of integrals that form a noncommutative Lie algebra. Then the system is "degenerate" (resonance) in the sense that its trajectories are everywhere dense in "few-dimensional tori." They are organized into "large" genuine Liouville tori (in the compact case). Consequently, an initial system is irrational only on small tori, but is not such on "large tori." In the four-dimensional case, however, the non-resonance nature is a "typical" situation.

THEOREM 6.1.1 (FOMENKO). *Let v be a Hamiltonian system, with a non-resonance Hamiltonian H, integrable by means of a certain Bott integral f on a compact nonsingular three-dimensional isoenergy surface Q. Then one can unqiely construct a certain graph $\Gamma(Q, f)$ with the following property: from the graph $\Gamma(Q, f)$ one can uniquely (with an accuracy to homeomorphism) reconstruct the whole topological picture of evolution and surgery (bifurcations) of the Liouville tori within the surface Q with varying integral f.*

Suppose $f: Q \to \mathbf{R}$ is a Bott integral, $\alpha \in R$ and $f_\alpha = f^{-1}(\alpha)$ is a connected component of the level surface of the integral (singular or nonsingular). If $\alpha = a$ is a regular (noncritical) value for f, then f_a is a union of a finite number of Liouville tori. Denote the critical values for f by c, the connected component of the critical level surface of the integral by f_c, and the set of critical points of the integral f on f_c by N_c. As proved in Chapter 2 of the present book (see also [282]—[284]), the connected components of sets N_c can be only of the following types. **Type I**—a minimax circle S^1 (a local minimum or maximum for f), then $N_c = f_c = S^1$. **Type II**—a minimax torus T^2 then $N_c = f_c = T^2$. **Type III**—a saddle critical circle S^1 with an orientable separatrix diagram, then $N_c = S^1 \neq f_c$. **Type IV**—a saddle

critical circle S^1 with a nonorientable separatrix diagram, then $N_c = S^1 \neq f_c$.
Type V—a minimax Klein bottle K^2, then $N_c = f_c = K^2$.

Denote a regular connected closed tubular neighbourhood of the component f_c in a manifold Q^3 by $U(f_c)$ (where c is a critical value of the integral). $U(f_c)$ may be assumed to be a connected three-dimensional manifold whose boundary consists of a disconnected union of tori. For $U(f_c)$ one can take a connected component of the manifold $f^{-1}[c-\varepsilon, c+\varepsilon]$. We may assume that $Q = \sum_c U(f_c)$, that is, Q is obtained from all the manifolds $U(f_c)$ by gluing together their boundaries through certain diffeomorphisms of boundary tori.

THEOREM 6.1.2 (FOMENKO). *Let Q be a compact nonsingular isoenergy surface of a system v, with a Hamiltonian H (not necessarily nonresonance), integrable by means of a certain Bott integral f. Then the manifolds $U(f_c)$ entering in the decomposition $Q = \sum_c U(f_c)$ admit the following representations depending on the type of the set N_c. Type I: $U(f_c) = P_c^1 \times S^1$, where $P_c^2 = D^2$ (disk); type II: $U(f_c) = P_c^1$, where $P_c^2 = S^1 \times D^1$ (cylinder); type III: $U(f_c) = P_c^2 \times S^1$, where P_c^2 is a certain two-dimensional surface with boundary; type IV: $U(f_c) = P_c^2 \dot\times S^1$, where P_c^2 is a certain two-dimensional surface with boundary and $P_c^2 \dot\times S^1$ is a space of a Seifert bundle with a base P^2 and a fibre S^1 (for a description see below); type V: $U(f_c) = P_c^2 \tilde\times S^1$, where $P_c^2 = \mu$ (Möbius strip), and $\mu \tilde\times S^1$ denotes a skew product (with a torus T^2 as a boundary).*

COROLLARY 6.1.1. *Each isoenergy surface Q (under conditions of Theorem 6.1.2) can be brought into one-to-one correspondence (with an accuracy to homeomorphism) with a certain closed two-dimensional surface $P^2(Q, f) = \sum_c P_c^2$ obtained by such gluing together of the surfaces P_c^2 which is induced by the gluing $Q = \sum_c U(f_c)$.*

THEOREM 6.1.3 (THE MAIN ONE; FOMENKO). *Suppose v is a Hamiltonian system integrable on Q by means of a Bott integral. Then there exists a single-valued (with an accuracy to homeomorphism) canonical embedding $k(Q, f)$ of the graph $\Gamma(Q, f)$ into the surface $P^2(Q, f)$. If a Hamiltonian H is nonresonance on Q, then the triplet (Γ, P, k) does not depend on the choice of a second integral f. Namely, if f and f' are any Bott integrals of the system v, then the corresponding graphs $\Gamma(Q, f)$ and $\Gamma(Q, f')$ and the surfaces $P(Q, f)$ and $P(Q, f')$ are homeomorphic, and the diagram*

$$k : \Gamma \to P$$
$$k' : \Gamma' \to P'$$

(with vertical homeomorphisms) is commutative.

COROLLARY 6.1.2. *In a nonresonance case the triplet (Γ, P, k) is a topological invariant of the integrable case (Hamiltonian) itself and makes it possible to classify integrable Hamiltonians by their topological type and complicacy.*

This triplet will be called a *topological portrait of an integrable Hamiltonian*. Division of the surface $P(Q)$ into regions (the division being determined by the graph $\Gamma(Q)$) is also a topological invariant. The surface $P(Q)$ need not necessarily be embedded into Q. Now let us define the manifold $P_c^2 \dot\times S^1$. A full torus obtained by gluing together two feet of a full cylinder $D^2 \times D^1$ through a mapping $(z, 0) \to (\exp(2\pi i b/a), 1)$, where $z \in \mathbb{C}, |z| \leqslant 1, a, b \in \mathbb{Z}$, and a and b are mutually simple is called a fibered full torus of type (a, b). The fibre is $\{(z,t), (\exp(2\pi i b/a)z, t), (\exp(4\pi i b/a)z, t), \ldots, (\exp(2\pi i (a-1)b/a)z, t), 0 \leqslant t \leqslant 1\}$. Then on the full torus $D^2 \times S^1$ a fibre bundle with a fibre S^1 above D^2 is defined, which is locally trivial for all nonzero points of the disk D^2. The circle fibre $\{(0,t), 0 \leqslant t \leqslant 1\}$ is singular if $a > 1$. Now consider the manifold P_c^2 with points $x_1, \ldots, x_m$ marked on it and encircle them with small disks $D_1^2, \ldots, D_m^2$. Then take the direct product $P_c^2 \times S^1$ and discard from it m full tori $D_i^2 \times S^1, 1 \leqslant i \leqslant m$. In the place of these full tori glue in fibered full tori of the type (2.1). Denote the manifold obtained by $P_c^2 \dot\times S^1$. It is a Seifert bundle and P_c^2 is its base.

We now present an explicit construction if invariants. First suppose for simplicity that on each critical level f_c there exists exactly one critical connected manifold N_c. For this case the construction of the graph $\Gamma(Q, f)$ was, in fact, described in Chapter 2 (see also [282]—[284]). Construct now the graph Γ in the general case. Now on one critical level may lie several critical manifolds. As distinguished from ordinary Morse functions, critical manifolds of a Bott integral which lie on one level cannot, generally speaking, be taken onto different levels by way of small perturbation of the integral (and remain in the class of integrals). The perturbation $\tilde{f}$ of the integral f may be not an integral. Let f_a be the level surface of an integral, that is, $f_a = f^{-1}(a)$. If a is a regular value, then f_a is a untion of a finite number of tori. Let us plot them as points in $\mathbb{R}^3$ on the level a, where the axis $\mathbb{R}$ has an upward direction. Varying a within the range of regular values, we make these points sweep some arcs—part of edges of a future graph Γ. Let N_c be a set of critical points f on f_c. We will mention two cases: a) $N_c = f_c$, b) $N_c \subset f_c$, where $N_c \neq f_c$. In the papers [282]—[284] (see also Chapter 2 of the present book) the author found all the posssibilities for N_c. Let us consider the case "a." Only three types of critical sets are possible here.

1. A "**minimax circle**". Here $N_c = f_c$ is homeomorphic to a circle on which f reaches its local minimum or maximum. Its tubular neighbourhood $S^1 \times D^2$ is homeomorphic to a full torus (in Q^3). As $a \to c$, nonsingular tori contract to the axis of the full torus and for $a = c$ they degenrate into S^1. This situation is shown schematically in Fig. 80, where the black point is the vertex of the graph into which (or from which) there goes one edge of the graph.

2. A "**torus**". Here $N_c = f_c$ and is homeomorphic to a torus T^2 on which f reaches its local minimum or maximum. Its tubular neighbourhood is homeomorphic to a cylinder $T^2 \times D^1$. The boundary of the cylinder is represented by two tori. As $a \to c$, they more towards each other and for $a = c$ they become one torus. We will depict this situation with a light circle (the vertex of the graph) into which (or from which) there go two edges of the graph (Fig. 80).

3. A "**Klein bottle**". Here $N_c = f_c$ and is homeomorphic to a Klein bottle K^2 on which f reaches its local minimum or maximum. Its tubular neighbourhood $K^2 \tilde\times D^1$ is homeomorphic to a skew product of K^2 by a segment. The boundary of $K^2 \tilde\times D^1$ is one torus. As $a \to c$, it tends to K^2 and is a two-fold covering when $a = c$. We will show this situation with a light circle with a point inside (the vertex of the graph) into which (or from which) there goes one edge of the graph (Fig.

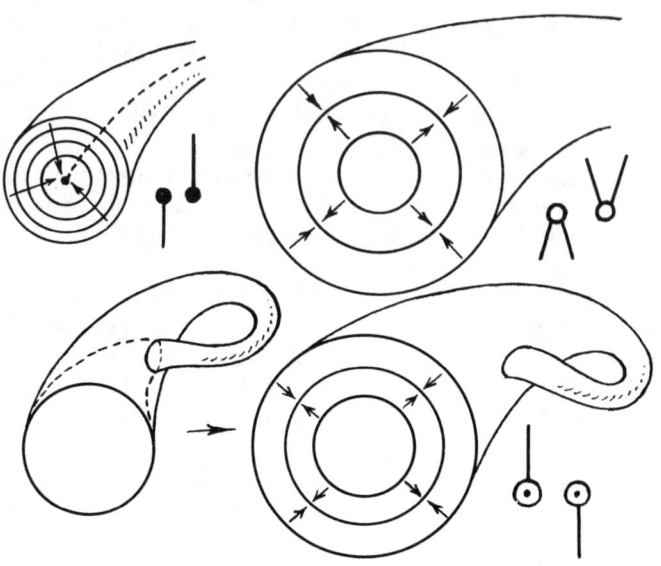

Figure 80

80).

Consider now the case "b." Here $N_c \subset f_c$ and $\dim N_c = 1$, $\dim f_c = 2$. Then (see [283] and Chapter 2) N_c is a disconnected union of non intersecting circles. Each of them is a saddle circle for f. The component f_c will also be called a saddle one. It may schematically be shown as a flat horizontal square lying on the level c in $\mathbb{R}^3$ (Fig. 81). Some edges of the graph stick into it from below (when $a \to c$ and $a < c$), the other edges go upward (when $a > c$). Thus, we have defined a certain graph A containing regular arcs (edges) some of which stick into squares and others end with vertices of the three types described above.

Fix the saddle value c and construct a graph T_c showing the exact manner in which the edges of the graph A interacting with f_c join with one another. Consider a vector field $w = \mathrm{grad}\, f$ on Q. Its trajectories entering into the critical points of the integral or going out of these points are called separatrices. Their union is a separatrix diagram of the critical submanifold. Let on the level f_c from each saddle critical circle S^1 there go its separatrix diagram. If it is orientable, it is obtained by gluing together two flat rings (cylinders) along the axial circle (see Chapter 2). If it is nonorientable, it is obtained by gluing together two Möbius strips along their axial circle. Consider non-critical values $c - \varepsilon$ and $c + \varepsilon$ close to c. The surfaces $f_{c+\varepsilon}$ and $f_{c-\varepsilon}$ consist of tori (Fig.82). The separatrix diagrams of the critical circles lying in f_c intersect these tori transversally along certain circles and divide the tori into unions of regions which we will call regular. In each region on a level $f_{c-\varepsilon}$ choose a point and let integral trajectories of the field w go from these

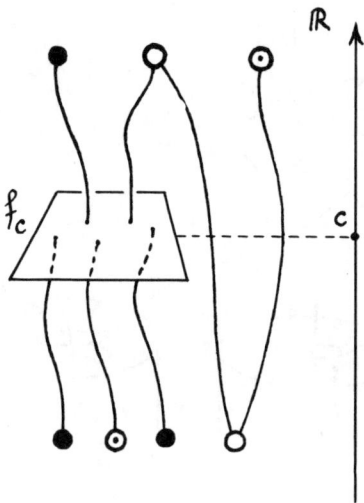

Figure 81

points. They will pass by critical circles on the level f_c and get into some other regular regions of tori which form $f_{c+\varepsilon}$. It is obvious that in this way we obtain a homeomorphism between the open regular regions from $f_{c-\varepsilon}$ and the open regular regions from $f_{c+\varepsilon}$.

First consider the orientable case, i.e., where all separatrix diagrams are orientable (i.e., there are no Möbius strips). Since each nonsingular torus is a point in the graph A, one can join the points on the level $f_{c-\varepsilon}$ and on the level $f_{c+\varepsilon}$ by arcs (segments) which stand for the bunch of the integral trajectories of the field w. We obtain a certain graph T_c. Its edges show us the motion of open regular regions of the tori. The tori fall into pieces which then rise (descend) and rearrange into new tori. Each upper torus is made of pieces of lower tori (and vice versa).

We now consider the nonorientable case, that is, when on f_c there is at least one critical circle with a nonorientable diagram. On each torus that comes up to f_c we mark with asterisks the regular regions incident with nonorientable separatrix diagrams (that is, with Möbius strips). Consequently, we may also mark with asterisks the corresponding edges of the graph under construction. Thus, we construct a graph by the scheme of the orientable case, after which mark with asterisks those of its edges which show the motion of askterisked regular regions. The graph obtained will be denoted by T_c (Fig. 83). It is clear that the ends of the edges of the graph T_c are identified with the ends of some edges of the graph A. We finally define the graph Γ as a union (gluing) $\Gamma = A + \sum_c T_c$, where $\{c\}$ are critical saddle values of the integral.

PROPOSITION 6.1.1. *Let f and f' be any two Bott integrals. Then under a homeomorphism $q(Q, f, f') : \Gamma(Q, f) \to \Gamma(Q, f')$ (see Theorem 6.1.3) the saddle*

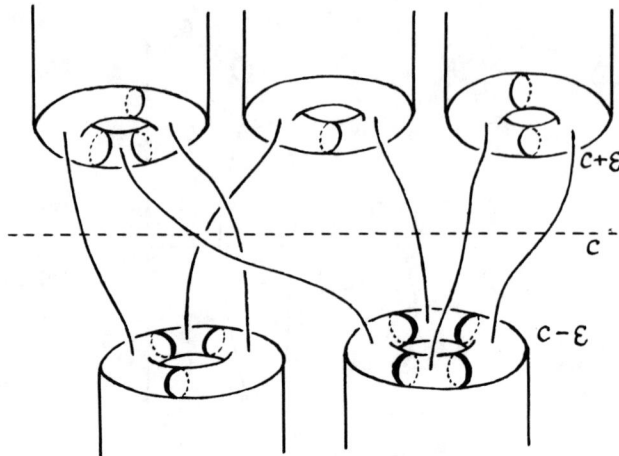

Figure 82

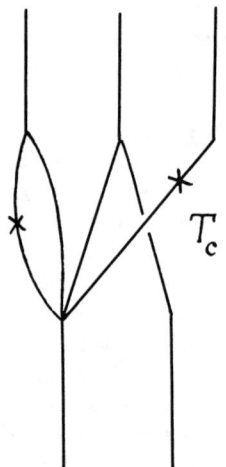

Figure 83

subgraphs T_c for the integral f homeomorphically pass over into saddle subgraphs T'_c for the integral f'. The asterisks of the graph Γ become asterisks of the graph Γ'. Vertices of the types "minimax circle" and "Klein bottle" of the graph Γ pass over into vertices of the same type (respectively) on the graph Γ'. The "torus" type vertices of the graph Γ can be mapped into ordinary interior points of edges of the graph Γ'. On the contrary, ordinary interior points of the edges of the graph Γ can be mapped into "torus" type vertices of the graph Γ'.

From the analytical point of view, the change of the type of vertex corresponds (locally) to the operation $f \to f^2$ or, vice versa, to the operation $f \to \sqrt{f}$.

DEFINITION 6.1.2: The triplet $\Gamma(Q), P(Q), k(Q)$ will be called a *topological invariant portrait* $I(H, Q)$ of an integrable Hamiltonian H on a given isoenergy surface Q. The set of all triplets $\{\Gamma(Q), P(Q), k(Q)\}$ over all Q will be called a *complete topological invariant portrait* $I(H)$. Homeomorphic triplets are of course assumed to be equivalent. The complete portrait $I(H)$ depends already on the Hamiltonian H only. Thus if two integrable systems have non-homeomorphic topological portraits, then the systems are not equivalent, for example, no isomorphism can be established between their trajectories. At the same time there exist certainly non-equivalent integrable systems with identical topological portraits $I(H)$.

Construct now a surface $P(Q, f)$. We will define it as a union (gluing) of the form $P(A) + \sum_c P(T_c)$, where $P(A)$ and $P(T_c)$ are two-dimensional surfaces with boundary. Now define $P(A) = (S^1 \times \text{Int } A)^{\cdot} + \Sigma D^2 + \Sigma \mu + \Sigma S^1 \times D^1$. Here Int A is a union of all open edges of the graph A. Consequently, $(S^1 \times \text{Int } A)$ is a union of open cylinders. The manifold $(S^1 \times \text{Int } A)^{\cdot}$ is obtained from it by adding boundary circles. Let an edge of the graph A end with a black vertex. The corresponding boundary circle on the boundary of the manifold $(S^1 \times \text{Int } A)^{\cdot}$ we will glue with a disk D^2. The gluing of such disks will be denoted by ΣD^2. Let two edges of the graph A meet at the light vertex. It determined two boundary circles on $(S^1 \times \text{Int } A)^{\cdot}$ which we will glue (join) with a cylinder $S^1 \times D^1$. This opration is denoted by $\Sigma S^1 \times D^1$. Let an edge of the graph A end with a light vertex with a point within it. Now glue the corresponding boundary circle on $(S^1 \times \text{Int } A)^{\cdot}$ with a Möbius strip μ. This operation is denoted by $\Sigma \mu$. Thus, ΣD^2, $\Sigma S^1 \times D^1$, $\Sigma \mu$ correspond to minimax circles, tori and Klein bottles. Now construct $P(T_c) = P_c$. First consider the orientable case where all critical saddle circles on f_c have orientable separatrix diagrams. As shown by the author in [208] (see also Chapter 2 of the book), f_c is homeomorphic to a direct product $K_c \times S^1$, where K_c is a certain graph obtained from several circles through identification of some pairs of points on them. Locally, from each vertex of the graph K_c there go exactly four edges. In the general case we have the following assertion.

PROPOSITION 6.1.2. *A complex f_c is obtained by gluing together several two-dimensional tori along circles that realize nonzero cycles on the tori. If there are several such circles on one torus, they do not intersect. The circles along which the tori entering in f_c are tangent, are critical for f. They are homologous and cut f_c into a union of several flat rings.*

Thus, the cycle γ is uniquely defined on f_c. Choose on each torus in f_c a certain circle (generator) α complementary of γ (intersecting with γ exactly at one point). We call it an oval. We may assume that ovals are tangent to one another at points lying on critical circles of the integral. In the orientable case, the union of ovals gives a graph K_c. It need not necessarily be flat. Let x be a tangent point of two ovals, that is, a critical point for f. Then segments of the integral trajectories

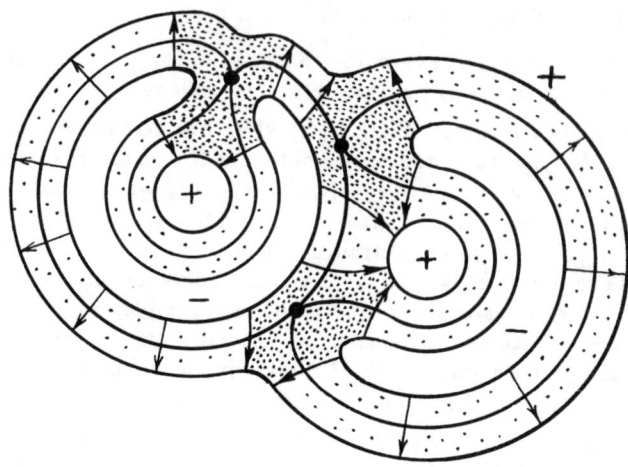

Figure 84

of the field w and the level lines of the function f determine (on a two-dimensional disk centred at the point x, lying in Q and orthogonal to the critical circle on which the point x lies) near the point x a "coordinate cross" on each end of which there is an arrow indicating the direction of w (Fig. 84). Let us construct such normal two-dimensional crosses in each vertex of the graph K_c. Different crosses are joined by segments which are parts of ovals. Join now the ends of the crosses by narrow strips going along arcs of ovals. These strips consist of segments of integral trajectories of the field w which orthogonally intersect the ovals (outside the critical points). Arcs of ovals go along the axis of these strips. As a results we obtain a smooth two-dimensional surface with boundary (Fig. 84). The sign " - " denotes the boundary circles corresponding to the tori which came up to f_c from below. The sign " + " stands for the boundary circles corresponding to the tori approaching f_c from above. As an example, Fig. 84 demonstrates three positive (upper) and two negative (lower) tori. The number of negative (positive) tori equals the number of the edges of the graph Γ approaching f_c from below (from above). The obtained surface will be denoted by $P(T_c) = P_c$. Its boundary circles are divided into two classes: lower (negative) and upper (positive). The graph K_c is uniquely (with an accuracy to surface homeomorphism) embedded into P_c^2.

We now construct P_c^2 in the nonorientable case. The scheme or arguments is basically the same. At first suppose for simplicity that exactly one critical circle with a nonorientable diagram lies on f_c. Then f_c has the form show in Fig. 85 (a particular case). Consequently, each oval α lying on such f_c must be doubled. Any other critical circle, with an orientable diagram, lying on this f_c encounters two copies of the cycle α at two points. This makes us double the cycles α also on those tori which are tangent to each other along circles with orientable diagrams but enter in the composition of a connected f_c that contains a circle with a nonorientable

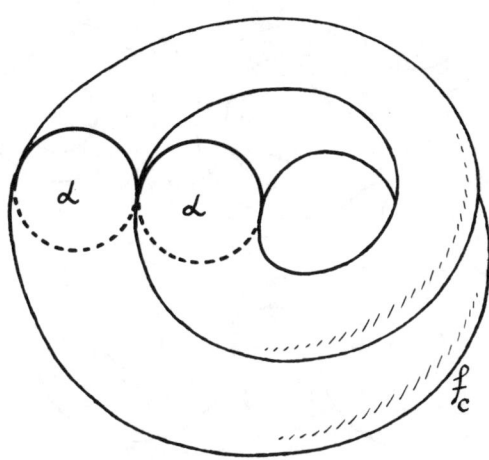

Figure 85

diagram. The number of circles with nonorientable diagrams on a connected f_c does not affect the doubling process, i.e., cycles should be doubled only once. The further constructions repeat the orientable case. As a result we obtain the surface $\tilde{P}_c$. The only distinction lies in that now each boundary circle of the surface $\tilde{P}_c$ (i.e., corresponding to one torus) is encountered exactly in two copies (it is doubled). Now construct a surface $P(T_c)$. From the construction of $\tilde{P}_c$ it is seen that on it a smooth action of the group $\mathbb{Z}_2$ (involution σ) is well defined; $\sigma(x) = x$ if and only if the point x belongs to a critical circle with a nonorientable separatrix diagram. Denote such points by $x_1, \ldots, x_m$ and the corresponding circles by $S_1, \ldots S_m$. Consider the factor-space $P_c = \tilde{P}_c/\mathbb{Z}_2$. It is clear that $P_c = P(T_c)$ is a two-dimensional manifold with boundary. Now each of its boundary circles corresponds exactly to one Liouville torus (upper or lower). It is readily seen that the points $x_1, \ldots x_m$ are interior points on the surface P_c. Mark them with asterisks. In the particular case where f_c has the form shown in Fig. 85 (that is, $m = 1$), the surface P_c is a two-dimensional cylinder with one asterisk (Fig. 86). Now we can construct the whole surface $P(Q, f)$.

It is clear that there exists a one-to-one correspondence between the boundary circles of the surface $P(A)$ and the boundary circles of the union of surfaces P_c. This correspondence is set by edges of the graph A. Identify the corresponding circles by means of certain homeomorphisms and uniquely obtain a closed two-dimensional surface $P(Q, f)$ (orientable or nonorientable). Clearly, the replacement of gluing homeomorphisms by some others induces only a homeomorphism of the surface onto itself. On $P(Q, f)$ there uniquely (with an accuracy to surface homeomorphism) exists a certain, generally speaking, disconnected graph, which we denote as $K(Q, f)$. Consider circles that cut in half the cylinders entering in the surface $S^1 \times \text{Int } A$. One may assume that each of them has the form $S^1 \times p$, where p is

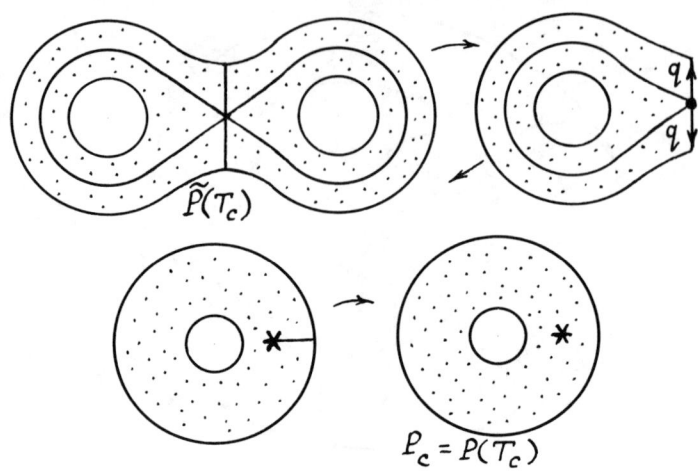

Figure 86

the middle of the corresponding edge of the graph A. Now define the graph K as a disconnected union of all graphs K_c and circles of the form $S^1 \times p$. The graph K has only vertices of multiplicity 4.

The construction of the surface $P(Q, f)$ is completed.

PROPOSITION 6.1.3. *If f and f' are any Bott integrals on Q, then $P(Q, f)$ is homeomorphic to $P(Q, f')$ (in a non-resonance case).*

Denote by K^* a graph conjugate to the graph K in the surface $P(Q, f)$. Its vertices are centres of the regions into which the graph K divides P, and edges are the arcs joining the vertices through the middles of the edges of the graph K.

PROPOSITION 6.1.4. *The graph $\Gamma(Q, f)$ coincides with the graph $K^*(Q, f)$. Therefore, the graph $\Gamma(Q, f)$ admits the embedding $k(Q, f) : \Gamma(Q, f) \to P(Q, f)$ which is uniquely (with an accuracy to surface homeomorphism) defined by an initial integrable nonresonance system. The graph K divides the surface P into regions homeomorphic either to a disk, or to a flat ring, or to a Möbius strip.*

Therefore, *each integrable Hamiltonian nonresonance system is naturally connected with a certain integer, which we will call "the genus of the system."* This is the genus of the surface $P(Q)$. Since according to Theorem 6.1.3 the graph Γ, the surface P and the embedding k do not depend on the choice of f, all these objects can be defined on the basis of only geometrical properties of foliation of Q^3 into Liouville tori, i.e., in the end, on the basis of the Hamiltonian properties only. We will not do this here and leave it to the reader who may, relying on the results of Chapter 2, restore the necessary arguments.

§2 Calculation of Topological Invariants of Certain Classical Mechanical Systems

We will now give examples of calculating topological portraits of certain classical integrable mechanical systems. The classical equations of motion of a heavy rigid body have the form:

$$A\dot\omega + \omega \times A\omega = \nu \times \operatorname{grad} \Pi, \quad \dot\nu = \nu \times \omega,$$

where $a \times b$ is a vector product, Π potential. In this example, the integrals of the system have the following form:

$$F = \nu_1^2 + \nu_2^2 + \nu_3^2 = 1,$$
$$G = A_1\omega_1\nu_1 + A_2\omega_2\nu_2 + A_3\omega_3\nu_3 = g,$$
$$H = \frac{1}{2}(A_1\omega_1^2 + A_2\omega_2^2 + A_3\omega_3^2) + \Pi(\nu) = h.$$

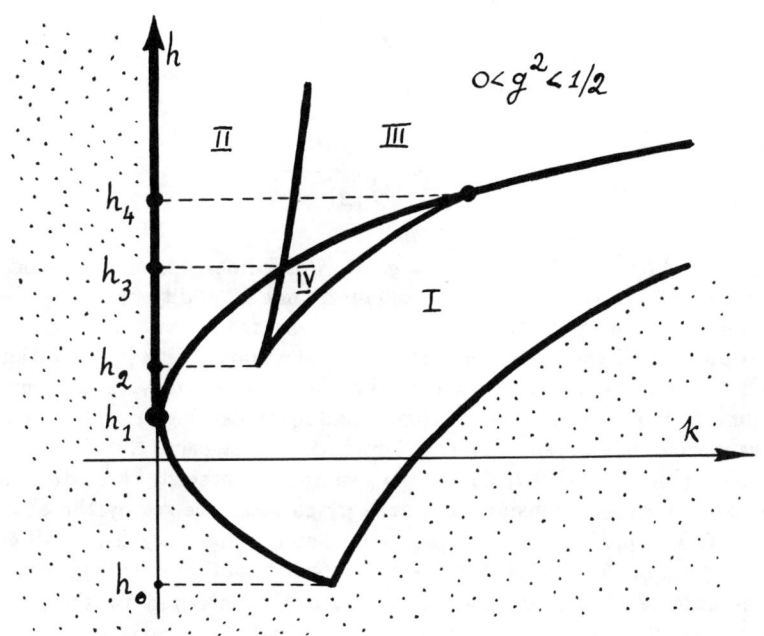

Figure 87 (1)

a) The case of Kovalevskaya is characterized by the fact that $A_1 = A_2 = 2A_3$, $\Pi = -\nu_1$. As is well known, there arises here another additional independent integral $f = (\omega_1^2 - \omega_2^2 + \nu_1)^2 + (2\omega_1\omega_2 + \nu_2)^2$. Consider a four-dimensional

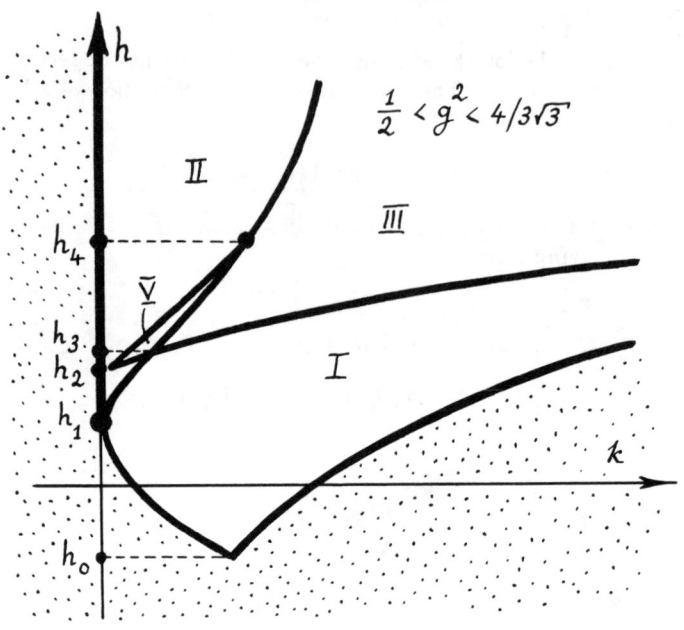

Figure 87 (2)

manifold $M^4 = F(\omega, \nu) = 1$; $G(\omega, \nu) = g$. It is diffeomorphic to a tangent bundle to a sphere T^*S^2. On M^4 a Hamiltonian H and an additional integral f are set. We can therefore apply the construction presented above and calculate the topological portrait of the integrable case of Kovalevskaya. This investigation was carried out by the author in collaboration with A. A. Oshemkov. Take a momentum mapping $m = H \times f : M^4 \to R^2$ corresponding to two integrals H and f. For different values of g one obtains different bifurcational diagrams $\Sigma \subset R^2$ calculated by Kharlamov (Figs. 87(1), 87(2)). In the example of interest, three-dimensional connected components of nonsingular isoenergy surfaces are set by the equations $Q^3 = \{F = 1, G = g, H = h\}$. It turned out that the complete list of all graphs $\Gamma(Q)$ encountered in the case of Kovalevskaya consists of 6 graphs (Fig. 88).

b) The case of Goryachev–Chaplygin. It is characterized by the fact that $A_1 = A_2 = 4A_3$, $\Pi = -\nu_1$. It is known that for $G = 0$ there exists an additional integral $f = 2\omega_3(\omega_1^2 + \omega_2^2) + 2\omega_1\nu_3$. Calculations show that here a complete list of graphs $\Gamma(Q)$ consists of only two graphs. The bifurcational diagram is shown in Fig. 89 and the topological portrait (graphs) in Fig. 90.

c) The solution of the Goryachev–Chaplygin case can be extended to the gyrostat case. In the equations of motion and in the integrals G and H one should replace $A\omega$ by $A\omega + \lambda$. Then the additional integral is known to be of the form $f = 2(\omega_3 - \lambda)(\omega_1^2 + \omega_2^2) + 2\omega_1\nu_3$. For $\lambda = 0$ we deal with the usual Goryachev–

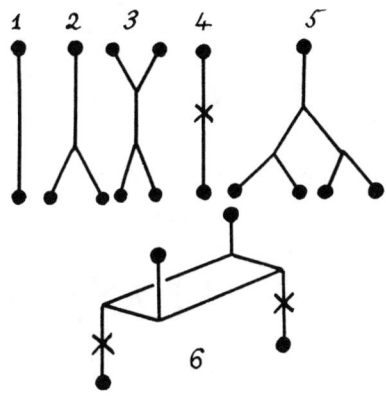

Figure 88

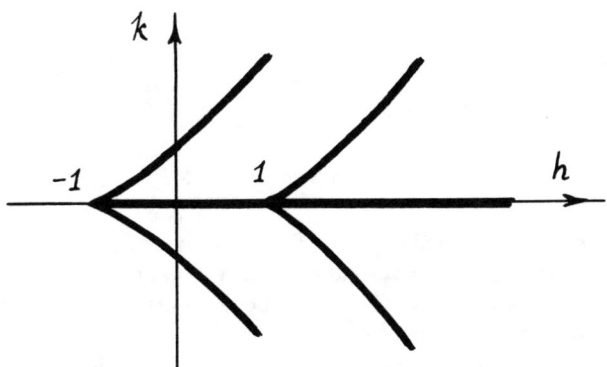

Figure 89

Chaplygin case. For $\lambda \neq 0$ it turns out that a complete list of graphs $\Gamma(Q)$ encoutered in this integrable case looks like the one shown in Fig. 91. Here again we deal with 6 types of graphs (Sretensky case).

Conclusion: complete topological portraits of the three above-mentioned integrable classical cases are not homeomorphic. Graphs 5 and 6 of the Kovalevskaya case are not homeomorphic to any graph of the Goryachev–Chaplygin case (gyrostat). Graphs 5 and 6 of the gyrostat case are not homeomorphic to any graph of the Kovalevskaya case. Thus, topological portraits serve for visual distinction between different integrable cases.

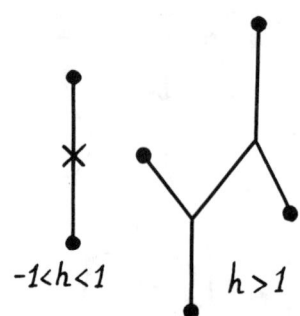

Figure 90

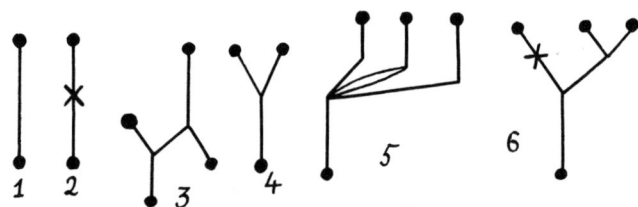

Figure 91

Oshemkov has constructed the graphs $\Gamma(Q)$ for an integrable case of the equations of motion of a four-dimensional rigid body (see below). A complete topological portrait of this case differs from all the three above-mentioned cases. In particular, among surfaces $P(Q)$ there appear tori.

We now present embeddings of the graphs $\Gamma(Q)$ into surfaces $P(Q)$. A detailed analysis of analytic expressions for integrals makes it possible to construct all these embeddings in an explicit form. For illustration, we first present the result for the case of Kovalevskaya (Fig. 92). The surface $P(Q)$ is seen here to be homeomorphic to a sphere in all cases. In Fig. 93 the graphs $\Gamma(Q)$ and $K(Q)$ of Kovalevskaya case are depicted on a plane obtained by puncturing one point from a sphere. Systems of ovals appear, the further analysis of which can be carried out within the classification theory of algebraic curves. In Figs. 92 and 93, graphs Γ are shown by solid arcs and graphs K by dashed arcs. It is obvious that $\Gamma = K^*$.

We also present an explicit form of the topological invariant for the case of integrability of the equations of motion of a four-dimensional rigid body (Oshemkov [301]).

Represent the Lie algebra so(4) as an algebra of skew-symmetric real matrices $X = (x_{ij})$, where $i,j = 1,2,3,4$. Two smooth functions (two polynomials) $f_1 =$

§2   Calculation of Topological Invariants

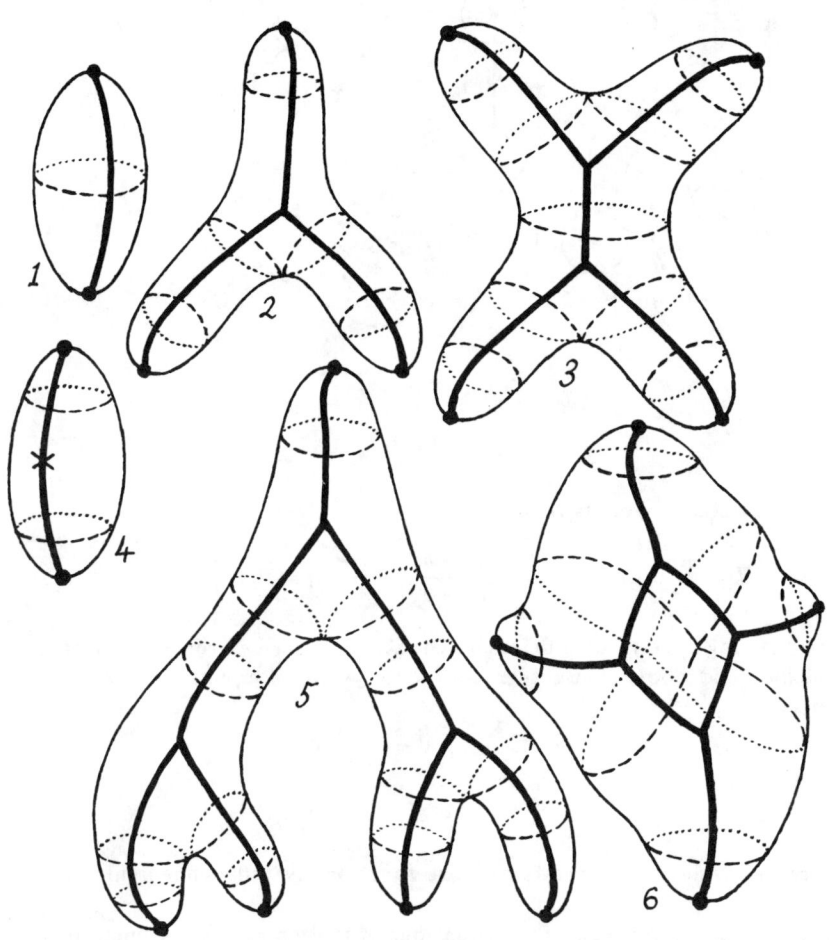

Figure 92

$\Sigma_{i<j} x_{ij}^2$ and $f_2 = x_{12}x_{34} - x_{13}x_{24} + x_{14}x_{23}$ distinguish orbits of the algebra (as common level surfaces of two polynomials on a six-dimensional Lie algebra). Under the condition $|2p_2| < p_1^2$ the two indicated polynomials have nonsingular common level surfaces of the form $S^2 \times S^2 = \{f_1 = p_1^2, f_2 = p_2\}$. These four-dimensional orbits are homeomorphic to a direct product of two spheres, $S^2 \times S^2$. On the orbits of the algebra there appears a Hamiltonian system of normal series (see above), $\dot{X} = [X, \varphi_{ab}(X)]$. The general form of the Hamiltonian of normal series

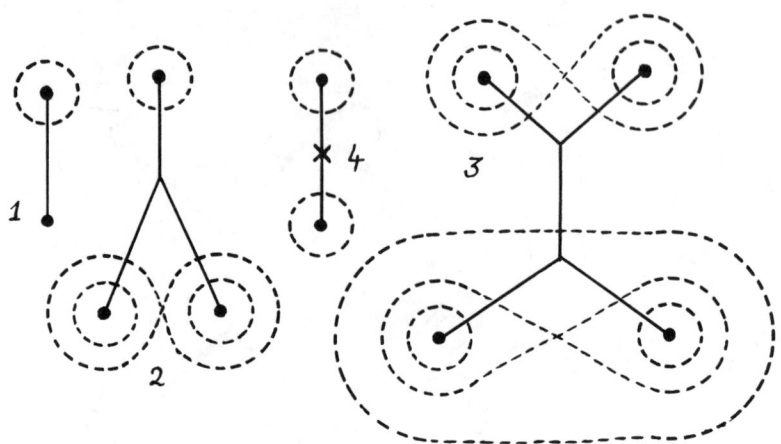

Figure 93 (1-4)

has been indicated above. It is as follows:

$$H = \sum_{i<j} \lambda_{ij} x_{ij}^2; \quad \lambda_{ij} = \frac{b_i - b_j}{a_i - a_j}, \quad \sum_i a_i = 0, \quad \sum_i b_i = 0.$$

A corresponding system of differential equations describes the motion of a four-dimensional rigid body. The integrals of this system of equations (see above) are of the form:

$$f_3 = \sum_{i<j} (a_i + a_j) x_{ij}^2$$

and

$$f_4 = \sum_{i<j} (a_i^2 + a_i a_j + a_j^2) x_{ij}^2.$$

They are in involution on orbits and are functionally independent almost everywhere.

The topological portrait of the equations of motion of a four-dimensional rigid body with a fixed centre of mass appears to contain *only black vertices* on graphs Γ. This means that all critical manifolds of the second integral are circles. These circles can be minimaxes, then Liouville tori close to them contract onto these circles. Next, critical circles can be saddles. In this case, on a critical level either two tori are transformed into one (or vice versa) or (as shown by calculations) sometimes two tori are transformed into two tori. In the latter case, the critical level surface of a Bott integral has the form $T \times S^1$, where the set T is homeomorphic to the cross-section of the sphere S^2 (embedded into $\mathbb{R}^3$ in a standard manner) with a pair of planes passing through its centre.

CLAIM 6.2.1. *Given a Hamiltonian system of a four-dimensional rigid body of normal series on the Lie algebra so(4) with a Hamiltonian f_3, nonsingular orbits*

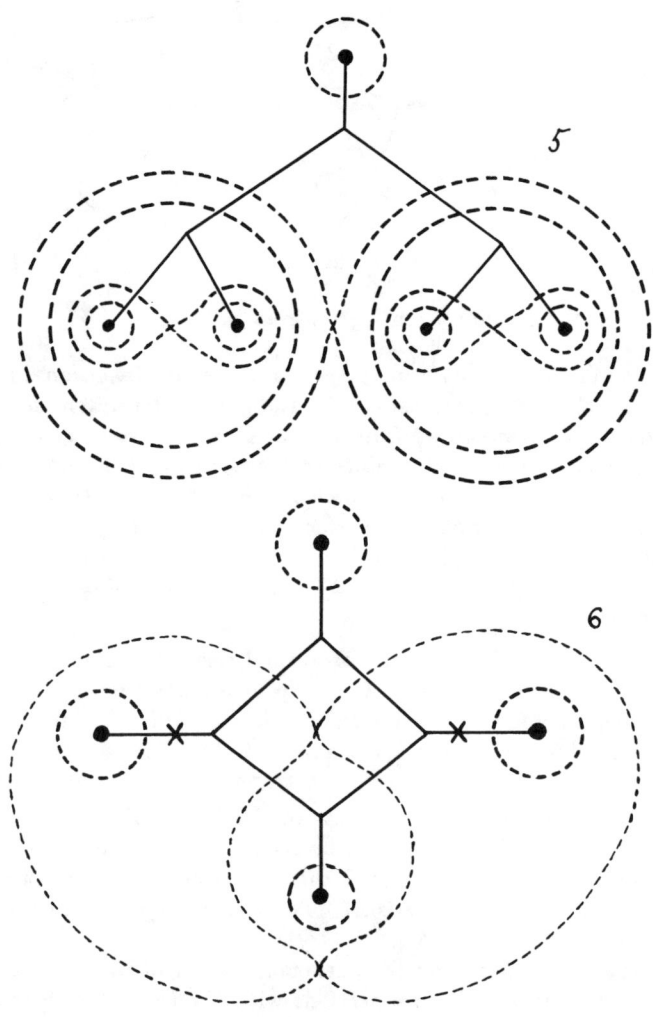

Figure 93 (5,6)

$S^2 \times S^2$ in the algebra so(4) fibre into three-dimensional isoenergy surfaces $Q^3 = \{f_3|_{S^2\times S^2} = \text{const}\}$. These surfaces are such that for all Q^3, except for a finite set, the function $f_4|_{S^2\times S^2}$ is a Bott integral. Depending on the values of p_1 and p_2, which set the orbit $S^2 \times S^2 = \{f_1 = p_1^2, f_2 = p_2\}$, the bifurcational diagrams for the momentum mapping $F = f_3 \times f_4|_{S^2\times S^2}$ can be of three types (see Fig. 94). Digits in Fig. 94 indicate the number if Liouville tori of which consists the complete

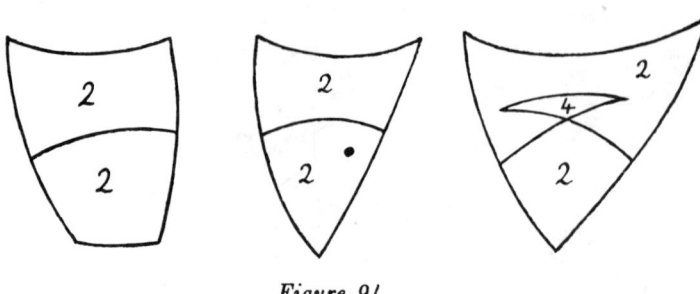

Figure 94

pre-image $F^{-1}(y)$ for points y from a given region.

THEOREM 6.2.1. Let a Hamiltonian system of a four-dimensional rigid body of normal series on the Lie algebra so(4) with the general Hamiltonian H mentioned above be given. For an integral f independent of H almost everywhere on orbits, choose f_3 or f_4. Then nonsingular orbits $S^2 \times S^2$ in so(4) are fibred into isoenergy surfaces $Q^3 = \{H|_{S^2 \times S^2} = \mathrm{const}\}$, which possess the property that the integral $f|_{S^2 \times S^2}$ is a Bott integral for all these surfaces Q^3 except for a finite set. A complete list of all connected graphs $\Gamma(Q)$ describing all nonsingular surfaces Q is presented in Fig. 95. All these graphs are realized for a certain Hamiltonian of the form H indicated above.

REMARK: An arbitrary Hamiltonian of normal series is represented in the form $H = k_1 f_1 + k_2 f_2 + k_3 f_3$, where k_1, k_2, k_3 do not depend on x_{ij}. Therefore, the bifurcational diagrams for a system with an aribtrary Hamiltonian of our form are obtained from the bifurcational diagrams for a system with a Hamiltonian H (see Fig. 94) by means if a non-degenrate linear transformation(Oshemkov).

Figure 95 shows topological portraits of the integrable case of normal series of a four-dimensional rigid body. This invariant is more complicated than those enumerated above. It should be noted that nonorientable edges of graphs $\Gamma(Q)$ are absent (i.e., there are no asterisks). The two-dimensional surfaces $P(Q)$ are homeomorphic either to a sphere or to a torus.

The surfaces constructed by the author are glued of basis surfaces of Seifert bundles. This offers interesting connections with the theory of three-dimensional manifolds. This material will be given elsewhere. All the enumerated results refer to each individual nonsingular isoenergy surface on which the system is integrable. Outside this surface the system can be non-integrable.

We will point to the connection of the graph $\Gamma(Q)$ with the bifurcational set Σ of the momentum mapping F of an integrable system. Here $F : M^4 \to R^2(H, f)$, where $F(x) = (H(x), f(x))$. Fixing a certain value of h for H, we obtain that the isoenergy surface $Q = (H = h)$ is a complete preimage (connected component) of a straight line with the equation $H = h$ on the plane $R^2(H, f)$. The restriction on F from M^4 to Q^3 coincides with the mapping f investigated above. Intersections of the straight line $H = h$ with the set Σ (which is an image, under the mapping F, of the set of critical points of the mapping F) correspond to the critical levels

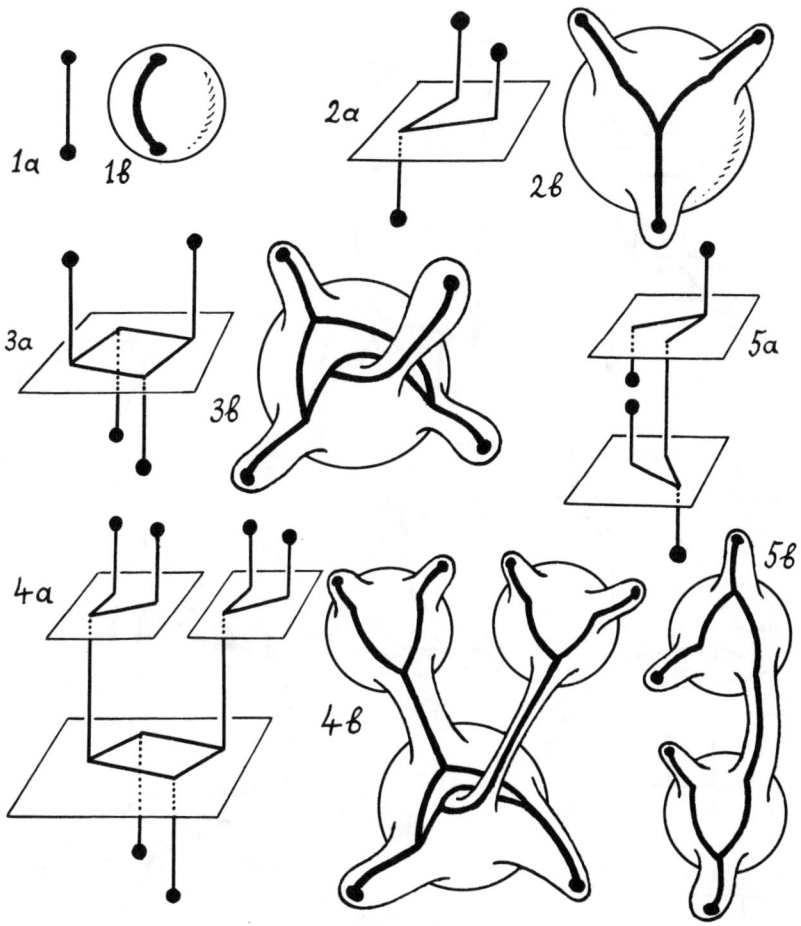

Figure 95 (1-5)

f_c of the integral f. It is readily seen that a replacement of the integral f by another integral f' affects, generally speaking, the set Σ. Consequently, Σ is not a topological invariant of an integrable system. The invariant constructed above, i.e., the triplet (Γ, P, k) is not restored completely from the geometry of the set Σ because it requires a more thorough analysis of critical sets of an integral and corresponding dynamic formulae.

Multidimensional analogues of the above-mentioned results also take place and we will present them elsewhere.

We now briefly dwell on the scheme of the proof of the above theorems. We

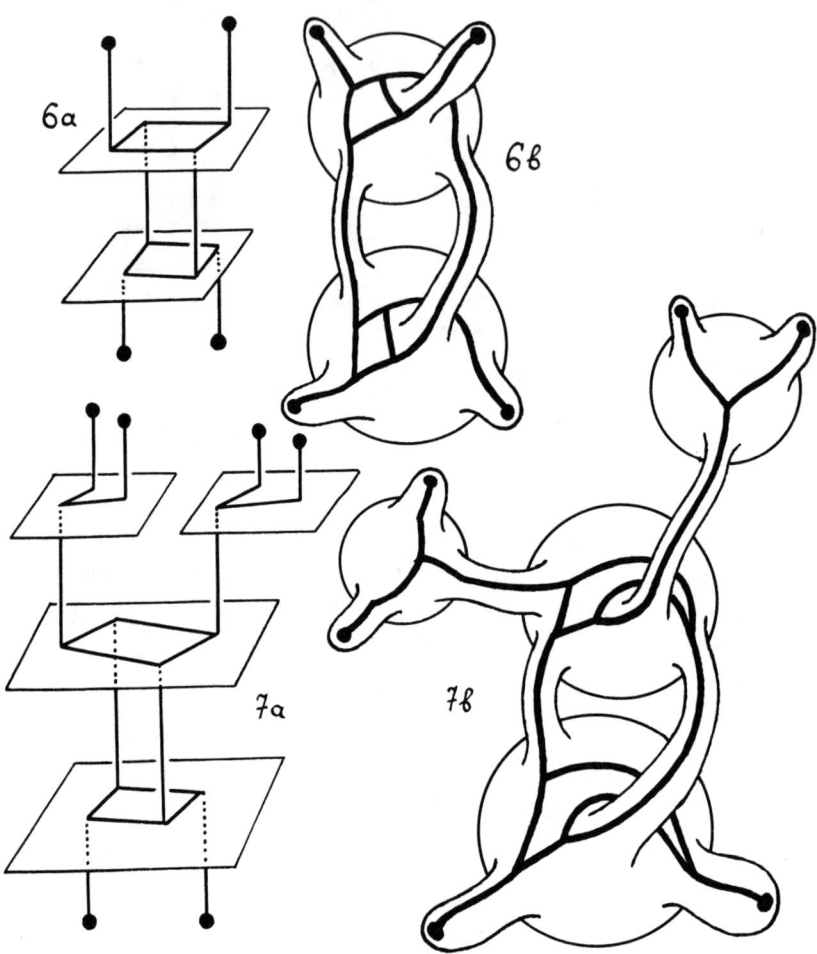

Figure 95 (6,7)

start with the proof of Proposition 6.1.1. Suppose f and f' are two Bott integrals on Q. Since the Hamiltonian is nonresonant, each nonsingular torus with an irrational winding is simultaneously a component of the level surface for each of the integrals because integrals are constant on the trajectories of the system v, and therefore constant on the closure of each trajectory. Each nonsingular torus with a rational winding is approximated arbitrarily closely by non-singular tori with irrational windings. Therefore, it is also a component of the level surface for each

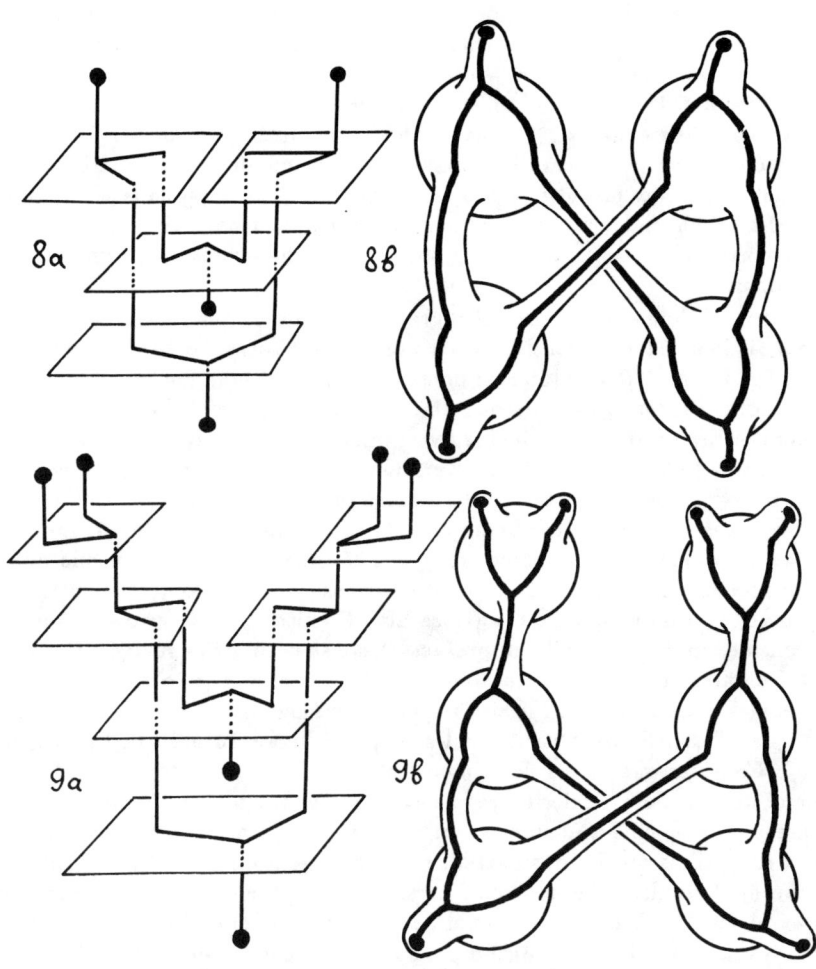

Figure 95 (8,9)

of the integrals. Consider fields $w = \text{grad } f$ and $w' = \text{grad } f'$. They define one and the same one-dimensional foliation of the surface Q into integral trajectories orthogonal (in an arbitrary metric) to all nonsingular level surfaces, i.e., to tori. The velocity of motion along trajectories (determined by the fields w and w') can be different and can become zero simultaneously on the whole torus for one of the fields (whereas the other field is nonzero). Since singular components of the level are approximated by non-singular tori (when they move along w and w'), then singular connected components of the level surface for f, which are other than a torus, coincide with analogous components for f'. If a level surface was a critical torus for f, it can become noncritical for f' (and vice versa). However, each singular component for f is a singular component for f' as well (and vice versa). Tori form a one-dimensional family in the neighbourhood of each nonsingular torus. Locally, considering operations $f \to f^2 = f'$ or $f \to \sqrt{f} = f'$, we can transform a noncirit- ical torus into a critical one and vice versa. All other functional operations locally preserving the Bott character reduce to these two because a smooth Morse func- tion on a one-dimensional manifold may have only quadratic singularities. From the geometrical point of view, the operation $f \to f^2$ corresponds to folding an edge of the graph and arising on it a minimum or a maximum. The operation $f \to \sqrt{f}$ corresponds to straightening the edge of the graph and eliminating the local minimum.

Construct a homeomorphism of a graph Γ onto a graph Γ'. To define it on a subgraph A, it suffices to consider each nonsingular torus for the integral f as a torus for the integral f' (singular or nonsingular). This mapping also continues to subgraphs of the form T_c because one-dimensional foliations of the surface Q which are generated by w and w', coincide everywhere except, maybe, some critical or noncritical tori (where the velocity of motion along w or w' becomes zero). As a consequence we note that total level surfaces of the integrals f and f' (contain- ing one and the same connected component) may comprise a different number of connected components. The values of f and f' may be distinct on one and the same connected component. This proves Proposition 1. Proposition 6.1.2 is, in fact, proved by the author in [283] (see also Chapter 2). To prove Proposition 6.1.3, consider in more detail the process of constructing the surface P. At first we restrict ourselves to the orientable case. A complex f_c is obtained by gluing several copies of tori $T_1, \ldots, T_N$ along cycles homologous to the cycle γ on f_c (see above). Let $S_1, \ldots S_p$ be critical circles of f_c. Choose on each of them a point $q_1, \ldots q_p$. Consider one of the flat rings r into which f_c is cut by circles S_i. Then on each of its boundary circles there lies exactly one point of the form q_α. Suppose these are points q_α and q_β. Join them with a smooth arc $\tau_{\alpha\beta}$ which wholly lies within a ring. This arc is constructed, of course, non-uniquely, but this does not affect the final result. The non-self-intersecting arc $\tau_{\alpha\beta}$ could be connected with the integral f. To this end, one should consider a field sgrad f on the ring r. This field can vanish only on the ring boundary. We might let out from the point q_α a short segment going inside the ring, and then from this ring let out an integral trajectory of the field sgrad f and move along it until we come sufficiently close to the opposite boundary of the ring on which the point q_β is situated. Then we turn from the integral trajectory and draw a short segment into the point q_β.

We repeat the process of constructing a smooth arc in the next ring starting from the point q_β, etc., until all the rings entering in the composition of f_c are over. As a result we obtain a system of circles–ovals. Each of them is composed of arcs of the form $\tau_{\alpha\beta}$. Consider in all interior points of the arc $\tau_{\alpha\beta}$ the normal segments along the integral trajectories of the field grad f. The ends of the obtained strip are determined by the coordinate crosses of the points q_α and q_β. Replacing the arc $\tau_{\alpha\beta}$ by the arc $\tau'_{\alpha\beta}$ which lies in r and joins q_α and q_β, we have a surface homeomorphic to the previous one. A replacement of f by f' leads to a homeomorphism of P

and may essentially affect the form of embedding of P_c into Q. Proposition 6.1.4 follows from the construction presented above. The graph K_c consists of arcs of the form $\tau_{\alpha\beta}$. Each of them determines its own ring of the form r on f_c. The proof of Theorem 6.1.2 follows from [283] as well as from the explicit construction of P_c described above. Consider in more detail the case $P_c \dot\times S^1$. Let $U(f_c) = \{x \in Q : c - \varepsilon \leqslant f(x) \leqslant c + \varepsilon\}$ be a regular neighbourhood of f_c, the latter containing critical circles $S_1, \ldots S_m$ with nonorientable diagrams. Let $U(S_i)$ be a small tubular neighbourhood of S_i (a full torus) in $U(f_c)$. Then $U(f_c) \backslash (U(S_1) \cup \ldots \cup U(S_m))$ is homeomorphic to a direct product $(P_c \backslash (x_1 \cup \ldots \cup x_m)) \times S^1$. In this case (see Fig. 86) each full torus $U(S_1)$ is a fibered full torus of the type (2.1).

An important comment. As has already been pointed out, the classical bifurcational diagram $\Sigma(H, f)$ of momentum mapping is not a topological invariant of an integrable system. The topological portrait $I(H)$ discovered by us is a more natural object from this point of view because it does not depend on the choice of a second integral f and is completely determined only by the nonresonance Hamiltonian H itself. In the case of general position, the bifurcational diagram Σ is shown by a certain curve (graph) with singularities which lies in a two-dimensional plane. Our invariant $I(H, Q)$ is also depicted by a certain one-dimensional graph but embedded into a certain closed two-dimensional surface. From this it is seen that invariance of the topological portrait $I(H)$ is reached due to a deeper penetration into the geometry of the integrable system. In other words, the invariant topological portrait $I(H)$ "knows" about the system of differential equations more than the non-invariant bifurcational diagram $\Sigma(H, f)$. The plane is replaced by a two-dimensional closed surface P^2, and the curve Σ is replaced by a graph Γ which has a "finer organization." The invariant $I(H) = \{\Gamma(Q), P(Q), k(Q)\}$ admits another interpretation. Fixing the value h of the enrgy H, we obtain one isoenergy manifold $Q_h^3 = (H = h)$. Varying h, we vary the manifold Q_h^3, and therefore vary the invariant $\Gamma(Q_h), P(Q_h), k(Q_h)$. The one-dimensional graph $\Gamma(Q_h)$ becomes deformed and sweeps a certain two-dimensional cell (simplicial) complex. Denote it by $\widetilde{\Gamma}(H)$. The two-dimensional surface $P(Q_h)$ sweeps a certain three-dimensional surface $\widetilde{P}(H)$ with varying h. Thus, we get a topological invariant $\widetilde{\Gamma}(H) \xrightarrow{\widetilde{k}(H)} \widetilde{P}(H)$. This object depends only on the symplectic representation of Abelian group R^2, which is generated by the integrals H and f. The object $\widetilde{I}(H) = (\widetilde{\Gamma}(H), \widetilde{P}(H), \widetilde{k}(H))$ does not depend on a concrete choice of H and f within the algebra generated by H and f (under nonresonance conditions).

Thus, each symplectic action (representation) of the group R^2 on M^4 defines in a unique manner the topological invariant $\widetilde{I}(\mathsf{R}^2)$. Here we assume that among the functions defining the representation of R^2 there exists at least one nonresonance integral.

The invariant $I(\mathsf{R}^2)$ admits another interpretation. Consider the momentum mapping $F : M^4 \to \mathsf{R}^2(H, f)$ for fixed H and f. If $a \in \mathsf{R}^2(H, f)$, then the complete pre-image $F^{-1}(a)$ consists of a certain number of closed connected components. If a is a regular value for the mapping F, then compact connected components of the pre-image $F^{-1}(a)$ are tori. Replacing each connected component by a point, we can consider the *space of connected components* of pre-images of a momentum mapping. Clearly, it coincides with $\widetilde{\Gamma}(\mathsf{R}^2)$ (Fig. 96).

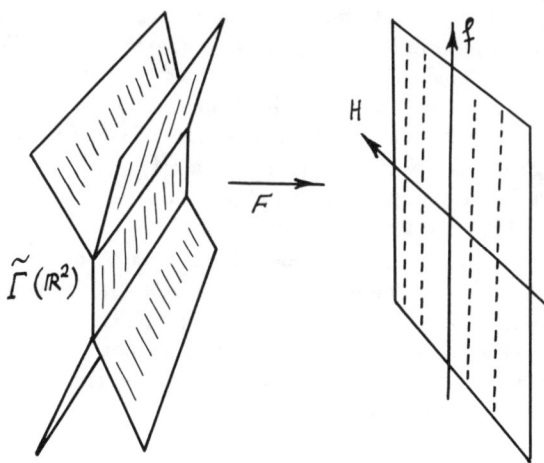

Figure 96

Cutting this space by "planes" $H = $ const, in the "cross-section" we obtain graphs $\Gamma(Q)$. The surface $\tilde{\Gamma}(R^2)$ is two-dimensional, and its singularities are of great interest. It is therefore relevant to formulate here the following problem:

> to describe singularities of the space of connected components of pre-images of the momentum mapping of an integrable system of differential equations.

The classification of singularities of codimension 1 has, in fact, been obtained above (in the nonresonance case). They are 5 types. It remains to classify singularities of codimension 2.

§3 Morse-Type Theory for Hamiltonian Systems Integrated by Means of Non-Bott Integrals

In Chapters 2 and 6 we have considered the case where a Hamiltonian system has an additional Bott (Morse) integral. As shown by concrete studies of real systems, this case is actually typical. But situations cometimes occur when a second additional integral is not a Bott one. It is natural to ask whether the above results are valid in this case. Of course, the integral should as before be imposed some reasonable limitations because smooth functions may have extremely complicated singularities. The study of real systems has shown that it suffices to require that the integral be "tame." The term is taken from topology where it denotes objects organized "rather well" as distinct from pathological "wild" spaces. An exact definition is given below. Many of the results that we have obtained prove to be valid also in the case when the smooth integral is only a tame, and not a Bott one.

Consider a four-dimensional symplectic smooth manifold and suppose that $v = \operatorname{sgrad} H$ is a Hamiltonian system with a smooth Hamiltonian H, which is Liouville-integrable by means of a certain smooth integral f on a nonsingular compact isoenergy surface $Q^3 = (H = \text{const})$. Restricting the integral f from the manifold M^4 (or from a neighbourhood of the submanifold Q^3) to the submanifold Q^3, we obtain a smooth mapping $f: Q \to R$.

DEFINITION 6.3.1. *A smooth integral f will be called tame if for any critical value $a \in R$ the set $f^{-1}(a)$ is tame, i.e., if there exists a homeomorphism of the whole manifold Q onto itself, which shifts this set to a polyhedron.*

Denote by (H^*) the class of three-dimensional compact closed manifolds which are isoenergy surfaces of Hamiltonian systems integrable by means of a tame integral (which is complementary of H, smooth, independent and in involution with H).

Since each Bott integral is apparently tame, a trivial inclusion $(H^*) \supset (H)$ takes place. In other words, extending the class of considered integrals, we have also extended the class of three-dimensional manifolds which are isoenergy surfaces. From the point of view of three-dimensional topology and its applications to Hamiltonian mechanics, the question is of interest whether or not the classes (H^*) and (H) coincide. To say it differently, to what extent the assumption about the Bott character is essential in many theorems of the Morse-type theory developed in Chapter 2.

THEOREM 6.3.1. (FOMENKO AND MATVEEV). *The classes (H^*) and (H) coincide.*

This result has many different interpretations. We will mention here only one of them. As we know from Chapter 2, the topology of the manifold Q of the class (H) shows the surgery in Liouville 2-tori with a varying value of the integral. We know already that all possible transformations in the Bott case are compostition of three basic types of surgery, I, II, III. On extending the class of integrals and introducing into consideration tame integrals, we have a priori extended also the store of possible modification of Liouville tori. The theorem formulated above asserts, in fact, that such an extension has not taken place. In other words, although the surgery in Liouville tori in the non-Bott tame case become, of course, much more complicated, nonetheless all of them are as before compositions of the same three types of surgery, I, II, III discovered by the author in Chapter 2. One should bear in mind that organizing these "tame transformations" into a composition of transformations of types I, II, III, we must, generally speaking, replace the initial integral by another, already a Bott one. The theorem formulated above states that such a replacement is always possible, i.e., "near" a tame integral there always exists a Bott integral (on the same manifold Q) which realizes decomposition of tame Liouville torus transformations into a composition of elementary transformations of types I, II, III.

REFERENCES

* In Russian

1. Arnold, V. "Sur la géométrie differentielle des groupes de Lie de dimension infinite et ses applications à l'hydrodynamique des fluides pur faits." *Annales de l'Institut Fourier* **16** (1966), No. 1, 319–361.
*2. Arnold, V. "Hamiltonian properties of the Euler equations of the dynamics of a rigid body in an ideal liquid." *Uspekhi Mat. Nauk* **26** (1969), 225–226.
3. Arnold, V. *Mathematical Methods of Classical Mechanics*. Springer-Verlag, New York (1978).
4. Adler, M., and Van Moerbeke, P. "Completely integrable systems, Euclidean Lie algebras and curves." *Advances in Math.* **38** (1980), No. 3, 267–317.
5. Adler, M., and Van Moerbeke, P. "Linearization of Hamiltonian systems, Jacobi varieties and representation theory." *Advances in Math.* **38** (1980), No. 3, 318–379.
6. Avez, A., Lichnerowich, A., and Dias-Miranda, A. "Sur l'algèbre des automorphises infinitisimaux d'une varieté symplectique." *J. of Differential Geometry* **9** (1974), No. 1, 1–40.
7. Adler, M. "On a trace functional for formal pseudodifferential operators and the symplectic structure for Korteweg–de Vries type equations." *Invent. Math.* **50** (1979), 219–248.
8. Adler, M. "Some finite-dimensional systems and their scattering behaviour." *Comm. Math. Phys.* **55** (1977).
9. Adler, M. "On a trace functional for formal pseudodifferential operators and the Hamiltonian structure of Korteweg–de Vries type equations." Lecture Notes in Math, v. 755 (1979), 1–16.
10. Adler, M., and Moser J. "On a class of polynomials connected with the Korteweg–de Vries equation." *Comm. Math. Phys.* (1978), 1–30.
11. Airault, H., McKean, H. P., and Moser, J. "Rational and elliptic solution of the Korteweg–de Vries equation and related many-body problem." *Comm. Pure Appl. Math* **30** (1977), 95–148.
*12. Arkhangelsky, A. A. "Completely integrable Hamiltonian systems on a group of triangular matrices." *Matem. Sbornik* **108** (1979), No. 1, 134–142.
*13. Belyaev, A. V. "On the motion of a multidimensional body with a fixed point in the gravity field." *Matem. Sbornik* **114** (1981), No. 3, 465–470.
14. Bogoyavlensky, O. I. "On perturbation of periodic toda lattice." *Comm. Math. Phys.* **51** (1976), 201–209.

*15. Bogoyavlensky, O. I. "New algebraic constructions of Euler equations." *Dokl. Akad. Nauk SSSR* **268** (1983), No. 2, 277–280.

*16. Bogoyavlensky, O. I. "The integrable Euler equations connected with filtrations of Lie algebras." *Matem. Sbornik* **121** (1983), No. 2, 233–242.

17. Burchnall, J. L., and Chaundy, T.W. "Commutative ordinary differential operators." *Proc. London Math. Soc.* **81** (1922), 420–440; *Proc. Royal Soc. London (A)* **118** (1928), 557–593; *Proc. Royal Soc. London (A)* **134** (1931), 471–485.

*18. Berezin, F. A., Perelomov, A. M. "Group-theoretical interpretation of Korteweg–de Vries equations." *Funkts. Analiz i yego Prilozhen.* **14** (1980), No. 2, 50–51.

*19. Brailov, A. V. "Involutive sets on Lie algebras and extension of the scalar ring." *Vestnik Mosk. Gos. Univers.*, ser. mat. mekh., No. 1 (1983), 7–15.

*20. Brailov, A. V. "Several cases of complete integrability of Euler equations and applications." *Dokl. Akad. Nauk SSSR* **268** (1983), No. 5, 1043–1046.

*21. Vishik, S. V., and Dolzhansky, F. V. "The analogues of Euler–Poisson and magnetic hydrodynamic equations connected with Lie groups." *Dokl. Akad. Nauk SSSR* **238** (1978), No. 5, 1032–1035.

*22. Vinogradov, A. M., and Kupershmidt, B. "The structure of Hamiltonian mechanics." *Uspekhi Mat. Nauk* **42** (1977), No. 4, 175–236.

23. Vergne, M. "La structure de Poisson sur l'algèbre symétrique d'une algèbre de Lie nilpotente." *Bull. Soc. Math. (France)* **100** (1972), 301–335.

*24. Veselov, A. P. "Finite-zoned potentials and integrable systems on a sphere with quadratic potential." *Funkts. Analiz i yego Prilozhen.* **14** (1980), No. 1, 48–50.

*25. Veselov, A. P. "On the Hamiltonian formalism for Novikov–Krichever equations for commutation of two operators." *Funkts. Analiz i yego Prilozhen.* **13** (1979), No. 1, 17.

*26. Vinogradov, A. M., and Krasil'shchik, I. S., "What is Hamiltonian formalism?" *Uspekhi Mat. Nauk* **30** (1975), No. 1, 173–198.

27. Wilson, G. "Commuting flows and conservation laws for Lax equations." *Math. Proc. Cambridge Philos. Soc.* **86** (1979), 31.

*28. Dubrovin, B. A., Novikov, S. P., and Fomenko, A. T. *Modern Geometry.* Nauka, Moscow (1979), Parts 1, 2; Nauka, Moscow (1984), Part 3. English translation: Dubrovin, B. A., Fomenko, A. T., and Novikov, S. P., *Modern geometry.* Springer Verlag, Part 1, GTM 93, (1984), Part 2, GTM 104, (1985).

*29. Gorr, G. V., Kudryashova, L. V., and Stepanova, L. A. *Classical Problems of Rigid-Body Dynamics.* Naukova Kumka, Kiev (1978).

*30. Gel'fand, I. M., and Dikii, L. A. "Fractional degrees of operators and Hamiltonian systems." *Funkts. Analiz i yego Prilozhen.* **10** (1976), No. 4, 13–29.

*31. Gel'fand, I. M., and Dorfman, I. Ya. "Hamiltonian operators and related algebraic structures." *Funkts. Analiz i yego Prilozhen.* **13** (1979), No. 4, 13–30.

*32. Gel'fand, I. M., and Cherednik, I. V. "Abstract Hamiltonian formalism for classical Yang–Baxter germs." *Uspekhi Mat. Nauk* **38** (1983), No. 3, 3–21.

33. Gardner, C. S. "Korteweg–de Vries equation and generalization. IV: "The Korteweg–de Vries equation as a Hamiltonian system." *J. Math. Phys.* **12**

(1971), No. 8, 1548–1551.
34. Gardner, C. S., Green, J. M., Kruskall, M. D., and Miura, R. M. "Korteweg–de Vries equation and generalization. VI: Methods for exact solution." *Comm. Pure Appl. Math.* **27** (1974), 97–133.
35. Godbillon, C. *Géométrie Differentielle et Mécanique Analitique.* Hermann, Paris (1969).
36. Dirac, P. "On generalized Hamiltonian dynamics." *Canadian Journ. of Mathematics* **11** (1950), 129–148.
*37. Dubrovin, B. A. "Theta-function and non-linear equations." *Uspekhi Mat. Nauk* **36** (1981), No. 2, 11-80.
*38. Dubrovin B. A. "Completely integrable Hamiltonian systems connected with matrix operators and Abelian manifolds." *Funkts. Analiz i yego Prilozhen.* **11** (1977), No. 4, 28–41.
*39. Dao Chong Thi."Integrability of Euler equations on homogeneous symplectic manifolds." *Matem. Sbornik* **106** (1978), No. 2, 154–161.
40. Duflo, M., and Vergne, M. "Une proprieté de la représentation coadjointe d'une algèbre de Lie." *C. R. Acad. Scient.* (Paris) **268** (1969), 583–585.
*41. Drinfeld, V. G., and Sokolov, V. V. "Korteweg–de Vries type equations and simple Lie algebras." *Dokl Akad. Nauk SSSR* **258** (1981), No. 1, 11–16.
42. Dickson, L. E. "Differential equations from the group standpoint." *Annals of Math.* **25** (1924), No. 4, 287–378.
*43. Dikii, L. A. "Note on Hamiltonian systems connected with rotation group." *Funkts. Analiz i yego Prilozhen.* **6** (1972), No. 4, 83–84.
*44. Dikii, L. A. "The Green function of differential operators and Hamiltonian systems." In *Nonlinear Waves.* Nauka, Moscow (1975), 36–45.
*45. Dubrovin, B. A., Matveev, V. B., and Novikov, S. P. "Nonlinear Korteweg–de Vries type equations, finite-zoned linear operators and Abelian manifolds." *Uspekhi Mat. Nauk* **31** (1976), No. 1, 55–136.
46. Dixmier, J., Duflo, M., and Vergne, M. "Sur la représentation coadjointe d'une algèbre de Lie." *Composito Math.* **25** (1974), 309–323.
*47. Zakharov, V. E., and Faddeev, L. D. "The Korteweg–de Vries equation, a completely integrable Hamiltonian system." *Funkts. Analiz i yego Prilozhen.* **5** (1971), No. 4, 18–27.
*48. Ilyushechkin, N. V. "On the variety of completely integrable left-invariant metrics on a semisimple Lie group." *Vestnik Mosk. Gos. Univers.*, ser. mat. mekh., No. 3 (1979), 35–37.
49. Cartan, E. *Leçons sur les Invariants Integraux.* Paris (1922).
*50. Krichever, I. M. "Algebraic curves and nonlinear differential equations." *Uspekhi Mat. Nauk* **33** (1978), No. 4, 215–216.
*51. Krichever, I. M. "The algebraic geometry methods in the theory of nonlinear equations." *Uspekhi Mat. Nauk* **32** (1977), No. 6, 183–208.
52. Kostant, B. "The solution to a generalized Toda lattice and representation theory." *Advances in Math.* **34** (1980), No. 3, 195–338.
*53. Kolesnikov, N. N. "Natural systems with a solvable symmetry group." *Vestnik Mosk. Gos. Univers.*, ser. mat. mekh., No. 5 (1978), 99–103.
*54. Kozlov, V. V., and Kolesnikov, N. N. "On integrability of Hamiltonian systems." *Vestnik Mosk. Gos. Univers.*, ser. mat. mekh., No. 5 (1979), 88–91.

55. Kupershmidt, B. A., and Wilson G. "Modifying Lax equations and the second Hamiltonian structure." *Invent. Math.* **62** (1981) 403–436.
*56. Kulish, P. P., and Reyman, A. G. "Hierarchy of symplectic forms for Schrödinger and Dirac equations on a straight line." *Zapiski Nauchnogo Seminara LOMI* **77** (1978), 134–147.
57. Kazhdan, D., Kostant, B., and Sternberg, S. "Hamiltonian group actions and dynamical systems of Calogero type." *Comm. Pure Appl. Math.* **31** (1978), No. 4, 481–507.
*58. Krasil'shchik, I. S. "Hamiltonian cohomologies of canonical algebras." *Dokl. Akad. Nauk SSSR* **251** (1980), No. 6, 136–1309.
59. Knörrer, H. "Geodesics on the ellipsoid." *Invent. Math.* **59** (1980), No. 2, 119-144.
60. Calogero, F. "Solution of the one-dimensional n-body problems with quadratic and/or inversely quadratic pair potentials." *J. Math. Phys.* **12** (1971), 419–436.
*61. Kozlov, V. V. "Integrability and nonintegrability in Hamiltonian mechanics." *Uspekhi Mat. Nauk.* **38** (1983), No. 1, 3–67.
*62. Kozlov, V. V. *Methods of Qualitative Analysis on Rigid-Body Dynamics*. Izdatel'stvo Mosk. Gos. Univers., Moscow (1980).
*63. Kolokol'tsov, V. N. "Geodesic flows on two-dimensional manifolds with an additional first integral polynomial in velocities." *Izvestiya Akad. Nauk SSSR*, ser. matem., No. 5 (1982), Vol. 46, 994–1010.
*64. Kirillov, A. A. *Elements of Representation Theory*. Nauka, Moscow (1972).
65. Langlois, M. "Contribution à l' étude du mouvement du coups rigide à n dimensions autour d'un point fixe." In *Thèse Presentée à la Faculté des Sciences de l'Université de Besançon*. Besançon (1971).
66. Lax, P. D. "Integrals of nonlinear equations of evolution and solitary waves." *Comm. Pure Appl. Math.* **21** (1968).
*67. Le Ngok Tyeuen. "Commutative sets of functions on orbits of general position of finite-dimensional Lie algebras." *Uspekhi Mat. Nauk* **38** (1983), No. 1, 179–180.
*68. Lebedev, D. R., and Manin, Yu. I. "The Hamiltonian operator of Gel'fand–Dikii and coadjoint representation of the Volterra group." *Funkts. Analiz i yego Prilozhen.* **13** (1980), No. 4, 40–46.
*69. Manin, Yu. I. "Algebraic aspects of nonlinear differential equations." In *Sovremennye Problemy Matematiki*, v. II (1978), 5–152.
*70. Manin, Yu. I. "Matrix solitons and fibre bundles over curves with singularities." *Funkts. Analiz i yego Prilozhen.* **12** (1978), No. 4, 53–63.
71. McKean, H. "Integrable systems and algebraic curves." In *Global Analysis*. Lecture Notes in Math., v. 755 (1976), 83–200. Springer, New York.
72. McKean, H., and Trubowitz, E. "Hill's operator and hyperelliptic function theory in the presence of infinitely many branch points." *Comm. Pure Appl. Math.* **29** (1976) 143–226.
73. Moser, J. "Various aspects of integrable Hamiltonian systems." In *Dynamical Systems*. Prog. in Math. 8. Birkhäuser-Boston: Cambridge, Mass. (1980).
74. Van Moerbeke, P., and Mumford, D., "The spectrum of difference operators and algebraic curves." *Acta Math.* **143** (1979), 93–154.

*75. Meshcheryakov, M. V. "Integration of the equations for geodesics of left-invariant metrics on semisimple Lie groups by means of special functions." *Matem. Sbornik* **117** (1982), No. 4, 481–493.

*76. Manakov, S. V. "Note on integration of Euler equations of the dynamics of an n-dimensional rigid body." *Funkts. Analiz i yego Prilozhen.* **10** (1976), No. 4, 93–94.

*77. Zakharov, V. E., Manakov, S. V., Novikov, S. P., and Pitayevsky, L. P. *Theory of Solitions. The Inverse Problem Method.* Edited by S. P. Novikov. Nauka, Moscow (1980).

78. Marsden, J., and Weinstein, A. "Reduction of symplectic manifolds with symmetry." *Reports on Math. Phys.* **5** (1974), No. 1, 121–130.

79. Moser, J. "Three integrable Hamiltonian systems connected with isospectral deformations." *Advances in Math.* **16** (1975), 197–220.

80. Moser, J. "Geometry of quadratic and spectral theory." In *The Chern Symposium. 1979.* Springer-Verlag (1980), 147–188.

81. Van Moerbeke, P. "The spectrum of Jacobi matrices." *Invent. Math.* **37** (1976), 45–81.

82. Moody, B. "A new class of Lie algebras." *J. Algebra* **10** (1968), 211–230.

*83. Mishchenko, A. S., and Fomenko, A. T. *A Course in Differential Geometry and Topology.* Nauka, Moscow (1980).

*84. Mishchenko, A. S. "Integrals of geodesic flows on Lie groups." *Funkts. Analiz i yego Prilozhen.* **4** (1970), No. 3, 73–78.

*85. Mishchenko, A. S. "Integration of geodesic flows on symmetric spaces." *Matematicheskiye Zametki* **31** (1982), No. 2, 257–262.

*86. Mishchenko, A. S. "Integration of geodesic flows on symmetric spaces." *Proceedings of a Seminar on Vector and Tensor Analysis*, Izdatel'stvo Mosk. Gos. Unives. **21** (1983), 13–22.

87. Magri, F. "A simple model of the integrable Hamiltonian equation." *J. Math. Phys.* **19** (1978), 1156–1162.

88. Mumford, D. "An algebra-geometric construction of commuting operators and of solution to the Toda lattice equation and related nonlinear equations." *Proc. Intern. Symp. on Algebraic Geometry, Kyoto* (1977), 115–153.

*89. Mishchenko. A. S., and Fomenko, A. T. "On integration of Euler equations on semisimple Lie algebras." *Dokl. Akad. Nauk SSSR* **231** (1976), No. 2, 536–538.

*90. Mishchenko, A. S., and Fomenko, A. T. "Generalized Liouville method of integrating Hamiltonian systems." *Funkts. Analiz i yego Prilozhen.* **12** (1978), 49–59.

*91. Mishchenko, A. S., and Fomenko, A. T. "Euler equations on finite-dimensional Lie groups." *Izvestiya Akad. Nauk SSSR*, ser. matem., **42** (1978), No. 2, 396–415.

*92. Mishchenko, A. S., and Fomenko, A. T. "Integrability of Euler equations on semisimple Lie algebras." In *Trudy Seminara po Vekt. i Tenz. Anal.*, issue 19 (1979), 3–94.

*93. Mishchenko, A. S., and Fomenko, A. T. "Integration of Hamiltonian systems with noncommutative symmetries." In *Trudy Seminara po Vect. i Tenz. Anal.*, issue 20 (1981), 5–54.

94. Mishchenko, A. S., and Fomenko, A. T. *Symplectic Lie Group Action.* Lecture Notes in Math., v. 763 (1979), 504–539. Springer-Verlag. "Algebraic Topology, Aarhus, 1978," *Proceedings.*
*95. Novikov, S. P. "Periodic Korteweg–de Vries problem." *Funkts. Analiz i yego Prilozhen.* **8** (1974), 54–66.
*96. Novikov, S. P. "Hamiltonian formalism and a multivalued analogue of the Morse theory." *Uspekhi Mat. Nauk* **37** (1982), No. 5, 3–49.
*97. Novikov, S. P., and Shmeltser, I. "Periodic solutions to the Kirchhoff equation of the free motion of the rigid body in an ideal liquid and an extended LMS theory. I." *Funkts. Analiz i yego Prilozhen.* **15** (1982), No. 3, 54–66.
*98. Novikov, S. P. "Variational methods and periodic solutions to Kirchhoff-type equations. II." *Funkts. Analiz i yego Prilozhen.* **15** (1982), No. 4, 3–49.
*99. Nekhoroshev, N. N. "Action-Angle Variables and Their Generalization". In: *Trudy MNO (Moscow Math. Soc.)* (1972), V. 26, 181–198.
*100. Nikolenko, N. V. "On complete integrability of a nonlinear Schrödinger equation." *Funkts. Analiz i yego Prilozhen.* **10** (1977), No. 3, 55–69.
101. Olshanetsky, M. A., and Perelomov, A. M. "Geodesic flows on symmetric zero-curvature spaces and explicity solutions of the generalized Calodgero model for the classical case." *Funkts. Analiz i yego Prilozhen.* **10** (1976), No. 3, 86–87.
102. Olshanetsky, M. A., and Perelomov, A. M. "Explicit solutions of the classical generalized Toda models." *Invent. Math.* **56** (1976), 261–269.
103. Olshanetsky, M. A., and Perelomov, A. M. "Completely integrable Hamiltonian systems connected with semisimple Lie algebras." *Invent. Math.* **37** (1976), 93–109.
104. Poincaré, H. "Les méthodes nouvelles de la mécanique céleste. I." Selected works. Moscow, Nauka. (1971).
105. Patera, J., Sharp, R. T., Winternitz, P., and Zassenhaus, H. "Invariant of real low-dimension Lie algebras." *J. Math. Phys.* **17** (1976), No. 6, 986–994.
*106. Pevtsova, T. A. "One way of constructing a commutative algebra of integrals on Lie algebras." *Vsesoyuznaya Konferentsiya po Sovremennym Problemam Geometrii.* Minsk (1979), 149.
*107. Perelomov, A. M. "Several notes on integration of the equations of motion of a rigid body in an ideal liquid." *Funkts. Analiz i yego Prilozhen.* **15** (1981), No. 2, 83–85.
108. Perelomov, A. M. "The simple relation between certain dynamical systems." *Math. Phys.* **63** (1978), 9–11.
109. Perelomov, A. M. " Lax representation for the systems of S. Kovalevskaya type." *Comm. Math. Phys.* **81** (1981), 239–241.
*110. Pidkuyko, S. I., and Stepin, A. M. "On the solution of one differential-functional equation." *Funkts. Analiz i yego Prilozhen.* **10** (1976), No. 2, 84–85.
*111. Pevtsova, T. A. "The symplectic structure of orbits of coadjoint representation of Lie algebras of the type $E \times_\rho G$." *Uspekhi Mat. Nauk* **37** (1982), No. 2, 225–226.
*112. Rashevsky, P. K. *Geometric Theory of Equations in Partial Derivatives.* Moscow, Leningrad (1947).
113. Reyman, A. G., and Semenov-Tian-Shansky, M. A. "Reduction of Hamiltonian systems, affine Lie algebras and Lax equations." *Invent. Math.* **54** (1979), No.

1, 81–100.
114. Reyman, A. G., and Semenov-Tian-Shansky, M. A. "Reduction of Hamiltonian systems, affine Lie algebras and Lax equations." *Invent. Math.* **63** (1981), No. 3, 423–432.
*115. Reyman, A. G., and Semenov-Tian Shansky, M. A. "Algebras of currents and nonlinear equations in partial derivatives." *Dokl. Akad. Nauk SSSR* **251** (1980), No. 6, 1310–1314.
*116. Reyman, A. G., Semenov-Tian-Shansky, M. A., and Frenkel, I. I. "Graduated Lie algebras and completely integrable dynamic systems." *Dokl. Akad. Nauk SSSR.* **247** (1979), No. 4, 802–805.
*117. Reyman, A. G. "Integrable Hamiltonian systems connected with graduated algebras." *Zapiski Nauchnykh Seminarov.* Leningrad (1980), v. 95, 3–54.
118. Ratiu, T. "The C. Neumann problem as a completely integrable system on an adjoint orbit." *Trans. of the Amer. Math. Soc.* **264** (1981), No. 2, 321–32.
119. Ratiu, T. "Euler–Poisson equations on Lie algebras and the n-dimensional heavy rigid body." *Amer. J. of Math.* **103** (1982), No. 3.
120. Ratiu, T., and Van Moerbeke, P. "The Lagrange rigid body motion." *Annales de l'Institut Fourier* **32** (1982), No. 1, 211–234.
121. Smale, S. "Topology and Mechanics: I." *Invent. Math* **10** (1970).
122. Sourian, J. M. *Structure des Systèmes Dynamiques.* Paris (1970).
*123. Stepin, A. M. "Integrable Hamiltonian systems, I. Methods of integration of Hamiltonian systems. II. Series of integrable systems." In *Qualitative Methods of Studying Nonlinear Differential Equations of Nonlinear Oscillations, Kiev, Math. Institut.* (1981).
124. Sternberg, S. *Lectures on Differential Geometry.* Prentice Hall, Englewood Cliffs, N.J. (1964).
*125. Tatarinov, Ya. V. *Lectures on Classical Mechanics.* Moscow, Moscow Univ. Press, (1984).
*126. Tatarinov, Ya. V. "Geometric formalism of classical dynamics; canonical originals." *Vestnik Mosk. Gos. Univers.*, ser. mat. mekh., No. 4 (1983), 85–95.
127. Tischler, D. "Closed 2-forms and an embedding theorem for symplectic manifolds." *J. Diff. Geometry* **12** (1977), 229–235.
128. Thimm, A. "Integrable geodesic flows on homogeneous spaces." *Ergod. Theory and Dynam. Syst.* **1** (1981), No. 4, 495–517.
129. Takiff, S. J. "Rings of invariant polynomials for a class of Lie algebras." *Trans. of the Amer. Math. Soc.* **160** (1971), 249–262.
*130. Trofimov, V. V. "Euler equations on Borel subalgebras of semisimple Lie algebras." *Izvestiya Akad. Nauk SSSR*, ser. matem. **43** (1979), No. 3, 714–732.
*131. Trofimov, V. V. "Euler equations on finite-dimensional solvable Lie Groups." *Izvestiya Akad. Nauk SSSR*, ser. matem. **44** (1980), No. 5, 1191–1199.
*132. Trofimov, V. V. "Finite-dimensional representations of Lie algebras and completely integrable systems." *Matem. Sbornik* **111** (1980), No. 4, 610–621.
*133. Trofimov, V. V. "On complete integrability of Euler equations on Borel subalgebras of simple Lie algebras." *VII All-Union Conference on Modern Problems in Geometry, Minsk* (1979), 201.
*134. Trofimov, V. V. "Group-theoretical interpretation of the equations of magnetic

hydrodynamics of an ideal liquid." In *Nonlinear Oscillations and Control Theory*. Izhevsk (1981), 118–124.

*135. Trofimov, V. V. "Completely integrable geodesic flows of left-invariant metrics on Lie groups connected with commutative graduated algebras with Poincaré duality." *Dokl. Akad. Nauk SSSR* **263** (1982), No. 4, 812–816.

*136. Trofimov, V. V. "Symplectic structures on groups of automorphisms of symmetric spaces." *Vestnik Mosk. Gos. Univers.*, ser. mat. mekh., No. 6 (1983).

*137. Trofimov, V. V. "On completely integrable geodesic flows on a group of motions of Euclidean space." In *Some Problems of Mathematics and Mechanics*. Moscow, Moscow Univ. Press, (1983), 8–9.

*138. Trofimov, V. V. "Commutative graduated algebras with Poincaré duality and Hamiltonian systems." In *Topological and Geometrical Methods in Mathematical Physics*. Voronezh, Voronezh Univ. Press, (1983), 128–132.

*139. Trofimov, V. V., and Fomenko, A. T. "Methods of constructing Hamiltonian flows on symmetric spaces and integrability of seveal hydrodynamic systems." *Dokl. Akad. Nauk SSSR*, **254** (1980), No. 6, 1349–1353.

*140. Trofimov, V. V., and Fomenko, A. T. "On the realization of mechanical systems on orbits of solvable Lie algebras." *Izvestiya Akad. Nauk SSSR*, ser. mekh. tverd. tela, no 3 (1981), 163.

*141. Trofimov, V. V., and Fomenko, A. T. "Dynamic systems on orbits of linear representations of Lie groups and complete integrability of some hydrodynamic systems." *Funkts. Analiz i yego Prilozhen.* **17** (1983), No. 1, 31–39.

*142. Trofimov, V. V., and Fomenko, A. T. "Group noninvariant symplectic structures and Hamiltonian flows on symmetric spaces." In *Trudy Seminara po Vect. i Tenz. Anal.*, issue 21, Moscow Univ. Press, (1983), 23–83.

*143. Fomenko, A. T. *Differential Geometry and Topology*. (Additional chapters.) Moscow, Moscow Univ. Press, (1983).

*144. Fomenko, A. T. "On symplectic structures and integrable systems on symmetric spaces." *Matem. Sbornik.* **115** (1981), No. 2, 263–280.

*145. Fomenko, A. T. "Group symplectic structures on homogeneous spaces." *Dokl. Akad. Nauk SSSR* **253** (1980), No. 5, 1062–1067.

*146. Fomenko, A. T. "Algebraic structure of several classes of completely integrable Hamiltonian systems on Lie algebas." In *Geometric Theory of Functions and Topology*. Kiev, Math. Institut, (1981).

*147. Fomenko, A. T. "Complete integrability of several classical Hamiltonian systems." In *Monogenic Functions and Mappings*. Kiev (1982), 3–19, Math. Institut.

*148. Fomenko, A. T. "Algebraic properties of several integrable Hamiltonian systems." *Tezisy Leningradskoy Mezhdunarodnoy Topologicheskoy Konferentsii*, Leningrad, Math. Institut, (1981).

*149. Fomenko, A. T. "Algebraic structure of several integrable Hamiltonian systems." In *Topological and Geometrical Methods in Mathematical Physics*. Voronezh, Voronezh Univ. Press, (1983), 84–110.

150. Flaschka, H. "Toda lattice II." *Prog. Theor. Phys.* **51** (1974), 703–716.

151. Iacob, A., and Sternberg, S. *Coadjoint Structures, Solutions and Integrability*. Lecture Notes in Phys., v. 120 (1980). Springer-Verlag.

152. Jacobi, C.G.J. *Vorlesungen über Dynamik*. Berlin (1884).

*153. Ibragimov, N. H. *Transformation Groups in Mathematical Physics*. Nauka, Moscow (1983).
154. Koblitz, A. H. *A Convergence of Lives of Sofia Kovalevskaia: Scientist, Writer, Revolutionary*. Birkhäuser Boston, Basel, Stuttgart. (1984).
155. Saveliev, M. V. "Integrable graded manifolds and nonlinear equations." *Comm. in Math. Phys.* **95** (1984), 199–216.
*156. Arkhangelsky, Yu. A. *Analytic Rigid-Body Dynamics*. Nauka, Moscow (1977).
157. Bogoyavlensky, O. I. "Integrable Euler equations on $SO(4)$ and their physical applications." *Comm. in Math. Phys.* **93** (1984), 417–436.
*158. Bogoyavlensky, O. I. "Rigid-body dynamics with n ellipsoid cavities filled with magnetic liquid." *Dokl. Akad. Nauk SSSR.* **272** (1983), No. 6, 1364–1367.
*159. Bogoyavlensky, O. I. "Periodic solutions in the pulsar rotation model." *Dokl. Akad. Nauk SSSR.* **276** (1984), No. 2, 343–347.
*160. Kozlov, V. V. "Two integrable problems of classical dynamics." *Vestnik Mosk. Gos. Univers.*, ser. mat. mekh., No. 4 (1981), 80–84.
*161. Kozlov, V. V. "Non-existence of an additional analytic integral in the problem of the motion of a non-symmetric heavy rigid body around a fixed point." *Vestnik Mosk. Gos. Univers.*, ser. mat. mekh., No. 1 (1975), 105–110.
162. Poincaré, H. "Les méthodes nouvelles de la mécanique céleste. II." *Selected Works*, Moscow, Nauka, (1972).
163. Fomenko, A. T. "Algebraic properties of some integrable Hamiltonian systems." *Topology, Proceedings, Leningrad 1982*. Lecture Notes in Math., v. 1060 (1984), 246–257. Springer-Verlag.
*164. Kozlov, V. V. "Separatrix splitting in the perturbed Euler–Poisson problem." *Vestnik Mosk. Gos. Univers.*, ser. mat. mekh., No. 6 (1976), 99–104.
*165. Ziglin, S. L. *Separatrix Splitting, Solution Branching and Nonexistence of Integral in Rigid-Body Dynamics*. Trudy Moskovskogo Matematicheskogo Obshchestva, Moscow (1980), V. 41, 287–303.
166. Galgani, L., Giorgilli, A., and Strelcyn, J. M. "Chaotic motions and transition to stochasticity in the classical problems of the heavy rigid body with a fixed point." Preprint (1980).
*167. Kozlov, V. V., and Onishchenko, D. A. "Nonintegrability of Kirchhoff equations." *Dokl. Akad. Nauk SSSR* **266** (1982), No. 6, 1298–1300.
*168. Gaydukov, E. V. "Asymptotic geodesics on a Riemannian manifold nonhomeomorphic to a sphere." *Dokl. Akad. Nauk SSSR* **169** (1966), No. 5, 999–1001.
*169. Bolotin, S. V. "Nonintegrability of the n-centre problem for $n > 2$." *Vestnik Mosk. Gos. Univers.*, ser. mat. mekh., no 3 (1984), 65–68.
170. Klingenberg, W. *Lectures on Closed Geodesics*. Berlin (1978).
*171. Anosov, D. V. "On typical properties of closed geodesics." *Izvestiya Akad. Nauk SSSR*, ser. mat. mekh. **46** (1982), No. 4.
172. Bott, R. "Non-degenerate critical manifolds." *Ann. of Math.* **60** (1954), 248–261.
*173. Palamodov, V. P. "On stability of equilibrium in a potential field." *Funkts. Analiz i yego Prilozhen.*, II, issue **4** (1977), 42–55.
*174. Gabrielov, A. M. "On projections of semi-analytic sets." *Funkts. Analiz i yego Prilozhen.*, II, issue **4** (1968), 18–30.

175. Lojasiewicz, S., "Triangulation of semi-analytic sets." *Annali della Scuola Normale Superiore di Pisa*, ser. III, **18** (1964), 449–474.
176. Preissmann, A. "Quelques propertiétés global des espaces de Riemann." *Comm. Math. Helv.*, **15** (1943), 175–216.
*177. Kharlamov, M. P. "Topologic analysis of classical integrable systems in rigid-body dynamics." *Dokl. Akad. Nauk SSSR*. **273** (1983), No. 6, 1322–1325.
*178. Pogosyan, T. I. *Critical Integral Surfaces in the Clebsch Problem*. Mekhanika Tverdogo Tela **16**, 19–24. Kiev.
*179. Pogosyan, T. I. *Construction of Bifurcation Sets in One Problem of Rigid-Body Dynamics*. Mekhanika Tverdogo Tela, Kiev, issue **12** (1980), 9–16.
*180. Tatarinov, Ya. V. "Portraits of classical integrals in the problem of rigid-body rotation around a fixed point." *Vestnik Mosk. Gos. Univers.*, ser. mat. mekh., No. 6 (1974).
181. Adler, M., and Van Moerbeke, P. "The algebraic integrability of geodesic flow on $SO(4)$." *Invent. Math.* **67** (1982), 297–331.
182. Fomenko, A. T. "The integrability of some Hamiltonian systems." *Ann. of Global Analysis and Geometry* **1**(2) (1983), 1–10.
183. Conley, C., and Zehnder, E. "Morse-type index theory for flows and periodic solutions for Hamiltonian equations." *Comm. Pure Appl. Math.* **37** (1984), No. 2, 207–255.
184. Waldhausen, F. "Eine Klasse von 3-dimensionalen Mannigfaltigkeiten." *I. Invent. Math.* **3** (1967), 308–333.
*185. Pogosyan, T. I., and Kharlamov, M. P. "Bifurcation set and integral manifolds of the problem of the rigid-body motion in a linear field of forces." *Prikl. Matem. i Mekh.* **43** (1979), 419–428.
186. Fomenko, A. T. *Algebraic Structure of Certain Integrable Hamiltonian Systems*. Lecture Notes in Math., v. 1108. Global Analysis: Studies and Appl. I. (1984), 103–127.
*187. Arnol'd, V. I., Varchenko, A. N., and Gusein-Zade, S. M. *Specific Features of Diffentiable Mappings*, v. I. Nauka, Moscow (1982).
*188. Trofimov, V. V., and Fomenko, A. T. "Liouville integrability of Hamiltonian systems on Lie algebras." *Uspekhi Mat. Nauk* **39** (1984), No. 2, 3–50.
189. Darboux, G. *Leçons sur la Théorie Générale des Surfaces et Les Applications Géométriques du Surfaces et les Applications géométriques du Calcul Infinitesimal*. Paris, Gautier-Villar (1981).
190. Zoll, O. "Über Flächen mit Scharen geschlossener geodätischen Linien." *Math. Ann.* **57** (1903), 108–133.
191. Besse, A. *Manifolds All of Whose Geodesics are Closed*. Berlin (1978).
192. Blaschke, W. *Einführung in die Differentialgeometrie*. Berlin (1950).
193. Funk, P. "Über Flächen mit lauter geschlossenen geodätischen Linien." *Math. Ann.* **74** (1913), 278–300.
194. Guillemin, V. "The Radon transform on Zoll surfaces." Preprint (1976).
195. Weinstein, A. "On the volume of manifolds all of whose geodesics are closed." *J. Diff. Geom.* **9** (1974), 513–517.
196. Green, L. W. "Auf Wiedersehenflächen." *Ann. Math* **78** (1963), 289–299.
*197. Kolokol'tsov, V. N. "New examples of manifolds all of whose geodesics are closed." *Vestnik Mosk. Gos. Univers.*, I, ser. mat. mekh, No. 4 (1984), 80–82.

198. Milnor, J. *Morse Theory*. Princeton University Press, Princeton, N.J. (1963).
199. Bernat, P., Conze, N., and Vergne, M. *Représentations des Groupes de Lie résolubles*. (A paraître dans la Collection "Monographies de la Société Matématique de France.") Dunod, Paris (1972).
200. Losco, L. "Intégrabilité en mécanique céleste." *J. Mécanique* **13** (1974), 197–223.
201. Golo, V. L. "Nonlinear regimes in spin dynamics of superfluid ^{3}He." *Letter in Math. Phys.* **5** (1981), 155–159.
202. Helgason, S. *Differential Geometry and Symmetric Spaces*. Academic Press, New York (1962).
203. Loos, O. *Symmetric Spaces*. W.A. Benjamin, New York (1969).
204. Dixmier, J. *Algèbres enveloppantes*. Bordas (1974).
205. Rais, M. "Indices of semi-direct product $E \times_\rho G$." R. C. Acad. Sci. (Paris), ser. A, **278**(4), (1978) 195–197.
206. Brailov, A. V. "Complete integrability of some geodesic flows and integrable systems with noncommuting integrals." *Dokl. Akad. Nauk SSSR* **271** (1983), No. 2, 273–276.
207. Brailov, A. V. "A series of completely integrable Hamiltonian systems on a semi-direct product $SO(n) \times R^n$." In *Geometry and Topology in Global Nonlinear Problems*. Voronezh, Voronezh Univ. Press, (1984), 145–148.
208. Brailov, A. V. "Complete integrability with noncommuting integrals of several Euler equations." In *Application of Topology in Modern Analysis*. Voronezh, Voronezh Univ. Press, (1985), 22–41.
*209. Veselov, A. P. *Dokl. Akad. Nauk SSSR* **270** (1983), No. 6, 1298–1300.
*210. Veselova, L. E. *Vestnik Mosk. Gos. Univers.*, ser. mat. mekh., No 2. (1985), 64–67.
*211. Moiseev, N. N., and Rumyantsev, V. V. *Dynamics of a Body with Cavities Containing Liquid*. Nauka, Moscow (1965).
*212. Veselov, A. P. *Dokl. Akad. Nauk SSSR* **276** (1984), No. 3, 590–593.
213. Stekloff, W. *Ann. Fac. Sci. Univ. de Toulouse*, ser. 3, **1** (1909), 145–256.
214. Kalnins, E. G., Miller, W., and Winternitz. *SIAM Journal of Appl. Math.* **30**(4) (1976), 630–634.
215. Adler, M., and Van Moerbeke, P. *Proc. Nat. Acad. of Sci. (USA)* **81** (1984), 4613–4616.
*216. Ziglin, S. L. *Funkts. Analiz i yego Prilozhen.* **16** (1982), 30–41.
*217. Ziglin, S. L. *Funkts. Analiz i yego Prilozhen.* **17** (1983), No. 1.
218. Landau, L. D., and Lifshitz, E. M. *Mekhanika*. Nauka, Moscow (1968).
219. Haine, L. *Comm. Math. Phys.* **94** (1984), 271–287.
220. Haine, L. *Math. Ann.* **263** (1983), 435–472.
*221. Veselov, A. P. *Dokl. Akad. Nauk SSSR*, **270** (5) (1983), 1094–1097.
*222. Perelomov, A. M. Preprint. Inst. Teoret. i Exper. Fiziki, 147 (1983).
*223. Baryakhtar, V. G., Belokolos, E. D., and Golod, P. I. Preprint. Inst. Teoret. Fiziki Uk. Akad. Nauk SSR, 84–128 P, (1984).
*224. Belavin, A. A., and Drinfeld, V. G. "On solutions of the classical Yang–Baxter equation for simple Lie algebras." *Funkts. Analiz i yego Prilozhen.* **16**, issue 3 (1982) 1–29.
225. Baldassarry, M. *Algebraic Varieties*. Berlin (1956).

226. Guillemin, V., and Sternberg, S. *Symplectic Technique in Physics.* Cambridge Univ. Press, Cambridge (1984).
227. Yau, S. T. "On the Ricci curvature of a compact Kähler manifold and the complex Monge–Ampère equations. I." *Comm. Pure Appl. Math.* **31** (1978) 339–411.
228. Yano, K., and Bochner, S. *Curvature and Betti Numbers.* Princeton, N.J. (1953).
229. Kobayashi, K. "First Chern class and holomorphic tensor fields." *J. Math. Soc. Japan* **32** (1980), No. 2, 325–329.
230. Kodaira, K. "On the structure of compact complex analytic surfaces, I, II, III, IV." *Amer. J. Math.* **86** (1964), 751–798; **88** (1966), 682–721; **90** (1968) 55–83, 170–192.
231. Beauville, A. "Variétes kähleriennes dont la première classe de Chern est nulle." *J. Diff. Geom.* **18** (1983), 755–782.
*232. *Algebraic surfaces.* Trudy Matemat. Inst. Steklova, v. 75. Nauka, Moscow (1965).
233. Burns, D., and Rapoport, M. "On the Torelli problems for kählerian $K3$ surfaces." *Ann. Sci. École Norm. Sup.* **8** (1975), 235–274.
234. Griffiths, P., and Harris, J. *Principles of Algebraic Geometry.* New York (1978).
235. Mumford, D. *Albelian Varieties.* Bombay (1968).
*236. Anosov, D. V. *Geodesic Flows on Closed Manifolds of Negative Curvature.* Trudy Matemat. Inst. Steklova, v. 90. Nauka, Moscow (1967).
237. Goto, M., and Grosshans, F. *Semisimple Lie Algebras.* New York (1978).
238. Koboyashi, Sh., and Nomidzu, K. *Foundations of Differential Geometry.* Interscience Publishers, New York, London (1963).
*239. Olshanetsky, M. A., and Perelomov, A. M. "Explicit solutions of several completely integrable Hamiltonian systems." *Funkts. Analiz i yego Prilozhen.* II, issue 1 (1977), 75–76.
*240. Meshcheryakov, M. V. "On the characteristic property of the inertia tensor of a multidimensional rigid body." *Uspekhi Mat. Nauk* **38**, issue 5 (1983), 201–202.
*241. Meshcheryakov, M. V. "Solutions of Euler equations on singular orbits of simple Lie groups." In *Geometry and Topology in Global Nonlinear Problems.* Voronezh, Voronezh Univ. Press, (1984), 158–162.
242. Kamalin, S. A., and Perelomov, A. M. "Construction of canonical coordinates on polarized coadjoint orbits of Lie groups." *Comm. Math. Phys.* **97** (1985), 553–568.
243. Lacomba, E., "Mechanical systems with symmetry on homogeneous spaces." *Trans. Amer. Math. Soc.* **185** (1973), 477–491.
244. Milnor, J. "Curvature of left-invariant metrics on Lie groups." *Advances in Math.* **21** (1976), 293–329.
245. Ebin, D. "Integrability of perfect fluid motion." *Comm. Pure Appl. Math.* **36** (1983), 37–54.
246. Kostant, B. "On differential geometry and homogeneous spaces." *Proc. Nat. Acad. Sci. (USA)* **46**(5) (1956).
*247. Meshcheryakov, M. V. "Notes on dynamic systems on semisimple Lie algebras." *Vestnik Mosk. Gos. Univer.*, ser. mat. mekh. (1980), 17–19.
248. Bourbaki, N., *Groupes et Algèbres de Lie.* Hermann, Paris (1968).

249. Wolf, J. A. *Representations Associated to Minimal Coadjoint Orbits*. Lecture Notes in Math., v. 676 (1978), Springer-Verlag, New York.
250. Kupershmidt, B. A., and Ratiu, T. "Canonical maps between semidirect products with applications to elasticity and superfluids." *Comm. Math. Phys.* **90** (1983), 235–250.
251. Holm, D. D., and Kupershmidt, B. A. "Poisson brackets and Clebsch representations for magnetohydrodynamics, multifluid plasmas, and elasticity." *Physica D. Nonlinear Phenomena* **6(3)** (1983), 347–363.
*252. Trofimov, V. V. "Extensions of Lie algebras and Hamiltonian system." *Izvestiya Akad. Nauk SSSR*, ser. matem. **47** (1983), 1303–1321.
*253. Trofimov, V. V. "The methods of constructing S-representations. *Vestnik Mosk. Gos. Univers.*, ser. mat. mekh. **1** (1984), 3–9.
254. Weinstein, A. "Symplectic Geometry." *Bull Amer. Math. Soc.* **5** (1981), 1–13.
255. Guillemin, V., and Sternberg, S. "Convexity properties of the moment mapping." *Invent. Math.* **67** (1982), 491–513.
256. Kac, V. *Infinite-Dimensional Lie Algebras*. Progr. in Math., v. 44. Birkhäuser Boston, Cambridge, Mass. (1983).
*257. Kapitsa, P. L. *Dokl. Acad. Nauk SSSR* **18** (1938), 1.
*258. Landau, L. D., and Lifshitz, E. M. *Statistical Physics*, Ch. 6. Nauka, Moscow (1964).
*259. Landau, L. D., and Lifshitz, E. M. *Mechanics of Continuous Media*. Fizmatgiz, Moscow (1956).
*260. Pitayevsky, L. P. *ZhETF* **37** (1959), 1974.
261. Osheroff, D. D., Richardson, R. C., and Lee, D. M. *Phys. Rev. Lett.* **44** (1972), 792.
262. Richardson, R. C., and Lee, D. M. *Proc. 24th Nobel Symp., 1973*. Academic Press, New York (1974), 84.
263. Anderson, P. W., and Morel, P. *Phys. Rev.* **123** (1961), 1911.
264. Balian, R., and Werthamer, N. R. *Phys. Rev.* **131** (1963), 1553.
265. Leggett, A. J. *Ann. Phys.* **85** (1974), 11. New York.
266. Dzyaloshinskii, I. E., and Volovik, G. E. *Ann. Phys.* **125** (1980), 67. New York.
267. Fomin, I. A. *Sov. Phys.* **3** (1981), 275.
268. Maki, K., and Ebisawa, H. *Phys. Rev. B.* **13** (1976), 2924.
269. Golo, V. L. *Lett. Math. Phys.* **5** (1981), 155.
*270. Poluektov, Yu. M. *Fizika Nizkikh Temper.* **10** (1984), 1013.
271. Vollhardt, D. *Proc. LT* **17**. Eckern, U., Schmid, A., Weber, W., and Wühl, H., eds.
272. Brinkman, W. F. *Phys. Lett. A* **49** (1974), 411.
273. Brinkman, W. F., and Smith, H. *Phys. Lett. A.* **53** (1975), 43.
274. Pohlmeyer, K. *Comm. Math. Phys.* **46** (1976), 207.
*275. Novikov, S. P. *Funkts. Analiz i yego Prilozhen.* **15**, issue 4 (1981), 50.
276. Orlik, P., Vogt, E., and Zieschang, H. "Zur Topologie gefaserter dreidimensionaler Mannigfaltigkeiten." *Topology*, **6(1)** (1967), 49–65.
*277. Bolsinov, A. V. "Completely integrable systems on contractions of Lie algebras." In *Trudy Seminara po Vekt. i Tenz. Anal.*, issue 22 (1985), 8–16. Moscow.

*278. Le Ngok Tyeuen. "Complete involutive sets of functions on extensions of Lie algebras connected with Frobenius algebras." In *Trudy Seminara po Vect. i Tenz. Anal.*, issue 22 (1985), 69–106. Moscow, Moscow Univ. Press.

279. Elashvili, A. G. "Frobenius Lie algebras, II." In *Trudy Tbilisskogo Matematicheskogo Instituta, Tbilisi*, v. 77 (1985), 127–137.

*280. Brailov, A. V., "Some constructions of complete families of functions in involution." In: *Trudy Seminara po Vect. i. Tenz. Anal.*, issue 22 (1985), 17–24, Moscow.

281. Perelomov, A. M., "Lax representation for systems of the type of S. Kovalevskaya." *Funkts. Analiz i yego Prilozhen* 16 (2) (1982), 80–81.

*282. Fomenko, A. T. "The Morse theory of integrable Hamiltonian systems." *Dokl. Akad. Nauk SSSR*, **287** (1986) No. 5, 1071–1075.

*283. Fomenko, A. T. "The topology of constant-energy surfaces of integrable Hamiltonian systems and obstacles on the way of integrability." *Izvestiya Akad. Nauk SSSR*, 50 (1986), No. 6, 1276–1307.

*284. Fomenko, A. T. "The topology of three-dimensional manifolds and integrable mechanical Hamiltonian systems." In: *The Vth Tiraspol' Symposium on General Topology and its Applications*. Kishinev, (1985), 235–237.

*285. Brailov, A. V., and Fomenko, A. T. "The topology of integrable manifolds of completely integrable Hamiltonian systems." *Matematicheskii sbornik*, (1987) No. 11.

*286. Fomenko, A. T., and Zieschang, H. "On the topology of three-dimensional manifolds arising in Hamiltonian mechanics." *Dokl. Akad. Nauk SSSR*, **294** (1987), No. 2.

*287. Matveev, S. V., Fomenko, A. T., and Sharko, V. V. "Round Morse functions and isoenergy level surfaces of Hamiltonian systems". *Matematicheskii sbornik*, (1988) (in print). (preprint published in 1986, Preprint 86, **76** *Kiev, Institut Matematiki Akad. Nauk Ukr. SSR*, (1986) 32 pages).

*288. Le Hong Van, and Fomenko, A. T. "Minimality criterion of Lagrangian submanifolds in Kählerian manifolds." *Matemat. Zametki*, 42, (1987), No. 4 559–571.

*289. Fomenko, A. T., and Zieschang, H. "On typical topological properties of integrable Hamiltonian systems." *Izvestiya Akad. Nauk SSSR*, (1987) (in print).

*290. Matveev, S. V., and Fomenko, A. T. "Isoenergy surfaces of Hamiltonian systems, enumeration of three-dimensional manifolds in the increasing order of their complicacy and computation of volumes of closed hyperbolic manifolds." *Usp. Mat. Nauk*, (1988) (in print).

*291. Matveev, S. V., and Fomenko, A. T. "Morse type theory for integrable Hamiltonian systems with tame integrals." *Matemat. Zametki*, (1988) (in print).

*292. Oshemkov, A. A. "Bott integrals of some integrable Hamiltonian systems." In: *Geometry, Differential Equations and Mechanics*. Moscow, izd. Mosk. Gos. Univers., (1986) 115–117.

*293. Brailov, A. V. "Construction of completely integrable geodesic flows on compact symmetric spaces." *Izv. Akad. Nauk SSSR, ser. matem.* **50** (1986) No. 4, 661–674

*294. Bolsinov, A. V. "New examples of completely integrable systems on Lie algebras." In: *Geometry, Differential Equations and Mechanics*. Moscow, izd.

Mosk. Gos. Univers., (1986) 54–58.

295. Bolsinov, A. V. "Complete integrability of Euler equations on orbits Ad of groups $U(n) \times_\varphi C^n$ and $SU(n) \times_\psi C^n$." *Vestnik Mosk. Gos. Univers., ser. matem. mekh.*, (1986) No. 4, 79–81.

*296. Le Hong Van, and Fomenko, A T. "Lagrangian manifolds and Maslov's index in the theory of minimal surfaces." *Dokl. Akad. Nauk SSSR*, (1987) (in print).

297. Fomenko, A. T., and Zieschang, H. "On the topology of three-dimensional manifolds arising in Hamiltonian mechanics." Preprint No. 55/1985, Ruhr-Universität Bochum, Germany, 1–10.

298. Fomenko, A. T., and Zieschang, H. "On typical topological properties of integrable Hamlton systems." Preprint No. 92/1987, Ruhr-Universität, Bochum, Germany, 1–47.

*299. Trofimov, V. V., and Fomenko, A. T. "The geometry of Poisson brackets and the methods of Liouville integration of systems on symetric spaces." *Itogi Nauki i Tekhniki. Sovr. Probl. Matem. Noveishiye Dost. VINITI, Moscow.* **29** (1986) 3–108.

*300. Taymanov, I. A. "Non-simply-connected manifolds with a nonintegrable geodesic flow." In: *Geometry, Differential Equations and Mechanics*, Moscow, Mosk. Gos. Univers., (1986) 119–120.

*301. Oshemkov, A. A. "The topology of isoenergy surfaces and bifurcation diagrams of integrable cases of the dynamics of a rigid body on so(4)." *Usp. Mat. Nauk*, (1988) (in print).

*302. Abrarov, D. L. "Topological obstacles for the existence of conditionally linear integrals." *Vestnik Mosk. Gos. Univers., ser. mat., mekh.*, (1984) No. 6, 72–75.

*303. Abrarov, D. L. "On topological obstacles for integrability of Hamiltonian systems." In: *Geometry, Diffrential Equations and Mechanics*, Moscow, Mosk. Gos. Univers., (1986), 25–31.

*304. Bolsinov, A. V. "Complete involutive families of functions on Lie algebras of the type $G +_\rho V$." *Usp. Mat. Nauk*, (1988) (in print).

*305. Bolsinov, A. V. "New constructions of integrable Hamiltonian systems." *Dokl. Akad. Nauk SSSR*, (1988) (in print).

*306. Belyaev, A. V. "On the motion of an n-dimensional rigid body with the group of symmetries $SO(k) SO(N-l)$ in a field with linear potential. Invariants of coadjoint representation of some Lie algebras." *Dokl. Akad. Nauk SSSR*, **282** (1985) No. 5, 1038–1042.

307. Birkhoff, G. D. "Dynamical systems." Amer. Math. Soc. Colloq. publ., **9** New York, Amer. Math. Soc., (1927). Revised ed. (1966).

*308. Bogoyavlensky, O. I. "Integrable Euler equations on Lie algebras, which arise in problems of mathematical physics." *Izvestiya Akad. Nauk SSSR, ser. matem.*, **48** (1984) No. 5, 883–983.

*309. Bogoyavlensky, O. I. "Some integrable cases of Euler equations." *Dokl. Akad. Nauk SSSR*, **287** (1986) No. 5, 1105–1108.

*310. Matveev, S. V., and Burmistrova, A. B. "The structure of S-functions on orientable 3-manifolds." *tezisy XI Vsesoyuz. Shkoly po Teorii Operatorov v Funkts. Prostr. Chelyabinsk*, (1986).

SUBJECT INDEX

A

Adjoint representation, Ch. 1, §4, p. 39
Affine Lie Algebra, Ch. 4, §1, p. 195
Algebraic dimension, Ch. 3, §4, p. 180
Angular velocity, Ch. 1, §1, p. 6
Annihilator of covector, Ch. 1, §4, p. 43

B

Beauville manifold, Ch. 3, §4, p. 184
Bifurcation diagram, Ch. 2, §1, p. 103
Bott integral, Ch. 2, §1, p. 56

C

Cartan subalgebra, Ch. 1, §4, p. 41
Canonical form, Ch. 1, §2, p. 13
Coadjoint representation, Ch. 1, §4, p. 39
Common level surface, Ch. 1, §2, p. 32
Commutative integration, Ch. 1, §2, p. 34
Commutator, Ch. 1, §4, p. 39
Complete commutative (involutive) set of integrals, Ch. 1, §2, p. 34
Complex integrability, Ch. 3, §4, p. 178
Compact form (subalgebra), Ch. 1, §4, p. 48
Constant-energy surface, Ch. 2, §1, p. 56
Contraction of Lie algebra, Ch. 4, §4, p. 243
Cylinder, Ch. 2, §1, p. 106

D

Darboux theorem, Ch. 1, §2, p. 19

Differential dimension, Ch. 3, §2, p. 161
Differential index, Ch. 3, §2, p. 162
Dissipative full torus, Ch. 2, §1, p. 105
Duplication of integrable equations, Ch. 4, §4, p. 231

E

Embeddings of dynamic systems, Ch. 4, §1, p. 187
Euler–Poisson equations, Ch. 1, §1, p. 1
Extension method of polynomial functions, Ch. 4, §4, p. 236
Exterior differential form, Ch. 1, §2, p. 15

F

First integral, Ch. 1, §1, p. 3
Frequency vector, Ch. 5, §1, p. 259
Frobenius Lie algebra, Ch. 4, §4, p. 234
Functions in involution, Ch. 1, §2, p. 32

G

Geodesic flow, Ch. 5, §4, p. 287
Geometrically simple set of integrals, Ch. 5, §3, p. 282
Globally Hamiltonian vector field, Ch. 1, §2, p. 21
Gyroscopic systems, Ch. 5, §2, p. 277

H

Hamiltonian field, Ch. 1, §2, p. 20

I

Inertia operator, Ch. 1, §1, p. 6

Subject Index

Integral of a vector field, Ch. 1, §2, p. 30
Invariant submanifold, Ch. 3, §2, p. 157
Involutive set of functions, Ch. 1, §2, p. 34
Isotropic plane, Ch. 1, §2, p. 14

J

Jacobi identity, Ch. 1, §2, p. 27
Jacobi multiplier, Ch. 1, §1, p. 4

K

Killing form, Ch. 1, §4, p. 41
Kirchhoff equations, Ch. 5, §1, p. 266
$K3$-type surfaces, Ch. 3, §4, p. 182

L

Lagrangian plane, Ch. 1, §2, p. 14
Legget Hamiltonian, Ch. 4, §5, p. 252
Lens space, Ch. 2, §1, p. 61
Lie algebra, Ch. 1, §4, p. 39
Lie group, Ch. 1, §4, p. 39
Liouville metric, Ch. 5, §4, p. 291
Liouville theorem, Ch. 1, §2, p. 32
Liouville tori, Ch. 1, §2, p. 32
Locally Hamiltonian vector field, Ch. 1, §2, p. 20

M

Many-body problem, Ch. 5, §2, p. 278
Maupertuis principle, Ch. 5, §1, p. 276
Maximal linear commutative subalgebra, Ch. 3, §1, p. 143
Maximal linear subalgebra, Ch. 3, §1, p. 146
Moment of inertia, Ch. 1, §1, p. 7
Momentum mapping, Ch. 2, §1, p. 103
Monodromy operator, Ch. 4, §3, p. 225
Morse function, Ch. 2, §1, p. 68
Morse lemma, Ch. 2, §1, p. 69

N

Noether's theorem, Ch. 1, §2, p. 31
Noncommutative algebra of integrals, Ch. 3, §1, p. 149
Noncommutative integration, Ch. 3, §1, p. 146
Normal form (subalgebra), Ch. 1, §4, p. 53
Nutation angle, Ch. 1, §1, p. 3

O

Orientable Bott integral, Ch. 2, §1, p. 60

P

Perturbation theory, Ch. 5, §1, p. 256
Poincaré method, Ch. 5, §1, p. 258
Poincaré lemma, Ch. 1, §2, p. 26
Poincaré set, Ch. 5, §1, p. 258
Poisson bracket, Ch. 1, §2, p. 26
Precession angle, Ch. 1, §1, p. 3
Projective space, Ch. 2, §1, p. 61

R

Rank of the algebra, Ch. 1, §4, p. 42
Real form (subalgebra), Ch. 1, §4, p. 48
Regular elements, Ch. 1, §4, p. 41
Rigid body, Ch. 1, §3, p. 34
Root (of Lie algebra), Ch. 1, §4, p. 42
Round handle, Ch. 2, §1, p. 71

S

Sectional operator, Ch. 4, §1, p. 215
Semisimple Lie algebra (group), Ch. 1, §4, p. 41
Separatrix diagram, Ch. 2, §1, p. 60
Separatrix splitting, Ch. 5, §1, p. 261
SC-property, Ch. 5, §4, p. 294
Shift of invariants, Ch. 4, §1, p. 190
Skew gradient, Ch. 1, §2, p. 18
Skew oprthogonal complement, Ch. 1, §2, p. 13
Skew symmetric scalar product, Ch. 1, §2, p. 12
Symmetric top, Ch. 1, §1, p. 8
Stable integral trajectory, Ch. 2, §1, p. 58
Symplectic group, Ch. 1, §2, p. 16

Symplectic manifold, Ch. 1, §2, p. 17
Symplectic structure, Ch. 1, §2, p. 12

T

Tangent space, Ch. 1, §2, p. 12
Topological surgery, Ch. 2, §1, p. 70
Toric saddle, Ch. 2, §1, p. 106

NOV 2 0 1990